Yeast Biotechnology: Diversity and Applications

Yeast Biotechnology: Diversity and Applications

Edited by Harvey Marshall

STATES
ACADEMIC PRESS
www.statesacademicpress.com

States Academic Press,
109 South 5th Street,
Brooklyn, NY 11249, USA

Visit us on the World Wide Web at:
www.statesacademicpress.com

ISBN: 978-1-63989-735-3

Cataloging-in-Publication Data

Yeast biotechnology : diversity and applications / edited by Harvey Marshall.
p. cm.
Includes bibliographical references and index.
ISBN 978-1-63989-735-3
1. Yeast fungi--Biotechnology. 2. Microbial biotechnology. 3. Biotechnology. I. Marshall, Harvey.
TP248.27.Y43 Y43 2023
660.62--dc23

Table of Contents

Preface

Yeast are single celled microorganisms which are classified as fungi. They are crucial in food chains, and are significant in sulfur, carbon and nitrogen cycles. They also have critical environmental applications, including the elimination of heavy metals from wastewater and bioremediation. The usage of yeasts in biotechnological processes is increasing due to their varied properties. They are used for the production of metabolites and recombinant proteins, as well as in-vivo biotransformations. Majority of yeast species are not harmful to animals or human beings, and have the potential to be utilized in a range of fields that are crucial to human activities. This book explores all the important aspects of yeast biotechnology in the present day scenario. It is a compilation of chapters that discuss the most vital concepts and emerging trends in this field. This book will serve as a reference to a broad spectrum of readers.

This book is the end result of constructive efforts and intensive research done by experts in this field. The aim of this book is to enlighten the readers with recent information in this area of research. The information provided in this profound book would serve as a valuable reference to students and researchers in this field.

At the end, I would like to thank all the authors for devoting their precious time and providing their valuable contribution to this book. I would also like to express my gratitude to my fellow colleagues who encouraged me throughout the process.

Editor

1

Effects of Medium Components on Isocitric Acid Production by *Yarrowia lipolytica* Yeast

Svetlana V. Kamzolova *, Vladimir A. Samoilenko, Julia N. Lunina and Igor G. Morgunov

Federal Research Center, Pushchino Scientific Center for Biological Research of the Russian Academy of Sciences, G.K. Skryabin Institute of Biochemistry and Physiology of Microorganisms of the Russian Academy of Sciences, Prospect Nauki 5, Pushchino, 142290 Moscow, Russia; samva@rambler.ru (V.A.S.); luninaju@rambler.ru (J.N.L.); morgunovs@rambler.ru (I.G.M.)

* Correspondence: kamzolova@rambler.ru

Abstract: The microbiological production of isocitric acid (ICA) is more preferable for its application in medicine and food, because the resulting product contains only the natural isomer—threo-D_S. The aim of the present work was to study ICA production by yeast using sunflower oil as carbon source. 30 taxonomically different yeast strains were assessed for their capability for ICA production, and *Y. lipolytica* VKM Y-2373 was selected as a promising producer. It was found that ICA production required: the limitation of *Y. lipolytica* growth by nitrogen, phosphorus, sulfur or magnesium, and an addition of iron, activating aconitate hydratase, a key enzyme of isocitrate synthesis. Another regulatory approach capable to shift acid formation to a predominant ICA synthesis is the use of inhibitors (itaconic and oxalic acids), which blocks the conversion of isocitrate at the level of isocitrate lyase. It is recommended to cultivate *Y. lipolytica* VKM Y-2373 under nitrogen deficiency conditions with addition of 1.5 mg/L iron and 30 mM itaconic acid. Such optimized nutrition medium provides 70.6 g/L ICA with a ratio between ICA and citric acid (CA) equal 4:1, a mass yield (Y_{ICA}) of 1.25 g/g and volume productivity (Q_{ICA}) of 1.19 g/L·h.

Keywords: microbial synthesis; (2*R*,3*S*)-isocitric acid (ICA); *Yarrowia lipolytica*; aconitate hydratase (AH); iron; isocitrate lyase (ICL); inhibitors; itaconic acid; oxalic acid

---

1. Introduction

Isocitric acid is a structural isomer of citric acid (CA), differing from the latter in the position of the hydroxyl group. Isocitric acid molecule has two asymmetric carbon atoms and hence exists in four stereoisomers, namely, threo-D_S-isocitric acid, threo-Ls-isocitric acid, erythro-D_S-isocitric acid, and erythro-L_S-isocitric acid [1], of which only threo-D_S-isocitric acid (or (2*R*,3*S*)-isocitric acid according to IUPAC Organic Nomenclature) (conventionally named as ICA) is an intermediate of the tricarboxylic acid (TCA) cycle, a common metabolite formed by microorganisms, plants, and animals [1].

ICA and its derivatives are widely used in industries producing pharmaceutical products, and cosmetics as surfactants, detergents, ion chelators, and biologically active additives to food products [2–10]. It was found that ICA surpasses the classical antioxidant ascorbic acid in the model of oxidative stress induced by the action of hydrogen peroxide and ecotoxicants (Cu, Pb, Zn, Cd) on infusorian cells [11]. In addition, ICA relieves the neurointoxication against the high concentration of lead and molybdenum salts, restores the memory and accelerates the learning [12]. It was also reported the positive influence of ICA on the spatial component of memory [13].

It is known that ICA accumulates in significant quantities in the juice of flowering showy stonecrop *Hylotelephium spectabile*, formerly called *Sedum spectabile*, which is sold at a price of 595 EUR per gram.

At present, microbiological production of ICA is more preferable for its applications in the medicine and food industries, because the resulting product contains only the natural isomer [2,4,5,9,10].

The effective processes for ICA production were developed on the cultivation of different wild, mutant and genetically engineered strains of *Yarrowia lipolytica* yeast on n-paraffin [14], ethanol [12,15], plant oils [3,4,6,16–19], and glycerol-containing waste of biodiesel industry [13,20–22]. It should be noted that CA accumulate simultaneously with ICA. The balance between CA and ICA can be shifted toward the overproduction of isocitrate by the activation of aconitate hydratase—enzyme, which isomerizes CA to ICA via *cis*-aconitate [14,16]. For the operation of aconitase, iron ions are required [14,15,17]. The ratio between citric acids toward the ICA production can be shifted by the inhibition of isocitrate lyase with itaconic and oxalic acid in *Y. lipolytica*, grown on rapeseed oil [23], while the genetically modified strain *Y. lipolytica* with the inactivated *ICL1* gene exhibited only a small increase in the synthesis of ICA in media with glucose and glycerol [24].

In recent years, sunflower oil has attracted increased interest as a renewable resource which can be obtained in any country. Sunflower oil is an excellent substrate for biotechnological processes because it provides for a high purity of the target product with the aim of its use in medicine and food industry. High ICA production from sunflower oil was revealed in wild and genetically modified yeast *Y. lipolytica* [3,4,6]. However, the knowledge on the physiology of isocitrate-producing strains *Y. lipolytica* and the fermentation conditions essential for ICA production from sunflower oil is still limited.

The aim of the present work was the selection of producer and the evaluation of critical factors affecting *Y. lipolytica* growth and ICA production for the optimal design of fermentation process.

2. Materials and Methods

All chemicals were of analytical grade (Sigma-Aldrich, St. Louis, MO, USA). Sunflower oil "Golden Seed Oil" was purchased from the "YugRossii" Processing Plant (Rostov, Russia). The fatty acid profile of the sunflower oil was (%, by mass): C14:0, 0.1; C16:0, 6.2; C18:0, 4.1; C18:1, 13.7; C18:2, 75.9.

For screening, 30 wild-type yeast strains were obtained from the Collection of the Laboratory of Aerobic Metabolism of Microorganisms, the Institute of Biochemistry and Physiology of Microorganisms, Russian Academy of Sciences and from the All-Russia Culture Collection of Microorganisms, Russian Academy of Sciences (VKM). The strains were maintained at 4 °C on agar slants with 1% paraffin as the carbon source. The yeast strains were maintained on agar medium with paraffin at +4 °C. Yeasts were cultivated on an orbital shaker (180 ± 10 rpm) at 29 °C during 96 h in 750-mL shaking flasks with 50 mL of the Reader medium containing (g/L): sunflower oil, 10; $(NH_4)_2SO_4$, 0.3; KH_2PO_4, 1.0; K_2HPO_4, 0.1; $MgSO_4 \cdot 7H_2O$, 0.7; $Ca(NO_3)_2 \cdot 4H_2O$, 0.4; NaCl, 0.5; trace element solution containing (mg/L): KJ, 0.1; $Na_2B_4O_7 \cdot 7H_2O$, 0.08; $MnSO_4 \cdot 5H_2O$ 0.05; $ZnSO_4 \cdot 7H_2O$ 0.04; $CuSO_4$, 0.04; Na_2MoO_4, 0.03; $FeSO_4(NH_4)_2SO_4 \cdot 6H_2O$, 0.35; yeast extract "Difco", 0.5. During cultivation, the pH of the medium was determined each 12 h and, if necessary, corrected by adding a certain volume of 10% solution of either KOH or H_2SO_4.

In experiments on optimization of ICA production by the selected strain *Y. lipolytica* VKM Y-2373, five mineral media were used: (1) Reader medium; (2) nitrogen-deficient; (3) phosphorus-deficient; (4) sulfur-deficient; (5) magnesium-deficient. The first medium contained (g/L): $(NH_4)_2SO_4$, 12.0; $MgSO_4 \cdot 7H_2O$, 1.4; $Ca(NO_3)_2$, 0.8; NaCl, 0.5; KH_2PO_4, 2.0; K_2HPO_4, 0.2; double volume of Burkholder microelement solution; yeast autolysate, 8 mL/L; thiamine, 100 µg/L. Limitation of the yeast growth was performed by decreasing concentration of the salt containing limiting component. The second medium contained nitrogen, phosphorus, sulfur and magnesium in concentrations of 630, 492, 362 and 280 mg/L, respectively. The third medium decreased (almost 30-fold) the amount of phosphorus; concentrations of nitrogen, phosphorus, sulfur and magnesium were 1960, 16.4, 362 and 280 mg/L. In the fourth medium the sulfur amount was decreased by almost 30 times; concentrations of nitrogen, phosphorus, sulfur and magnesium were 1960, 492, 12 and 280 mg/L, respectively. The fifth medium decreased (almost 50-fold) the amount of magnesium; concentrations of nitrogen, phosphorus, sulfur and magnesium were 1960, 492, 362 and 5.6 mg/L. The initial concentration of oil was 20 g/L; pulsed additions of oil (by 20 g/L) at 24, 48 and 72 h. The fermentation conditions were maintained automatically at the constant level: temperature (29 °C); dissolved oxygen concentration (pO_2) (55–60% from air saturation); pH 6.0 was adjusted with 20% KOH. These fermentation parameters were selected on the basis of data on optimization of the ICA synthesis from rapeseed oil by *Y. lipolytica* [17]. Cultivation was performed for 96 h.

The effect of inhibitors on the metabolization and production of ICA and CA was studied in experiments carried out in test tubes and in a fermenter. In the first case, *Y. lipolytica* VKM Y-2373 were cultivated on an orbital shaker (180 ± 10 rpm) at 29 °C during 48 h in large test tubes (20 cm long, 2 cm in diameter) with 5 mL Reader medium containing sunflower oil (5 g/L) as carbon substrate. In the second case, *Y. lipolytica* was cultivated in the 10-L fermenter with 5 L of medium containing (g/L): $(NH_4)_2SO_4$, 3; $MgSO_4{\cdot}7H_2O$, 1.4; $Ca(NO_3)_2$, 0.8; NaCl, 0.5; KH_2PO_4, 2.0; K_2HPO_4, 0.2; double volume of Burkholder microelement solution; yeast autolysate, 8 mL/L; thiamine, 100 µg/L and $FeSO_4{\cdot}7H_2O$, 7.5 mg/L. The fermentation conditions were maintained automatically at the constant level: temperature (29 °C); dissolved oxygen concentration (pO_2) (55–60% from air saturation); pH 6.0 was adjusted with 20% KOH. Cultivation was performed for 96 h.

The amount of biomass, oil was determined as described earlier [17].

Concentration of organic acids (pyruvic acid, citric acid, α-ketoglutaric acid, fumaric acid, malic acid, oxaloacetic acid, itaconic acid, oxalic acid) was determined using high-performance liquid chromatography (HPLC) with an HPLC chromatograph (Pharmacia, LKB, Uppsala, Sweden) on an Inertsil ODS-3 reversed-phase column (250 × 4 mm, Elsiko, Moscow, Russia) at 210 nm; 20 mM phosphoric acid was used as a mobile phase with the flow rate of 1.0 mL/min; the column temperature was maintained at 35 °C. Acids were also identified using the standard solutions (Roche Diagnostics GmbH. Germany). Glucose 6-phosphate was determined by enzymatic assay with glucose-6-phosphate dehydrogenase. Fructose 1,6-bisphosphate was determined by fructose based on the Seliwanoff color reaction.

For the enzyme assay, yeast cells were collected by centrifugation at 3000× *g* for 10 min (4 °C) and washed with an ice-cold 0.9% NaCl solution. The cell pellet was suspended in a proportion of 1:10 in 100 mM phosphate buffer (pH 7.4) supplemented with 1 mM EDTA. Cells in the suspension were disrupted with Ballotini glass beads (d = 150–250 µ) on a planetary mixer for 3 min at 1000 rpm (0 °C). The cell homogenate was centrifuged at 5000 g for 30 min (4 °C), and the supernatant was used for the assay of citrate synthase (CS), aconitate hydratase (AH), NAD-isocitrate dehydrogenase (NAD-ICDH) and isocitrate lyase (ICL). CS was assayed in the reaction mixture containing 0.25 mM oxaloacetate, 0.25 mM acetyl CoA, 0.1 mM 5.5-dithiobis(2-nitrobenzoate), and 100 mM Tris-HCl buffer (pH 8.5). AH was assayed in the reaction mixture containing 5 mM ICA and 50 mM potassium phosphate buffer (pH 7.5). NAD-ICDH was assayed in the reaction mixture containing 0.25 mM ICA, 4 mM NAD, 0.5 mM AMP, 10 mM $MgCl_2$, 2 mM antimycin, and 50 mM Tris-HCl buffer (pH 7.5). ICL was assayed in the reaction mixture containing 4 mM ICL, 8 mM phenylhydrazine-HCl, 4 mM cysteine-HCl, 10 mM $MgCl_2$, and 75 mM potassium phosphate buffer (pH 6.85). The amount of enzyme catalyzing conversion of 1 µmol of substrate per min per mg of protein was taken as one unit of enzyme activity (U/mg of protein).

The mass yield coefficient of ICA production (Y_{ICA}), expressed in g of ICA per g of sunflower oil consumed, was calculated from:

$$Y_{ICA} = P/S$$

The volume productivity (Q_{ICA}), expressed in g/L·h, was calculated from:

$$Q_{ICA} = P/(V \cdot t)$$

where P is the total amount of ICA in the culture liquid at the end of cultivation (g), S is the total amount of sunflower oil consumed (g), V is the initial volume of culture liquid (L), t is the fermentation duration (h).

All the data presented are the means values of three experiments and two measurements for each experiment; standard deviations were calculated (S.D. < 10%). The ICA/CA ratio, the mass yield coefficient of ICA production (Y_{ICA}) and the volume productivity (Q_{ICA}) are calculated using the mean value of ICA.

3. Results and Discussion

3.1. Selection ICA Producers

Screening of yeasts actively producing ICA from sunflower oil was carried out 30 natural strains belonging to 18 species. Yeasts were cultivated under limitation of cell growth by nitrogen (67 mg/L) and excess of sunflower oil (10 g/L). It should be noted that in this study, for the first time, a wide range of yeast species was tested for their ability to produce ICA from sunflower oil, whereas until now, the studies on ICA production were carried out mainly with Yarrowia lipolytica (syn. Candida lipolytica) [3,4,6]. On other substrates (n-alkanes, glucose, rapeseed oil), the acid-producing activity was found in yeast species *Meyerozyma guilliermondii* (syn. *Candida guilliermondii*), *Candida ravautii*, *Diutina catenulata* (syn. *Candida brumptii*, *C. catenulata*) and *Y. lipolytica* [17,25–27].

As seen from Table 1, all strains grew well in a medium with sunflower oil as only carbon source. However, the strains were heterogeneous in ICA production. Thirteen strains belonging to species *Aciculoconidium aculeatum*, *Babjeviella inositovora*, *Candida intermedia*, *C. parapsilosis*, *C. tropicalis*, *C. utilis*, *Diutina catenulata*, *D. rugosa*, *Pichia besseyi*, *P. media*, *P. membranifaciens*, *Sugiyamaella paludigena* and *Wickerhamomyces anomalus* not produced ICA. Strains belonging to species *Blastobotrys adeninivorans*, *Candida zeylanoides*, *Meyerozyma guilliermondii* and *Y. lipolytica* produced ICA in the amounts from 0.14 to 6.3 g/L. The strain *Y. lipolytica* VKM Y-2373 synthesized the maximum amount of ICA (6.3 g/L). It should be noted that, acid-producing yeasts synthesized also a by-product, CA; a ratio of ICA to CA varied from 1.3: 1 to 1: 2.9, and was the best (1.3:1) in *Y. lipolytica* VKM Y-2373.

Out of 30 strains studied, we selected strain *Y. lipolytica* VKM Y-2373 for further optimization of ICA production.

3.2. Effect Limiting Factors on ICA Production

Numerous studies have noted that the ability of the yeast *Y. lipolytica* to produce citric acids was found under nitrogen deficiency [2,4,5,9,10,28–30].

The present study deals with ICA synthesis in the selected strain by using various growth-limiting components such as nitrogen, phosphorus, sulfur and magnesium. As seen from Table 2, when *Y. lipolytica* VKM Y-2373 was grown in a Reader medium without limitation, no acid formation occurred; under these conditions the cells grew well, and the biomass reached 27.7 g/L. When growth of *Y. lipolytica* VKM Y-2373 was limited by nitrogen, phosphorus, sulfur and magnesium, then concentration of these elements was selected so that the biomass did not exceed 10–12 g/L, since at a higher biomass, the main factor limiting growth is the oxygen deficiency. To fulfill this condition, the nitrogen concentration was decreased 4-fold, phosphorus and sulfur—30-fold, magnesium—50-fold. Under nitrogen limitation, *Y. lipolytica* VKM Y-2373 synthesized 46.97 g/L of ICA and 35.67 g/L of CA; a ratio of ICA to CA was 1.3:1. Similar data were found with a deficiency of phosphorus and sulfur; a ratio of ICA to CA varied from 1.1:1 to 1.2:1. The limitation of *Y. lipolytica* VKM Y-2373 growth by magnesium led to the predominant accumulation of ICA (a ratio of a ratio of ICA to CA was 3:1); however, the amount of ICA was decreased in 2.3 times compared with one under nitrogen deficiency.

Table 1. Screening of yeasts producing of isocitric acid (ICA) from sunflower oil.

Strains	Biomass (g/L)	ICA (g/L)	CA (g/L)	ICA/CA Ratio
Aciculoconidium aculeatum VKM Y-1301	4.43 ± 0.35	0	0.53	n.d.
Babjeviella inositovora VKM Y-2494	4.60 ± 0.53	0	0	n.d.
Blastobotrys adeninivorans VKM Y-2676	4.90 ± 0.56	0.60 ± 0.10	1.64 ± 0.07	1:2.7
C. intermedia	4.20 ± 0.60	0	0	n.d.
C. parapsilosis	4.10 ± 0.66	0	0	n.d.
C. tropicalis 303	3.00 ± 0.56	0	0	n.d.
C. utilis VKM Y-33	1.47 ± 0.50	0	0	n.d.
C. zeylanoides VKM Y-6	2.21 ± 0.50	0.61 ± 0.12	1.22 ± 0.12	1:2
C. zeylanoides VKM Y-14	2.20 ± 0.20	1.11 ± 0.20	2.22 ± 0.13	1:2
C. zeylanoides VKM Y-2324	2.57 ± 0.21	1.12 ± 0.12	1.90 ± 0.20	1:1.7
Diutina catenulata VKM Y-5	4.70 ± 0.20	1.30 ± 0.20	2.60 ± 0.20	1:2
D. catenulata VKM Y-36	2.30 ± 0.20	1.10 ± 0.20	1.1 ± 0.20	1:1
D. rugosa VKM Y-67	2.03 ± 0.35	0	0	n.d.
Meyerozyma guilliermondii	5.20 ± 0.56	0.80 ± 0.10	2.35 ± 0.10	1:2.9
Pichia besseyi VKM Y-2084	2.15 ± 0.15	0	0	n.d.
P. media VKM Y-1381	2.55 ± 0.15	0	0	n.d.
P. membranifaciens VKM Y-292	3.01 ± 0.40	0	0	n.d.
Sugiyamaella paludigena VKM Y-2443	2.65 ± 0.35	0	0	n.d.
Wickerhamomyces bisporus VKM Y-1065	2.90 ± 0.25	0	0	n.d.
Y. lipolytica VKM Y-2412	2.00 ± 0.21	2.10 ± 0.35	1.10 ± 0.16	1:1
Y. lipolytica VKM Y-2373	1.70 ± 0.17	6.30 ± 0.61	4.84 ± 0.39	1.3:1
Y. lipolytica 68	2.37 ± 0.23	0.14 ± 0.01	0.16 ± 0.02	1:1.1
Y. lipolytica 69	2.60 ± 0.22	2.90 ± 0.26	2.21 ± 0.21	1.3:1
Y. lipolytica 374/4	2.40 ± 0.24	2.46 ± 0.40	2.66 ± 0.29	1:1.1
Y. lipolytica 571	3.50 ± 0.29	1.1 ± 0.18	1.07 ± 0.10	1:1
Y. lipolytica 581	2.40 ± 0.21	1.26 ± 0.24	1.73 ± 0.27	1:1.4
Y. lipolytica 585	1.95 ± 0.20	1.20 ± 0.19	1.52 ± 0.25	1:1.3
Y. lipolytica 607	2.20 ± 0.20	1.80 ± 0.60	1.7 ± 0.44	1:1.1
Y. lipolytica 684	2.78 ± 0.40	2.10 ± 0.60	2.0 ± 0.24	1.1:1
Zygoascus hellenicus VKM Y-2007	2.70 ± 0.20	0	0	n.d.

n.d.—not determined.

Table 2. Effect of limiting component on the growth of *Y. lipolytica* VKM Y-2373 and ICA production.

Parameter	Limiting Component (mg/L)				
	Full Medium	N	P	S	Mg
	2545 (N) 492 (P) 362 (S) 280 (Mg)	630 (N) 492 (P) 362 (S) 280 (Mg)	1960 (N) 16.4 (P) 362 (S) 280 (Mg)	1960 (N) 492 (P) 12 (S) 280 (Mg)	1960 (N) 492 (P) 362 (S) 5.6 (Mg)
Time (h)	48	96	96	96	96
Biomass (g/L)	27.7 ± 3.3	9.83 ± 0.15	8.23 ± 0.15	8.40 ± 0.20	9.80 ± 0.20
ICA (g/L)	1.37 ± 0.12	46.97 ± 2.73	38.0 ± 1.1	40.36 ± 1.86	20.83 ± 0.57
CA (g/L)	1.37 ± 0.10	35.67 ± 1.05	32.36 ± 2.25	38.46 ± 0.31	6.40 ± 0.36
ICA/CA ratio	1:1	1.3:1	1.2:1	1.1:1	3.3:1
Y_{ICA} (g/g)	n.d.	0.83	0.67	0.71	0.37
Q_{ICA} (g/L·h)	n.d.	0.79	0.64	0.68	0.35

n.d.—not determined.

The obtained data demonstrate that limitation of *Y. lipolytica* growth by biogenic macroelements (N, P, S and Mg) is the effective method for regulation of the acid production. At the same time, it is noted that the requirement of limiting this components in the nutrition medium can be a problem for citric acids production using carbon sources enriched by these elements [30]. All further experiments were carried out under nitrogen deficiency because it showed the maximum values of the ICA yield (Y_{ICA} = 0.83 g/g) and volume productivity (Q_{ICA} = 0.79 g/L·h).

3.3. *Effect of Iron*

Aconitate hydratase (AH) is a key enzyme which involved in the synthesis of isocitrate in TCA cycle. It was reported the genetically modified strain *Y. lipolytica* with the superexpressed *ACO1* gene cultivated on sunflower oil showed a higher content of ICA (66–71%) in comparison with the wild-type strain *Y. lipolytica* H222 (35–49% ICA) [16]. This transformant produced 68.4 g/L ICA (75.6% of total acids) with the yield Y_{ICA} = 0.64 g/g [6].

It is known that iron is an essential element for active functioning AH [2,9]. As can be seen from Table 3, the cell growth was observed in iron-free medium that apparently due to a low amount of iron available from the sunflower oil. An iron concentration in the range 0–0.05 mg/L was growth-limiting. This is confirmed by a direct correlation between iron concentration and biomass density. An increase in iron concentration from 0.05 to 6 mg/L showed no effect on biomass level. An increase iron concentration up to 10 mg/L led to a forced growth and the biomass production at level of 15.73 g/L. Earlier it was reported that an increased iron content (10–50 mg/L) promotes the accumulation of biomass in the erythritol-producing *Y. lipolytica* grown on glycerol [31].

Table 3. Effect of iron on the growth of *Y. lipolytica* VKM Y-2373 and ICA production.

Parameter	Concentration of Iron (mg/L)					
	Without Fe	**0.05**	**1.5**	**3.0**	**6.0**	**10.0**
Biomass (g/L)	2.83 ± 0.15	9.83 ± 0.15	10.67 ± 0.15	11.10 ± 0.36	10.73 ± 0.78	15.73 ± 0.15
ICA (g/L)	10.97 ± 0.73	46.97 ± 2.73	51.50 ± 1.5	48.90 ± 3.57	48.50 ± 0.50	35.67 ± 2.04
CA (g/L)	10.67 ± 1.25	35.67 ± 1.05	30.23 ± 0.60	28.60 ± 0.62	28.73 ± 1.15	20.58 ± 2.31
ICA/CA ratio	1:1	1.3:1	1.7:1	1.7:1	1.7:1	1.7:1
Y_{ICA} (g/g)	0.19	0.83	0.91	0.86	0.86	0.63
Q_{ICA} (g/L·h)	0.18	0.79	0.87	0.80	0.81	0.62
Enzyme activity (U/mg protein)						
CS	2.20 ± 0.30	2.25 ± 0.35	2.66 ± 0.20	2.62 ± 0.40	2.70 ± 0.30	2.75 ± 0.25
AH	0.20 ± 0.02	0.25 ± 0.03	0.95 ± 0.15	0.93 ± 0.25	0.85 ± 0.16	0.88 ± 0.07
NAD-ICDH	0.02 ± 0.01	0.02 ± 0.01	0.03 ± 0.01	0.03 ± 0.01	0.03 ± 0.01	0.03 ± 0.01
ICL	0.11 ± 0.01	0.12 ± 0.01	0.12 ± 0.01	0.11 ± 0.01	0.11 ± 0.01	0.12 ± 0.01

As seen from Table 3, high ICA concentration (46.97–51.5 g/L) and the mass yield (Y_{ICA}) (0.83–0.91 g/g) were found at iron concentration between 0.05 and 0.6 mg/g. An increase in iron from 6 to 10 mg/L imposed a decrease in ICA production for 25%.

As seen from Table 3, under iron limitation in the range 0–0.05 mg/L, AH was insignificant (0.2–0.25 U/mg protein). An increase in iron concentration from 0.05 to 10 mg/L resulted in AH activity in 3.4–3.8 times. In this case, CS activity increased only for 22%. NAD-ICDH remained practically constant at low level (0.02–0.03 U/mg protein). ICL which is involved in the metabolism of fatty acids, was high (0.11–0.12 U/mg protein) in all variants of experiment.

Further experiments were carried out with the optimal concentration of iron in the cultivation medium equal to 1.5 mg/L.

3.4. *Inhibition of Isocitrate Lyase*

It is known that the ratio between citric acids toward the ICA production can be shifted by the inhibition of isocitrate lyase (ICL) [2,13–15,23,24]. Yeast ICL is a constitutive enzyme subject to catabolite repression [32,33]. ICL is induced when the yeast grows on acetate, n-alkanes, ethanol and fatty acids [21,32–38]. At the same time, glucose in the cultivation medium suppresses ICL [36,37,39]. If the medium contains a mixture of hexadecane and glucose, ICL is suppressed until glucose is exhausted and then is induced providing the consumption of hexadecane [37]. In contrast, the glyoxylate cycle in *Candida tropicalis* functions even in the presence of glucose in the medium; this yeast first transforms carbohydrates to fatty acids, which then are utilized through β-oxidation and the glyoxylate cycle [40]. The study of the genetic mechanisms of ICL in *Escherichia coli* showed that the *aceA* gene of the acetate operon by the special repressor protein IclR [41]. The ICL gene responsible for the synthesis of ICL was isolated and cloned from the yeast *Saccharomyces cerevisiae* [42], the fungus *Aspergillus nidulans* [43], the yeast *Y. lipolytica* [44] and other microorganisms. The genetic studies of *S. cerevisiae* showed that glucose suppresses ICL with the aid of ten amino acids inside the polypeptide chain with the involvement of the cAMP-dependent protein kinase [45,46]. The promoter region necessary for the activation of ICL synthesis is common to the genes encoding the enzymes involved in gluconeogenesis [47]. The deletion of the *ICL1* gene from the genome of *Y. lipolytica* not only inhibits the assimilation of acetate, ethanol and fatty acids, but also reduces the rate of growth of this yeast on glucose [44].

The task of this section was to find an inhibitor of ICL, which could not be metabolized by cells. The published data on ICL inhibitors are discussed in detail below. ICL of *C. guillermondii* was found to be inhibited by low concentrations of glucose-6-P (Ki = 0.8 mM), the degree of inhibition being decreased by high concentrations of ICA. Glucose-6-P also inhibits the ICL of *C. tropicalis* but Ki is considerably higher (Ki = 4–5 mM) [48]. The effect of glucose-6-P on ICL is explained by the inhibition of the first enzyme of gluconeogenesis from C2-compounds by the end product, and hence can be considered as retroinhibition [49]. The strong inhibitor of yeast ICL is 6-phosphogluconate, whose inhibitory action is complete and competitive. This fact is explained by the structural and steric similarity of 6-phosphogluconate and ICA [49]. Fructose-1,6-bisphosphate is an inhibitor of ICL; Ki for *C. tropicalis* is equal to 5 mM [48]. Phosphoenolpyruvate is a strong allosteric inhibitor of ICL. When yeasts are cultivated in the medium with glucose or intermediates of gluconeogenesis, phosphoenolpyruvate can be an intermediate of glycolysis provided that the yeast cells utilize noncarbohydrate carbon sources [32]. The inhibitor constant (Ki) for phosphoenolpyruvate is low: 0.22 mM for *E. coli* [50], and 0.6 mM for *Phycomyces blakesleeanus* [51]. On the other hand, ICL of *Ashbya gossypii* is weakly (by 55%) inhibited by 10 mM phosphoenolpyruvate [52]. The data available in the literature for yeasts show that phosphoenolpyruvate is a noncompetitive inhibitor of ICL, the inhibitor constant Ki being different for different yeasts. For example, Ki for *S. cerevisiae* is equal to 1–4 mM [49], for *C. guilliermondii*, 1.6 mM [53], for *C. tropicalis*, 5 mM [48]. Pyruvate is also a strong inhibitor of ICL, Ki for ICL from *C. guilliermondii* being equal to 0.8 mM [53], and that from *C. tropicalis*, 5.0 mM [48]. ICL is also inhibited by TCA cycle intermediates: citrate (≥5 mM for *C. tropicalis* [48] and 10 mM for *C. guilliermondii* [53]); α-ketoglutarate (≥5 mM for *C. tropicalis* [48]); fumarate (9 mM for *C. guilliermondii* [53]) and malate 9 mM for *C. guilliermondii* [53]).The ICL of *C. guilliermondii* is strongly inhibited by 0.1 mM oxaloacetate (Ki = 0.05 mM) [53]. But in *C. tropicalis*, ICL is also inhibited by oxaloacetate, but to a lower degree (Ki = 5.0 mM) [48]. ICL is inhibited by the excess of its substrate and products (ICA, glyoxylate and succinate) [33,48,53–56].

The inhibitory action of itaconate (a structural analogue of succinate) and oxalate (a structural analogue of glyoxylate) is well known. Experiments with the permeabilized cells of *Pseudomonas indigofera* showed that itaconic acid specifically inhibits ICL and is not metabolized by the cells [55]. Ki for ICL from different cultures was found to be very different. For example, Ki = 100 mM for the fungus *Mucor racemosus* grown on acetate [57]. Oxalate and itaconate strongly inhibited the ICL of *Ashbya gossypii* (Ki = 4 μM and 127 μM, respectively) [52] and that of the basidiomycete *Fomitopsis palustris* (Ki = 37 μM and 68 μM, respectively) [58]. The inhibitory action of itaconate on ICL and the formation of ICA were found in the wild strain *C. lipolytica* grown on n-alkanes [2,59]. We reported that itaconate shifts of acid formation toward the preferential synthesis of ICA from rapeseed oil and increased the ratio ICA/CA by 6 times [23]. However, this inhibitor elevated the ICA/CA ratio only by 1.5 times for *Y. lipolytica*, grown on biodiesel waste [13].

Up to yet the metabolization of all the aforementioned inhibitors by *Y. lipolytica* were not studied in details.

We carried out experiments to study the metabolization of putative inhibitors noted by *Y. lipolytica* VKM Y-2373.

As can be seen from Table 4, when inhibitors were added to the medium in an amount of 0.2–2.1 g/L simultaneously with the main source—sunflower oil (5 g/L), putative ICL inhibitors were quickly assimilated and included in the metabolism. However, *Y. lipolytica* VKM Y-2373 does not metabolize only two compounds—itaconic and oxalic acids. These acids introduced into the medium at the beginning of cultivation are found in the medium after the complete consumption of sunflower oil.

Table 4. The metabolization of putative inhibitors by *Y. lipolytica* VKM Y-2373, grown on sunflower oil.

Inhibitor	Concentration in Cultural Medium (g/L)	
	Initial	Residual
Glucose 6-phosphate	1.1	0
Fructose 1,6-bisphosphate	2.1	0
Pyruvic acid	0.2	0
Citric acid	0.6	0
α-Ketoglutaric acid	0.4	0
Fumaric acid	0.3	0
Malic acid	0.5	0
Oxaloacetic acid	0.4	0
Itaconic acid	0.4	0.4
Oxalate	0.4	0.4

In subsequent experiments we examined the effect of itaconate, oxalate, as well as their combined effect on the growth and ICA production of *Y. lipolytica* VKM Y-2373.

As seen from Table 5, the concentration of itaconate in the medium should not exceed 40 mM, since at higher concentrations (50 mM), the biomass accumulation decreased by 1.3 times, as compared to control experiment, when the inhibitor was not added to the medium. With the addition of 30 mM itaconic acid, ICA production increased from 51.5 to 70.60 g/L. A further increase in the concentration of itaconate to 50 mM and more led to inhibition of ICA synthesis in 2.3 times. At 30 mM itaconate *Y. lipolytica* VKM Y-2373 shifted towards ICA (ICA/CA = 4:1 in comparison with a ratio of 1.7:1 without inhibitor. The maximum values of the ICA yield (Y_{ICA} = 1.25 g/g) and volume productivity (Q_{ICA} = 1.19 g/L·h) were obtained at 30 mM itaconic acid. The study of enzyme activity revealed that an addition of 20–50 mM itaconic acid resulted in a 12-fold decrease in activity of ICL, while CS, AH and NAD-ICDH slowly down.

Table 5. Effect of itaconic acid on the growth of *Y. lipolytica* VKM Y-2373 and ICA production.

Parameter	Concentration of Itaconic Acid (mM)				
	0	20	30	40	50
Biomass (g/L)	10.67 ± 0.15	10.1 ± 1.15	10.85 ±1.15	9.7 ± 1.10	7.7 ± 1.20
ICA (g/L)	51.50 ± 1.5	62.70 ± 4.45	70.60 ± 2.15	51.00 ± 3.45	31.00 ± 6.15
CA (g/L)	30.23 ± 0.60	23.50 ± 2.15	17.50 ± 1.15	12.75 ± 2.25	7.75 ± 2.15
ICA/CA ratio	1.7:1	2.8:1	4:1	4:1	4:1
Y_{ICA} (g/g)	0.91	1.11	1.25	0.90	0.55
Q_{ICA} (g/L·h)	0.87	1.06	1.19	0.86	0.52
Enzyme activity (U/mg protein)					
CS	2.66 ± 0.20	2.30 ± 0.35	2.20 ± 0.35	2.20 ± 0.35	2.10 ± 0.20
AH	0.95 ± 0.15	0.85 ± 0.07	0.85 ± 0.07	0.80 ± 0.17	0.75 ± 0.10
NAD-ICDH	0.03 ± 0.01	0.02 ± 0.01	0.02 ± 0.01	0.02 ± 0.01	0.02 ± 0.01
ICL	0.12 ± 0.01	0.01 ± 0.01	0.01 ± 0.01	0.01 ± 0.01	0.01 ± 0.01

As seen from Table 6, oxalic acid had a similar, but even weaker, effect for ICA production. An addition of 30 mM oxalic acid resulted in inhibition of isocitrate lyase (8-fold), whereas a shift in ratio between formed ICA and CA consisted of 2.85:1 and ICA production was only 23% higher than in the absence of the inhibitor. It can be assumed that the weaker effect of oxalate in comparison with itaconate on *Y. lipolytica* was associated with the different permeability of the cell wall for these two anions.

Table 6. ICA production by *Y. lipolytica* cultivated on sunflower oil with and without inhibitors of ICL.

Parameter	Without Inhibitor	With Inhibitor		
		Itaconic Acid 30 mM	Oxalic Acid 30 mM	Itaconic Acid + Oxalic Acid
Biomass (g/L)	10.67 ± 0.15	10.85 ± 1.15	9.80 ± 0.75	6.10 ± 0.65
ICA (g/L)	51.50 ± 1.5	70.60 ± 2.15	62.80 ± 4.75	68.00 ± 2.20
CA (g/L)	30.23 ± 0.60	17.50 ± 1.15	22.20 ± 2.15	17.00 ± 2.15
ICA/CA ratio	1.7:1	4:1	2.9:1	4:1
Y_{ICA} (g/g)	0.91	1.25	1.11	1.2
Q_{ICA} (g/L·h)	0.87	1.19	1.05	1.15
Enzyme activity (U/mg protein)				
CS	2.66 ± 0.20	2.20 ± 0.35	2.70 ± 0.20	2.20 ± 0.25
AH	0.95 ± 0.15	0.85 ± 0.07	0.78 ± 0.10	0.80 ± 0.10
NAD-ICDH	0.03 ± 0.01	0.02 ± 0.01	0.03 ± 0.02	0.03 ± 0.01
ICL	0.12 ± 0.01	0.01 ± 0.01	0.01 ± 0.01	0.01 ± 0.01

The combined effect of itaconic and oxalic acids on the production of ICA by *Y. lipolytica* VKM Y-2373) was first studied in this work. As evident from the comparison of biotechnological parameters shown in Table 6, the combined effect of inhibitors practically do not affect ICA production.

The presented results demonstrate that ICA synthesis is associated with high activities of citrate synthase and aconitate hydratase and low activities of NAD-isocitrate dehydrogenase and isocitrate lyase. Recently it was reported the genetically modified strain with the superexpressed *CIT1* and *CIT2* genes produced citric and isocitric acid from vegetable oil in a ratio close to 1:1while CA/ICA ratio for wild-type strain was 4.12:1 [19].

Y. lipolytica VKM Y-2373 is recommended to cultivate under nitrogen deficiency conditions with addition of iron (1.5 mg/L) and 30 mM itaconic acid. Such nutrition medium provides 70.6 g/L ICA with a ratio of 4:1, a mass yield (Y_{ICA}) of 1.25 g/g and volume productivity (Q_{ICA}) of 1.19 g/L·h.

4. Conclusions

Thus, *Y. lipolytica* VKM Y-2373 is as a perspective producer of ICA from sunflower oil. The intensive ICA production occurred only under the limitation of cell growth by nitrogen, phosphorus, sulfur or magnesium and the excess sunflower oil. The concentration of iron (activator of AH) and itaconic acid (inhibitor of ICL) are effective factors controlling the ICA synthesis in *Y. lipolytica*. Under chosen concentration of iron (1.5 mg/L) and itaconic acid (30 mM), the selected strain produced up to 70.6 g/L ICA with a ratio between ICA and CA equal 4:1, a mass yield (Y_{ICA}) of 1.25 g/g and volume productivity (Q_{ICA}) of 1.19 g/L·h.

Author Contributions: Conceptualization, I.G.M. and V.A.S.; investigation, S.V.K., J.N.L. and I.G.M.; methodology, S.V.K. and I.G.M.; resources, I.G.M. and V.A.S.; supervision, I.G.M.; writing—original draft, S.V.K. All authors have read and agreed to the published version of the manuscript.

References

1. Vickery, H.B. A suggested new nomenclature for the isomers of isocitric acid. *J. Biol. Chem.* **1962**, *237*, 1739–1741.
2. Finogenova, T.V.; Morgunov, I.G.; Kamzolova, S.V.; Chernyavskaya, O.G. Organic acid production by the yeast *Yarrowia lipolytica*: A review of prospects. *Appl. Biochem. Microbiol.* **2005**, *41*, 418–425. [CrossRef]
3. Heretsch, P.; Thomas, F.; Aurich, A.; Krautscheid, H.; Sicker, D.; Giannis, A. Syntheses with a chiral building block from the citric acid cycle: (2R,3S)-isocitric acid by fermentation of sunflower oil. *Angew Chem. Int. Ed. Engl.* **2008**, *47*, 1958–1960. [CrossRef]
4. Aurich, A.; Specht, R.; Müller, R.A.; Stottmeister, U.; Yovkova, V.; Otto, C.; Holz, M.; Barth, G.; Heretsch, P.; Thomas, F.A.; et al. Microbiologically produced carboxylic acids used as building blocks in organic synthesis.

In *Reprogramming Microbial Metabolic Pathways. Subcellular Biochemistry*; Wang, X., Chen, J., Quinn, P., Eds.; Springer: Dordrecht, The Netherlands, 2012; Volume 64, pp. 391–423. [CrossRef]

5. Otto, C.; Holz, M.; Barth, G. Production of Organic Acids by *Yarrowia lipolytica*. In *Yarrowia lipolytica*; Microbiology Monographs; Barth, G., Ed.; Springer: Berlin/Heidelberg, Germany, 2013; pp. 137–151. [CrossRef]
6. Aurich, A.; Hofmann, J.; Oltrogge, R.; Wecks, M.; Glaser, R.; Blömer, L.; Mauersberger, S.; Roland, A.; Müller, R.A.; Sicker, D.; et al. Improved isolation of microbiologically produced (2R,3S)-isocitric acid by adsorption on activated carbon and recovery with methanol. *Org. Process Res. Dev.* **2017**, *21*, 866–870. [CrossRef]
7. Giannis, A.; Mousavizadeh, F. On Artemisinin, Cyclopamine, D-Isocitric acid, Hyperforin, Epigenetics, Sialic Acid, and More. *Synlett* **2019**, *30*, 1401–1418. [CrossRef]
8. Bullin, K.; Hennig, L.; Herold, R.; Krautscheid, H.; Richter, K.; Sicker, D. An optimized method for an (2R,3S)-isocitric acid building block. *Monatsh. Chem.* **2019**, *150*, 247–253. [CrossRef]
9. Kamzolova, S.V.; Morgunov, I.G. Microbial production of (2R,3S)-isocitric acid: State of the arts and prospects (Review). *Appl. Microbiol. Biotechnol.* **2019**, *103*, 9321–93333. [CrossRef]
10. Fickers, P.; Cheng, H.; Sze, K.; Lin, C. Sugar alcohols and organic acids synthesis in *Yarrowia lipolytica*: Where Are We? *Microorganisms* **2020**, *8*, 574. [CrossRef]
11. Morgunov, I.G.; Karpukhina, O.V.; Kamzolova, S.V.; Samoilenko, V.A.; Inozemtsev, A.N. Investigation of the effect of biologically active threo-Ds-isocitric acid on oxidative stress in *Paramecium caudatum*. *Prep. Biochem. Biotechnol.* **2018**, *48*, 1–5. [CrossRef]
12. Morgunov, I.G.; Kamzolova, S.V.; Karpukhina, O.V.; Bokieva, S.V.; Inozemtsev, A.N. Biosynthesis of isocitric acid in repeated-batch culture and testing of its stress-protective activity. *Appl. Microbiol. Biotechnol.* **2019**, *103*, 3549–3558. [CrossRef]
13. Morgunov, I.G.; Kamzolova, S.V.; Karpukhina, O.V.; Bokieva, S.V.; Lunina, J.N.; Inozemtsev, A.N. Microbiological production of isocitric acid from biodiesel waste and its effect on spatial memory. *Microorganisms* **2020**, *8*, 462. [CrossRef]
14. Finogenova, T.V. Biosynthesis of Organic Acids by Yeast Organisms and its Regulation. Ph.D. Thesis, USSR Academy of Sciences, Institute of Biochemistry and Physiology of Microorganisms, Pushchino, Russia, 1982.
15. Kamzolova, S.V.; Shamin, R.V.; Stepanova, N.N.; Morgunov, G.I.; Lunina, J.N.; Allayarov, R.K.; Samoilenko, V.A.; Morgunov, I.G. Fermentation conditions and media optimization for isocitric acid production from ethanol by *Yarrowia lipolytica*. *BioMed Res. Int.* **2018**, 2543210. [CrossRef]
16. Holz, M.; Förster, A.; Mauersberger, S.; Barth, G. Aconitase overexpression changes the product ratio of citric acid production by *Yarrowia lipolytica*. *Appl. Microbiol. Biotechnol.* **2009**, *81*, 1087–1096. [CrossRef]
17. Kamzolova, S.V.; Dedyukhina, E.G.; Samoilenko, V.A.; Lunina, J.N.; Puntus, I.F.; Allayarov, R.K.; Chiglintseva, M.N.; Mironov, A.A.; Morgunov, I.G. Isocitric acid production from rapeseed oil by *Yarrowia lipolytica* yeast. *Appl. Microbiol. Biotechnol.* **2013**, *97*, 9133–9144. [CrossRef]
18. Kamzolova, S.V.; Lunina, Y.N.; Puntus, I.F.; Allayarov, R.K.; Laptev, I.A.; Samoilenko, V.A.; Morgunov, I.G. Biosynthesis of isocitric acid by the yeast *Yarrowia lipolytica* and its regulation. *Appl. Biochem. Microbiol.* **2015**, *51*, 249–254. [CrossRef]
19. Hapeta, P.; Rakicka-Pustułka, M.; Juszczyk, P.; Robak, M.; Rymowicz, W.; Lazar, Z. Overexpression of citrate synthase increases isocitric acid biosynthesis in the *Yarrowia lipolytica*. *Sustainability* **2020**, *12*, 7364. [CrossRef]
20. Da Silva, L.V.; Tavares, C.B.; Amaral, P.F.F.; Coehlo, M.A.Z. Production of citric acid by *Yarrowia lipolytica* in different crude oil concentrations and in different nitrogen sources. *Chem. Eng. Trans.* **2012**, *27*, 199–204.
21. Morgunov, I.G.; Kamzolova, S.V. Physiologo-biochemical characteristics of citrate-producing yeast *Yarrowia lipolytica* grown on glycerol-containing waste of biodiesel industry. *Appl. Microbiol. Biotechnol.* **2015**, *99*, 6443–6450. [CrossRef]
22. Rzechonek, D.A.; Dobrowolski, A.; Rymowicz, W.; Mirończuk, A.M. Aseptic production of citric and isocitric acid from crude glycerol by genetically modified *Yarrowia lipolytica*. *Bioresour. Technol.* **2019**, *271*, 340–344. [CrossRef]
23. Kamzolova, S.V.; Allayarov, R.K.; Lunina, J.N.; Morgunov, I.G. The effect of oxalic and itaconic acids on threo-Ds-isocitric acid production from rapeseed oil by *Yarrowia lipolytica*. *Bioresour. Technol.* **2016**, *206*, 128–133. [CrossRef]

24. Förster, A.; Jacobs, K.; Juretzek, T.; Mauersberger, S.; Barth, B. Overexpression of the *ICL1* gene changes the product ratio of citric acid production by *Yarrowia lipolytica*. *Appl. Microbiol. Biotechnol.* **2007**, *77*, 861–869. [CrossRef]
25. Abe, M.; Tabuchi, T.; Tanaka, M. Studies on organic acid fermentation in yeast. III. Accumulation of isocitric acid in cultures of yeast. *J. Agric. Chem. Soc. Jpn.* **1970**, *44*, 493–498.
26. Oogaki, M.; Inoue, M.; Kaimaktchiev, A.C.; Nakahara, T.; Tabuchi, T. Production of isocitric acid from glucose by *Candida ravautii*. *Agric. Biol. Chem.* **1984**, *48*, 789–795. [CrossRef]
27. Nakahara, T.; Kaimaktchiev, A.C.; Oogaki-Chino, M.; Uchida, Y.; Tebuchi, T. Isocitric acid production from n-alanes by *Candida catenulate*. *Agric. Biol. Chem.* **1987**, *51*, 2111–2116.
28. Moeller, L.; Strehlitz, B.; Aurich, A.; Zehnsdorf, A.; Bley, T. Optimization of citric acid production from glucose by *Yarrowia lipolytica*. *Eng. Life Sci.* **2007**, *7*, 504–511. [CrossRef]
29. Papanikolaou, S.; Galiotou-Panayotou, M.; Fakas, S.; Komaitis, M.; Aggelis, G. Citric acid production by *Yarrowia lipolytica* cultivated on olive-mill wastewater-based media. *Bioresour. Technol.* **2008**, *99*, 2419–2428. [CrossRef]
30. Carsanba, E.; Papanikolaou, S.; Fickers, P.; Erten, H. Screening various *Yarrowia lipolytica* strains for citric acid production. *Yeast* **2019**, *36*, 319–327. [CrossRef]
31. Tomaszewska, L.; Rakicka, M.; Rymowicz, W.; Rywińska, A. A comparative study on glycerol metabolism to erythritol and citric acid in *Yarrowialipolytica* yeast cells. *FEMS Yeast Res.* **2014**, *14*, 966–976. [CrossRef] [PubMed]
32. Kornberg, H.L. The role and control of the glyoxylate cycle in *Escherichia coli*. *Biochem. J.* **1966**, *99*, 1–11. [CrossRef]
33. Eprintsev, A.T.; Salnikov, A.V.; Haba, A.M.; Zaichikova, M.V. Isocitrate lyase isozymes and their role in organisms with different levels of organization. *Biol. Bull. Rev.* **2014**, *4*, 323–334. [CrossRef]
34. Evans, C.T.; Ratledge, C. The physiological significance of citric acid in the control of metabolism in lipid-accumulating yeasts. *Biotechnol. Genet. Eng. Rev.* **1985**, *3*, 85–111. [CrossRef]
35. Holdsworth, J.E.; Veenhuis, H.; Ratledge, C. Enzyme activities in oleaginous yeasts accumulating and utilizing exogenous or endogenous lipids. *J. Gen. Microbiol.* **1988**, *134*, 2907–2915. [CrossRef] [PubMed]
36. De Jong-Gubbels, P.; Vanrolleghem, P.; Heijnen, S.; Van Dijken, J.P.; Pronk, J.T. Regulation of carbon metabolism in chemostat cultures of *Saccharomyces cerevisiae* grown on mixtures of glucose and ethanol. *Yeast* **1995**, *11*, 407–418. [CrossRef] [PubMed]
37. Kamzolova, S.V.; Lunina, J.N.; Morgunov, I.G. Biochemistry of citric acid production from rapeseed oil by *Yarrowia lipolytica* yeast. *J. Am. Oil Chem. Soc.* **2011**, *88*, 1965–1976. [CrossRef]
38. Papanikolaou, S.; Aggelis, G. Lipids of oleaginous yeasts. Part II: Technology and potential applications. *Eur. J. Lipid Sci. Technol.* **2011**, *113*, 1052–1073. [CrossRef]
39. Lopez-Boado, Y.S.; Herrero, P.; Gascon, S.; Moreno, F. Catabolite inactivation of isocitrate lyase from *Saccharomyces cerevisiae*. *Arch. Microbiol.* **1987**, *147*, 231–234. [CrossRef]
40. Cioni, M.; Pinzauti, G.; Vanni, P. Comparative biochemistry of the glyoxylate cycle. *Comp. Biochem. Physiol.* **1981**, *70*, 1–26. [CrossRef]
41. Cortay, J.C.; Negre, D.; Galinier, A.; Duclos, B.; Perriere, G.; Cozzone, A.J. Regulation of the acetate operon in *Escherichia coli*: Purification and functional characterization of the IclR repressor. *EMBO J.* **1991**, *10*, 675–679. [CrossRef]
42. Schöler, A.; Schüller, H.J. Structure and regulation of the isocitrate lyase gene ICL1 from the yeast *Saccharomyces cerevisiae*. *Curr. Genet.* **1993**, *23*, 375–381. [CrossRef]
43. McCullough, W.; Shanks, A. Properties of genes involved in the control of isocitrate lyase production in *Aspergillus nidulans*. *J. Gen. Microbiol.* **1993**, *139*, 509–511. [CrossRef]
44. Barth, G.; Scheuber, T. Cloning of the isocitrate lyase gene (*ICL1*) from *Yarrowia lipolytica* and characterization of the deduced protein. *Mol. Gen. Genet.* **1993**, *241*, 422–430. [CrossRef]
45. Ordiz, I.; Herrero, P.; Rodicio, R.; Moreno, F. Glucose-induced inactivation of isocitrate lyase in *Saccharomyces cerevisiae* is mediated by an internal decapeptide sequence. *FEBS Lett.* **1995**, *367*, 219–222. [CrossRef]
46. Ordiz, I.; Herrero, P.; Rodicio, R.; Moreno, F. Glucose-induced inactivation of isocitrate lyase in *Saccharomyces cerevisiae* is mediated by the cAMP-dependent protein kinase catalytic subunits Tpk 1 and Tpk2. *FEBS Lett.* **1996**, *385*, 43–46. [CrossRef]

47. Schöler, A.; Schüller, H.J. A carbon source-responsive promoter element necessary for activation of the isocitrate lyase gene ICL1 is common to genes of the gluconeogenic pathway in the yeast *Saccharomyces cerevisiae*. *Mol. Cell. Biol.* **1994**, *14*, 3613–3622. [CrossRef]
48. Tanaka, A.; Nabeshima, S.; Tokuda, M.; Fukui, S. Studies on the physiology and metabolism of hydrocarbon-utilizing microorganisms. XX. Partial purification of isocitrate lyase from *Candida tropicalis* and some kinetic properties of the enzyme. *Agric. Biol. Chem.* **1977**, *41*, 795–801. [CrossRef]
49. Hanozet, G.M.; Guerritore, A. Role of phosphoenolpyruvate and 6-phosphogluconate in the short-term control of yeast isocitrate lyase. *Arch. Biochem. Biophys.* **1972**, *149*, 127–135. [CrossRef]
50. LaPorte, D.C.; Walsh, K.; Koshland, D.E., Jr. The branch point effect. Ultrasensitivity and subsensitivity to metabolic control. *J. Biol. Chem.* **1984**, *259*, 14068–14075.
51. Rúa, J.; de Arriaga, D.; Busto, F.; Soler, J. Isocitrate lyase from *Phycomycesblakesleeanus*. The role of Mg^{2+} ions, kinetics and evidence for two classes of modifiable thiol groups. *Biochem. J.* **1990**, *272*, 359–367. [CrossRef]
52. Schmidt, G.; Stahmann, K.P.; Sahm, H. Inhibition of purified isocitrate lyase identified itaconate and oxalate as potential antimetabolites for the riboflavin overproducer *Ashbya gossypii*. *Microbiology* **1996**, *142*, 411–417. [CrossRef]
53. Hildebrandt, W.; Weide, H. Isocitrate lyase of *Candida guilliermondii*, strain H17. 3. Regulation by intermediates of alkane catabolism and general model of regulation. *Z. Allg. Mikrobiol.* **1974**, *14*, 47–52. [CrossRef]
54. Hoyt, J.C.; Johnson, K.E.; Reeves, H.C. Purification and characterization of *Acinetobacter calcoaceticus* isocitrate lyase. *J. Bacteriol.* **1991**, *173*, 6844–6848. [CrossRef]
55. McFadden, B.A.; Purohit, S. Itaconate, an isocitrate lyase-directed inhibitor in *Pseudomonas indigofera*. *J. Bacteriol.* **1977**, *131*, 136–144. [CrossRef]
56. Krátký, M.; Vinšová, J. Advances in mycobacterial isocitrate lyase targeting and inhibitors. *Curr. Med. Chem.* **2012**, *19*, 6126–6137. [CrossRef]
57. O'Connell, B.T.; Paznokas, J.L. Glyoxylate cycle in *Mucor racemosus*. *J. Bacteriol.* **1980**, *143*, 416–421. [CrossRef]
58. Munir, E.; Hattori, T.; Shimada, M. Purification and characterization of isocitrate lyase from the wood-destroying basidiomycete *Fomitopsis palustris* grown on glucose. *Arch. Biochem. Biophys.* **2002**, *399*, 225–231. [CrossRef]
59. Karklin, R.; Peltzmane, I.; Raminya, L.; Korde, G. Overproduction of isocitric acid by wild strain of *Candida lipolytica*. In *Metabolism of n-Alkanes and Oversynthesis of Products by Microorganisms*; Finogenova, T.V., Sharyshev, A.A., Eds.; USSR Academy of Sciences, Institute of Biochemistry and Physiology of Microorganisms Pushchino: Pushchino, Russia, 1991; pp. 143–146.

2

Saccharomyces arboricola and its Hybrids' Propensity for Sake Production: Interspecific Hybrids Reveal Increased Fermentation Abilities and a Mosaic Metabolic Profile

Matthew J. Winans [1,2,*], **Yuki Yamamoto** [1], **Yuki Fujimaru** [1], **Yuki Kusaba** [1], **Jennifer E. G. Gallagher** [2] and **Hiroshi Kitagaki** [1]

1 Graduate School of Advanced Health Sciences, Saga University, 1, Honjo, Saga city, Saga 840-8502, Japan; 18555010@edu.cc.saga-u.ac.jp (Y.Y.); yukidenkan37@gmail.com (Y.F.); 19626004@edu.cc.saga-u.ac.jp (Y.K.); ktgkhrs@cc.saga-u.ac.jp (H.K.)

2 Biology Department, West Virginia University, 53 Campus Drive, Morgantown, WV 26506-6057, USA; jegallagher@mix.wvu.edu

* Correspondence: mwinans@mix.wvu.edu

Abstract: The use of interspecific hybrids during the industrial fermentation process has been well established, positioning the frontier of advancement in brewing to capitalize on the potential of *Saccharomyces* hybridization. Interspecific yeast hybrids used in modern monoculture inoculations benefit from a wide range of volatile metabolites that broaden the organoleptic complexity. This is the first report of sake brewing by *Saccharomyces arboricola* and its hybrids. *S. arboricola* x *S. cerevisiae* direct-mating generated cryotolerant interspecific hybrids which increased yields of ethanol and ethyl hexanoate compared to parental strains, important flavor attributes of fine Japanese ginjo sake rice wine. Hierarchical clustering heatmapping with principal component analysis for metabolic profiling was used in finding low levels of endogenous amino/organic acids clustered *S. arboricola* apart from the *S. cerevisiae* industrial strains. In sake fermentations, hybrid strains showed a mosaic profile of parental strains, while metabolic analysis suggested *S. arboricola* had a lower amino acid net uptake than *S. cerevisiae*. Additionally, this research found an increase in ethanolic fermentation from pyruvate and increased sulfur metabolism. Together, these results suggest *S. arboricola* is poised for in-depth metabolomic exploration in sake fermentation.

Keywords: sake; fermentation; hybrid; ginjo; ethanol; ethyl hexanoate; isoamyl acetate; yeast; *Saccharomyces*; metabolism

---

1. Introduction

As currently defined, the budding yeast clade known as *Saccharomyces* has eight members: *S. cerevisiae*, *S. paradoxus* (syn. *S. cerevisiae* var. *tetraspora*, *S. cerevisiae* var. *terrestris*), *S. uvarum* (syn. *S. bayanus* var. *uvarum*), *S. mikatae*, *S. kudriavzevii*, *S. arboricola* (syn. *S. arboricolus*), *S. eubayanus*, and *S. jurei* [1–9]. Additionally, *S. pastorianus* (syn. *S. carlsbergensis*, *S. monacensis*) and *S. bayanus* are two natural hybrids found in this group [10–12]. *S. arboricola* was first isolated from the tree bark of the *Fagaceea* family in the west and southwest of China [2] in 2008 and later also isolated in western Taiwan [5] and northern New Zealand [13]. The New Zealand isolate was reported to be highly divergent from the Chinese isolates used in this study, supporting the far East Asian origin theory for *Saccharomyces* yeast. The Chinese *S. arboricola* has a high quality annotated genomic sequence that when compared to *S. cerevisiae* shows a reciprocal translocation on chromosome (chr.) IV and XIII, inversion on chr.VI, high presence of non-syntenic genes near telomeric regions, lack of introgression, and four novel genes [14]. Between *S. cerevisiae* and other Saccharomyces members there is a 15–30% divergence, but the divergence between S. cerevisiae strains is 0.1–0.6% [15,16]. Additionally, subpopulations of *S. paradoxus* show a 1.5–4.5% divergence, while the divergence between *S. eubayanus* and *S. uvarum* is 6% [7,17]. The nucleotide divergence of yeast within *Saccharomyces* is near the man to bird comparison, but with the advantage of a physiological and structural similarity of man to chimp [18]. Favorable properties of different *Saccharomyces* yeast can be brought together by hybridization that often results in transgressive phenotypes. Typical environmentally isolated *Saccharomyces* strains are diploid while laboratory strains are often used in the haploid form to reduce variability and industrial yeast are frequently found to be polyploidy, potentially a result of favored fermentation environment. Although interspecies hybrids of the *Saccharomyces* genus are viable, the they do not produce viable gametes or ascospores in yeast resulting in reproductive isolation as a sterility barrier [9]. Hybridization of phenotypically diverse yeast has been exploited for the improvement of industrial strains for specific purposes.

Fermenting yeasts, specifically *S. cerevisiae,* has been unwittingly domesticated throughout humanity's existence via the production of various foods and beverages, including bread, beer, wine, and sake. One factor in *S. cerevisiae* selection is its unique metabolic capabilities as a Crabtree positive yeast; they amass ethanol in the presence of oxygen. The commonly preferred eukaryotic metabolic pathway in the presence of oxygen is oxidative respiration for its high ATP yield. But in the high concentrations of six-carbon carbohydrates, such as glucose, this pathway is repressed by the Crabtree effect. Under these conditions, the energy for growth comes from the glycolysis pathway until exhaustion of six-carbon molecules causes a diauxic shift to oxidation of the two-carbon ethanol molecules into CO_2 [19,20]. Crabtree positive microbes are beneficial in the production of alcoholic beverages which have been produced for at least 9000 years, and the earliest known recording of Japanese sake is mentioned ca. 285 AD in an ancient Chinese document called *Gishiwajinden* [21–23]. Sake is the alcoholic drink generated by fermenting rice and beer is that of barley, making their development specific to regions dominated by these grains. Rice and barley diverged 50 million years ago and rice was introduced to Japan from the Chinese Yangzi region around 400 BC, many differences exist between these two grains [24,25]. During the 15th century in Bavaria, lager-brewing arose [26] and is the most popular alcoholic beverage in the world today. Since the lager beer yeast, *S. pastorianus* was discovered to be a hybrid of *S. eubayanus* x *S. cerevisiae,* interest has sparked in the use of interspecific hybrids in fermentation [27]. Originally the *S. eubayanus* parent was identified in Patagonia and the question of lager yeast's origin has generated controversy [28–31]. Yeasts originated multiple times from various fungal lineages of which one gave rise to the budding yeasts [32]. The genomic diversity with geographical boundaries and many other lines of evidence support a Far East Asia origin for *Saccharomyces* [33–35].

Beer, wine, and sake are complex chemical matrixes resulting from many metabolic and chemical reactions in which raw ingredients determine initial sugar, assimilable nitrogen, and fatty acid profiles [25,36]. The volatile metabolic profile obtained during fermentation of different carbohydrates is reflective of the raw ingredients composition [37,38]. Wine is perhaps one of the oldest biotechnologies as yeast are often found on damaged grapes and crushing of the fruit facilitates the release of the fermentable sugars [39,40]. Obtaining fermentable sugars for beer and sake is quite different. Production of beer involves a saccharification of malted barley by the grain's own enzymes to convert their starches to maltose, glucose, and maltotriose found in descending concentrations. Boiling of the wort results in less than 3% of the malt lipids available to yeast during fermentation [41]. Fermentation of green beer is accomplished in 3 days to 10 days [42] and results in an average alcohol content of 5% (vol/vol) with some brews tipping the scales around 10% (vol/vol) or more.

Immature sake is generally produced over 41 days and involves several steamed rice additions at temperatures between 12–20 °C applying high osmotic pressure on the yeast from the rice mash ratios [43]. The highest alcohol yield in the world, 17–20% (vol/vol) without distillation, is obtained through this process [25]. *Aspergillus oryzae*, colloquially known as koji, is a keystone organism in the production of sake from polished rice [44]. By traditional Japanese sake mash methods, koji performs a continuous saccharification of highly milled rice starches to the fermentable sugars and the budding yeast, *S. cerevisiae*, ferments these sugars into ethanol [45]. Free sugars such as glucose and oligomers are produced by the koji enzymes glucoamylase and α-amylase during the sake mash [23]. Recent work has shown that koji secretes mannitol 2-dehydrogenase which converts fructose to mannitol in the sake mash. Interestingly, *S. arboricola* efficiently catabolizes mannitol, an abundant natural energy storage molecule that *Aspergillus* sp. produce in high concentrations [46,47]. This phenotype is superior to *S. cerevisiae* metabolism of mannitol [14]. Considering the temperature tolerance and growth capabilities on mannitol, it was hypothesized that *S. arboricola* and interspecific hybrids are uniquely fit for sake fermentation with *A. oryzae* because of their metabolic access to the underutilized sugar-alcohol and temperature preference which affects the flavor active metabolites.

The aim of this study was to characterize the metabolic capabilities of *S. arboricola* and its hybrids with *S. cerevisiae* in Japanese rice wine. For the first time, sake fermentation profiling of *S. arboricola* and newly generated *S. arboricola* x *S. cerevisiae* interspecific hybrids is reported. The wine and beer brewing industries have often employed interspecific hybrids for their fermentations. *S. kudriavzevii* x *S. cerevisiae* are known to ferment wines, *S. mikatae* x *S. cerevisiae* hybrids have been actively produced for wine, and the *S. cerevisiae* x *S. eubayanus* hybrid produces the most popular beer in the world [33,48,49]. Limited research has been conducted on the metabolic capabilities of *S. arboricola* and these results suggest *S. arboricola* and the hybrids' propensity for fermentation may be suited for sake production.

2. Materials and Methods

2.1. Yeast Strains

Yeast strains of *Saccharomyces cerevisiae* included Italian wine isolate M22 [50], West African wine isolate DBVPG6044 [51], sake isolate Kyokai No. 7 known as haploid 868K7 [52], sake isolate TCR7 [53], American brewing isolate WLP001, and American brewing isolate HT01. Yeast strains of *Saccharomyces arboricola* included SA350 (syn. AS 2.3317), SA351 (syn. AS 2.3318), SA352 (syn. AS 2.3319 [14]. Hybrids SAM8c and SAM9a were between *S. arboricola* 350 x *S. cerevisiae* M22 (Table 1). A minimum of three biological replicates were used in each analysis.

Table 1. Yeast Strains Used During this Study.

Strain	Species	Isolation Source	Reference
SA350 (syn. AS 2.3317)	*S. arboricola*	Tree bark-China	Liti, G., et al. (2013)
SA351 (syn. AS 2.3318)	*S. arboricola*	Tree bark-China	Liti, G., et al. (2013)
SA352 (syn. AS 2.3319)	*S. arboricola*	Tree bark-China	Liti, G., et al. (2013)
M22	*S. cerevisiae*	Wine-Italy	Capece, A., et al. (2012)
DBVPG6044	*S. cerevisiae*	Billi wine-West Africa	Song, G., et al. (2015)
Kyokai No. 7 (syn. 868K7/ K7)	*S. cerevisiae*	Sake-Japan	Katou, T., et. al. (2008)
TCR7	*S. cerevisiae*	Sake-Japan	Horie, K., et al. (2010)
WLP001	*S. cerevisiae*	Beer, American ale	Commercially available from White Labs, USA
HT01	*S. cerevisiae*	Beer, American ale	Current study
SAM8c	Hybrid (*S. arboricola x S. cerevisiae*)	Lab, spore to spore mating	Current study
SAM9a	Hybrid (*S. arboricola x S. cerevisiae*)	Lab, spore to spore mating	Current study

2.2. *Yeast Isolation*

Saccharomyces cerevisiae HT01 was the only strain isolated for this study and was isolated as described previously [54] with slight modification. Briefly, beer was harvested from a can sanitized on the outside with 70% ethanol into an enrichment media (3 g yeast extract, 3 g malt extract, 5 g peptone, 10 g sucrose, 76 mL EtOH, 1 mg chloramphenicol, and 1 mL of 1-M HCl per liter). Samples were transferred to sterile tubes and incubated until signs of yeast appeared. Cultures were then streaked onto YPD media and single colonies were grown and stocked. From these stocks, the yeast culture's genomic DNA was extracted, the ITS region was amplified via polymerase chain reaction (PCR) [55], Sanger sequenced at the West Virginia University Genomic Core, and the isolated yeast was confirmed to be *S. cerevisiae* via utilizing NCBI Blast.

2.3. *Serial Spot Dilutions*

Serial spot dilution growth assays were conducted as described previously [56] with some modification. Briefly, OD_{600nm} growth readings were taken from fresh, overnight cultures of each yeast grown at 23 °C. For each yeast assayed, a total of 3.2×10^5 yeast cells were transferred to a well in the first column of a 96 well plate. Each yeast assayed was serially diluted 4 times in a 1:10 ratio, using sterile distilled and MilliQ (MQ) filtered water, across the rows which resulted in 5 wells each. A handheld 48 pin plate was used to place the serial dilution spots onto solid YPD agar media. Plates were then incubated at 4 °C for ten days, 10 °C for seven days, and 15 °C for four days. Growth was assessed and photographed.

2.4. *Sporulation and Spore Staining*

Fresh liquid yeast cultures were grown in YPD (1% yeast extract, 2% peptone, 2% dextrose) overnight at 23 °C in a shaking water bath. Isolates were streaked out onto YEP(EG) buffered agar plates (1% succinic acid, 1% yeast extract, 2% peptone, 2% glycerol, 2% agar, potassium hydroxide to pH 5.5, 2.5% ethanol) for 2–3 days. Single colonies were thinly patched onto Snack-Spo Zn agar plates (1% potassium acetate, 0.1% yeast extract, 0.05% dextrose, 0.005% zinc acetate, 2% agar) and incubated for 6 days. To measure spore formation, a modified Schaeffer-Fulton endospore staining technique was used. Briefly, cultures from the sporulation plates were suspended in 1ml distilled water and recovered by pipette. The suspended culture was fixed to a glass microscope slide via flame. Malachite green (10%, 0.1 g/mL) was then used to stain the spore walls for 15 min (minute/s) before decoloring and rinsing with water. Safranin (0.25%, 2.5 mg/mL) was used as a counterstain for non-sporulated cells for 1 min before rinsing with water and mounting in glycerol. Malachite green cells were counted as spores and safarin red cells were counted as non-sporulated yeast cells [57]. Images were taken under a light microscope and sporulation efficiency for each strain was calculated, reported in the results section, to select the best possibility of a successful hybridization.

2.5. *Interspecific Hybridization*

The ascus of the spore was digested over 1 hr. with Zymolase (25 mg/mL) at 37 °C. The digested spores were placed dropwise on YPD agar at an angle to facilitate a central streak of liquid across the plate. Micromanipulation of spores from *S. arboricola* strain 350 and *S. cerevisiae* M22 spores was performed by fiber optic needle under a light microscope using the SporePlay (Singer Instruments, Roadwater, England) dissection microscope by Singer Instruments and yielded 40 spore to spore mattings. Visual observations were conducted on the shmooing process during incubation at 25 °C. Single colonies were selected, streaked out for purification, and stocked at −80 °C with cryogenic media (1% yeast extract, 2% peptone, 15% glycerol).

2.6. *Genomic DNA Extraction*

Alkaline lysis genomic DNA extraction was performed on fresh, overnight cultures by harvesting (0.3×10^7) cells via cold centrifugation at 7500× *g* for 3 min. Cells were washed twice with phosphate-buffered solution (PBS) before NaOH (0.02 M) addition and subsequent incubation at 100 °C for 10 min. The cellular solution underwent cold centrifugation at 4 °C for 3 min before 250 μL of supernatant was mixed with 50 μL acetic acid (2.5 M) and then 300 μL of 2-propanol. Cold centrifugation was performed again for 15 min before the supernatant was discarded. Ethanol (80% v/v) was added to the DNA pellet to lightly mix the solution before a 5 min cold centrifugation. The DNA was dried at room temperature and resuspended in TE buffer.

2.7. *Polymerase Chain Reaction*

DNA concentrations were measured by OD_{260nm} via a glass black cell cuvette. PCR confirmed *S. arboricola* 350 and *S. cerevisiae* M22 (SAM hybrids) interspecific hybridization. Two multiplexed species-specific primer sets were used, detailed previously [58]. The oligonucleotide sequences used for primers were as follows; for *S. arboricola* sarbF1 (GGC ACG CCC TTA CAG CAG CAA TCG) and sarbR2 (TCG TCG TAC AGA TGC TGG TAG GGC GCT); for *S. cerevisiae* scerF2 (GCG CTT TAC ATT CAG ATC CCG AG) and scerR2 (TAA GTT GGT TGT CAG CAA GAT TG ATC). A master mix for the PCR reactions was made on ice and consisted of PCR reagents from kit: KOD-Plus-Neo (product code KOD-401) ordered from TOYOBO, Japan. The PCR Touchdown [59] thermocycle protocol was used. During gel electrophoresis [60], the *S. cerevisiae* band generated at 150bp (base pairs) and *S. arboricola* generated at 349 bp.

2.8. *Sake Fermentations*

Freshly grown cells (2.0×10^9) were measured by OD_{600nm}, harvested from YPD via centrifugation at 3200× *g* for 3 min, and washed in MilliQ water twice. The harvested yeast, rice, koji, lactate, and water were mixed together in mason jars covered with loose lids. Incubation occurred at 15 °C for 2 weeks with an initial remixing of the mash occurring after the first 24 h. Fermentation was monitored daily by mass reduction of the sake mash, which corresponded to CO_2 production. Fermentation trials were performed in triplicate. Post-fermentation, the sake mash was harvested into conical tubes by centrifuged at 4 °C and 3200× *g*. The supernatant was transferred to storage at 4 °C for further analysis. Statistical analysis of ethanol content was validated by one-way ANOVA with a two-tailed student's *t*-test post-hoc analysis and ordered difference report for each day of fermentation.

2.9. *Chromatography Analysis*

2.9.1. Gas Chromatography-Flame Ionization Detection

For gas chromatography-flame ionization detection (GC-FID) of the sake supernatant, ribitol (syn. adonitol) (0.2 mg/mL) was supplied as an internal standard. Samples were freeze-dried via liquid nitrogen bath followed by vacuum chamber drying for 6 h. The freeze-dried samples were mixed with 100 μL of methoxylamine diluted with pyridine (20 mg/mL] and incubated at 30 °C for 90 min with agitation (1500 rpm). The sample was mixed with 50 μL of N-methyl-N-TMS-trifluoroacetamide (MSTFA) incubated for 30 min at 37 °C with agitation and transferred into a glass vial for chromatography metabolome analysis. A Cp-sil8CB column from Agilent, model #CP8751 (30 mm, 0.25 mm, 0.25 μm) was used to assess the sake fermentation samples. The injection was at 230 °C with a 1 mL/min helium carrier gas flow rate. The column temperature was held at 80 °C for 2 min before rising 10 °C/min until 320 °C was achieved, the temperature was held for 12 min before reaching a 320 °C detector. The Shimadzu Corporation (Shimadzu Manufacturing Company, Kyoto, Japan) model GC-2014AFSC 100 V was used in performing the GC-FID analysis.

2.9.2. Head Space-Solid Phase Microextraction

For the head space-solid phase microextraction (HS-SPME), the sake supernatant samples were prepared with NaCl and methyl hexanoate (40 mg/100 mL) was supplied as an internal standard. In a glass vial, the solution was agitated under 80 °C heat with a stir bar (160 rpm) for 15 min. A syringe was inserted into the vial to expose a SPME fiber (100 μm PDMS) for 15 min. The fiber was retracted into the syringe before removal from the vial and inserted into the injector for a 5 min exposure. A DB wax column from Agilent, model #122-7032 (30 mm, 0.25 mm, 0.25 μm) was used. The injection was at 200 °C with a 0.95 mL/min helium carrier gas flow rate. The column temperature was held at 60 °C for 1 min before rising 3 °C/min until 100 °C was achieved, the rate increased to 15 °C/min until 200 °C was reached and held for 4 min before reaching a 320 °C detector. The Shimadzu Corporation (Shimadzu Manufacturing Company, Kyoto, Japan) model GC-2014AFSC 100 V was used in performing the HS-SPME analysis.

2.9.3. Gas Chromatography–Mass Spectroscopy

For gas chromatography–mass spectroscopy (GC-MS) of endogenous metabolites, stationary phase yeast cultures were grown in YPD liquid media. Stationary phase was defined at an OD_{600nm} of 2.0. Cells (4×10^7) were harvested, washed, and freeze-dried. An extraction solution (chloroform:methanol:MQ [2:5:2]) with ribitol (0.2 mg/mL) as the internal standard was added to the freeze-dried samples and incubated for 30 min at 37 °C under agitation (1500 rpm). The samples were centrifuged, and the supernatant was harvested. MQ water was added to the supernatant and centrifuged. The supernatant was concentrated under vacuum centrifugation (−100 kPa) at 37 °C. MQ water was added, and the water-soluble metabolites were freeze-dried. Methoxyamine (20 mg/mL) was added to the samples and incubated for 90 min at 30 °C under agitation (1500 rpm). MSTFA (*N*-methyl-*N*-trimethylsilyl trifluoroacetoamide) was added to the samples and incubated for 30 min at 37 °C with agitation (1500 rpm). Samples were supplied in glass vials for GC-MS analysis using an Agilent DB-5MS+DG (30 m, 0.25 mm, 0.25 um) column, splitless injection, and helium carrier gas (1 mL/min). The column temperature rose from 80 °C at a rate of 13 °C/min until reaching a 320 °C detector. The Agilent Technologies, Inc. (Agilent Technologies, Santa Clara, California, United States of America) model 7890 B was used for the GC system, model 5977 A was used for the MS system, and the National Institute for Standards and Technology (NIST) mass spectral library (United States Department of Commerce, Gaithersburg, MD, USA) was used.

2.9.4. Data Analysis

The peak area of identified compounds was divided by the peak area of the internal standard for normalization of the data. Normalized peak areas of the compounds and ethanol concentrations measured during sake fermentations were auto-scaled and used as independent variables in the loading plots for construction of the principal component analysis (PCA). Statistical analysis was performed by soft independent modeling by class analogy (SIMCA) methodology by SIMCA version 13.0.3.0 (Umetrics, Umeå, Sweden) computer software. Analysis of variance (ANOVA), post-hoc students t-test, and post hoc Tukey-Kramer honest significance test (Tukey HSD), $p < 0.05$ was used for statistical significance. Deviations from this statistical analysis procedure, if occurred, was reported in individual results or methods sections. Heatmap was generated through MetaboAnalyst 4.0 [61]. Heatmaps were based on data normalization by internal reference feature, cubed root transformation, pareto/range data scaling for endogenous/sake metabolites, clustering by Ward's method, and Euclidean distance measurements.

3. Results

Endogenous metabolome characterization of *S. arboricola* was put in context of industrially relevant *S. cerevisiae* yeast. Two yeast strains were chosen to represent either beer, wine, or sake fermentation industries. WLP001 and HT01 represented beer, M22 and DBVPG6044 represented wine, K7 and TCR7 represented sake production. SA350, SA351, and SA352 were used to represent wild *S. arboricola* (Table 1). Endogenous metabolite profiling was visualized in a hierarchical clustering heatmap that showed clustering of yeast by industry and species (Figure 1A). Ward's linkage clustering algorithm created two main clusters separated by species with the exception of West African bili wine strain DBVPG6044 that grouped within *S. arboricola*. The yeast in this cluster, located in the right half of the heatmap, all showed a low concentration of endogenous amino acids. Beer strains, sake strains, and wine strain M22 composed the other cluster on the left half. Beer strains paired together, and sake strains paired together with Italian wine strain M22. These industrial strains had similar but varying metabolic profiles displaying high endogenous levels of many amino and organic acids.

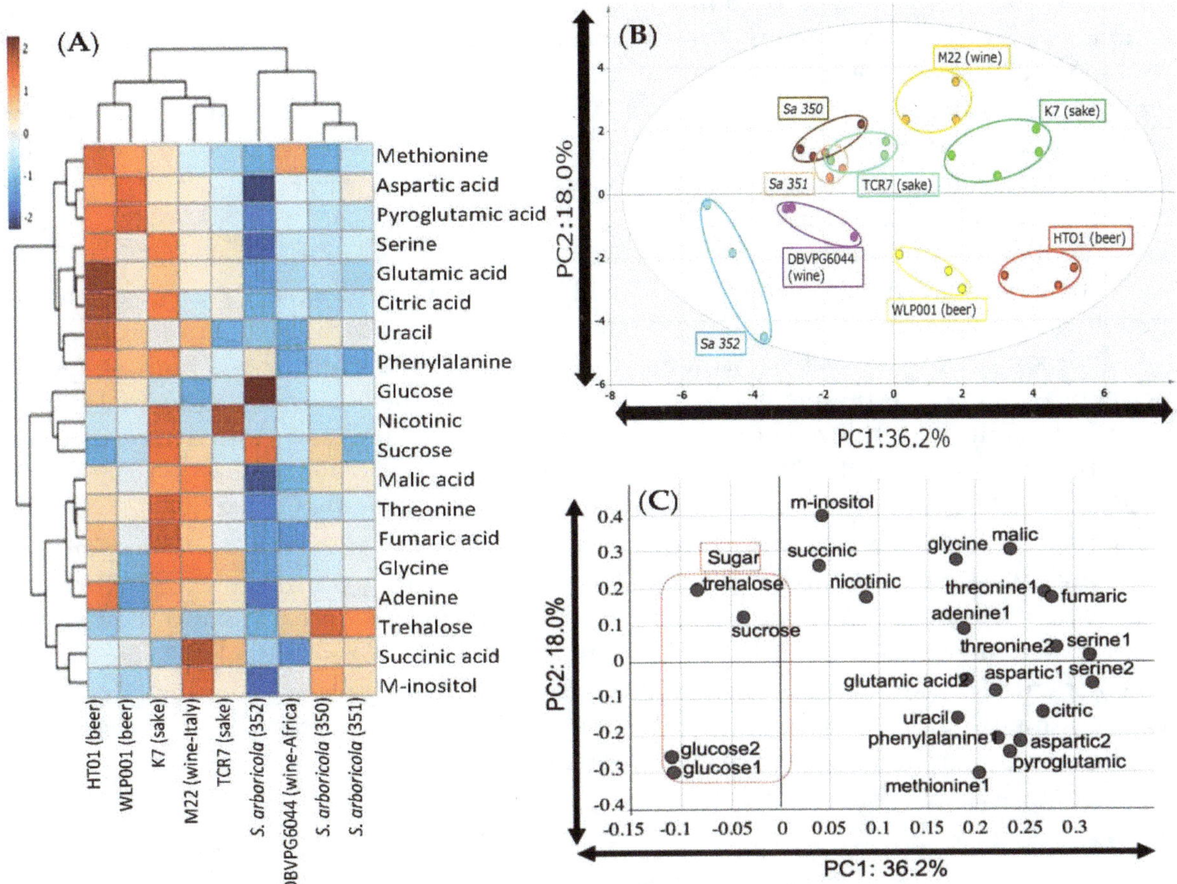

Figure 1. Yeasts endogenous metabolic profiles. (**A**) Hierarchical clustering heatmap of endogenous metabolite profiles between strains and species. Relatively high compound concentrations were shown in varying degrees of red. Relatively low compound concentrations were shown in varying degrees of blue. Shown as average concentration per sample groupings of three replicates. (**B**) Principal component analysis (PCA) score plot of *Saccharomyces* yeast explaining 17.4% of the variance on PC2 and 31.6% of the variance on PC1. (**C**) PCA loading plot. Saccharide content is circled to show importance in the PC1 variance. PCA and loading plot generated through SIMCA analysis. Three replicates were used for each strain.

A score plot generated through principal component analysis separated the strains into general industry-specific groups (Figure 1B). A total of 19 metabolites identified by GC-MS were used as variables in the construction of the loading plot (Figure 1C). Sugar utilization played a major role in separation of the strains. *S. arboricola* strains were grouped to the left while beer and sake strains grouped opposite to the right on principal component (PC1) explaining 36.2% of the sample variance and separating the groups by species. On the axis of PC2, sake yeast grouped high, wine yeasts were neutral/variable, and beer yeasts were low explaining 18.0% of sample variance and separating the yeast by their industrial niche domestication. Glucose, sucrose, and trehalose negatively contributed to PC1, which exemplified sugar uptake differences and was inferred to translate to fermentation aptitude. Beer brewing strains were found to be the most inclined yeasts in this regard and *S. arboricola* were found to be the least adept. A third component, explaining 12.7% was used to create a 3D PCA plot of the endogenous metabolites (Figure S1). Citric acid, malic acid, fumaric acid, glutamic acid, nicotinic acid, succinic acid, aspartic acid, threonine, serine, glycine, phenylalanine, pyroglutamine, methionine, uracil, inositol, and adenine positively contributed to PC1. A loading matrix was constructed for the variables (Table S1). Glycine, threonine, malic acid, fumaric acid, nicotinic acid, succinic acid, sucrose, trehalose, inositol, and adenine positively contributed to PC2. Phenylalanine, pyroglutamine, methionine, uracil, aspartic acid, and glucose contributed negatively to PC2, which distinguished the beer brewing strains from the sake strains. These two groups are represented in the heatmap's industrial cluster. Factors from the heatmap show the difference between sake and beer yeast include the endogenous inositol levels; which were high in sake yeast and affect membrane fluidity during osmotic stress such as an industrial sake mash.

The potential of *Saccharomyces* hybridization to broaden metabolic capabilities and sensory complexity from volatile metabolites was explored using *S. arboricola* and *S. cerevisiae*. Sporulation efficiency was used in selection of strains for direct-mating hybridization as industrial strains are typically poor sporulators. *S. arboricola* 350 and *S. cerevisiae* M22 yielded the highest sporulation efficiency at 69% and 91% respectively (Figure S2A). Selection of the *S. cerevisiae* industrial strain relied on sporulation results to optimize spore-spore mattings and given M22 grouped similar to the sake strain metabolism in Figure 1B, M22 was choosen. Other *S. arboricola* strains sporulated at 68% for 351 and 20% for 352. Interspecific SAM (*S. arboricola* 350 and *S. cerevisiae* M22) hybrids, SAM8c and SAM9a, generated double bands representing genomic input from both species. Two spore-to-spore mating colonies were advanced to lab-scale sake fermentation after confirmation of hybrid state by species-specific multiplexed PCR (Figure S2B). Cryotolerance of *S. arboricola* and the hybrids was shown as substantial growth at 4 and 10 °C after ten and seven days, respectively (Figure 2). At 15 °C, there was no perceivable difference between yeast growth. Thermosensitivity was not detected using the serial spotting dilution assay in temperatures up to 30 °C, slow growth of *S. arboricola* cultures during extreme temperature highs in the summer months was observed.

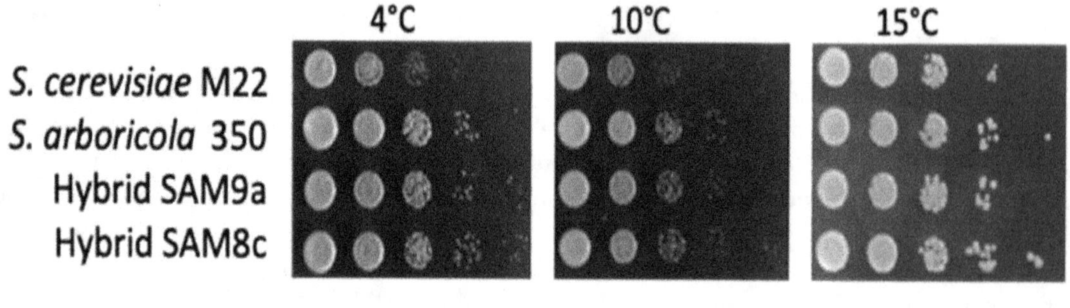

Figure 2. Cryotolerance of parental and hybrid yeasts. Serial spot dilution growth assay of parental *S. cerevisiae* M22, parental *S. arboricola* 350, and the SAM hybrids. Three different temperatures of 4 °C, 10 °C, and 15 °C was used in assessment of their cryotolerance. Yeast were incubated on solid YPD agar for ten days at 4 °C, seven days at 10 °C, and four days at 15 °C.

To determine if important sake flavor molecules changed with hybridization, ethyl hexanoate, isoamyl acetate, and ethanol fermentation was measured. In fact, it was found that *S arboricola*, *S. cerevisiae*, and the hybrids metabolite profiles after lab-scale sake mash fermentation trials were different. The hybrids had a more vigorous fermentation, yielding a higher alcohol content than *S. cerevisiae* by day seven ($p = 0.031$) of the 14 day trial fermentation (Figure 3A). A one-way ANOVA with a two-tailed student's t-test post-hoc analysis and ordered difference report for each day of fermentation was generated in determine ethanol content significance (Table S2). The gap between fermentation rates declined over time trending towards non-significance between all samples if more time was allotted. The fermentations were stopped as the fermentation rate slowed down and a fair amount of ethanol 10%+ was produced. The fermentation trials were designed with time constraints as this is often a major factor in production settings and attenuation was secondary. The final alcohol yield at day 14 was significant between *S. cerevisiae* and the hybrids ($p = 0.038$), while *S. arboricola* was not significantly different than either hybrid nor *S. cerevisiae*. Hybrid SAM9a had the most productive fermentation rate, followed by hybrid SAM8c, *S. arboricola* 350, and *S. cerevisiae* M22. The mean final alcohol content of sake produced was 12.92% for the hybrids, 12.15% for *S. arboricola* 350, and 11.46% for *S. cerevisiae* M22.

Detection of volatile aromatic compounds allows insight into flavor and sensory attributes of the final sake produced and HS-SPME technology in the gas chromatography system was utilized. Sake fermented with *S. arboricola* contained significantly less ethyl hexanoate (syn. ethyl caproate) (0.207 mg mL^{-1}, st error 0.016) than *S. cerevisiae* (0.316 mg mL^{-1}, st error 0.019, $p = 0.005$), or the hybrids (0.308 mg mL^{-1}, st error 0.014, $p = 0.003$) (Figure 3B). Sake fermented by *S. arboricola* contained a concentration of isoamyl acetate (0.368 mg mL^{-1}, st error 0.064) that was high compared to *S. cerevisiae* (0.259 mg mL^{-1}, st error 0.064, $p = 0.25$) and the hybrids (0.252 mg mL^{-1}, st error 0.045, $p = 0.17$), but fell short of a significant *p*-value (Figure 3C). Ethyl hexanoate imparts apple or aniseed and isoamyl acetate imparts a banana-like attribute to human sensory receptors. Whole metabolite heatmap by sample was created showing a separation of *S. arboricola* from the hybrids and M22 individual samples when volatile compounds were used (Figure S3A) or a separation of all three when non-volatile metabolites were used (Figure S3B).

Sake fermentation profiles were visualized in a hierarchical clustering heatmap with hybrids clustering in closer proximity to the *S. arboricola* yeast (Figure 4A). Parental strain *S. cerevisiae* M22's sake metabolite profile contained high amounts of glucose, malic acid, threonine, and valine while parental *S. arboricola*'s profile contained high amounts of citric acid, aspartic acid, glutamic acid, leucine, isoleucine, alanine, phenylalanine, inositol, and glycerol. Parental profiles almost completely opposed one another with only serine, succinic acid, and urea showing similar parental concentrations. M22 was the most distant parent to the hybrids' profile. The heatmap of the hybrid strains showed a mosaic blending of the two parental species. Metabolomic analysis using principal components clearly separated interspecific hybrids from each parent based in a PCA score plot (Figure 4B). A total of 20 metabolites, including 17 non-volatile compounds, were identified by GC-FID and 3 volatile compounds identified by HS-SPME These were used as variables in construction of the loading plot (Figure 4C). Principal component (PC1) separated samples by their species and explained 43.2% of the variance between the samples. Fermentation profiles showed a separation of the two *Saccharomyces* species on opposite ends of PC1 and the hybrid yeast grouped to the middle. PC2 separated the hybrid from the parental samples by the novel hybrid fermentation features that explained 23.1% of the sample variance. Threonine, valine, malic acid, glucose, ethyl hexanoate, and dimethyl trisulfur (DMTS) positively contributed to PC1. Phenylalanine, isoleucine, leucine, alanine, aspartic acid, citric acid, glutamic acid, succinic acid, inositol, glycerol, ethanol, and isoamyl acetate were amino acids, organic acids, and other molecules that negatively contributed to PC1. Ethanol, DMTS, phenylalanine, ethyl hexanoate, glycerol, isoleucine, glutamic acid, valine, and threonine negatively contributed to PC2. Serine, succinic acid, malic acid, glucose, urea, isoamyl acetate, aspartic acid, inositol, and alanine positively contributed to PC2. A loading matrix was constructed for the variables (Table S3).

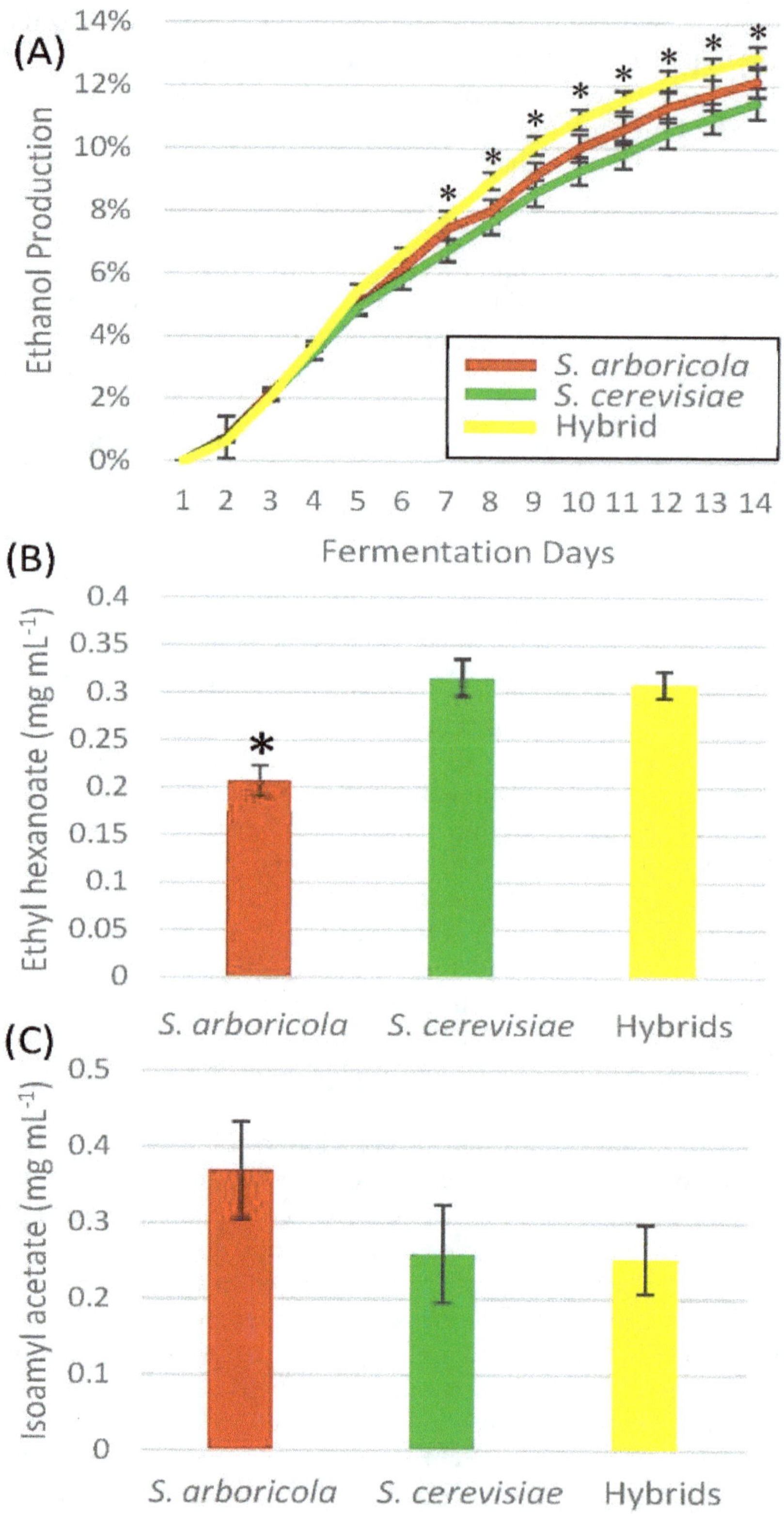

Figure 3. Phenotypic characteristics of hybrids and parental yeast. (**A**) Mean alcoholic fermentation rates of *S. arboricola* 350, *S. cerevisiae* M22, and the SAM hybrids are shown as a line graph representing the alcohol content over 14 days in time. Standard error bars shown in line graph (**B**) Bar chart depicting the production of ethyl hexanoate in mg mL^{-1}. (**C**) Bar chart depicting the production of isoamyl acetate in mg mL^{-1}. Statistical analysis performed by ANOVA, post-hoc students *t*-test, and post-hoc Tukey-HSD, $p < 0.05$.

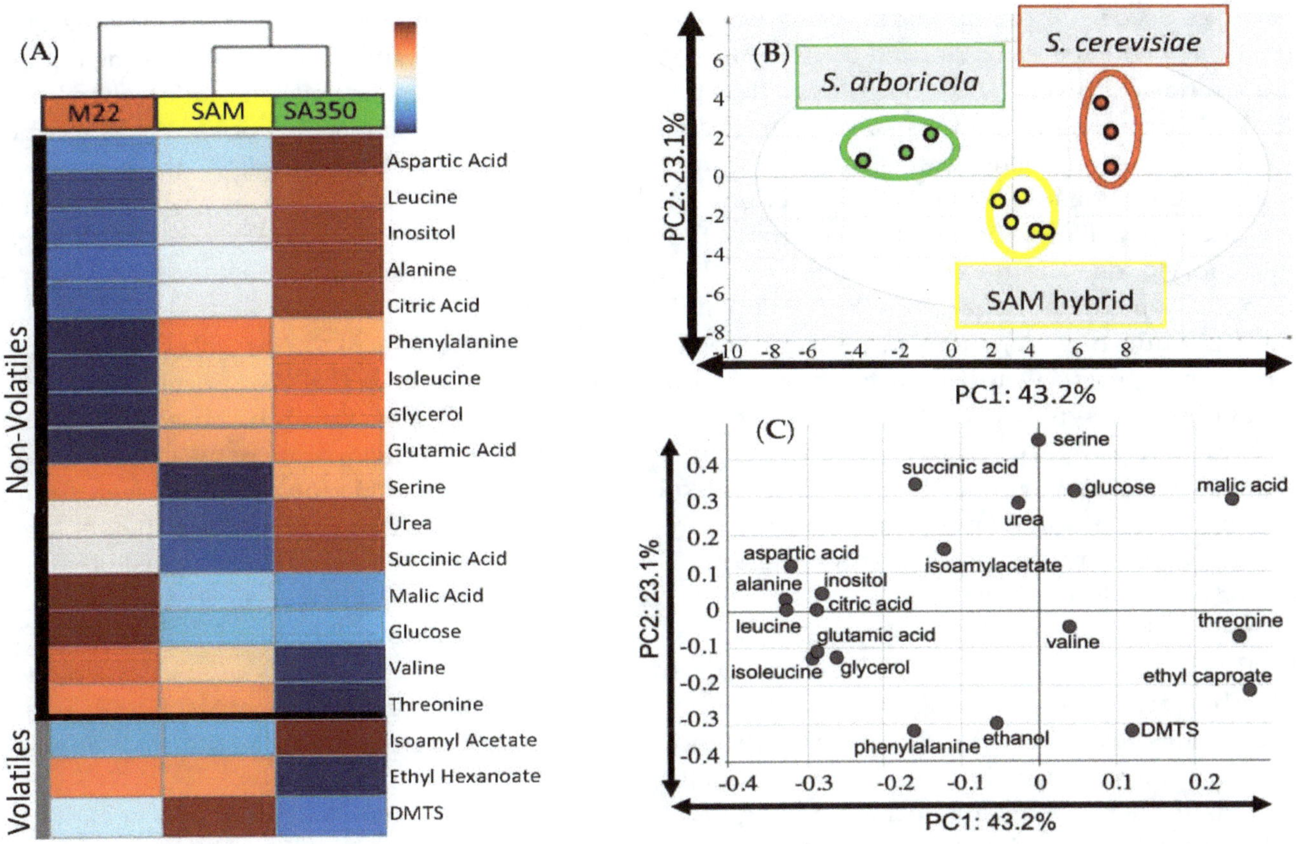

Figure 4. Sake fermentation profiles of *S. arboricola* interspecific hybrids and parental strains. (**A**) Hierarchical clustering heatmap of sake fermentation profiles between strains and species. Relatively high compound concentrations were shown in varying degrees of red. Relatively low compound concentrations were shown in varying degrees of blue. Shown as average concentration per sample groupings of three replicates each. (**B**) PCA score plot of sake fermentation profiles obtained for *S. cerevisiae*, *S. arboricola*, and the SAM hybrids. (**C**) PCA loading plot of sake fermentation profiles obtained for *S. cerevisiae*, *S. arboricola*, and the SAM hybrids. PC1 explained 43.2% and PC2 explained 23.1% of sample variance. PCA and loading plot generated through SIMCA analysis. Three replicates were used for each strain.

4. Discussion

This is the first report of sake brewing by *Saccharomyces arboricola* and its hybrids. The exploratory metabolic analysis of *S. arboricola*, a neglected *Saccharomyces* yeast lacking metabolic evaluation, adds to the novelty of this study. Through direct mating of *S. arboricola* to *S. cerevisiae*, interspecific hybrids gained a higher final ethanol content over *S. cerevisiae* and a higher ethyl hexanoate over *S. arboricola*. It was found that the low levels of endogenous amino/organic acids clustered *S. arboricola* apart from the industrial strains by principal component analysis and hierarchical clustering heatmapping. In sake fermentations, hybrid strains showed a mosaic profile of parental strains except where parental profiles were similar. This metabolic analysis suggested *S. arboricola* had higher transportation activity of pyruvate to the mitochondria and a lower amino acid net uptake than *S. cerevisiae*. Additionally, hybrid metabolic analysis suggested an increase in ethanolic fermentation from pyruvate and increased sulfur metabolism. Together, these results suggest *S. arboricola* is poised for in-depth and focused metabolomic exploration.

Endogenous metabolite comparison of nine industrially relevant and diverse yeast by GC-MS provided a separation of metabolomic profiles where industry and species appeared to be drivers of distinction in both heat map clustering and PCA. The internal composition is characteristic of their metabolic tendencies and influences the environment in which many endogenous metabolic reactions occur. Of the industrial yeast, sake yeast separated from beer yeast by PC2 of the PCA (Figure 1B) which is characterized by the glucose to fructose ratio, inositol level, as well as the organic and amino acid profile. The endogenous inositol levels were high in sake yeast and affect membrane fluidity during osmotic stress such as an industrial sake mash. Sugar utilization played a major role in separation of the strains as evident in the loading plot which highlighted the importance of sugar metabolism. Metabolic distinctions between the glucose to fructose ratio of beer to sake or wine yeasts can be inferred to resemble the kinetic differences of hexokinase-mediated sugar phosphorylation. The glucose to fructose ratio between wine strains was correlated with the kinetics of hexokinase-mediated sugar phosphorylation and the inherent properties of each unique strain [62,63]. High amounts of endogenous inositol and trehalose were characteristic of *S. arboricola*. Furthermore, serine and glycine were found to be low in the *S. arboricola* cluster and are important glycolytic metabolites biosynthesized from 3-phosphoglycerate or alternatively through the gluconeogenic pathway utilizing glyoxylate [64]. Organic and amino acid metabolism shift between alternative biosynthesis pathways and their byproducts likely play a role in the difference seen between strains. Additional genetic and environmental factors are also suspected to play a role in strain metabolomic profiles as seen by increased copy number of *MAL* (maltose) genes in sake and beer yeast, yet wine strains tend to lose this attribute because their typical industrial fermentation sugar availability lacks maltose [65]. Copy number variations (CNV) are concentrated in the sub telomeric regions, with chr. I, VII, VIII, IX, X, XII, XV, and XVI being more susceptible to variation [65]. Gene ontology related to carbon and nitrogen metabolism, ion transport, and flocculation are prone to CNV [66,67] which often results in deletions rather than amplifications of genetic material [65]. The CNVs are thought to underline niche adaptation and domestication [65,66].

Flavor-active elements formed during yeast fermentation can be divided into five key groups: sulfur-containing molecules, organic acids, higher alcohols, carbonyl compounds, and volatile esters [68–71]. Of these classifications, volatile esters are the largest and are responsible for the highly desired fruity character in several alcoholic beverages [72]. Production of ethyl hexanoate and isoamyl alcohol is maximized during fermentations at 15 °C [73,74]. Ethyl hexanoate imparts apple or aniseed and isoamyl acetate imparts a banana-like attribute to human sensory receptors. Ethyl hexanoate, a MCFA (medium chain fatty acids) ester with a 0.17–0.21 mg mL^{-1} flavor threshold, is an important flavor molecule for ginjo sake [68,75]. The fruity ester, formed from hexanoic acid or hexanoyl CoA with ethanol, imparts a light, sweet, apple complexity to the delicate and fragrant beverage. Ginjo sake is produced from rice with the outer 40% of the grain polished away to remove the hull, bran, and part of the starchy endosperm, leaving 60% of the rice to create the sake mash. During the sake fermentation trials, *S. arboricola* produced significantly less ethyl hexanoate than either the hybrid or *S. cerevisiae* wine strain M22. Increased inositol concentrations have been shown to decrease production of ethyl hexanoate during fermentation [76]. This relationship was seen in PC1 of the hybrids and suggests *S. arboricola*'s inositol biosynthesis pathway is more active, in turn producing low ethyl hexanoate levels seen in this study. Additionally, *S. arboricola*'s production of isoamyl acetate showed a trend towards a significant increase over the hybrid or *S. cerevisiae* M22 (Figure 3C). Isoamyl acetate, an acetate ester with a 0.6–1.2 mg mL^{-1} flavor threshold, imparts a distinct banana character typical of hefeweizen yeast [68]. Hybrid strains produced a pleasant aromatic profile containing both ethyl hexanoate and isoamyl acetate. Ethanol hexanoyl transferase, Eht1, is believed to catalyze the formation of ethyl hexanoate [68]. Alcohol acetyltransferase genes Atf1 and Atf2 are responsible for the total biosynthesis of isoamyl alcohol and divergence in the Atf proteins has been suggested to influence the whole ester aroma profile of yeast strains [72]. Protein identity between *S. arboricola* and *S. cerevisiae* for Atf1 was 81%, Atf2 was 36%, and Eht1 was 84%. Together the aromatic ester profile of *S. arboricola* was unique in the combination of low ethyl hexanoate and isoamyl alcohol concentrations are speculated to be owing from inositol lipid activity and variation in factors affecting Atf activity. This area of research is merits further investigation.

Interspecific hybrids are commonly employed in agricultural and horticultural settings where hybrid vigor provides an increased or transgressive phenotype [77]. In oenology *Saccharomyces* hybrids, such as *S. cerevisiae* x *S. kudrazevii*, are often employed [78,79]. Many times *Saccharomyces* hybrids are able to yield more ethanol during fermentation under lab controlled environments [80], and these results support previous findings. The use of *S. arboricola* in sake production is novel and no other study had *S. arboricola's* metabolism as the primary focus of the manuscript. To date several labs have investigated *Saccharomyces* hybrids, one study included *S. arboricola* in a beer brewing context while one other included *S. arboricola* among their hybrids in a biofuel context [80,81]. This research showed that in a sake mash, SAM interspecific hybrids produced a more vigorous fermentation in a short amount of time which may be beneficial if used or further developed in a production context. Of the range of temperatures for *Saccharomyces* growth, *S. arboricola* is cryotolerant and thermosensitive [2,80,82]. Cryotolerance is a common attribute of yeast hybrids as seen in the lager yeast *Saccharomyces pastorianus*, which is a hybrid between *S. cerevisiae* x *S. eubayanus* [7,83–85]. This phenotype has proven beneficial for the cold fermentations typical of lagering beer or brewing sake. The ability to ferment at colder temperatures allows for more volatile compounds to be retained and a decrease in bitter or acidic taste [86]. The mechanism by which thermal phenotypes are determined has not been fully elucidated, but mitotype, ploidy, and lipid composition are factors [80,86–91]. Lipid profiles with a higher number of MCFA attributes to cryotolerance of *S. cerevisiae* x *S. kudriavzevii* hybrids [92]. At colder temperatures, increasing the MCFA composition and reducing saturated and unsaturated fatty acid composition allows increased fluidity of the membranes.

Previous reports show that wine yeast produce lower amino acidity and less succinic acid, but more malic acid than sake yeast [93]. Yeast strain has a major impact on metabolic profiles as seen in this study's two wine strains. During fermentation, organic acid production is attributed to the metabolism of pyruvate in the mitochondria and the reverse tricarboxylic acid cycle (TCA) cycle in the cytoplasm [94,95]. Most succinic acid and malic acid are produced through mitochondrial TCA cycle, however reductive TCA can supplement their production during sake fermentation [95]. These amino and organic acidity profiles are metabolically unique to individual strains and these may contribute a favorable fresh acidity such as malate while succinate contributes an unfavorable acidity to sake. During sake metabolic analysis, PC1 was inferred to represent the amino acid net uptake and mitochondrial transportation of pyruvate which contributes to the amino and organic acidity profiles (Figure 4C). Of the sake metabolites detected, 75% of the compounds were produced in high abundance in the sake by *S. arboricola* as compared to *S. cerevisiae* which produced 44% of the compounds in high amounts. *S. arboricola* was considered to have a lower amino acid net uptake than *S. cerevisiae* strains in this study. The abundance of residual amino acids detected in the sake medium and the low abundance of endogenous metabolites were considered in these findings. Moreover, membrane fluidity and amino acid and organic acid transport of *S. arboricola* has not been investigated and merits further research.

The generated interspecific hybrids were considered to have vigorous glycolysis and fermentation of pyruvate to ethanol compared to parental strains. The reduced malic acid, succinic acid, and glucose (Figure 4A) compared to the increased ethanol yield (Figure 3A) supported these claims. The hybrid condition was distinctly exemplified in principal component 2 of the sake metabolite score plot. In this study, where parental metabolic profiles were similar, the hybrids revealed an inversion of these compounds in the heatmap as seen in DMTS, serine, urea, and succinic acid. DMTS is associated with the stale aroma of Japanese sake called hineka. Recycling of sulfur via the methionine salvage pathway produces DMTS-P1, a precursor of this aromatic compound [96]. Mercaptans (thiols) typically have a low sensory threshold and the hybrids were considered to have an active sulfur metabolism. Additionally, urea is a yeast waste product from arginase activity and can be catabolized to form succinate in the cell [97]. Mechanistic elucidation of these metabolic variances is beyond the scope of this initial report of sake fermentation by *S. arboricola* and the hybrids, further research is required.

Research for *S. arboricola* yeast is very limited and one other report [80], to the author's knowledge, has explored their use and their interspecific hybrids' use in beverage fermentation. During their wort fermentation trials, *S. arboricola* produced high levels of ethyl hexanoate at 0.35 mg mL^{-1}, while this study revealed low production at 0.207 mg mL^{-1} by substituting sake mash for wort. Yeast strain and raw ingredients certainly play a role in the final fermentation product qualities. The chemistry of sake and beer have some similarities, but there are also many differences including raw materials such as rice or barley and the presence of *A. oryzae* in the sake mash. These results indicate koji was an integral part of the sake mash converting rice carbohydrate into fermentable sugars. *Aspergillus* spp. produce mannitol at high levels [46,47] and *S. arboricola* has great growth utilizing mannitol [14], but the bioavailability of this sugar for fermentation by *S. arboricola* remains unexplored. This metabolic capability likely facilitated the increase in fermentation performance seen in this study. *S. arboricola* and their hybrids may be uniquely fit for use in sake fermentations.

In two recent large phylogenetic analysis of 157 and 1011 *Saccharomyces cerevisiae* strains, an Asian cluster was found basal to the root of the trees [34,65] suggesting little divergence from their wild ancestors in comparison to the highly diverged beer1 clade. Although 1011 genome study was a great step in unmasking the lost ecology and domestication of natural *Saccharomyces* yeast, it had an overrepresentation of beer brewing strains. In order to dive deeper into the Out of Asia Origin Hypothesis, many more Asian and diverse strains will need to be sequenced and compared phylogenetically. The investigation into *S. arboricola*'s biotechnological value yielded promising results, but the dynamics between *A. oryzae*'s mannitol production and *S. arboricola*'s mannitol utilization merits further research. Exploration of the biotechnical potential of the *Saccharomyces* clade is only now beginning and understanding the diversity of yeast facilitates humanity's exploitation of *Saccharomyces*.

5. Conclusions

This study explored the novel use of *S. arboricola* in Japanese rice wine known as sake, traditionally produced from polished rice with *A. oryzae* during winter months. The metabolites in the sake produced from *S. arboricola*, *S. cerevisiae*, and their interspecific hybrids were distinct by PCA analysis. *S. arboricola* had a lower amino acid net uptake than *S. cerevisiae* and the hybrids showed a mosaic profile of parental strains. Hybrid strains had a more vigorous ethanolic fermentation in the sake mash than *S. cerevisiae* but had a higher ethyl hexanoate yield than *S. arboricola*. In the ginjo style of Japanese sake, ethanol and ethyl hexanoate are important sensory metabolic compounds. *S. arboricola* differed from *S. cerevisiae* industrial strains because of lower levels of endogenous amino/organic acids, suggesting a lower amino acid net uptake. The endogenous metabolite study highlighted the impact of strain selection which industry niche has unwittingly had an influence in each strains metabolism. The diversity of the *Saccharomyces* clade of budding yeast is expansive and future studies focused metabolic phenotypes of new strains and novel applications will most likely surface in the following years to come.

Author Contributions: Conceptualization, M.J.W., J.E.G.G., and H.K.; methodology, M.J.W.; validation, M.J.W., Y.Y., and Y.F.; formal analysis, M.J.W., Y.Y., and Y.F.; Investigation, M.J.W., Y.F., and Y.Y.; resources, H.K., Y.Y., Y.F., and Y.K.; data curation, Y.Y., Y.F., and Y.K.; writing—original draft preparation, M.J.W.; writing—review and editing, M.J.W., J.E.G.G., and H.K.; visualization, M.J.W., Y.F, and Y.Y.; supervision, H.K.; project administration, M.J.W. and H.K.; funding acquisition, M.J.W. and H.K. All authors have read and agreed to the published version of the manuscript.

Acknowledgments: Thank you Jenny Bellon from the Austrian Wine Research Institute who aided with experimental design, Gianni Liti of Institute for Research on Cancer and Aging (IRCAN) who provided the *Saccharomyces arboricola* strains, Barbra Dunn of Stanford University who provided advice and materials, Angela Chu/Joe Horecka of Stanford University who shared their sporulation protocol, and Tetsuya Goshima/Takeshi Akao of the Japanese National Research Institute of Brewing (NRIB) who provided access to the micromanipulator and materials/provided the K7/868 haploid.

References

1. Martini, A.V.; Martini, A. Three newly delimited species of Saccharomyces sensu stricto. *Antonie Van Leeuwenhoek* **1987**, *53*, 77–84. [CrossRef]
2. Wang, S.A.; Bai, F.Y. Saccharomyces arboricolus sp. nov., a yeast species from tree bark. *Int. J. Syst. Evol. Microbiol.* **2008**, *58*, 510–514. [CrossRef] [PubMed]
3. Naumova, E.S.; Roberts, I.N.; James, S.A.; Naumov, G.I.; Louis, E.J. Three new species in the Saccharomyces sensu stricto complex: Saccharomyces cariocanus, Saccharomyces kudriavzevii and Saccharomyces mikatae. *Int. J. Syst. Evol. Microbiol.* **2015**, *50*, 1931–1942. [CrossRef]
4. Naumov, G.I.; Naumova, E.S.; Hagler, A.N.; Mendonça-Hagler, L.C.; Louis, E.J. A new genetically isolated population of the Saccharomyces sensu stricto complex from Brazil. *Antonie Van Leeuwenhoek* **1995**, *67*, 351–355. [CrossRef]
5. Naumov, G.I.; Lee, C.-F.; Naumova, E.S. Molecular genetic diversity of the Saccharomyces yeasts in Taiwan: Saccharomyces arboricola, Saccharomyces cerevisiae and Saccharomyces kudriavzevii. *Antonie Van Leeuwenhoek* **2013**, *103*, 217–228. [CrossRef]
6. Naumov, G.I.; Naumova, E.S.; Louis, E.J. Two New Genetically Isolated Popoulations of the Saccharomyces Sensu Stricto Complex from Japan. *J. Gen. Appl. Microbiol.* **1995**, *41*, 499–505. [CrossRef]
7. Libkind, D.; Hittinger, C.T.; Valério, E.; Gonçalves, C.; Dover, J.; Johnston, M.; Gonçalves, P.; Sampaio, J.P. Microbe domestication and the identification of the wild genetic stock of lager-brewing yeast. *Proc. Natl. Acad. Sci. USA* **2011**, *108*, 14539–14544. [CrossRef] [PubMed]
8. Boynton, P.J.; Greig, D. The ecology and evolution of non-domesticated Saccharomyces species Primrose. *Yeast* **2014**, *31*, 449–462. [PubMed]
9. Sipiczki, M. Interspecies hybridisation and genome chimerisation in Saccharomyces: Combining of gene pools of species and its biotechnological perspectives. *Front. Microbiol.* **2018**, *9*, 3071. [CrossRef] [PubMed]
10. Masneuf, I.; Hansen, J.; Groth, C.; Piskur, J.; Dubourdieu, D. New hybrids between Saccharomyces sensu stricto yeast species found among wine and cider production strains. *Appl. Environ. Microbiol.* **1998**, *64*, 3887–3892. [CrossRef]
11. Querol, A.; Bond, U. The complex and dynamic genomes of industrial yeasts: Minireview. *FEMS Microbiol. Lett.* **2009**, *293*, 1–10. [CrossRef]
12. Nguyen, H.V.; Legras, J.L.; Neuvéglise, C.; Gaillardin, C. Deciphering the hybridisation history leading to the lager lineage based on the mosaic genomes of Saccharomyces bayanus strains NBRC1948 and CBS380 T. *PLoS ONE* **2011**, *6*, e25821. [CrossRef] [PubMed]
13. Gayevskiy, V.; Goddard, M.R. Saccharomyces eubayanus and Saccharomyces arboricola reside in North Island native New Zealand forests. *Environ. Microbiol.* **2016**, *18*, 1137–1147. [CrossRef] [PubMed]
14. Liti, G.; Nguyen Ba, A.N.; Blythe, M.; Müller, C.A.; Bergström, A.; Cubillos, F.A.; Dafhnis-Calas, F.; Khoshraftar, S.; Malla, S.; Mehta, N.; et al. High quality de novo sequencing and assembly of the Saccharomyces arboricolus genome. *Bmc Genom.* **2013**, *14*, 69. [CrossRef] [PubMed]
15. Cliften, P.; Sudarsanam, P.; Desikan, A.; Fulton, L.; Fulton, B.; Majors, J.; Waterston, R.; Cohen, B.A.; Johnston, M. Finding functional features in Saccharomyces genomes by phylogenetic footprinting. *Science* **2003**, *301*, 71–76. [CrossRef] [PubMed]
16. Birren, B.; Lander, E.S.; Kellis, M.; Endrizzi, M.; Patterson, N. Sequencing and comparison of yeast species to identify genes and regulatory elements. *Nature* **2003**, *423*, 241–254.
17. Liti, G.; Carter, D.M.; Moses, A.M.; Warringer, J.; Parts, L.; James, S.A.; Davey, R.P.; Roberts, I.N.; Burt, A.; Koufopanou, V.; et al. Population genomics of domestic and wild yeasts. *Nature* **2009**, *458*, 337–341. [CrossRef]
18. Dujon, B. Yeasts illustrate the molecular mechanisms of eukaryotic genome evolution. *Trends Genet.* **2006**, *22*, 375–387. [CrossRef]
19. Postma, E.; Verduyn, C.; Scheffers, A.; Van Dijken, J.P. Enzymic analysis of the Crabtree Effect in glucose-limited chemostat cultures of Saccharomyces cerevisiae. *Appl. Environ. Microbiol.* **1989**, *55*, 468–477. [CrossRef]
20. Galdieri, L.; Mehrotra, S.; Yu, S.; Vancura, A. Transcriptional regulation in yeast during diauxic shift and stationary phase. *OMI. J. Integr. Biol.* **2010**, *14*, 629–638. [CrossRef]

21. McGovern, P.E.; Zhang, J.; Tang, J.; Zhang, Z.; Hall, G.R.; Moreau, R.A.; Nuñez, A.; Butrym, E.D.; Richards, M.P.; Wang, C.-S.; et al. Fermented beverages of pre- and proto-historic China. *Proc. Natl. Acad. Sci. USA* **2004**, *101*, 17593–17598. [CrossRef]
22. McGovern, P.E. *Ancient Wine: The Search for the Origins of Viniculture*; Princeton Publishing: Princeton, NJ, USA, 2003; pp. 1–15, ISBN 0691070806.
23. Kitagaki, H.; Kitamoto, K. Breeding Researches of Sake Yeasts in Japan: History, Recent Technological Advances, and Future Perspectives. *Food Sci. Technol.* **2012**, *4*, 215–235.
24. Charles, M.; Tang, H.; Belcram, H.; Paterson, A.; Gornicki, P.; Chalhoub, B. Sixty million years in evolution of soft grain trait in grasses: Emergence of the softness locus in the common ancestor of pooideae and ehrhartoideae, after their divergence from panicoideae. *Mol. Biol. Evol.* **2009**, *26*, 1651–1661. [CrossRef]
25. Arendt, E.; Zannini, E. Rice. In *Cereal Grains for the Food and Beverage Industries*; Woodhead Publishing: Cambridge, UK, 2013; pp. 114–154. ISBN 9780857098924.
26. Corran, H.S. *A History of Brewing*; David & Charles: London, UK, 1975.
27. Sampaio, J.P. Microbe profile: Saccharomyces eubayanus, the missing link to lager beer yeasts. *Microbiolog* **2018**, *164*, 1069–1071. [CrossRef] [PubMed]
28. Peris, D.; Langdon, Q.K.; Moriarty, R.V.; Sylvester, K.; Bontrager, M.; Charron, G.; Leducq, J.B.; Landry, C.R.; Libkind, D.; Hittinger, C.T. Complex Ancestries of Lager-Brewing Hybrids Were Shaped by Standing Variation in the Wild Yeast Saccharomyces eubayanus. *PLoS Genet.* **2016**, *12*, e1006155. [CrossRef] [PubMed]
29. Nespolo, R.F.; Villarroel, C.A.; Oporto, C.I.; Tapia, S.M.; Vega, F.; Urbina, K.; De Chiara, M.; Mozzachiodi, S.; Mikhalev, E.; Thompson, D.; et al. An Out-of-Patagonia dispersal explains most of the worldwide genetic distribution in Saccharomyces eubayanus. *bioRxiv* **2019**, 709253. [CrossRef]
30. Langdon, Q.K.; Peris, D.; Eizaguirre, J.I.; Opulente, D.A.; Buh, K.V.; Sylvester, K.; Jarzyna, M.; Rodríguez, M.E.; Lopes, C.A.; Libkind, D.; et al. Genomic diversity and global distribution of Saccharomyces eubayanus, the wild ancestor of hybrid lager-brewing yeasts. *bioRxiv* **2019**, 709535.
31. Eizaguirre, J.I.; Peris, D.; Rodríguez, M.E.; Lopes, C.A.; De Los Ríos, P.; Hittinger, C.T.; Libkind, D. Phylogeography of the wild Lager-brewing ancestor (Saccharomyces eubayanus) in Patagonia. *Environ. Microbiol.* **2018**, *20*, 3732–3743. [CrossRef]
32. Nagy, L.G.; Ohm, R.A.; Kovács, G.M.; Floudas, D.; Riley, R.; Gácser, A.; Sipiczki, M.; Davis, J.M.; Doty, S.L.; De Hoog, G.S.; et al. Latent homology and convergent regulatory evolution underlies the repeated emergence of yeasts. *Nat. Commun.* **2014**, *5*, 4471. [CrossRef]
33. Bing, J.; Han, P.J.; Liu, W.Q.; Wang, Q.M.; Bai, F.Y. Evidence for a far east asian origin of lager beer yeast. *Curr. Biol.* **2014**, *24*, R380–R381. [CrossRef]
34. Peter, J.; De Chiara, M.; Friedrich, A.; Yue, J.X.; Pflieger, D.; Bergström, A.; Sigwalt, A.; Barre, B.; Freel, K.; Llored, A.; et al. Genome evolution across 1,011 Saccharomyces cerevisiae isolates. *Nature* **2018**, *556*, 339–344. [CrossRef]
35. Duan, S.F.; Han, P.J.; Wang, Q.M.; Liu, W.Q.; Shi, J.Y.; Li, K.; Zhang, X.L.; Bai, F.Y. The origin and adaptive evolution of domesticated populations of yeast from Far East Asia. *Nat. Commun.* **2018**, *9*, 2690. [CrossRef] [PubMed]
36. He, Y.; Dong, J.; Yin, H.; Zhao, Y.; Chen, R.; Wan, X.; Chen, P.; Hou, X.; Liu, J.; Chen, L. Wort composition and its impact on the flavour-active higher alcohol and ester formation of beer—A review. *J. Inst. Brew.* **2014**, *120*, 157–163. [CrossRef]
37. Engan, S. Wort Composition and Beer Flavour. II. The Influence of Different Carbohydrates on the Formation of Some Flavour Components During Fermentation. *J. Inst. Brew.* **1971**, *78*, 169–173. [CrossRef]
38. Younis, O.S.; Stewart, G.G. Sugar uptake and subsequent ester and higher alcohol production by Saccharomyces cerevisiae. *J. Inst. Brew.* **1998**, *104*, 255–264. [CrossRef]
39. Mortimer, R.; Polsinelli, M. On the origins of wine yeast. *Res. Microbiol.* **1999**, *150*, 199–204. [CrossRef]
40. Bisson, L.F. Gene Expression in Yeasts During Wine Fermentation. In *Yeasts in The Production of Wine*; Romano, P., Ciani, M., Fleet, G., Eds.; Springer: New York, NY, USA, 2019; pp. 165–209.
41. Kühbeck, F.; Back, W.; Krottenthaler, M. Influence of lauter turbidity on wort composition, fermentation performance and beer quality—A review. *J. Inst. Brew.* **2006**, *112*, 215–221. [CrossRef]
42. White, C.; Zainasheff, J. *Yeast: The Practical Guide to Beer Fermentation (Brewing Elements)*; Brewers Publications: Boulder, CO, USA, 2010.

43. Kodama, K. Sake Yeast. In *The Yeasts*; Rose, H., Harrison, J.S., Eds.; Academic Press: London, UK, 1993; pp. 129–168.
44. Ferdouse, J.; Yamamoto, Y.; Taguchi, S.; Yoshizaki, Y.; Takamine, K.; Kitagaki, H. Glycosylceramide modifies the flavor and metabolic characteristics of sake yeast. *PeerJ* **2018**, *10*, e4768. [CrossRef]
45. Zhang, B.; Guan, Z.B.; Cao, Y.; Xie, G.F.; Lu, J. Secretome of Aspergillus oryzae in Shaoxing rice wine koji. *Int. J. Food Microbiol.* **2012**, *155*, 113–119. [CrossRef]
46. Song, S.H.; Vieille, C. Recent advances in the biological production of mannitol. *Appl. Microbiol. Biotechnol.* **2009**, *84*, 55–62. [CrossRef]
47. Smiley, K.L.; Cadmus, M.C.; Liepins, P. Biosynthesis of D-mannitol from D-glucose by Aspergillus candidus. *Biotechnol. Bioeng.* **1967**, *9*, 365–374. [CrossRef]
48. Bellon, J.R.; Schmid, F.; Capone, D.L.; Dunn, B.L.; Chambers, P.J. Introducing a New Breed of Wine Yeast: Interspecific Hybridisation between a Commercial Saccharomyces cerevisiae Wine Yeast and Saccharomyces mikatae. *PLoS ONE* **2013**, *8*, e62053. [CrossRef] [PubMed]
49. González, S.S.; Barrio, E.; Gafner, J.; Querol, A. Natural hybrids from Saccharomyces cerevisiae, Saccharomyces bayanus and Saccharomyces kudriavzevii in wine fermentations. *FEMS Yeast Res.* **2006**, *6*, 1221–1234. [CrossRef] [PubMed]
50. Capece, A.; Romaniello, R.; Siesto, G.; Romano, P. Diversity of Saccharomyces cerevisiae yeasts associated to spontaneously fermenting grapes from an Italian "heroic vine-growing area". *Food Microbiol.* **2012**, *31*, 159–166. [CrossRef] [PubMed]
51. Song, G.; Dickins, B.J.A.; Demeter, J.; Engel, S.; Dunn, B.; Cherry, J.M. AGAPE (Automated Genome Analysis PipelinE) for pan-genome analysis of Saccharomyces cerevisiae. *PLoS ONE* **2015**, *10*, e0120671. [CrossRef]
52. Katou, T.; Kitagaki, H.; Akao, T.; Shimoi, H. Brewing characteristics of haploid strains isolated from sake yeast Kyokai No. *Yeast* **2008**, *25*, 799–807. [CrossRef]
53. Horie, K.; Oba, T.; Motomura, S.; Isogai, A.; Yoshimura, T.; Tsuge, K.; Koganemaru, K.; Kobayashi, G.; Kitagaki, H. Breeding of a Low Pyruvate-Producing Sake Yeast by Isolation of a Mutant Resistant to Ethyl α-Transcyanocinnamate, an Inhibitor of Mitochondrial Pyruvate Transport. *Biosci. Biotechnol. Biochem.* **2010**, *74*, 843–847. [CrossRef]
54. Sniegowski, P.D.; Dombrowski, P.G.; Fingerman, E. Saccharomyces cerevisiae and Saccharomyces paradoxus coexist in a natural woodland site in North America and display different levels of reproductive isolation from European conspecifics. *FEMS Yeast Res.* **2002**, *1*, 299–306.
55. Schoch, C.L.; Seifert, K.A.; Huhndorf, S.; Robert, V.; Spouge, J.L.; Levesque, C.A.; Chen, W.; Consortium, F.B.; Bolchacova, E.; Voigt, K.; et al. Nuclear ribosomal internal transcribed spacer (ITS) region as a universal DNA barcode marker for Fungi. *Proc. Natl. Acad. Sci. USA* **2012**, *109*, 1–6. [CrossRef]
56. Rong-Mullins, X.; Winans, M.J.; Lee, J.B.; Lonergan, Z.R.; Pilolli, V.A.; Weatherly, L.M.; Carmenzind, T.W.; Jiang, L.; Cumming, J.R.; Oporto, G.S.; et al. Proteomic and genetic analysis of the response of S. cerevisiae to soluble copper leads to improvement of the antimicrobial function of cellulosic copper nanoparticles. *Metallomics* **2017**, *9*, 1304–1315. [CrossRef]
57. Shimwell, J.L. A Simple Staining Method for the Detection of Ascospores in Yeasts. *J. Inst. Brew.* **1938**, *44*, 474. [CrossRef]
58. Muir, A.; Harrison, E.; Wheals, A. A multiplex set of species-specific primers for rapid identification of members of the genus Saccharomyces. *FEMS Yeast Res.* **2011**, *11*, 552–563. [CrossRef] [PubMed]
59. Korbie, D.J.; Mattick, J.S. Touchdown PCR for increased specificity and sensitivity in PCR amplification. *Nat. Protoc.* **2008**, *3*, 1452–1456. [CrossRef] [PubMed]
60. Lee, P.Y.; Costumbrado, J.; Hsu, C.-Y.; Kim, Y.H. Agarose Gel Electrophoresis for the Separation of DNA Fragments. *J. Vis. Exp.* **2012**, *20*, e3923. [CrossRef]
61. Xia, J.; Sinelnikov, I.V.; Han, B.; Wishart, D.S. MetaboAnalyst 3.0-making metabolomics more meaningful. *Nucleic Acids Res.* **2015**, *43*, W251–W257. [CrossRef]
62. Berthels, N.J.; Cordero Otero, R.R.; Bauer, F.F.; Thevelein, J.M.; Pretorius, I.S. Discrepancy in glucose and fructose utilisation during fermentation by Saccharomyces cerevisiae wine yeast strains. *FEMS Yeast Res.* **2004**, *4*, 683–689. [CrossRef]
63. Berthels, N.J.; Cordero Otero, R.R.; Bauer, F.F.; Pretorius, I.S.; Thevelein, J.M. Correlation between glucose/fructose discrepancy and hexokinase kinetic properties in different Saccharomyces cerevisiae wine yeast strains. *Appl. Microbiol. Biotechnol.* **2008**, *77*, 1083–1091. [CrossRef]

64. Melcher, K.; Entian, K.-D. Genetic analysis of serine biosynthesis and glucose repression in yeast. *Curr. Genet.* **1992**, *21*, 295–300. [CrossRef]
65. Gallone, B.; Steensels, J.; Baele, G.; Maere, S.; Verstrepen, K.J.; Prahl, T.; Soriaga, L.; Saels, V.; Herrera-Malaver, B.; Merlevede, A.; et al. Domestication and Divergence of Saccharomyces cerevisiae Beer Yeasts. *Cell* **2016**, *166*, 1397–1410.e16. [CrossRef]
66. Dunn, B.; Richter, C.; Kvitek, D.J.; Pugh, T.; Sherlock, G. Analysis of the Saccharomyces cerevisiae pan-genome reveals a pool of copy number variants distributed in diverse yeast strains from differing industrial environments. *Genome Res.* **2012**, *22*, 908–924. [CrossRef]
67. Bergström, A.; Simpson, J.T.; Salinas, F.; Barré, B.; Parts, L.; Zia, A.; Nguyen Ba, A.N.; Moses, A.M.; Louis, E.J.; Mustonen, V.; et al. A high-definition view of functional genetic variation from natural yeast genomes. *Mol. Biol. Evol.* **2014**, *31*, 872–888. [CrossRef] [PubMed]
68. Verstrepen, K.J.; Derdelinckx, G.; Dufour, J.P.; Winderickx, J.; Thevelein, J.M.; Pretorius, I.S.; Delvaux, F.R. Flavor-active esters: Adding fruitiness to beer. *J. Biosci. Bioeng.* **2003**, *96*, 110–118. [CrossRef]
69. Pisarnitskii, A.F. Formation of wine aroma: Tones and imperfections caused by minor components (review). *Appl. Biochem. Microbiol.* **2001**, *37*, 552–560. [CrossRef]
70. Pretorius, I.S.; Lambrechts, M.G. Yeast and its importance to wine aroma: A review. *S. Afr. J. Enol. Vitic.* **2000**, *21*, 97–129.
71. Quilter, M.G.; Hurley, J.C.; Lynch, F.J.; Murphy, M.G. The Production of Isoamyl Acetate from Amyl Alcohol by Saccharomyces cerevisiae. *J. Inst. Brew.* **2003**, *109*, 34–40. [CrossRef]
72. Verstrepen, K.J.; Van Laere, S.D.M.; Vanderhaegen, B.M.P.; Derdelinckx, G.; Dufour, J.-P.; Pretorius, I.S.; Winderickx, J.; Thevelein, J.M.; Delvaux, F.R. Expression Levels of the Yeast Alcohol Acetyltransferase Genes ATF1, Lg-ATF1, and ATF2 Control the Formation of a Broad Range of Volatile Esters. *Appl. Environ. Microbiol.* **2003**, *69*, 5228–5237. [CrossRef]
73. Norstedt, C.; Bengtsson, A.; Bennet, P.; Lindström, I.; Äyräpää, T. Technical Measures to Control the Formation of Esters During Beer Fermentation. In *Proceedings of the European Brewery Convention: Proceedings of the 15th Congress*; Elsevier Scientific Publishing Company: Amsterdam, The Netherlands, 1975; pp. 581–600.
74. Hammond, J.R.M. Yeast Technology. In *Brewer's Yeast*; Rose, A.H., Harrison, J.S., Eds.; Academic Press: San Diego, CA, USA, 1993; Volumne 5, pp. 8–56.
75. Takahashi, T.; Ohra, Y.; Sueno, K. Breeding of a sake yeast mutant with enhanced ethyl caproate productivity in sake brewing using rice milled at a high polishing ratio. *J. Biosci. Bioeng.* **2017**, *123*, 707–713. [CrossRef]
76. Furukawa, K.; Yamada, T.; Mizoguchi, H.; Hara, S. Increased Ethyl Caproate Production by Inositol Limitation in Saccharomyces cerevisiae. *J. Biosci. Bioeng.* **2003**, *95*, 448–454. [CrossRef]
77. Dittrich-Reed, D.R.; Fitzpatrick, B.M. Transgressive Hybrids as Hopeful Monsters. *Evol. Biol.* **2013**, *40*, 310–315. [CrossRef]
78. Bisson, L.F. Yeast Hybrids in Winemaking. *Catal. Discov. Pract.* **2017**, *1*, 27–34. [CrossRef]
79. Morales, L.; Dujon, B. Evolutionary Role of Interspecies Hybridization and Genetic Exchanges in Yeasts. *Microbiol. Mol. Biol. Rev.* **2012**, *76*, 721–739. [CrossRef]
80. Nikulin, J.; Krogerus, K.; Gibson, B. Alternative Saccharomyces interspecies hybrid combinations and their potential for low-temperature wort fermentation. *Yeast* **2018**, *35*, 113–127. [CrossRef]
81. Peris, D.; Moriarty, R.V.; Alexander, W.G.; Baker, E.C.; Sylvester, K.; Sardi, M.; Langdon, Q.K.; Libkind, D.; Wang, Q.M.; Bai, F.Y.; et al. Hybridization and adaptive evolution of diverse Saccharomyces species for cellulosic biofuel production. *Biotechnol. Biofuels* **2017**, *10*, 1–19. [CrossRef] [PubMed]
82. Salvadó, Z.; Arroyo-López, F.N.; Guillamón, J.M.; Salazar, G.; Querol, A.; Barrio, E. Temperature adaptation Markedly Determines evolution within the genus Saccharomyces. *Appl. Environ. Microbiol.* **2011**, *77*, 2292–2302. [CrossRef]
83. Krogerus, K.; Magalhes, F.; Vidgren, V.; Gibson, B. New lager yeast strains generated by interspecific hybridization. *J. Ind. Microbiol. Biotechnol.* **2015**, 769–778. [CrossRef] [PubMed]
84. Magalhães, F.; Krogerus, K.; Vidgren, V.; Sandell, M.; Gibson, B. Improved cider fermentation performance and quality with newly generated Saccharomyces cerevisiae × Saccharomyces eubayanus hybrids. *J. Ind. Microbiol. Biotechnol.* **2017**, *44*, 1203–1213. [CrossRef] [PubMed]
85. Magalhães, F.; Krogerus, K.; Castillo, S.; Ortiz-Julien, A.; Dequin, S.; Gibson, B. Exploring the potential of Saccharomyces eubayanus as a parent for new interspecies hybrid strains in winemaking. *FEMS Yeast Res.* **2017**, *17*, 1–10. [CrossRef]

86. Torija, M.J.; Beltran, G.; Novo, M.; Poblet, M.; Guillamón, J.M.; Mas, A.; Rozès, N. Effects of fermentation temperature and Saccharomyces species on the cell fatty acid composition and presence of volatile compounds in wine. *Int. J. Food Microbiol.* **2003**, *85*, 127–136. [CrossRef]
87. Krogerus, K.; Magalhães, F.; Vidgren, V.; Gibson, B. Novel brewing yeast hybrids: Creation and application. *Appl. Microbiol. Biotechnol.* **2017**, *101*, 65–78. [CrossRef]
88. Fay, J.C.; Hittinger, C.T.; Sia, E.A.; Peris, D.; Li, X.C. Mitochondria-encoded genes contribute to evolution of heat and cold tolerance in yeast. *Sci. Adv.* **2019**, *5*, eaav1848.
89. Beltran, G.; Novo, M.; Guillamón, J.M.; Mas, A.; Rozès, N. Effect of fermentation temperature and culture media on the yeast lipid composition and wine volatile compounds. *Int. J. Food Microbiol.* **2008**, *121*, 169–177. [CrossRef] [PubMed]
90. Redón, M.; Guillamón, J.M.; Mas, A.; Rozès, N. Effect of lipid supplementation upon Saccharomyces cerevisiae lipid composition and fermentation performance at low temperature. *Eur. Food Res. Technol.* **2009**, *228*, 833–840. [CrossRef]
91. Baker, E.P.; Peris, D.; Moriarty, R.V.; Li, X.C.; Fay, J.C.; Hittinger, C.T. Mitochondrial DNA and temperature tolerance in lager yeasts. *Sci. Adv.* **2019**, *5*, eaav1869. [CrossRef] [PubMed]
92. Tronchoni, J.; Rozès, N.; Querol, A.; Guillamón, J.M. Lipid composition of wine strains of Saccharomyces kudriavzevii and Saccharomyces cerevisiae grown at low temperature. *Int. J. Food Microbiol.* **2012**, *155*, 191–198. [CrossRef] [PubMed]
93. Iwase, T.; Takada, A.; Watanade, T.; Fukuda, H.; Sasaki, K.; Yoshitake, M. Breeding of Sake Yeast by Protoplast Fusion of Sake Yeast and Wine Yeast. *J. Brew. Soc. Jpn.* **1995**, *90*, 137–142. [CrossRef]
94. Motomura, S.; Horie, K.; Kitagaki, H. Mitochondrial activity of sake brewery yeast affects malic and succinic acid production during alcoholic fermentation. *J. Inst. Brew.* **2012**, *118*, 22–26. [CrossRef]
95. Asano, T.; Kurose, N.; Hiraoka, N.; Kawakita, S. Effect of NAD+-dependent isocitrate dehydrogenase gene (IDH1, IDH2) disruption of sake yeast on organic acid composition in sake mash. *J. Biosci. Bioeng.* **1999**, *88*, 258–263. [CrossRef]
96. Wakabayashi, K.; Isogai, A.; Watanabe, D.; Fujita, A.; Sudo, S. Involvemwnt of methionine salvage pathway genes of Saccharomyces cerevisiae in the production of precursor compounds of dimethyl trisulfide (DMTS). *J. Biosci. Bioeng.* **2013**, *116*, 475–479. [CrossRef]
97. Schneider, B.L.; Kiupakis, A.K.; Reitzer, L.J. Arginine catabolism and the arginine succinyltransferase pathway in Escherichia coli. *J. Bacteriol.* **1998**, *180*, 4278–4286. [CrossRef]

3

Kluyveromyces marxianus: Current State of Omics Studies, Strain Improvement Strategy and Potential Industrial Implementation

Dung Minh Ha-Tran [1,2,3], Trinh Thi My Nguyen [2] and Chieh-Chen Huang [2,4,*]

1 Molecular and Biological Agricultural Sciences Program, Taiwan International Graduate Program, Academia Sinica and National Chung Hsing University, Taipei 11529, Taiwan; hatranminhdung@gmail.com
2 Department of Life Sciences, National Chung Hsing University, Taichung 40227, Taiwan; mytrinhnguyen0410@gmail.com
3 Graduate Institute of Biotechnology, National Chung Hsing University, Taichung 40227, Taiwan
4 Innovation and Development Center of Sustainable Agriculture, National Chung Hsing University, Taichung 40227, Taiwan
* Correspondence: cchuang@dragon.nchu.edu.tw

Abstract: Bioethanol is considered an excellent alternative to fossil fuels, since it importantly contributes to the reduced consumption of crude oil, and to the alleviation of environmental pollution. Up to now, the baker yeast *Saccharomyces cerevisiae* is the most common eukaryotic microorganism used in ethanol production. The inability of *S. cerevisiae* to grow on pentoses, however, hinders its effective growth on plant biomass hydrolysates, which contain large amounts of C_5 and C_{12} sugars. The industrial-scale bioprocessing requires high temperature bioreactors, diverse carbon sources, and the high titer production of volatile compounds. These criteria indicate that the search for alternative microbes possessing useful traits that meet the required standards of bioethanol production is necessary. Compared to other yeasts, *Kluyveromyces marxianus* has several advantages over others, e.g., it could grow on a broad spectrum of substrates (C_5, C_6 and C_{12} sugars); tolerate high temperature, toxins, and a wide range of pH values; and produce volatile short-chain ester. *K. marxianus* also shows a high ethanol production rate at high temperature and is a Crabtree-negative species. These attributes make *K. marxianus* promising as an industrial host for the biosynthesis of biofuels and other valuable chemicals.

Keywords: bioethanol; *Kluyveromyces marxianus*; omics technologies; gTME; CRISPR-Cas9

---

1. Introduction

Saccharomyces cerevisiae plays an extremely important role for millennia in food and beverage productions, and is the most studied yeast species [1]. The importance of *S. cerevisiae* in bioethanol production is unquestionable, as it is the most common microorganism, being used in the first generation bioethanol production from sugar or starch crops [2]. In addition to the well-known yeast *S. cerevisiae*, however, the demand for other non-conventional yeasts that possess advantageous characters, such as thermotolerance or pentose metabolism for industrial application, is continuously rising. The thermotolerant yeast *K. marxianus* has many good traits to be used as cell factory in food and biotechnology [3]. These advantages include the fastest growth rate (with the maximum growth rate of 0.80 h^{-1}) among any eukaryotic microbes [4,5], the ability to assimilate a wide range of sugars (e.g., glucose, lactose, galactose, xylose, inulin, and arabinose), thermo (up to 52 °C) and toxin (furaldehyde) tolerance, a wide range of pH values (pH 2.5–9), high ethanol yield at elevated temperatures, production of value-added aromatic chemicals (e.g., 2-phenylethylethanol and 2-phenylethyl acetate), and secretion of lytic enzymes [6–8].

This review aims to focus on the latest progress in Omics studies of *K. marxianus*, especially the recent transcriptomic and proteomic studies of *K. marxianus* grown on specific substrates (e.g., Jerusalem artichokes) or in stress conditions (heat, ethanol stress, or toxic compounds). The review also updates the current state of *K. marxianus* strain improvement using advanced molecular biology techniques. The third part emphasizes the role of *K. marxianus* as a complementary microbe in microbial co-culture system, and also highlights the use of *K. marxianus* in different configurations of substrate hydrolysis and fermentation for bioethanol production. The fourth part of the review discusses some studies on the Crabtree effect in *K. marxianus* and *S. cerevisiae*.

2. Omics Studies in *K. marxianus* upon Stress Conditions

Contrary to the conventional yeast *S. cerevisiae*, *K. marxianus* cannot tolerate high ethanol concentration. Under ethanol stress, e.g., 6% (*v*/*v*) ethanol, Diniz et al. [9] found that the metabolic flow through the central metabolic pathways was impaired. Genes encoding heat shock proteins were upregulated and ribosome biogenesis-related genes were down regulated, indicating the harmful effect of ethanol on *K. marxianus* growth and cell proliferation machinery. Upon ethanol stress, *S. cerevisiae* increases the degree of unsaturated fatty acids and ergosterol of plasma membrane to maintain the membrane stability [10]. In *K. marxianus* CCT 7735, however, genes encoding unsaturated fatty acid biosynthesis were downregulated at high ethanol concentration [9].

In addition, the fatty acid profile of *K. marxianus* showed that the degree of unsaturated fatty acid did not increase upon ethanol stress. This finding is in agreement with the study of Alvim et al. [11] as they found that after 12 h of ethanol exposure, the concentration of ergosterol decreased compared to that of 1 h, 4 h, and 8 h after ethanol exposure. Similarly, in the study of Wang et al. [12], various genes encoding fatty acid and ergosterol metabolism were downregulated under multiple inhibitors stress such as phenols, furfural, HMF, and acetic acid. These consistent findings might explain differences in ethanol tolerance capability between *S. cerevisiae* and *K. marxianus*. In the work of Fu et al. [13], in contrast, at high temperature (45 °C), genes related to lipid metabolism of the plasma membrane were upregulated and *K. marxianus* DMKU3-1042 also produced more ergosterol than it did in the normal condition. Differences in the expression patterns of lipid metabolism-encoding genes and/or ergosterol profiles between the studies of Diniz et al. [9], Wang et al. [12], Alvim et al. [11], and Fu et al. [13] might be explained by the differences in *K. marxianus* strains used in these experiments (e.g., CCT 7735, YHJ010, CCT 7735, and DMKU3-1042, respectively), since they used the same method [14] to measure ergosterol content.

The downregulation of genes involved in central carbon metabolism are consensus between ethanol stress [9,11], high temperature [13], and mixed inhibitors stress [12] (Table 1). In 1995, Piper stated that many changes induced by ethanol stress were similar to those triggered by heat stress and the synergistic effects of heat and ethanol stresses were recorded [15]. Intriguingly, these present reports reconfirmed Piper's statement as various genes related to central carbon metabolic pathways were found to exhibit the low expression levels upon heat or ethanol exposure (Table 1).

Table 1. A brief overview of Omics studies in *K. marxianus*.

Omics Technologies	Growth Conditions	Important Findings	Sources
RNA-seq, HiSeq 2000 system (Illumina, San Diego, CA 92121, USA)	Yeast strain: Y179 Treatments: - 120 g/L inulin without aeration (120–N–24) - 230 g/L inulin without aeration (230–N–72) (control) - 230 g/L inulin with aeration (ORP -130 mV) (230–130 mV–36) Differentially expressed genes (DEGs) analysis: Module 1: 230–130 mV–36 vs. 230–N–72 Module 2: 120–N–24 vs. 230–N–72	Result 1: Module 1 had 1840 DEGs, module 2 had 2658 DEGs Conclusion 1: Inulin concentrations had greater effect on transcriptome profiles than aeration and hypoxic condition. Result 2: More genes related to ethanol metabolism and transcriptional factors upregulated in 120–N–24 relative to 230–N–72 Conclusion 2: High inulin loading inhibited yeast metabolism Downregulated genes in 230–130 mV–36: GPM1 Downregulated genes in 120–N–24: *GPM1* Upregulated genes in 230–130 mV–36: - Central carbon metabolic pathways: *INU1, HXK1, GLK1, MDH1p, PDC1, ADH3, GPD1p, TRXR, GPX, KMALLA2475, TPO1, HSP31* - Upregulated genes in 120–N–24: - Central carbon metabolic pathways: *INU1, HXK1, GLK1* - Upregulated genes in 230–N–72: PDC1, MIG1, ATG8	[16]
RNA-seq, (SOLiD 5500 XL sequencer, Thermo Fisher Scientific, Waltham, MA, USA)	Yeast strain: CCT 7735 Treatments: High ethanol exposure 6% (*v*/*v*) DEGs analysis: Module 1: 1 h vs. 0 h Module 2: 4 h vs. 0 h Module 3: 1 h vs. 4 h	Downregulated genes in ethanol treatment: - Unsaturated fatty acid and ergosterol biosynthesis: *FEN1, SUR4, FAS1, SCS7, KLMA-40623, ERG25, ERG3, SUR2, OLE1, KLMA_20527, KLMA_10244, KLMA_20392* - Central carbon metabolic pathway: *RAG5, GLK1, RAG2, FBA1, GAP3, GAP1, PGK, GPM1, ENO, PYK1, LAT1, PYC2, ACO2b, LSC2* - Leloir pathway: *GAL1, GAL7, GAL10* - Fermentation pathway: *LAT1, ACS2, ADH, ADH1, ADH2, ADH3, ADH4b* - Translation initiation factors: *eIF3a, eIF3e, eIF5A* Upregulated genes in ethanol treatment: - Central metabolic pathway: *ZWF, KLMA70303, PYC2* - Heat shock protein: *HSP26, HSP60, HSP78*	[9]

Table 1. *Cont.*

Omics Technologies	Growth Conditions	Important Findings	Sources
RNA-seq, Illumina HiSeq 4000 instrument (Illumina, San Diego, CA 92121, USA)	Yeast strain: YHJ010 Treatments: Mixed inhibitors (0.7 g/L furfural + 0.7 g/L HMF + 3 g/L acetate acid + 0.28 g/L phenols (4-hydroxybenzaldehyde, syringaldehyde, catechol and vanillin, 0.07 g/L each) DEGs analysis: Mixed inhibitors treatment vs. Control (without stress)	Downregulated genes in mixed inhibitors: - Central carbon metabolism: HXK1, GND1, PGI1, PFK1, PFK2, FBA1, TPI1, TDH1, TDH3, PGK1, GPM1, GPM2, ENO1, PYK1, PDC, ADH2, DAK1 - Fatty acid and ergosterol metabolism: OLE1, SCS7, FAS2, DUG3, LipA, ERG25, LTA4H, ERG1, ATH1, ERG20 - B1 & B6 metabolism: KMAR_30698, KMAR_30699, KMAR_30041, KMAR_20540, KMAR_40549, KMAR_30339 - Transporters: KMAR_50344, KMAR_10529, KMAR_10514, KMAR_10360, KMAR_10458, KMAR_10759, KMAR_20313, KMAR_70169, KMAR_20003, KMAR_70277, KMAR_30323, KMAR_40422, KMAR_60332, KMAR_50593 - Transcription factors: KMAR_40216, KMAR_40526, KMAR_70129, KMAR_10730, KMAR_60223 Upregulated genes in mixed inhibitors: - Central carbon metabolism: FBP1, TDH2, ADH3, ADH4, ADH6, ALD6, GUT2, MAE1, CIT1, ACO1, ACO2, IDH1, IDH2, KGD1, KGD2, SDH1, SDH2, SDH3, SDH4, MDH2, PCK1, ICL1, MLS1, GDH1 - Transcription factors: KMAR_30570, KMAR_50272, KMAR_30474, KMAR_30246, KMAR_60382, KMAR_50274, KMAR_40048 - Mitochondrial respiratory chain: NDI1, SDH1, SDH2, SDH3, SDH4, QCR1, QCR2, QCR9, RIP1, CYT1, ATP1, ATP16, ATP14, ATP6C - ROS detoxification: KMAR_70075, KMAR_20527, KMAR_40107, KMAR_80342, KMAR_40185, KMAR_50400 - Transporters: KMAR_80370, KMAR_30579, KMAR_80266, KMAR_50347, KMAR_20602, KMAR_70126, KMAR_10531, KMAR_40029, KMAR_50130, KMAR_80409, KMAR_60406, KMAR_10004, KMAR_40093, KMAR_10790, KMAR_20248, KMAR_40425, KMAR_60075, KMAR_30642, KMAR_30337, KMAR_40188, KMAR_40156, KMAR_70262, KMAR_10802, KMAR_80400, KMAR_40340, KMAR_20004, KMAR_30588, KMAR_70319	[12]
RNA-seq, HiSeq 4000 system (Illumina Inc., San Diago, CA 92121, USA)	Yeast strain: DMKU3-1042 Treatments: High temperature 45 °C DEGs analysis: 45 °C–14 h vs. 30 °C–14 h 45 °C–22 h vs. 30 °C–22 h	Downregulated genes at 45 °C vs. 30 °C: - Central carbon metabolic network: GLK1, RAG2, PFK1, GPD1, FBA1, TDH1, TDH3, RHR2, TPI1, PGK, ADH2, GPM1, PDX1, LAT1, ACS2, ALD4, CIT1, MDH1, MDH3, FUM1, LSC2, ACO2b, IDP1, KGD1 - BCAA biosynthesis: LEU1, LEU2, LEU4, SDL1, ILV3, ILV6 Upregulated genes at 45 °C vs. 30 °C: - Mitochondrial respiratory chain: COX5A, COX7, COX12, RIP, QCR2 - Glycerol and acetic acid generation: GPD2, ALD6	[13]
	DEGs analysis: 45 °C–16 h vs. 45 °C–14 h 45 °C–18 h vs. 45 °C–14 h 45 °C–20 h vs. 45 °C–14 h 45 °C–22 h vs. 45 °C–14 h	Downregulated genes at 45 °C (16, 18, 20, 22 h vs. 14 h): - Central carbon metabolic network: HXK, ZWF, GPD1, FBA1, TDH1, TDH3, PGK, ADH1, ADH2, PDC1, ENO, ALD6, MDH1, MDH2, MDH3, SDH1, LSC2, KGD1 Upregulated genes at 45 °C (16, 18, 20, 22 h vs. 14 h): - Central carbon metabolic network: GLK1, RAG2, FBP1, GPD2, PFK1, RHR2, TPI1, GPM1, GPM3, PYK1, ADH3, ADH4b, ALD4, LAT1, PDX1, ACS2, FUM1, ACO2b, IDP1	

Table 1. *Cont.*

Omics Technologies	Growth Conditions	Important Findings	Sources
MALDI-TOF/TOF (Ultraflex III, Bruker, Daltonics, Bremen, Germany)	Yeast strain: CCT 7735 Treatments: High ethanol exposure 6% (*v*/*v*) Protein abundance analysis: 1 h and 4 h after ethanol exposure vs. 1 h and 4 h (absence of ethanol) (control)	Less abundant at 1 h (ethanol stress) vs. control - Central carbon metabolism: Enolase_8, Enolase_9, Triosephosphate isomerase_2, Triosephosphate isomerase_3, Phosphoglycerate mutase 1_3, NAD(P)H-dependent D-xylose reductase_1, Pyruvate kinase, Fructose-bisphosphate aldolase_1, Phosphoglycerate kinase, Transaldolase_1, Transaldolase_2, Triosephosphate isomerase_1 - Heat shock proteins: HSP SSA3_9, HSP SSA2 - Translational proteins: 40S ribosomal protein S14, 40S ribosomal protein S18 More abundant at 1 h (ethanol stress) vs. control - Central carbon metabolism: Enolase_1, Enolase_2, Enolase_4, Enolase_5, Enolase_6, Enolase_7, Enoate reductase 1_1, Enoate reductase 1_3, Enoate reductase 1_4, Hexokinase, Phosphoglycerate mutase 1_2, Phosphoglycerate mutase 1_4, Malate dehydrogenase, Alcohol dehydrogenase 1, Alcohol dehydrogenase 2_2 - Heat shock proteins: HSP SSA3_1, HSP SSA3_2, HSP SSA3_3, HSP SSA3_4, HSP SSA3_5, HSP SSA3_6, HSP SSA3_7, HSP SSA3_8 Less abundant at 4 h (ethanol stress) vs. control - Central carbon metabolism: Phosphoglycerate kinase, Fructose-bisphosphate aldolase_1, Enolase_8, Enolase_9, Transaldolase_1, Triosephosphate isomerase_1, Enolase_1, Enolase_2, Enolase_3, Enolase_4, Enolase_5, Enoate reductase 1_2 - Heat shock proteins: HSP SSA2, HSP104 More abundant at 4 h (ethanol stress) vs. control - Central carbon metabolism: Enolase_6, Enolase_7, Enoate reductase 1_1, Enoate reductase 1_3, Enoate reductase 1_4, Fructose-bisphosphate aldolase_2, Transaldolase_2, NAD(P)H-dependent D-xylose reductase_1 - Heat shock proteins: HSP78, HSP26	[11]

Abbreviation of gene/enzyme in the Table 1: RAG5, hexokinase; RAG2, glucose-6-phosphate isomerase; FBA1, fructose-bisphosphate aldolase; GAP3, glyceraldehyde-3-phosphate dehydrogenase 3; GAP1, glyceraldehyde-3-phosphate dehydrogenase 1; PGK, phosphoglycerate kinase; GPM1, phosphoglycerate mutase 1; GPM2, probable phosphoglycerate mutase YOR283W; GPM3, phosphoglycerate mutase 3; ENO, enolase; PYK1, pyruvate kinase; LAT1, acetyltransferase component of pyruvate dehydrogenase complex; PYC2, pyruvate carboxylase; ACO1, aconitate hydratase; ACO2b, aconitate hydratase; LSC2, succinyl-CoA ligase subunit β; PFK1, phosphofructokinase 1; GPD1, glycerol-3-phosphate dehydrogenase 1; GPD2, glycerol-3-phosphate dehydrogenase 2; TDH1, glyceraldehyde-3-phosphate dehydrogenase 1; TDH2, glyceraldehyde-3-phosphate dehydrogenase 2; TDH3, glyceraldehyde-3-phosphate dehydrogenase 3; RHR2, glycerol-3-phosphatase 1; TPI1, triose phosphate isomerase; PDX1, pyruvate dehydrogenase; ACS2, acetyl-CoA synthetase 2; ALD4, aldehyde dehydrogenase; CIT1, citrate synthase; MDH1, malate dehydrogenase 1; MDH2, malate dehydrogenase 2; MDH3, malate dehydrogenase 3; FUM1, fumarate hydratase; IDP1, isocitrate dehydrogenase 1; KGD1, 2-oxoglutarate dehydrogenase E1 component; KGD2, dihydrolipoyllysine-residue succinyltransferase component of 2-oxoglutarate dehydrogenase complex; GAL1, galactokinase; GAL7, galactose-1-phosphate uridylyltransferase; GAL10, bifunctional protein; ADH, alcohol dehydrogenase; ADH1, alcohol dehydrogenase 1; ADH2, alcohol dehydrogenase 2; ADH3 alcohol dehydrogenase 3; ADH4b, alcohol dehydrogenase 4; ADH6, NADP-dependent alcohol dehydrogenase 6; ZWF, glucose-6-phosphate 1-dehydrogenase; KLMA_70303, 6-phosphofructo-2-kinase; PYC2, pyruvate carboxylase; ALD6, magnesium-activated aldehyde dehydrogenase; COX5A, cytochrome c oxidase polypeptide 5A; COX7, cytochrome c oxidase subunit 7; COX12, cytochrome c oxidase subunit 6B; RIP, cytochrome b-c1 complex subunit Rieske; QCR2, cytochrome b-c-1 complex subunit 2; LEU1, 3-isopropylmalate dehydratase; LEU2, 3-isopropylmalate dehydrogenase; LEU4, 2-isopropylmalate synthase; SDL1, L-serine dehydratase; ILV3, dihydroxy-acid dehydratase; ILV6, acetolactate synthase small subunit; BCAA, bacterial branched-chain amino acid biosynthesis, HXK, hexokinase; PDC1, pyruvate decarboxylase 1; SDH1, succinate dehydrogenase 1; SDH2, succinate dehydrogenase 2; SDH3, succinate dehydrogenase 3; SDH4, succinate dehydrogenase 4; FBP1, fructose 1,6-bisphosphatase; HXK1, hexokinase; GND1, 6-phosphoglunonate dehydrogenase; PGI1, glucose-6-phosphate isomerase; PFK1, 6-phosphofructokinase subunit alpha; PFK2, 6-phosphofructokinase subunit beta; PGK1, phosphoglycerate kinase; ENO1, enolase 1; DAK1, dihydroxyacetone kinase 1; GUT2, glycerol-3-phosphate dehydrogenase; MAE1, NAD-dependent malic enzyme; IDH1, isocitrate dehydrogenase [NAD]; IDH2, isocitrate dehydrogenase [NAD]; PCK1, phosphoenolpyruvate carboxykinase [ATP]; ICL1, isocitrate lyase; MLS1, malate synthase 1; GDH1, NADP-specific glutamate dehydrogenase; NDI1, rotenone-insensitive NADH-ubiquinone oxidoreductase; SDH1, succinate dehydrogenase 1; SDH2, succinate dehydrogenase 2; SDH3, succinate dehydrogenase 3; SDH4, succinate dehydrogenase 4; QCR1, cytochrome b-c1 complex subunit 1; QCR2, cytochrome b-c1 complex subunit 2; QCR9, c reductase complex; RIP1, cytochrome b-c1 complex subunit Rieske; CYT1, cytochrome c1; ATP1, ATP synthase subunit alpha; ATP16, ATP synthase subunit delta; ATP14, ATP synthase subunit H; ATP6C, v-Type proton ATPase subunit C; OLE1, acyl-CoA desaturase 1; SCS7, inositolphosphorylceramide-B C-26 hydroxylase; FAS2, fatty acid synthase subunit alpha; DUG3, probable glutamine amidotransferase DUG3; LipA, lipoyl synthase; ERG25, c-4 methylsterol oxidase; LTA4H, leukotriene A-4 hydrolase; ERG1, squalene monooxygenase; ATH1, vacuolar acid trehalase; ERG20, farnesyl pyrophosphate synthetase; INU1, inulinase1; GLK1, glucokinase 1; GPD1p, glycerol-3-phosphate dehydrogenase; TRXR, thioredoxin reductase; GPX, glutathione peroxidase.

Industrially relevant substrate loading caused overflow metabolism and growth cessation in *Hungateiclostridium thermocellum*, a robust thermophilic, cellulolytic bacterium [17,18]. This phenomenon also happens to *K. marxianus* when growing at high concentration of inulin-containing materials like Jerusalem artichokes. In the study of Gao et al. [16], two modules of treatments were used to investigate the effects of inulin loadings and aeration conditions on sugar consumption and ethanol fermentation of *K. marxianus* Y179. The first module consists of treatment with 230 g/L inulin, with micro-aeration by oxidation-reduction potential (ORP) controlling at −130 mV and the yeast sample was collected at 36 h (230–130 mV–36) vs. treatment with 230 g/L inulin, without aeration, and the sample was collected at 72 h (230–N–72). Similarly, the second module composes the treatment with low substrate loading 120 g/L inulin, without aeration (120–N–24) vs. 230–N–72. In the first module, micro-aeration condition (230–130 mV–36) promoted inulin consumption and ethanol fermentation. In the second module, genes related to ethanol metabolism and transcriptional factors were upregulated in 120–N–24, thus, suggesting the inhibitory effect of high inulin loading on *K. marxianus* metabolism. Accordingly, gene *HXK1* encodes hexokinase, gene *GLK1* encodes glucokinase and gene *INU1* encodes inulinase were upregulated in 230–130 mV–36 and 120–N–24 relative to those in 230–N–72. In a previous study [19], under carbon deprivation (when ethanol is the sole carbon source in the medium), the gene *GPM1* encoding phosphoglycerate mutase in *S. cerevisiae* was upregulated, and ethanol was used to generate ATP (through oxidative phosphorylation) and sugar phosphates for nucleotide biosynthesis, cell wall construction and storage carbohydrates biosynthesis. Therefore, the function of *GPM1* gene is inferred to be associated with respiratory growth on non-fermentable substrates like ethanol. This gene might be a good candidate for further gene silencing strategy to prevent the consume of ethanol as a substrate, thus, enhancing ethanol productivity. The *GPM1* gene in *K. marxianus* was found to be downregulated in low inulin loading (120–N–24) and micro-aeration (230–130 mV–36), suggesting the abundance of carbon source and the favored growth condition for *K. marxianus*. The high expression levels of *PDC1* gene encoding pyruvate decarboxylase and *ADH3* encoding alcohol dehydrogenase in 230–130 mV–36 assumed the carbon flux towards fermentative pathways, which enhance the respiration and regulate reduction reactions. Glycerol-encoding gene *KmGPD1* was upregulated in high inulin concentration to maintain the high cell viabilities during ethanol fermentation. Moreover, two reactive oxygen species (ROS) stress-related genes thioredoxin reductase *TRXR* and glutathione peroxidase *GPX* were upregulated in 230–130 mV–36 treatment, suggesting their important roles in helping cells defend better against ROS damages. Altogether, the regulation patterns of key genes in this study indicated that the micro-aeration in high substrate loading system is suitable for ethanol fermentation using inulin as the starting material.

In summary, the rapid development of Omics technologies helps to gain insight into transcriptomic and proteomic profiles of *K. marxianus* in response to stress conditions such as high temperature, high ethanol concentration or furfural, phenol inhibitors. Based on the gene expression patterns and/or protein abundance upon these harsh circumstances, best candidate genes could be selected for further detailed study or metabolic engineering to develop industrially relevant phenotypes.

3. Advanced Techniques in *Kluyveromyces marxianus* Strain Improvement

K. marxianus can transport various types of sugar, such as glucose [20], lactose [21], fructose [22], galactose [23], xylose [24], cellobiose, and arabinose [8], and organic acids, such as lactic acid [25] and malic acid [26], into the cells. However, the ability of *K. marxianus* to digest cellobiose was still very poor [8]. To improve the capability of metabolizing cellobiose for *K. marxianus* KY3, Chang et al. [8] transformed a rumen fungal β-glucosidase gene from *Neocallimastix* sp. W5 into its genome. Consequently, the transformant *K. marxianus* KY3-NpaBGS strain was able to use cellobiose better and produced approximately 1 g/L ethanol when growing on YP medium supplemented with 20 g/L cellobiose. In contrast, *K. marxianus* SSSJ-0, a native kefir yeast strain that possesses β-glucosidase enzyme, could only use cellobiose for cell growth, but was unable to convert cellobiose into ethanol.

Lignocellulosic biomass is an abundant and renewable resource for the production of biofuels and other value-added compounds [27]. Therefore, many efforts have been made to combine the high ethanol yield and the robust lignocellulose degradability into a single host cell for a consolidated bioprocessing (CBP). In this concept, the *K. marxianus* KY3 was engineered to be an artificial cellulolytic microbe with five cellulase genes including two exoglucanases (from *Trichoderma reesei*), two endoglucanases (from *Aspergillus niger*) and one β-glucosidase (from *Neocallimastix patriciarum*) transformed into the yeast genome using the Promoter-based Gene Assembly and Simultaneous Overexpression (PGASO) technique [28,29]. In addition, to facilitate the import of cellodextrin into the cells, a fungal cellodextrin transporter gene from the red bread mold *Neurospora crassa* was selected for genetic transformation. Consequently, the ethanol yield of the recombinant *K. marxianus* KR7 strain in YP medium with 10% (*w/v*) Avicel as the sole carbon source was ~0.6 g/L. These foreign genes functioned properly in the host cell, reflecting via cellulolytic enzyme assay, cellodextrin transport, cellobiose digestion, and ethanol production. Although the conversion of Avicel to ethanol was not that efficient, the PGASO method proved its potential for practical applications, as it could assemble multiple exogenous genes into *K. marxianus* genome in one single step to facilitate enzyme combinations or to construct *de novo* desired pathways in *K. marxianus* host cell [28].

Hungateiclostridium thermocellum cellulosome, nature's largest cellulolytic machinery, accounts for the fastest growth rate of any bacterium on crystalline cellulose [30]. A cellulosomal enzyme contains a type I dockerin, which could interact with the type I cohesin of the central nonenzymatic scaffolding subnit CipA via type I dockerin-type I cohesin interaction. Due to the Lego-like architecture of cellulosome, each scaffolding subunit CipA, with nine type I cohesins on its structure, can carry simultaneously nine different cellolosomal enzymes. In turn, CipA, with its type II dockerin, enables the interaction with one of three surface anchoring proteins SdbA, Orf2p, or OlpB via type II dockerin-type II cohesin modules. Since the anchoring protein OlpB has seven type II cohesins, the interaction between CipA-OlpB can accommodate up to $9 \times 7 = 63$ cellolosomal enzymes in a single cellulosome complex. Up to now, several research groups have been sought to design cellulosome microbes that can express a full size of cellulosome structure instead of some individual cellulosomal genes called mini-cellulosomes [31–36]. Recently, the group of Anandharaj et al. [37] succeeded in developing an engineered *K. marxianus* host that can express a full size of *H. thermocellum* cellulosome on its cell surface. The engineered yeast, with its *de novo* powerful cellulosome, could efficiently degrade Avicel and phosphoric acid-swollen cellulose (PASC) to produce 3.09 g/L and 8.61 g/L of ethanol, respectively. This result could be recorded as the highest ethanol titer of any constructed yeast cellulosome thus far [37].

The thioredoxin/thioredoxin reductase (Trx/TrxR) system is widely present in yeast mitochondria and plays important roles in protecting yeast from ROS damages [16]. In the study of Gao et al. [16], they found that the gene encoding thioredoxin reductase (*KmTrxR*) in *K. marxianus* was upregulated under high substrate loading and aerobic conditions. To confirm the protective functions of *K. marxianus* Trx/TrxR system in other yeasts, two genes *KmTRX* and *KmTrxR* were transformed into the *S. cerevisiae* 280 host cell to create the *KmTRX* overexpression strain, the *KmTrxR* overexpression strain, and the double *KmTRX-KmTrxR* overexpression strain. The results showed that although the overexpression of a single *KmTRX* gene in *S. cerevisiase* 280 had adverse effect on the host cell, the overexpression of *KmTrxR*, in contrast, aided the host cell tolerate to lignocellulose-derived inhibitors, such as acetic and formic acids. Moreover, the double overexpression of two genes *KmTRX* and *KmTrxR*, with their synergistic effects, could improve ethanol yield, and shorten the lag phase of *S. cerevisiae* cell under the inhibitory effects of mixed chemicals such as acetic and formic acids and furfural (FAF) [38]. Additionally, in the study of Gao et al. [16], the *KmTPX1* gene, which encodes peroxiredoxin, was found greatly upregulated under aerobic conditions and high inulin concentration. The gene *KmTPX1* is homologous to *Tsa1p* gene in *S. cerevisiae*, which is involves in redox reactions to remove excess ROS like peroxides, to regulate the concentration of peroxides to protect cells from DNA damage and cell death [39]. Taking advantage of their previous finding, Gao et al. [40] constructed an overexpression vector which contained *KmTPX1* gene and transformed it into *S. cerevisiae* cell. As expected, the overexpression of *KmTPX1* in the transformant *S. cerevisiae* strain helped the yeast tolerate better to both oxidative stress and inhibitory compounds released from the degradation of lignocellulose. Consequently, the enhanced tolerance of *S. cerevisiae* to oxidative stress and furfural led to the overall higher rates of glucose consumption and ethanol fermentation in the transformant *KmTPX1* strain compared with the control.

Based on the stress-related transcription factor (TF) profiles in *S. cerevisiae* in a prior study [41], Li et al. [42] performed a protein-protein BLAST to determine the stress-related TFs in *K. marxianus*. Subsequently, they carried out the genetic transformation of exogeneous stress-related TF derived from *K. marxianus* DMKU3-1042 into *S. cerevisiae* TSH3 cell to enhance the thermotolerance, growth and ethanol yield of *S. cerevisiae* TSH3. As a consequence, at elevated temperature (43 °C) and 104.8 g/L glucose, the transformant KmHSF1 and KmMSN2 *S. cerevisiae* strains yielded the final ethanol concentrations of 27.2 ± 1.4 g/L and 27.6 ± 1.2 g/L, respectively, much higher than the control with 18.9 ± 0.3 g/L ethanol. When looking into details, the transcriptomic profiles of these transgenic *S. cerevisiae* strains revealed that the *KmHsf1* gene improved ethanol production by regulating transporter-related genes in the host cell to limit the excessive ATP consumption and by promoting glucose uptake, whereas the *KmMsn2* gene might aid in regulating glucose metabolism and glycolysis/gluconeogenesis. In addition, *KmMsn2* promoted the host cell tolerate better to high temperature by regulating genes involved in lipid metabolism, thereby changing membrane fluidity. These above studies exemplify excellently a straightforward procedure from transcriptomic or proteomic studies to the selection of candidate genes for genetic transformation or other technologies for the improvement of microbial biofuel microorganisms.

Recently, in the study of da Silveira et al. [43], ethanol-tolerant *K. marxianus* CCT 7735 strains were developed using the Adaptive laboratory evolution (ALE) strategy [44]. Briefly, hundreds of generations of *K. marxianus* were exposed to 4% (*v/v*) ethanol, and the trained yeast cells were considered "ethanol tolerant" when a significant increase (>50%) in the specific growth rate was observed. In the evolved ethanol-tolerant *K. marxianus* ETS4 strain, the intracellular amine/amide compounds and organic acids abundance were higher than those in its parent strain P4 under ethanol stress. The membrane fatty acid and ergosterol, an important sterol in yeast membranes, which is responsible for ethanol tolerance trait [45], were more abundant in the evolved strain ETS4 than in the P4 strain. This phenotype was in accordance with a INDEL mutation in the upstream region of the coding sequence (CDS) detected in the *RRI1* gene which is involved in the positive regulation of ergosterol biosynthesis. Likewise, two genes *KLMA_10136* and *PXA*, which are associated with lipid metabolic process, had mutations as follows: INDEL in the upstream region of *KLMA_10136* CDS and INDEL in the downstream region of *PXA* CDS, respectively. Additionally, the accumulations of valine and metabolites of the TCA cycle such as isocitric acid, citric acid, and cis-aconitric acid were recorded only in the ETS4 strain when exposed to ethanol. This might contribute to an increase in ethanol tolerance of the evolved strain.

The TATA-binding protein (TBP) Spt15, one of the components of the general factor RNA polymerase II (RNA Pol II) transcription factor D (TFIID), is the most common target of yeast for global transcription machinery engineering (gTME) technique [46,47]. This technique could induce the global perturbations of the transcriptome through mutagenesis of key proteins that regulate the global transcriptome, thereby improving complex phenotypes quickly and effectively [47]. In the study of Li et al. [48], the *SPT15* gene was subjected to error-prone PCR, cloned into an expression vector and, then, pooled recombinant plasmids were transformed into *K. marxianus* to construct a random mutagenesis library in its cells. The results of mutant screening under 6% (*v/v*) ethanol stress showed that two mutant strains M2 and M10 demonstrated faster growth rates than others. Regarding ethanol productivity, M2 strain performed better compared with M10 and control strain (i.e., M2 produced 57.29 ± 1.96 g/L ethanol, which was 23.74% and 22.05% higher than those of M10 and the control, respectively). Moreover, the M2 strain also tolerated to high ethanol concentration better than M10 and the control, e.g., its ethanol inhibition concentration (EIC) value was 57 g/L, much higher than that of M10 and the control with 46 and 47 g/L, respectively. As a global transcriptome regulator, a non-synonymous (Non-Syn) mutation (Lys was substituted by Glu[31]) in the *Spt15* gene could influence the expression patterns of hundreds of genes including those involved in the central carbon metabolism, amino acid transport, long-chain fatty acid biosynthesis and MAPK signaling pathway (upregulated) and also ribosome biosynthesis, translation and protein synthesis (downregulated). From this perspective, the gTME method could be used for the improvement of other complex phenotypes, such as furfural tolerance or thermotolerance in *K. marxianus*.

Despite several advantageous traits for industrial applications, however, genetic engineering approaches for *K. marxianus* strain improvement have been still limited since the genome-editing tools and stable heterologous expression systems for this yeast species have not well-established yet [49]. In the study of Löbs et al. [50], CRISPR-Cas9 system, which was adapted from *Streptococcus pyogenes*, was used to create functional disruptions to alcohol dehydrogenase (ADH) and alcohol-O-acetyltransferase (ATF) genes in *K. marxianus*. The study aimed to investigate the metabolic pathways that are involved in the ethyl acetate and ethanol biosynthesis. In industry, ethyl acetate is used as a solvent and as flavor and fragrance compound and its worldwide demand is ~1.7 million tons per year [51]. The data from Löbs et al. report showed that the knockout of *KmAtf* gene reduced the production of ethyl acetate by 15%, whereas the disruption of *KmAdh2* gene almost entirely abolished the production of ethanol, resulting in the accumulation of acetaldehyde. The data obtained from KmADH2 and KmATF knock-out strains indicated the fundamental role of *KmAdh2* gene in ethanol production in both aerobic and anaerobic conditions. In regard to ethyl acetate biosynthesis, KmADH2 played a role in providing ethanol as a substrate for the reaction of Atf-catalyzed condensation with acetyl-CoA. Since the disruption of *KmAtf* gene only reduced a little amount of ethyl acetate, it suggested that probable alternative metabolic routes might take responsibility for the biosynthesis of ethyl acetate in *K. marxianus*.

4. Mono-, Co-Culture Systems and Other Fermentation Process Configurations in Bioethanol Production Using *K. marxianus*

Many fermentation approaches have been widely investigated to improve the productivity of bioethanol, thereby reducing the cost of industrial operation [28,35,37,48,52–75] (Table 2). A compatible co-culturing strategy could leverage the useful features from different microbes, thereby improving the productivity relative to monocultures [76]. Since an ideal microbe for consolidated bioprocessing (CBP) still remains to be found, the co-culture of engineered microorganisms, which confer newly advantageous genetic traits on microbes, would be a good approach for biofuels production. As numerous studies have been published, we just took few examples to clarify this concept. In the study of Ho et al. [77], a recombinant cellulosomal *Bacillus subtilis* which carried eight genes from *H. thermocellum*, namely one scaffolding protein gene (*cipA*), one cell-surface anchoring gene (*sdbA*), two exoglucanase genes (*celK* and *celS*), two endoglucanase genes (*celA* and *celR*), and two xylanase genes (*xynC* and *xynZ*) was cultured with a recombinant *K. marxianus* KY3-NpaBGs carrying a β-glucosidase gene from rumen fungus in the YP medium supplemented with 20 g/L Napier grass as the sole carbon source. At 42 °C, the dual-microbe co-culturing yielded 3.28 g/L, indicating the potential of *K. marxianus* as a complementary partner for bioprocessing. In this dual *K. marxianus*-*B. subtilis* system, the engineered *B. subtilis* was responsible for cellulolytic hydrolysis via its complex heterologous cellulosomal enzymes and the engineered *K. marxianus*, in turn, helps to convert the resultant cellobiose into glucose via secretory β-glucosidase. The study of Guo et al. [54], used cheese whey powder (CWP), a by-product of cheese industry, which contains high concentration of lactose and other essential nutrients for co-culturing *S. cerevisiase* and *K. marxianus*. As *S. cerevisiase* cannot ferment lactose but *K. marxianus* can, the co-culturing strategy was applied to make use of the carbon source and nutrient availability in CWP to produce ethanol. In addition, the mixed and alginate-immobilized cells produced higher ethanol yield relative to the free cell cultures. To enhance ethanol production and thermotolerance of yeast cells, the immobilized cocultures of *K. marxianus* DMKU 3-1042 and *S. cerevisiae* M30 on thin-shell silk cocoons (TSC) and alginate-loofa matrix (ALM) were carried out by Eiadpum et al. [55]. At high temperatures (range of 40–45 °C), both monoculture and coculture performed better than the monoculture of *S. cerevisiae* in producing ethanol. TSC and ALM functioned as yeast cell carriers and might protect cells from adverse conditions like high concentration of inhibitors or elevated temperatures [78]. On average, TSC-immobilized cell system yielded 16% higher ethanol production than ALM-immobilized cell system. This might be due to the high biocompatibility, high mechanical strength, light weight, high surface area, and proper porous structure of TSC that provided a convenient growth environment for yeast cells to live and to produce ethanol [78]. However, in a mixed culture, the cells–cells interaction between different strains is an important issue that should be taken into consideration. Differences in growth rates, nutrient uptake rates and secreted metabolites might be probable factors that affect cell viability [79]. In addition, killer toxins and extracellular proteases synthesized by yeasts may be another matter of mixed fermentation, as these toxic compounds might function against their coculture partners [80]. As both *S. cerevisiae* and *K. marxianus* could produce killer toxins [80], they might exclude each other in specific circumstances. In the study by Lopez et al. [79], the viability loss of *K. marxianus* was recorded in mixed culture conditions. Moreover, in the direct contact mixed culture, *S. cerevisiae* was also unfavorably affected.

Table 2. Monoculture, co-culture of *K. marxianus* with other microbes and other fermentation processes for bioethanol production.

Strain	Growth Condition	Theoretical Ethanol Yield (%)	Ethanol Yield (g eth/g sugar)	Maximum Ethanol (g/L)	Sources
	Monoculture and direct fermentation				
UFV-3	Aerobic: 30 °C, 250 rpm, whey permeate (240 g/L lactose)	-	0.35	57	[56]
	Hypoxia: 30 °C, 40 rpm, whey permeate (170 g/L lactose)	-	0.53	76	
	Anoxia: 30 °C, whey permeate (170 g/L lactose)	-	0.51	80	
DMKU 3-1042	37 °C, sugarcane juice (22% total sugars)	77.5	-	8.7	[57]
KD-15	30 °C, 90 rpm, 60 h, saccharified flour mixed with cheese whey (99.3 g/L glucose, 59.4 g/L lactose)	-	0.45 ± 0.027	71.4 ± 2.6	[58]
	30 °C, 90 rpm, 60 h, saccharified potato tubers mixed with cheese whey (137 g/L glucose, 19.1 g/L lactose)	-	0.44 ± 0.05	69.1 ± 3.9	
Kluyveromyces sp. IIPE453	45 °C, pH 4.5, 16 h, pretreated sugarcane bagasse pith (40 g/L total sugar)	88	-	17.4	[59]
DMB3-7	30 °C, 40 g/L xylose, 96 h	-	0.187 ± 0.01	6.9	[60]
-	30 ± 2 °C, enzyme-hydrolyzed henequen leaf juice (74.4 ± 3.29 g/L reducing sugar)	80.04 ± 5.29	-	16.5 ± 0.56	[52]
Engineered TATA-binding protein Spt15 strain	45 °C, 100 rpm, 200 g/L glucose	-	-	58	[48]

Table 2. *Cont.*

Strain	Growth Condition	Theoretical Ethanol Yield (%)	Ethanol Yield (g eth/g sugar)	Maximum Ethanol (g/L)	Sources
UFV-3	48 °C, 100 rpm, 10 g/L glucose	-	0.4 ± 0.01	-	[61]
PW	30 ± 1 °C, pH 6, 4% (*w*/*v*) NaCl, 10% (*v*/*v*) molasses	39.1	-	7.92	[62]
OFF1	30 °C, *A. angustifolia* juice (140 g/L reducing sugar)	-	0.38	52.27	[53]
-	30 °C, 24 h, 100 rpm, pomegranate peels (100 g/L~16.83 g/L reducing sugar)	83.1	0.48	7.2	[63]
		Coculture			
K. marxianus (isolated from the henequen plant) & *S. cerevisiae* (commercial strain) (25% Km/75% Sc)	35 ± 2 °C, henequen leaf juice + molasses (69.4 g/L total sugar)	-	-	41.2	[64]
K. marxianus TY-3 & *S. cerevisiae* AY-5	30 °C, alginate-immobilized cells, cheese whey powder (100 g/L total sugar)	-	0.43	41.8	[54]
K. marxianus DMKU 3-1042 & *S. cerevisiae* M30	37 °C, thin-shell silk cocoon-immobilized cells (IC-TSC), sugarcane juice or blackstrap molasses (220 g/L total sugar)	-	0.41	81.4	[55]
	40 °C, thin-shell silk cocoon-immobilized cells (IC-TSC), sugarcane juice or blackstrap molasses (220 g/L total sugar)	-	0.43	77.3	
		Other process configurations			
IMB3	SSF: 45 °C, 168 h, hydrothemolysis pretreated switchgrass 8% (*w*/*v*) + 0.7 mL Accellerase 1500/g glucan	86.3		22.5	[65]
β-glucosidase-producing strain YG1027	SSF: 45 °C, air ventilation (3 L/min), 100 g/L cellobiose, 48 h	51		29.5	[66]
CECT 10875	SSF: 42 °C, 72 h, 150 rpm, 50 mM sodium citrate buffer + 15 FPU cellulase/g substrate + 15 IU β-glucosidase/g substrate	10.8	0.06	10.8	[67]
	PSSF: Pre-saccharification 50 °C, 8 h, 150 rpm, + 15 FPU cellulase/g substrate + 15 IU beta-glucosidase/g substrate, followed by SSF, 42 °C, 72 h	10.7	0.05	10.7	
	LSSF: 50 °C, 8 h, 150 rpm + 10 IU laccase/g substrate, followed by SSF, 42 °C, 72 h	69.2	0.35		
	LPSSF: Pre-saccharification 50 °C, 8 h, 150 rpm, + 10 IU laccase/g substrate + 15 FPU cellulase/g substrate + 15 IU beta-glucosidase/g substrate, followed by SSF 42 °C, 72 h	70.9	0.36	10.7	

Table 2. *Cont.*

Strain	Growth Condition	Theoretical Ethanol Yield (%)	Ethanol Yield (g eth/g sugar)	Maximum Ethanol (g/L)	Sources
K213	PSSF: pretreated carrot pomace, 50 °C, 84 h, 15 FPU Accellerase TM 1000/g dry carrot pomace + 52.3 U pectinase/g dry carrot pomace, followed by SSF with 10% (*w*/*v*) resultant carrot pomace, 15 FPU Accellerase^{TM} 1000/g dry carrot pomace + 52.3 U pectinase/g dry carrot pomace, 42 °C, pH 5, 680 rpm	-	0.18	18	[68]
K. marxianus UFV-3	PSSF: 8% (*w/v*) pretreated sugarcane bagasse, 50 °C, 72 h, 15 FPU cellulase/g substrate, 180 rpm, followed by SSF, 37 °C	-	0.28	22.62	[69]
S. cerevisiae CAT-1	PSSF: 8% (*w/v*) pretreated sugarcane bagasse, 50 °C, 72 h, 15 FPU cellulase/g substrate, 180 rpm, followed by SSF 42 °C	-	0.29	22.84	
Km UOFS Y-2791 *Sc* UOFS Y-0528	SHF: Pretreated slurry of *O. ficus-indica* cladode + 15 FPU cellulase/g substrate + 15 IU beta-glucosidase/g substrate + 100 U pectinase/g substrate, 50 °C, 300 rpm, 48 h. SHF: non aeration (*S. cerevisiae* 35 °C, 36 h; *K. marxianus*, 40 °C, 48 h)	-	0.4; 0.42	19.6; 19.5	[70]
	SSF: Pretreated slurry of *O. ficus-indica* cladode + 15 FPU cellulase/g substrate + 15 IU beta-glucosidase/g substrate + 100 U pectinase/g substrate, non-aeration (*S. cerevisiae* 35 °C, 36 h; *K. marxianus* 40 °C, 48 h)	70; 64	-	20.6; 19.3	
CCT 7735	SSF: 39.5 °C, 72.5 rpm, pH 5.05, 72 h, 22.5 FPU cellulase/g substrate, saccharified sugarcane bagasse (80 g/L) + ricotta whey 5% (*w/v*), hipoxia	-	-	49.65	[71]
K213	SHF: NaOH/H_2O_2-pretreated water hyacinth, 52.29 FPU/g substrate, 50 °C, incubated 3 days, 150 rpm.	-	0.13	6.41	[72]
	SSF: 42 °C, 20 mL fermentation medium, 52.29 FPU cellulase, 1 g NaOH/H_2O_2-pretreated water hyacinth	-	0.16	7.34	
KR9 (glycoside hydrolase from *A. niger*, *T. reesei*, *N. patriciarum*)	37 °C, 200 rpm, saccharified rice straw (~60 g/L glucose)	90	-	50	[35]
CCT 7735	PSSF: Pre-saccharification of pretreated elephant grass (16%, *w/v*) + 60 FPU cellulase/mL substrate, 50 °C, 72 h, gentle agitation, followed by SSF, 38 °C, pH 4.8, 50 rpm	-	-	45.5	[73]
	Consolidated bioprocessing (CBP)				
Engineered *K. marxianus* (*T. reesei* endoglucanase, *A. aculeatus* β-glucosidase)	48 °C, 10 g/L β;-glucan, 12 h	92.2	0.47	4.24	[74]

Table 2. *Cont.*

Strain	Growth Condition	Theoretical Ethanol Yield (%)	Ethanol Yield (g eth/g sugar)	Maximum Ethanol (g/L)	Sources
Engineered *K. marxianus* KR5 (*T. reesei* endoglucanase, exoglucanase, cow rumen fungus beta-glucosidase)	37 °C, 120 rpm, 2% (*w/v*) cellobiose, 168 h	93	-	8.5	[28]
	37 °C, 120 rpm, 2% (*w/v*) β-glycan, 168 h	74	-	5.4	
Inulinase-producing strain Y179	Aeration (0.025 vvm): 33 °C, pH 4.7, 250 rpm, Jerusalem artichoke tuber meal (210 g/L total sugars), 48 h	77.1	0.4 ± 0.01	75.6 ± 1.6	[75]
	Without aeration: 33 °C, pH 4.7, Jerusalem artichoke tuber meal (210 g/L total sugars), 84 h	86.9	0.45 ± 0.01	83.1 ± 1.5	
Engineered *K. marxianus* (*H. thermocellum* largest cellulosome complex OlpB)	1% (*w/v*) Avicel, 37 °C, 300 rpm	-	-	3.09	[37]
	1% (*w/v*) PASC, 37 °C, 300 rpm	-	-	8.61	

Abbreviation: LSSF, laccase treatment, simultaneous saccharification and fermentation; LPSSF, laccase treatment, pre-saccharification, simultaneous saccharification and fermentation; PSSF, pre-saccharification, simultaneous saccharification and fermentation; SSF: simultaneous saccharification and fermentation; IU, international unit; FPU, filter paper unit; SHF, separate hydrolysis and fermentation.

5. Studies of Crabtree Effect in *K. marxianus*

Crabtree effect is the repression of respiration in aerobic glucose excess conditions and this effect is believed to play roles in a competition mechanism as it allows yeasts to growth rapidly and produce ethanol in such conditions [81]. This evolution feature promotes the rapid use of glucose and the production of ethanol, an antimicrobial chemical, resulting in the advantage of Crabtree-positive species over other microorganisms in its ecological niche. However, in the context of microbial production of biofuels and chemicals, this feature also hinders yields when the Crabtree-positive yeasts would be used as cell factory platforms to produce other chemicals than ethanol [82]. Consequently, it is of interest in abolishing this effect in Crabtree-positive species. The disruption of genes encoding puruvate decarboxylase in *S. cerevisiae* completely eliminated the Crabtree effect, however, it caused the growth deficiency of mutant strains in excess glucose condition [83]. Recently, Dai et al. [81] succeeded in turning the Crabtree-positive property of *S. cerevisiae* into Crabtree-negative by using systematic engineering. In the study of Sakihama et al. [84], under anaerobic condition and 5.5 g/L glucose, the Crabtree-negative species *K. marxianus* had significantly increased metabolite abundances in glycolysis (e.g., phosphoenol pyruvate, isocitrate, 2-ketoglutarate, succinate, malate, and fumarate), especially pyruvate with 4.5-fold higher than that in aerobic condition. Furthermore, under anaerobic condition, the transcript abundances of genes involved in glycolysis in *K. marxianus* were higher than those in aerobic condition, in accordance with their metabolic profiles. In contrast, the pool sizes of metabolites in aerobic condition (e.g., fructose 6-phosphate, 3-phosphoglycerate, phosphoenolpyruvate, acetyl-CoA, isocitrate, fumarate, fructose 1,6-bisphosphate, glucose 6-phosphate, and dihydroxyacetone phosphate) in *S. cerevisiae*, a Crabtree-positive yeast, were higher than those under anaerobic condition, suggesting the incline of these metabolites toward glycolytic route in the presence of oxygen. Concerning biomass production, under aeration, *K. marxianus* cell density was 2.9-fold higher than that of itself in anaerobic condition and 2.2-fold higher than that of *S. cerevisiae* in aerobic cultivation. The cell density of *S. cerevisiae* in aerobic culture (OD_{600} ~ 54), however, was not much higher than that in the anaerobic culture (OD_{600} ~ 46), suggesting a slow growth rate when the Crabtree effect occurred. Regarding ethanol and acetate productions, under aerobic conditions, *S. cerevisiae* reached an ethanol titer of 22.1 g/L and acetate titer of 1.3 g/L, while *K. marxianus* only produced 5 g/L ethanol and 0.5 g/L acetate in the same circumstances. These data were consistent with the previous study of Wardrop et al. [85], as they also found that *K. marxianus* yielded higher biomass than *S. cerevisiae* but produced a lower ethanol concentration (0.4 g/L vs. 6 g/L of *S. cerevisiae*) in glucose pulse treatment (sudden increased glucose from 1 g/L to 50 g/L). In addition, the oxygen uptake in *S. cerevisiae* immediately declined after glucose upshift, whereas the increase in oxygen uptake in such circumstances was recorded in *K. marxianus*, indicating the maintenance of respiratory activity in the Crabtree-negative yeast. These data indicate the fundamental differences between Crabtree-negative and Crabtree-positive species in aerobic culture with high concentration of glucose

6. Conclusions

The non-conventional yeast *K. marxianus* has been proved to be a promising eukaryotic microbe for bioethanol production and other food and environmental applications. Although having several useful traits that are suitable for bioethanol production at an industrial scale, its genetic drawbacks, such as the sensitivity to high concentration of ethanol or the incapability of growing on polysaccharides should be improved to meet the demands of industrial fermenting yeast strains. In addition, despite a lot of efforts having been deployed for constructing a robust *K. marxianus* strain appropriate for CBP, the ethanol production of these current engineered strains was still modest. Up to now, the highest ethanol concentration produced by an engineered *K. marxianus* is only 8.61 g/L, too low for any practical consideration.

Author Contributions: D.M.H.-T. and T.T.M.N. wrote the manuscript. C.-C.H. supervised the work. All authors have read and agreed to the published version of the manuscript.

References

1. Radecka, D.; Mukherjee, V.; Mateo, R.; Stojiljkovic, M.; Foulquie-Moreno, M.; Thevelein, J. Looking beyond Saccharomyces: The potential of non-conventional yeast species for desirable traits in bioethanol fermentation. *FEMS Yeast Res.* **2015**, *15*, 1–13. [CrossRef] [PubMed]
2. Jansen, M.; Bracher, J.; Papapetridis, I.; Verhoeven, M.; de Bruijn, H.; de Waal, P.; van Maris, A.; Klaassen, P.; Pronk, J. Saccharomyces cerevisiae strains for second-generation ethanol production: From academic exploration to industrial implementation. *FEMS Yeast Res.* **2017**, *17*. [CrossRef] [PubMed]
3. Karim, A.; Gerliani, N.; Aider, M. Kluyveromyces marxianus: An emerging yeast cell factory for applications in food and biotechnology. *Int. J. Food Microbiol.* **2020**, *333*. [CrossRef] [PubMed]
4. Groeneveld, P.; Stouthamer, A.; Westerhoff, H. Super life-how and why "cell selection" leads to the fastest-growing eukaryote. *FEBS J.* **2009**, *276*, 254–270. [CrossRef] [PubMed]
5. Mo, W.; Wang, M.; Zhan, R.; Yu, Y.; He, Y.; Lu, H. Kluyveromyces marxianus developing ethanol tolerance during adaptive evolution with significant improvements of multiple pathways. *Biotechnol. Biofuels* **2019**, *12*. [CrossRef] [PubMed]
6. Fonseca, G.; Heinzle, E.; Wittmann, C.; Gombert, A. The yeast Kluyveromyces marxianus and its biotechnology potential. *Appl. Microbiol. Biotechnol.* **2008**, *79*, 339–354. [CrossRef] [PubMed]
7. Lane, M.; Morrissey, J. Kluyveromyces marxianus: A yeast emerging from its sister's shadow. *Fungal. Biol. Rev.* **2010**, *24*, 17–26. [CrossRef]
8. Chang, J.; Ho, C.; Mao, C.; Barham, N.; Huang, Y.; Ho, F.; Wu, Y.; Hou, Y.; Shih, M.; Li, W.; et al. A thermo- and toxin-tolerant kefir yeast for biorefinery and biofuel production. *Appl. Energy* **2014**, *132*, 465–474. [CrossRef]
9. Diniz, R.; Villada, J.; Alvim, M.; Vidigal, P.; Vieira, N.; Maceiras, M.; Cerdan, M.; Gonzalez-Siso, M.; Lahtvee, P.; da Silveira, W. Transcriptome analysis of the thermotolerant yeast Kluyveromyces marxianus CCT 7735 under ethanol stress. *Appl. Microbiol. Biotechnol.* **2017**, *101*, 6969–6980. [CrossRef]
10. Vanegas, J.; Contreras, M.; Faller, R.; Longo, M. Role of unsaturated lipid and ergosterol in ethanol tolerance of model yeast biomembranes. *Biophys. J.* **2012**, *102*. [CrossRef]
11. Alvim, M.; Vital, C.; Barros, E.; Vieira, N.; da Silveira, F.; Balbino, T.; Diniz, R.; Brito, A.; Bazzolli, D.; de Oliveira-Ramos, H.; et al. Ethanol stress responses of Kluyveromyces marxianus CCT7735 revealed by proteomic and metabolomic analyses. *Antonie Van Leeuwenhoek* **2019**, *112*, 827–845. [CrossRef] [PubMed]
12. Wang, D.; Wu, D.; Yang, X.; Hong, J. Transcriptomic analysis of thermotolerant yeast Kluyveromyces marxianus in multiple inhibitors tolerance. *RSC Adv.* **2018**, *8*, 14177–14192. [CrossRef]
13. Fu, X.; Li, P.; Zhang, L.; Li, S. Understanding the stress responses of Kluyveromyces marxianus after an arrest during high temperature ethanol fermentation based on integration of RNA-seq and metabolite data. *Appl. Microbiol. Biotechnol.* **2019**, *103*, 2715–2729. [CrossRef] [PubMed]

14. Lahtvee, P.; Kumar, R.; Hallstrom, B.; Nielsen, J. Adaptation to different types of stress converge on mitochondrial metabolism. *Mol. Biol. Cell* **2016**, *27*. [CrossRef] [PubMed]
15. Piper, P. The heat shock and ethanol stress responses of yeast exhibit extensive similarity and functional overlap. *FEMS Microbiol. Lett.* **1995**, *134*, 121–127. [CrossRef] [PubMed]
16. Gao, J.; Yuan, W.; Li, Y.; Xiang, R.; Hou, S.; Zhong, S.; Bai, F. Transcriptional analysis of Kluyveromyces marxianus for ethanol production from inulin using consolidated bioprocessing technology. *Biotechnol. Biofuels* **2015**, *8*. [CrossRef]
17. Holwerda, E.; Thorne, P.; Olson, D.; Amador-Noguez, D.; Engle, N.; Tschaplinski, T.; van Dijken, J.; Lynd, L. The exometabolome of Clostridium thermocellum reveals overflow metabolism at high cellulose loading. *Biotechnol. Biofuels* **2014**, *7*, 155. Available online: http://www.biotechnologyforbiofuels.com/content/7/1/155 (accessed on 20 October 2020). [CrossRef] [PubMed]
18. Thompson, R.; Trinh, C. Overflow metabolism and growth cessation in Clostridium thermocellum DSM1313 during high cellulose loading fermentations. *Biotechnol. Bioeng.* **2017**, *114*. [CrossRef]
19. Papini, M.; Nookaew, I.; Scalcinati, G.; Siewers, V.; Nielsen, J. Phosphoglycerate mutase knock-out mutant Saccharomyces cerevisiae: Physiological investigation and transcriptome analysis. *Biotechnol. J.* **2010**, *5*, 1016–1027. [CrossRef]
20. Gasnier, B. Characterization of low- and high-affinity glucose transports in the yeast Kluyveromyces marxianus. *Biochim. Biophys. Acta* **1987**, *903*, 425–433. [CrossRef]
21. Carvalho-Silva, M.; Spencer-Martins, I. Modes of lactose uptake in the yeast species Kluyveromyces marxianus. *Antonie Van Leeuwenhoek* **1990**, *57*, 77–81. [CrossRef] [PubMed]
22. Postma, E.; van den Broek, P. Continuous culture study of the regulation of glucose and fructose transport in Kluyveromyces marxianus CBS 6556. *J. Bacteriol.* **1990**, *172*, 2871–2876. [CrossRef] [PubMed]
23. van Leeuwen, C.; Postma, E.; van den Broek, P.; van Steveninck, J. Proton-motive force-driven D-galactose transport in plasma membrane vesicles from the yeast Kluyveromyces marxianus. *J. Biol. Chem.* **1991**, *266*, 12146–12151.
24. Stambuk, B.; Franden, M.; Singh, A.; Zhang, M. D-xylose transport by Candida succiphila and Kluyveromyces marxianus. *Appl. Biochem. Biotechnol.* **2003**, *105–108*, 255–263. [CrossRef]
25. Fonseca, A.; Spencer-Martins, I.; van Uden, N. Transport of lactic acid in Kluyveromyces marxianus: Evidence for a monocarboxylate uniport. *Yeast* **1991**, *7*, 775–780. [CrossRef]
26. Queiros, O.; Casal, M.; Althoff, S.; Moradas-Ferreira, P.; Leao, C. Isolation and characterization of Kluyveromyces marxianus mutants deficient in malate transport. *Yeast* **1998**, *14*, 401–407. [CrossRef]
27. Kucharska, K.; Rybarczyk, P.; Holowacz, I.; Lukajtis, R.; Glinka, M.; Kaminski, M. Pretreatment of lignocellulosic materials as substrates for fermentation processes. *Molecules* **2018**, *23*, 2937. [CrossRef]
28. Chang, J.; Ho, C.; Ho, F.; Tsai, T.; Ke, H.; Wang, C.; Chen, H.; Shih, M.; Huang, C.; Li, W. PGASO: A synthetic biology tool for engineering a cellulolytic yeast. *Biotechnol. Biofuels* **2012**, *5*, 1–12. Available online: http://www.biotechnologyforbiofuels.com/content/5/1/53 (accessed on 20 October 2020). [CrossRef]
29. Chang, J.; Ho, F.; Ho, C.; Wu, Y.; Hou, Y.; Huang, C.; Shih, M.; Li, W. Assembling a cellulase cocktail and a cellodextrin transporter into a yeast host for CBP ethanol production. *Biotechnol. Biofuels* **2013**, *6*, 1–13. Available online: http://www.biotechnologyforbiofuels.com/content/6/1/19 (accessed on 20 October 2020). [CrossRef]
30. Demain, A.; David Wu, J. Cellulase, Clostridia, and Ethanol. *Microbiol. Mol. Biol. Rev.* **2005**, *69*, 124–154. [CrossRef]
31. Wen, J.; Sun, J.; Zhao, H. Yeast surface display of trifunctional mini-cellulosomes for simultaneous saccharification and fermentation of cellulose to ethanol. *Appl. Environ. Microbiol.* **2010**, *76*, 1251–1260. [CrossRef] [PubMed]
32. Fan, L.; Zhang, Z.; Yu, X.; Xue, Y.; Tan, T. Self-surface assembly of cellulosomes with two miniscaffoldins on Saccharomyces cerevisiae for cellulosic ethanol production. *Proc. Natl. Acad. Sci. USA* **2012**, *109*, 13260–13265. [CrossRef] [PubMed]
33. Tsai, S.; DaSilva, N.; Chen, W. Functional display of complex cellulosomes on the yeast surface via adaptive assembly. *ACS Synth. Biol.* **2013**, *2*, 14–21. [CrossRef] [PubMed]
34. Liang, Y.; Ang, E.; Zhao, H. Engineered Pentafunctional Minicellulosome for Simultaneous Saccharification and Ethanol Fermentation in Saccharomyces cerevisiae. *Appl. Environ. Microbiol.* **2014**, *80*, 6677–6684. [CrossRef] [PubMed]

35. Chang, J.; Lin, Y.; Lay, C.; Thia, C.; Wu, Y.; Hou, Y.; Huang, C.; Li, W. Constructing a cellulosic yeast host with an efficient cellulase cocktail. *Biotechnol. Bioeng.* **2018**, *115*. [CrossRef]
36. Chang, J.; Anandharaj, M.; Ho, C.; Tsuge, K.; Tsai, T.; Ke, H.; Lin, Y.; Ha Tran, M.; Li, W.; Huang, C. Biomimetic strategy for constructing Clostridium thermocellum cellulosomal operons in Bacillus subtilis. *Biotechnol. Biofuels* **2018**, *11*. [CrossRef]
37. Anandharaj, M.; Lin, Y.; Rani, R.; Nadendla, E.; Ho, M.; Huang, C.; Cheng, J.; Chang, J.; Li, W. Constructing a yeast to express the largest cellulosome complex on the cell surface. *Proc. Natl. Acad. Sci. USA* **2020**, *117*, 2385–2394. [CrossRef]
38. Gao, J.; Yuan, W.; Li, Y.; Bai, F.; Jiang, Y. Synergistic effect of thioredoxin and its reductase from Kluyveromyces marxianus on enhanced tolerance to multiple lignocellulose-derived inhibitors. *Microb. Cell Factories* **2017**, *16*, 1–13. [CrossRef]
39. Iraqui, I.; Kienda, G.; Soeur, J.; Faye, G.; Baldacci, G.; Kolodner, R.; Huang, M. Peroxiredoxin Tsa1 is the key peroxidase suppressing genome instability and protecting against cell death in Saccharomyces cerevisiae. *PLoS Genet.* **2009**, *5*. [CrossRef]
40. Gao, J.; Feng, H.; Yuan, W.; Li, Y.; Hou, J.; Zhong, S.; Bai, F. Enhanced fermentative performance under stresses of multiple lignocellulose-derived inhibitors by overexpression of a typical 2-Cys peroxiredoxin from Kluyveromyces marxianus. *Biotechnol. Biofuels* **2017**, *10*. [CrossRef]
41. Shui, W.; Xiong, Y.; Xiao, W.; Qi, X.; Zhang, Y.; Lin, Y.; Guo, Y.; Zhang, Z.; Wang, Q.; Ma, Y. Understanding the Mechanism of Thermotolerance Distinct From Heat Shock Response Through Proteomic Analysis of Industrial Strains of Saccharomyces cerevisiae. *Mol. Cell. Proteomics* **2015**, *14*, 1885–1897. [CrossRef] [PubMed]
42. Li, P.; Fu, X.; Zhang, L.; Zhang, Z.; Li, J.; Li, S. The transcription factors Hsf1 and Msn2 of thermotolerant Kluyveromyces marxianus promote cell growth and ethanol fermentation of Saccharomyces cerevisiae at high temperatures. *Biotechnol. Biofuels* **2017**, *10*. [CrossRef] [PubMed]
43. Da Silveira, F.; de Oliveira-Soares, D.; Bang, K.; Balbino, T.; de Moura-Ferreira, M.; Diniz, R.; de Lima, L.; Brandao, M.; Villas-Boas, S.; da Silveira, W. Assessment of ethanol tolerance of Kluyveromyces marxianus CCT 7735 selected by adaptive laboratory evolution. *Appl. Microbiol. Biotechnol.* **2020**, *104*, 7483–7494. [CrossRef] [PubMed]
44. Dragosits, M.; Mattanovich, D. Adaptive laboratory evolution-principles and applications for biotechnology. *Microb. Cell Factories* **2013**, *12*, 1–17. Available online: http://www.microbialcellfactories.com/content/12/1/64 (accessed on 20 October 2020). [CrossRef]
45. Caspeta, L.; Chen, Y.; Ghiaci, P.; Feizi, A.; Buskov, S.; Hallstrom, B.; Petranovic, D.; Nielsen, J. Altered sterol composition renders yeast thermotolerant. *Science* **2014**, *346*, 75–78. [CrossRef]
46. Alper, H.; Moxley, J.; Nevoigt, E.; Fink, G.; Stephanopoulos, G. Engineering Yeast Transcription Machinery for Improved Ethanol Tolerance and Production. *Science* **2006**, *314*, 1565–1568. [CrossRef]
47. Lam, F.; Hartner, F.; Fink, G.; Stephanopoulos, G. Chapter 20-enhancing stress resistance and production phenotypes through transcriptome engineering. In *Methods in Enzymology-Guide to Yeast Genetics: Functional Genomics, Proteomics, and Other Systems Analysis*; Weissman, J., Guthrie, C., Fink, G., Eds.; Elsevier Inc.: Amsterdam, The Netherlands, 2010; Volume 470, pp. 509–532. ISBN 978-0-12-375171-3.
48. Li, P.; Fu, X.; Li, S.; Zhang, L. Engineering TATA-binding protein Spt15 to improve ethanol tolerance and production in Kluyveromyces marxianus. *Biotechnol. Biofuels* **2018**, *11*. [CrossRef]
49. Heo, P.; Yang, T.; Chung, S.; Cheon, Y.; Kim, J.; Park, J.; Koo, H.; Cho, K.; Seo, J.; Park, J.; et al. Simultaneous integration of multiple genes into the Kluyveromyces marxianus chromosome. *J. Biotechnol.* **2013**, *167*, 323–325. [CrossRef]
50. Löbs, A.; Engel, R.; Schwartz, C.; Flores, A.; Wheeldon, I. CRISPR–Cas9-enabled genetic disruptions for understanding ethanol and ethyl acetate biosynthesis in Kluyveromyces marxianus. *Biotechnol. Biofuels* **2017**, *10*. [CrossRef]
51. Posada, J.; Patel, A.; Roes, A.; Blok, K.; Faaij, A.; Patel, M. Potential of bioethanol as a chemical building block for biorefineries: Preliminary sustainability assessment of 12 bioethanol-based products. *Bioresour. Technol.* **2013**, *135*, 490–499. [CrossRef]
52. Villegas-Silva, P.; Toledano-Thompson, T.; Canto-Canché, B.; Larqué-Saavedra, A.; Barahona-Perez, L. Hydrolysis of Agave fourcroydes Lemaire (henequen) leaf juice and fermentation with Kluyveromyces marxianus for ethanol production. *BMC Biotechnol.* **2014**, *14*, 14. Available online: http://www.biomedcentral.com/1472-6750/14/14 (accessed on 20 October 2020). [CrossRef] [PubMed]

53. Alcazar-Valle, M.; Gschaedler, A.; Gutierrez-Pulido, H.; Arana-Sanchez, A.; Arellano-Plaza, M. Fermentative capabilities of native yeast strains grown on juices from different Agave species used for tequila and mezcal production. *Biotechnol. Ind. Microbiol.* **2019**, *50*, 379–388. [CrossRef] [PubMed]
54. Guo, X.; Zhou, J.; Xiao, D. Improved Ethanol Production by Mixed Immobilized Cells of Kluyveromyces marxianus and Saccharomyces cerevisiae from Cheese Whey Powder Solution Fermentation. *Appl. Biochem. Biotechnol.* **2010**, *160*, 532–538. [CrossRef]
55. Eiadpum, A.; Limtong, S.; Phisalaphong, M. High-temperature ethanol fermentation by immobilized coculture of Kluyveromyces marxianus and Saccharomyces cerevisiae. *J. Biosci. Bioeng.* **2012**, *114*, 325–329. [CrossRef]
56. Silveira, W.; Passos, F.; Mantovani, H.; Passos, F. Ethanol production from cheese whey permeate by Kluyveromyces marxianus UFV-3: A flux analysis of oxido-reductive metabolism as a function of lactose concentration and oxygen levels. *Enzyme Microb. Technol.* **2005**, *36*, 930–936. [CrossRef]
57. Limtong, S.; Sringiew, C.; Yongmanitchai, W. Production of fuel ethanol at high temperature from sugar cane juice by a newly isolated Kluyveromyces marxianus. *Bioresour. Technol.* **2007**, *98*, 3367–3374. [CrossRef] [PubMed]
58. Nakamura, K.; Shinomiya, N.; Orikasa, Y.; Oda, Y. Efficient Production of Ethanol from Saccharified Crops Mixed with Cheese Whey by the Flex Yeast Kluyveromyces marxianus KD-15. *Food Sci. Technol. Res.* **2012**, *18*, 235–242. [CrossRef]
59. Dasgupta, D.; Suman, S.; Pandey, D.; Ghosh, D.; Khan, R.; Agrawal, D.; Jain, R.; Vadde, V.; Adhikari, D. Design and optimization of ethanol production from bagasse pith hydrolysate by a thermotolerant yeast Kluyveromyces sp. IIPE453 using response surface methodology. *SpringerPlus* **2013**, *2*, 1–10. Available online: http://www.springerplus.com/content/2/1/159 (accessed on 20 October 2020). [CrossRef] [PubMed]
60. Goshima, T.; Negi, K.; Tsuji, M.; Inoue, H.; Yano, S.; Hoshino, T.; Matsushika, A. Ethanol fermentation from xylose by metabolically engineered strains of Kluyveromyces marxianus. *J. Biosci. Bioeng.* **2013**, *116*, 551–554. [CrossRef] [PubMed]
61. Madeira, J.V., Jr.; Gombert, A. Towards high-temperature fuel ethanol production using Kluyveromyces marxianus: On the search for plug-in strains for the Brazilian sugarcane-based biorefinery. *Biomass Bioenergy* **2018**, *119*, 217–228. [CrossRef]
62. Karatay, S.; Demiray, E.; Donmez, G. Bioethanol production by newly isolated halotolerant Kluyveromyces marxianus strains. *Environ. Prog. Sustain. Energy* **2019**, *38*, 542–547. [CrossRef]
63. Demiray, E.; Karatay, S.; Donmez, G. Efficient bioethanol production from pomegranate peels by newly isolated Kluyveromyces marxianus. *Energy Sources Part Recovery Util. Environ. Eff.* **2020**, *42*, 709–718. [CrossRef]
64. Caceres-Farfan, M.; Lappe, P.; Larque-Saavedra, A.; Magdub-Mendez, A.; Barahona-Perez, L. Ethanol production from henequen (Agave fourcroydes Lem.) juice and molasses by a mixture of two yeasts. *Bioresour. Technol.* **2008**, *99*, 9036–9039. [CrossRef] [PubMed]
65. Pessani, N.; Atiyeh, H.; Wilkins, M.; Bellmer, D.; Banat, I. Simultaneous saccharification and fermentation of Kanlow switchgrass by thermotolerant Kluyveromyces marxianus IMB3: The effect of enzyme loading, temperature and higher solid loadings. *Bioresour. Technol.* **2011**, *102*, 10618–10624. [CrossRef] [PubMed]
66. Matsuzaki, C.; Nakagawa, A.; Koyanagi, T.; Tanaka, K.; Minami, H.; Tamaki, H.; Katayama, T.; Yamamoto, K.; Kumagai, H. Kluyveromyces marxianus-based platform for direct ethanol fermentation and recovery from cellulosic materials under air-ventilated conditions. *J. Biosci. Bioeng.* **2012**, *113*, 604–607. [CrossRef] [PubMed]
67. Moreno, A.; Ibarra, D.; Ballesteros, I.; Gonzalez, A.; Ballesteros, M. Comparing cell viability and ethanol fermentation of the thermotolerant yeast Kluyveromyces marxianus and Saccharomyces cerevisiae on steam-exploded biomass treated with laccase. *Bioresour. Technol.* **2013**, *135*, 239–245. [CrossRef]
68. Yu, C.; Jiang, B.; Duan, K. Production of Bioethanol from Carrot Pomace Using the Thermotolerant Yeast Kluyveromyces marxianus. *Energies* **2013**, *6*, 1794–1801. [CrossRef]
69. Costa, D.; de Souza, C.; Costa, P.; Rodrigues, M.; dos Santos, A.; Lopes, M.; Genier, H.; Silveira, W.; Fietto, L. Physiological characterization of thermotolerant yeast for cellulosic ethanol production. *Appl. Microbiol. Biotechnol.* **2014**, *98*, 3829–3840. [CrossRef]
70. Kuloyo, O.; du Preez, J.; Garcıa-Aparicio, M.; Kilian, S.; Steyn, L.; Görgens, J. Opuntia ficus-indica cladodes as feedstock for ethanol production by Kluyveromyces marxianus and Saccharomyces cerevisiae. *World J. Microbiol. Biotechnol.* **2014**, *30*, 3173–3183. [CrossRef]

71. Ferreira, P.; da Silveira, F.; dos Santos, R.; Genier, H.; Diniz, R.; Junior, J.; Fietto, L.; Passos, F.; da Silveira, W. Optimizing ethanol production by thermotolerant Kluyveromyces marxianus CCT 7735 in a mixture of sugarcane bagasse and ricotta whey. *Food Sci. Biotechnol.* **2015**, *24*, 1421–1427. [CrossRef]
72. Jan, J.; Wei, Z.; Wang, Q.; He, M.; Li, S.; Irbis, C. Bioethanol production from sodium hydroxide/hydrogen peroxide-pretreated water hyacinth via simultaneous saccharification and fermentation with a newly isolated thermotolerant Kluyveromyces marxianu strain. *Bioresour. Technol.* **2015**, *193*, 103–109. [CrossRef]
73. Campos, B.; Diniz, R.; da Silveira, F.; Junior, J.; Fietto, L.; Machado, J.; da Silveira, W. Elephant grass (Pennisetum purpureum Schumach) is a promising feedstock for ethanol production by the thermotolerant yeast Kluyveromyces marxianus CCT 7735. *Braz. J. Chem. Eng.* **2019**, *36*, 43–49. [CrossRef]
74. Yanase, S.; Hasunuma, T.; Yamada, R.; Tanaka, T.; Ogino, C.; Fukuda, H.; Kondo, H. Direct ethanol production from cellulosic materials at high temperature using the thermotolerant yeast Kluyveromyces marxianus displaying cellulolytic enzymes. *Appl. Microbiol. Biotechnol.* **2010**, *88*, 381–388. [CrossRef] [PubMed]
75. Yuan, W.; Chang, B.; Ren, J.; Liu, J.; Bai, F.; Li, Y. Consolidated bioprocessing strategy for ethanol production from Jerusalem artichoke tubers by Kluyveromyces marxianus under high gravity conditions. *J. Appl. Microbiol.* **2012**, *112*, 38–44. [CrossRef] [PubMed]
76. Goers, L.; Freemont, P.; Polizzi, K. Co-culture systems and technologies: Taking synthetic biology to the next level. *J. R. Soc. Interface* **2014**, *11*. [CrossRef]
77. Ho, C.; Chang, J.; Lee, S.; Chin, T.; Shih, M.; Li, W.; Huang, C. Development of cellulosic ethanol production process via co-culturing of artificial cellulosomal Bacillus and kefir yeast. *Appl. Energy* **2012**, *100*, 27–32. [CrossRef]
78. Rattanapan, A.; Limtong, S.; Phisalaphong, M. Ethanol production by repeated batch and continuous fermentations of blackstrap molasses using immobilized yeast cells on thin-shell silk cocoons. *Appl. Energy* **2011**, *88*, 4400–4404. [CrossRef]
79. Lopez, C.; Beaufort, S.; Brandam, C.; Taillandier, P. Interactions between Kluyveromyces marxianus and Saccharomyces cerevisiae in tequila must type medium fermentation. *World J. Microbiol. Biotechnol.* **2014**, *30*, 2223–2229. [CrossRef]
80. Abranches, J.; Morais, P.; Rosa, C.; Mendonca-Hagler, L.; Hagler, A. The incidence of killer activity and extracellular proteases in tropical yeast communities. *Can. J. Microbiol.* **1997**, *43*, 328–336. [CrossRef]
81. Dai, Z.; Huang, M.; Chen, Y.; Siewers, V.; Nielsen, J. Global rewiring of cellular metabolism renders Saccharomyces cerevisiae Crabtree negative. *Nat. Commun.* **2018**, *9*. [CrossRef]
82. Nielsen, J.; Larsson, C.; van Maris, A.; Pronk, J. Metabolic engineering of yeast for production of fuels and chemicals. *Curr. Opin. Biotechnol.* **2013**, *24*, 398–404. [CrossRef] [PubMed]
83. Flikweert, M.; van der Zandens, L.; Janssen, W.; Steensma, H.; van Dijken, J.P.; Pronk, J. Pyruvate Decarboxylase: An Indispensable Enzyme for Growth of Saccharomyces cerevisiae on Glucose. *Yeast* **1996**, *12*, 247–257. [CrossRef]
84. Sakihama, Y.; Hidese, R.; Hasunuma, T.; Kondo, A. Increased flux in acetyl-CoA synthetic pathway and TCA cycle of Kluyveromyces marxianus under respiratory conditions. *Sci. Rep.* **2019**, *9*. [CrossRef] [PubMed]
85. Wardrop, F.; Liti, G.; Cardinali, G.; Walker, G. Physiological responses of Crabtree positive and Crabtree negative yeasts to glucose upshifts in a chemostat. *Ann. Microbiol.* **2004**, *54*, 103–114.

4

Antioxidant Properties of Fermented Green Coffee Beans with *Wickerhamomyces anomalus* (Strain KNU18Y3)

Mesfin Haile [1] and Won Hee Kang [1,2,*]

[1] Department of Horticulture, Kangwon National University, Chuncheon 24341, Korea; mesfinhaile97@gmail.com
[2] Convergence Program of Coffee Science, Kangwon National University, Chuncheon 24341, Korea
* Correspondence: whkang@kangwon.ac.kr

Abstract: A few yeast species have been tested frequently to improve the tastes, flavors, and other important quality parameters of coffee. However, continuing evaluations of different yeast species for fermenting green coffee beans will have a significant positive contribution to the coffee industry. This experiment was conducted to evaluate the antioxidant properties, total phenol content (TPC), total flavonoid content (TFC), total tannin content (TTC), and the consumer acceptability of fermented green coffee beans with *Wickerhamomyces anomalu*. The coffee beans were roasted at different roasting conditions (light, medium, and dark). There was no significant ($p > 0.05$) difference between the yeast-fermented and non-fermented coffee with regard to the oxygen radical absorbance capacity (ORAC) values in medium and dark roasted coffee. Similarly, the superoxide dismutase-like (SOD)-like activity did not significantly differ in all roasting conditions. However, the SOD-like activity was significantly different ($p < 0.05$), particularly within light roasted and medium roasted, and between light roasted and dark roasted in both the control and fermented coffee extracts. The 2, 2-diphenyl-1-picrylhydrazyl (DPPH) radical scavenging assay and ferric reducing antioxidant power (FRAP) were improved in fermented coffee beans. There was a significant ($p \leq 0.05$) difference between the yeast-fermented and non-fermented coffee with respect to the TPC and TFC in all roasting types and the TTC in the light and dark roasting conditions. The fermentation of green coffee beans with *W. anomalus* increased the TPC and TFC. However, the TTC was lower in the fermented coffee beans compared to the non-fermented coffee beans in medium and dark roasted coffee. In general, fermentation of green coffee beans with *W. anomalus* has the potential to improve the functionality of coffee beans.

Keywords: antioxidant; coffee; fermentation; *W. anomalus*

---

1. Introduction

Fermentation has the potential to improve the functionality of foods. Fermentation is primarily relied upon during the wet processing of coffee to remove mucilage. However, it is also improves coffee's sensory quality attributes [1]. Fermentation has been extensively applied in the food, chemical, and pharmaceutical industries to aid in the manufacturing, extraction, and modification of bioactive compounds [2,3]. Microbial fermentation is an interesting biotechnological processing system that can improve the total phenolic content of foods and herbs by liberating their insolubly bound phenolics and hence boost their nutritional value [4,5]. Numerous research papers provide a comparison of the antioxidant properties of popular drinks such as coffee, tea, and cocoa [6–8]. Phenolic compounds are found in a broad range of regularly consumed plant foods, such as vegetables, fruits, legumes, and cereals, and in beverages of plant origin, such as coffee, tea, and wine [9,10]. Flavonoids naturally exist in plants with a variable phenolic structure. Functional hydroxyl groups in flavonoids mediate their antioxidant effects by chelating metal ions and scavenging free radicals [11,12]. Flavonoids in food are commonly responsible for the color, the taste, the prevention of fat oxidation, and the protection of enzymes and vitamins [13].

Wickerhamomyces anomalus (formerly *Pichia anomala*) is an ascomycetous heterothallic yeast of the family *Wickerhamomycetaceae* that propagates sexually through the development of hat-shaped ascospores and asexually by budding [14]. *W. anomalus* strains are found in several environments and have been isolated from cereal grains, fruits, maize silage, wine, and high-sugar food products [14]. *W. anomalus* is categorized as a biosafety level-one organism; it is deemed harmless for healthy individuals and can grow under severe environmental stress conditions, such as high and low pH, high osmotic pressure, anaerobic conditions, and low water activity [15]. As mentioned by Comitini et al. [16], *W. anomalus* toxins have been examined as antimicrobial agents against some spoilage yeasts such as *Dekkera/Brettanomyces*. Some researchers have reported that the wild species of *W. anomalus* strains have the potential to use maltose and grow better than other commercial brewing yeasts [17]. However, other studies have referred to its clear inability to metabolize maltose [18]. Considering these findings, *W. anomalus* is preferred as a good starter culture in sequence or inoculated together with other yeasts in winemaking [19]. *p. anomala*-mixed starters have been evaluated to improve the final quality of cider [18]. Due to its positive effect on sensory quality, a mixed starter of *W. anomalus* and *Saccharomyces cerevisiae* has been proposed for making Chinese baijiu [20].

Recently, studies have been conducted to improve the functionality of coffee through green bean fermentation with selected microorganisms. In these studies, the *Saccharomyces* species [21,22], *Saccharomycopsis fibuligera* [22], *Rhizopus oligosporus* [23], and *Yarrowia lipolytica* [24], have been used to ferment green coffee beans. The fermentation of green coffee beans with *R. oligosporus* significantly enhanced the compositions of aroma precursors [23]. As reported by Kwak et al. [21] and Mesfin and Kang [22], the fermentation of green coffee beans improves antioxidant activity, total phenol content (TPC), and total flavonoid content (TFC), and it reduces the total tannin content (TTC), which is mostly responsible for coffee's astringency. *W. anomalus* strain KNU18Y3 has the ability to produce a pectinase enzyme, it was selected as a starter culture for coffee fermentation in the wet processing method [25]. Since this yeast has not been tested previously to ferment green coffee beans, this study was conducted to evaluate the antioxidant activities of fermented coffee beans with *W. anomalus* (strain KNU18Y3).

2. Materials and Methods

2.1. Chemicals

Disodium phosphate (Na_2HPO_4), aluminium chloride ($AlCl_3{\cdot}6H_2O$), monosodium phosphate (NaH_2PO_4), sodium carbonate (Na_2CO_3), D-glucose, and sodium nitrite ($NaNO_2$) were supplied by Dae-Jung Chemicals & Metals Co., Ltd., (Jeongwang-dong, Shiheung-city, Gyeonggi-do, Republic of Korea). Pyrogallol, 2,2′-azobis (2-amidinopropane) dihydrochloride (AAPH), and fluorescein sodium were purchased from Santa Cruz Biotechnology, Inc. (Dallas, Texas, USA). NaCl, ethylenediaminetetraacetic acid, yeast peptone dextrose, gallic acid, Folin-Ciocalteu's phenol reagent, quercetin, Trolox, 2,2-diphenyl-1-picrylhydrazyl, potassium ferricyanide($C_6N_6FeK_3$), trichloroacetic acid ($C_2HC_{13}O_2$), and ferric chloride ($FeCl_3$) were purchased from Sigma Aldrich LLC (St. Louis, MO, USA). Methanol was supplied by Merck KGaA (Darmstadt, Germany), and HCl was purchased from Junsei Chemical Co., Ltd. (4-4-16 Nihonbashi-honcho, Chuo-ku, Tokyo, Japan).

2.2. Fermentation of Green Coffee Beans

A freeze-dried *W. anomalus* (strain KNU18Y3) yeast cells were cultivated in a sterile yeast peptone dextrose broth (0.5% *w/v* yeast extract, 2 % *w/v* dextrose and 1% *w/v* casein peptone) for 48 h at 30 °C. The pH of the media was calibrated to 5.0 with 1 M HCl. Then, the yeast cells were collected via centrifugation at 8000× *g*. Finally, the collected yeast cells were washed twice using the deionized distilled water (ddH_2O) and mixed in a similar volume of 100 mM phosphate-buffered saline solution.

The green coffee beans (*Coffea arabica* L.) that imported from Kenya were used for this study. The coffee beans (600 g) saturated in ddH_2O at 4 °C for 24 h and then steamed at 80 °C for 40 min to kill native yeasts that exist on the coffee beans. Next, the steamed coffee beans then allowed to cool at 20–23 °C. The heat-treated green coffee beans were divided into two lots: fermented through inoculation (1.0×10^4 colony-forming unit (CFU)/g of coffee beans) and non-inoculated (control). Then, each lot was divided into three parts and underwent a different roasting treatment. Fermentation of green coffee beans and roasting were done in three replications. The fermentation was carried out for 24 h at 30 °C. After that, the coffee beans were washed thoroughly using a ddH_2O. The fermented coffee beans were dried using oven at 45 °C until the moisture content reached between 10% and 12%.

The pH of the fermented solution (the water that used to ferment the coffee beans) was recorded using a pH meter at 0 h and 24 h. A sample solution was taken after 24 h fermentation and serially diluted and plated onto yeast peptone dextrose agar (YPDA) plates. The plates were incubated for 24 h at 30 °C to estimate the growth of yeast cells. The pH measurements and microbial counts were made in five replications.

2.3. *Coffee Roasting, Grinding, Extraction*

The green coffee beans were roasted at 245 °C for 11.5 (light roast), 13.5 (medium roast) and 16 (dark roast) min using a coffee roaster (GeneCafe CR-100 coffee roaster, Genesis, Ansan, Korea) to ensure uniform roasting conditions. Each roasting was done in triplicate. After roasting, the beans were allowed to cool for 30 min and then ground by adjusting in a medium level on a coffee grinder (Latina 600N electric grinder). Hot brew extraction made using a coffee maker (HD7450, Philips, Nanjing, China) by mixing 36 g of coffee powder with 500 mL of water. A filter paper was used during brewing. After each brewing, the coffee machine was cleaned prior to brewing the next sample. The extraction was made in three replications. The analysis of antioxidant activities, TPC, TFC, and TTC were measured from this extract.

2.4. *Color Parameters*

Chroma Meter (CR-400 Chroma Meter, Tokyo, Japan) was used to estimate the color of the roasted coffee powder. A small coffee powder (5 g) was kept into a Petri dish for reading, and a* (redness), b* (yellowness) and L* (lightness) were assessed. The color reading was made in triplicate.

2.5. *Antioxidant Activity*

2.5.1. Oxygen Radical Absorbance Capacity

The oxygen radical absorbance capacity (ORAC) was estimated with some modifications using the protocol described by Ou et al. [26]. Fluorescein powder was dissolved using a phosphate buffer (NaH_2PO_4-Na_2HPO_4, 10 mM, pH 7.0). The coffee extract (50 µL) was mixed with 25 mM of the fluorescein solution (150 µL) and incubated in a dark environment for about 10 min. Then, 25 µL of 120 mM AAPH solution was mixed to the coffee extract and fluorescein mixture. Meanwhile, 10 mM phosphate buffer was substituted instead of coffee extract and used as a control. The absorbance was measured using a UV/visible spectrophotometer (U-2900, Hitachi High-Tech Corporation, Tokyo, Japan). The readings were recorded at one minute interval for 30 min (excitation wavelength: 485 nm; emission wavelength: 535 nm). The absorbance readings for ORAC estimation were made in triplicate and expressed as a µM Trolox equivalent/mL of coffee (µM TE/mL). The ORAC estimated using the formula stated below;

$$\text{ORAC } (\mu\text{M TE/g}) = (C_{\text{Trolox}} \times (AUC_{\text{Sample}} - AUC_{\text{Blank}}) \times k)/AUC_{\text{Trolox}} - AUC_{\text{Blank}} \quad (1)$$

where C_{Trolox}, k, and AUC were the concentrations of Trolox (5 μM), the sample dilution factor, and the area under the curve, respectively. AUC was computed according to the formula presented below;

$$\text{AUC} = 1 + \sum_{n=1}^{30} \frac{fn}{f0} \tag{2}$$

where f_n was the fluorescence at time n (min).

2.5.2. Superoxide Dismutase-Like Activity

The protocol suggested by Marklund and Marklund [27] used with some modifications to estimate the SOD-like activity. Coffee extract (400 μL), Tris-HCl buffer (600 μL, 50 mM tris (hydroxymethyl) aminomethane and 10 mM ethylenediaminetetraacetic acid, pH 8.0), and 7.2 mM pyrogallol (40 μL) were mixed together and kept at 25 °C for 10 min. The reaction was terminated by mixing 0.1N HCl (20 μL). Absorbance readings were done at 420 nm using a UV/visible spectrophotometer (U-2900, Hitachi High-Tech Corporation, Tokyo, Japan). The absorbance readings for the estimation of the SOD-like activity were performed in triplicate. SOD-like activity was computed with the formula presented below:

$$\text{SOD-like activity (\%)} = (1 - A/B) \times 100 \tag{3}$$

where A indicates the sample absorbance and B indicates the absorbance of the blank (control).

2.5.3. 2.,2-diphenyl-1-picrylhydrazyl (DPPH) Radical Scavenging Assay

The DPPH activity was measured using the protocol described by Pataro et al. [28], with some modifications. DPPH was dissolved in methanol (0.1 mM) and 3.9 mL of it transferred into a cuvette, then immediately the absorbance was measured at 515 nm (used as a control). The coffee extracts (800 μL) from each sample were mixed with 3.2 mL of methanol-DPPH solution and placed in a dark condition for 30 min. Then the absorbance was recorded. Methanol was used as a blank. The DPPH assay was done in three replications. The DPPH inhibition percentage was calculated using the following formula;

$$\text{DPPH inhibition (\%)} = (A_{control} - A_{sample})/\ A_{control} \times 100 \tag{4}$$

where $A_{control}$ is the absorbance of control reaction (DPPH and Methanol), and A_{sample} is the absorbance in the presence of coffee extract.

2.5.4. Ferric Reducing Antioxidant Power (FRAP)

FRAP of the coffee extract was assessed using the method described by Oyaizu [29]. A 2.5 mL of coffee extract was mixed with 2.5 mL of 200 mM sodium phosphate buffer (pH 6.6) 2.5 mL of 1% potassium ferricyanide. The mixture was incubated for 20 min at 50 °C following 2.5 mL of 10% trichloroacetic acid was added. Then, this mixture was centrifuged (2000× *g* for 10 min) and 5 mL of the supernatant was taken and mixed with 5 mL of water and finally added 1 mL of 0.1% ferric chloride. The absorbance was measured at 700 nm using a spectrophotometer (U-2900, Hitachi High-Tech Corporation, Tokyo, Japan) and the FRAP was subsequently determined using ascorbic acid as standard. The FRAP analysis was done in three replications.

2.6. Total Phenol Content, Flavonoid Content, and Tannin Content

The total phenol content of the coffee extract was measured using a protocol described by Singleton's method [30], with some adjustments. The coffee extract (20 μL) was diluted with 1580 μL ddH_2O. Diluted coffee extract (160 μL) and Ciocalteu's phenol reagent (10 μL) were mixed and kept for 8 min. Then, 30 μL of 20% Na_2CO_3 solution was mixed and incubated in a dark environment for 2 h. The coffee extract was substituted with distilled water and used as a blank. The absorbance readings were determined at 765 nm using a UV/visible spectrophotometer (U-2900, Hitachi High-Tech Corporation, Tokyo, Japan). Gallic acid solutions (0–1 mg/mL) were used to generate a standard curve ($r^2 = 0.997$). The estimation of the TPC was done in triplicate. Results were expressed as mg gallic acid equivalent/mL (mg GAE/mL) of coffee extract.

A protocol stated by Dewanto [31], applied with some modification to estimate the total flavonoid content of the coffee extract. The mixture was prepared using 250 µL coffee extract, 1 mL of distilled water and 75 µL of 5% $NaNO_2$. The 5 min later, 10% $AlCl_3$ $6H_2O$ solution (150 µL) was mixed and incubated for 6 min. Finally, 1N NaOH (500 µL) was added and incubated for 11 min. The ddH_2O used as a blank sample. The absorbance was measured at 510 nm using a UV/visible spectrophotometer (U-2900, Hitachi High-Tech Corporation, Tokyo, Japan). The standard solution was prepared using a quercetin solution (0–1 mg/mL) to make a standard curve ($r^2 = 0.999$). The estimation of the TFC was done in triplicate. The total flavonoids in the coffee were expressed as mg quercetin equivalent/mL of coffee extract.

The total tannin content (TTC) was measured using the Folin-Ciocalteu method, with some modifications [32]. About 100 µL of the coffee extract was added to a tube (10 mL) containing 7500 µL of distilled water, 500 µL of Folin-Ciocalteu phenol reagent, and 1000 µL of 35% sodium carbonate solution and was then diluted to 10,000 µL with distilled water. The mixture was well-mixed and placed at room temperature for 30 min. A set of standard solutions of tannic acid (20, 40, 60, 80, 100 µg/mL) were made as a reference. Absorbance for the test and standard solutions was measured with a UV/Visible spectrophotometer (U-2900, Hitachi High-Tech Corporation, Tokyo, Japan) against the blank (distilled water) at 700 nm. The estimation of the TTC was done in triplicate. The tannin content was expressed in mg/mL of tannic acid in the coffee extracts.

2.7. Sensory Evaluation

The consumer responses of the coffee made from fermented and non-fermented coffee beans were evaluated with 50 people (students and staff members of the horticulture department) at Kangwon National University, Republic of Korea. The medium roasted coffee beans were selected for testing the consumer acceptability since the Korean people widely used the medium roasted coffee. The coffee was brewed using a coffee maker (HD7450, Philips, Nanjing, China) using 36 g of coffee powder with 500 mL of water. The prepared coffee was poured into a thermos to maintain the temperature. Each consumer was provided with coffee sample (20 mL, 60 ± 2 °C) in a paper cup (150 mL) and a glass of water (24 ± 2 °C) for cleansing palate after each taste. The scale was 1 to 7 and labeled as 1 = dislike very much, 2 = dislike moderately, 3 = dislike slightly, 4= neither like nor dislike, 5 = like slightly, 6 = like moderately, 7 = like very much.

2.8. Statistical Analysis

The results were compiled using Microsoft Excel 2013. Analysis of variance (ANOVA) was performed using SAS 9.4 software (SAS Institute, 100 Campus Drive, Cary, Raleigh, North Carolina, USA), to classify significant variations among samples.

3. Results and Discussion

3.1. pH and Colony Count (log CFU/mL)

The pH of the solution (the water that was used to ferment the coffee beans) was measured right after the inoculation of the yeast (0 h) and after 24 h of fermentation. A pH reduction was observed in both the yeast-inoculated and non-yeast-inoculated (control) treatments. However, the pH reduction rate was higher in the yeast-inoculated sample. The pH in the control treatment was 6.02 at 0 h and 5.78 after 24 h (Figure 1A). In the fermented treatment, the pH was reduced from 6.05 to 5.46 within 24 h of fermentation (Figure 1A). The pH of the yeast-inoculated solution showed a higher reduction compared to the non-yeast-inoculated solution. The alkalinity and acidity of the solution are expressed based on the pH values. The pH continues to decrease while fermentation takes place for various reasons. Fermentation is responsible for the production of organic acids and the absorption of basic amino acids, which substantially reduces the pH [33]. However, a small pH reduction was found in the non-yeast-inoculated treatment. This reduction was expected because the water-soluble organic acids leaked from the coffee beans into the water [21,34] and/or because of bacteria that naturally exist on the coffee beans [21]. The yeast populations were measured; initially, they were 7.66 log CFU/mL, and after 24 h, they reached 9.62 log CFU/mL (Figure 1B).

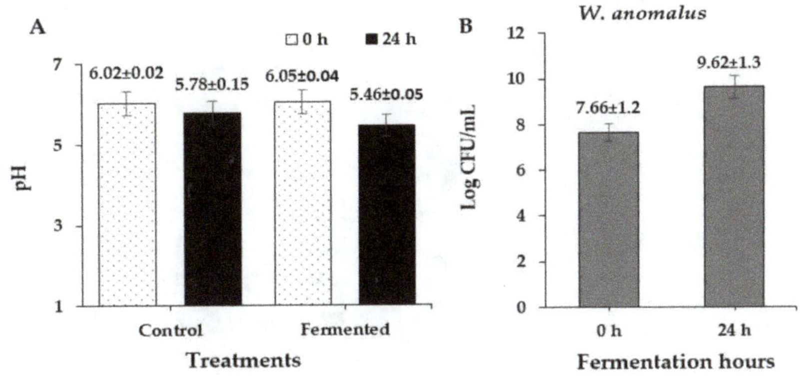

Figure 1. The pH in yeast inoculated and non-inoculated treatments (**A**) and yeast population (**B**).

3.2. Colorimeter Data

The fermented and non-fermented ground coffee after roasting is shown in Figure 2. A significant ($p < 0.05$) differences were observed between the yeast-fermented and non-fermented treatments of ground roasted (light roasting level) coffee with respect to the colorimetric parameter (L* and a*) (Figure 3A,B). The L* and a* of ground roasted coffee did not significantly differ between the fermented and control treatment in both medium and dark roasting condition. However, we observed that the L* value increased and the *a** value decreased as the roasting time increased. The b* value was significantly higher in the fermented treatment at the light, medium and dark roasting level (Figure 3C). Supporting reports showed that fermentation of green coffee beans with *S. cerevisiae* strains [21,22], and *S. fibuligera* [22] resulted in high L*, a*, and b* values compared to non-fermented coffee beans. During cocoa fermentation, gradually, a color change was observed as a result of a high amount of phenolic compounds and flavonoids, which act as a substrate of polyphenol oxidase in the presence of oxygen [35].

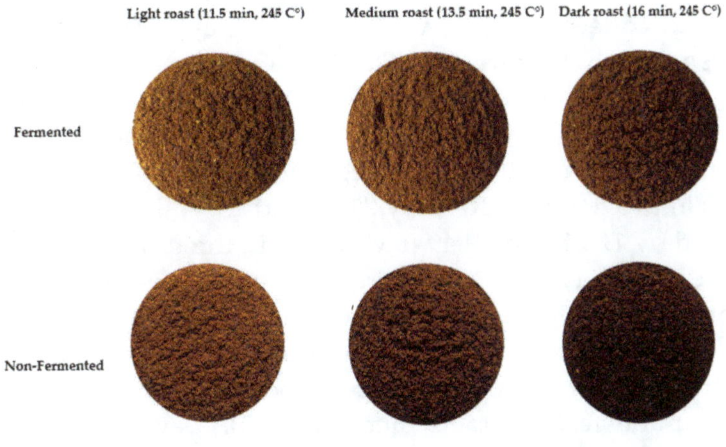

Figure 2. The fermented and non-fermented ground coffee that roasted at a different level.

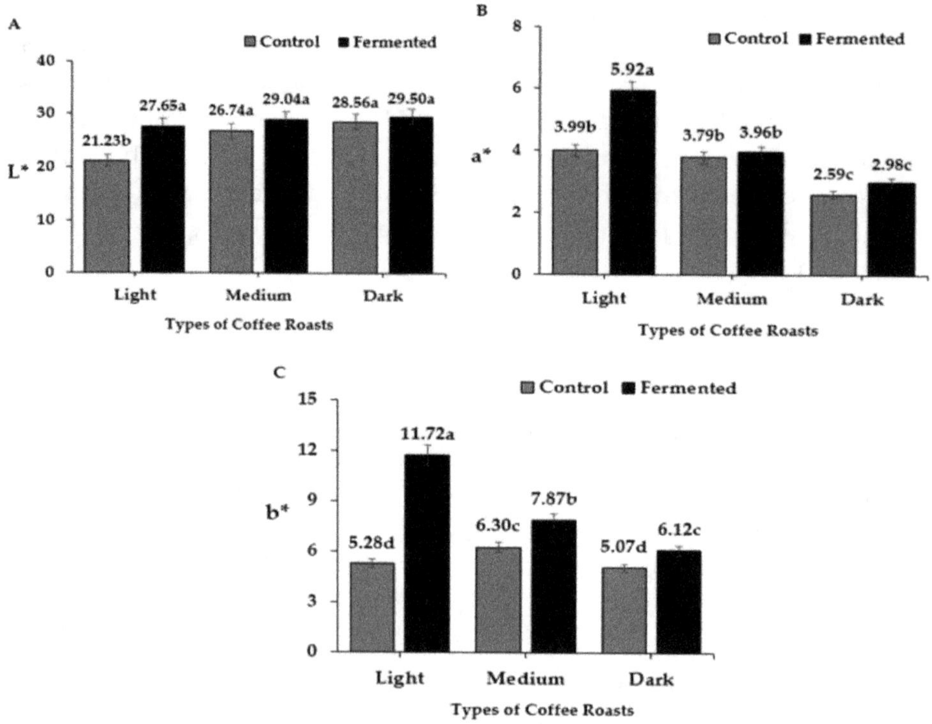

Figure 3. Colorimeter (L*, a* and b*) values of fermented roasted and ground coffee (**A**–**C**). Bars with different letters indicate a significant difference ($p < 0.5$) among treatments.

3.3. Antioxidant Activity

3.3.1. SOD-Like and ORAC

The antioxidants of fermented and non-fermented coffee extract were evaluated in terms of the oxygen radical absorbance capacity (ORAC) and superoxide dismutase (SOD)-like activity. The peroxyl radical scavenging potentials of aqueous soluble components present in the coffee could be estimated more accurately with a modified ORAC assay [36,37]. There was no significant ($p > 0.05$) difference between the yeast-fermented and non-fermented coffee extracts with respect to the ORAC value in medium (Figure 4A) and dark roasted coffee. The SOD-like activity was not significantly differed between fermented and non-fermented coffee in all roasting conditions (Figure 4B). However, both ORAC and SOD-like activity were significantly lower in both treatments in the light roasting condition. The means of ORAC value for the yeast-fermented coffee extracts was 28.33, 46.08, and 47.82 µ MTE/mL in the light, medium and dark roasted coffee, respectively (Figure 4A). The means of the SOD-like activity of yeast-fermented coffee extracts was 59.56, 87.31, and 89.05% at the light, medium and dark roasting levels, respectively (Figure 4B). The ORAC value was higher by 3.96 µ MTE/mL in fermented coffee compared to non-fermented coffee in the light roasting condition (Figure 4A). We found a significant improvement in the ORAC value and SOD-like activity in the medium and dark conditions compared to the light roasted coffee regardless of the fermentation. The medium and dark roasting conditions improved the ORAC by 21.93 and 23.94 µ MTE/mL in the control treatment. However, it was increased by 17.75 and 19.49 µ MTE/mL in the fermented coffee extracts compared to light roasted coffee, respectively. This is in agreement with previous reports on the influence of roasting on antioxidant activity. Nicoli et al. [38] reported higher antioxidant activity for medium and dark roast brews. Similarly, the elevated antioxidant activity for higher degrees of roasting has been confirmed in studies [39–41]. As the roasting temperature increased the antioxidant activities of the kernel, cashew nut and testa extracts increased [42]. Several publications have described variations in coffee antioxidant contents as a cause of roasting conditions [43,44]. Del Castillo et al. [43] noticed a higher antioxidant activity in the medium roasted coffee brew. However, Perrone et al. [45]

found a greater antioxidant activity for light roast brews. The results from various investigations regarding the antioxidant activity of coffee have been considerably diverse. The causes of such variable research results can be associated with the development of melanoidin compounds, which increase the antioxidant capacity despite the reductions in the polyphenol content. Nevertheless, other studies have not detected any significant effects of roasting on antioxidant activity [46], nor any indications that roasting for a longer duration at high temperature reduce antioxidant activity [44,47,48] when polyphenol degradation is not compensated by melanoidin formation [49,50]. However, amongst other coffees, this phenomenon was noticed for dark roasted Robusta coffee [40].

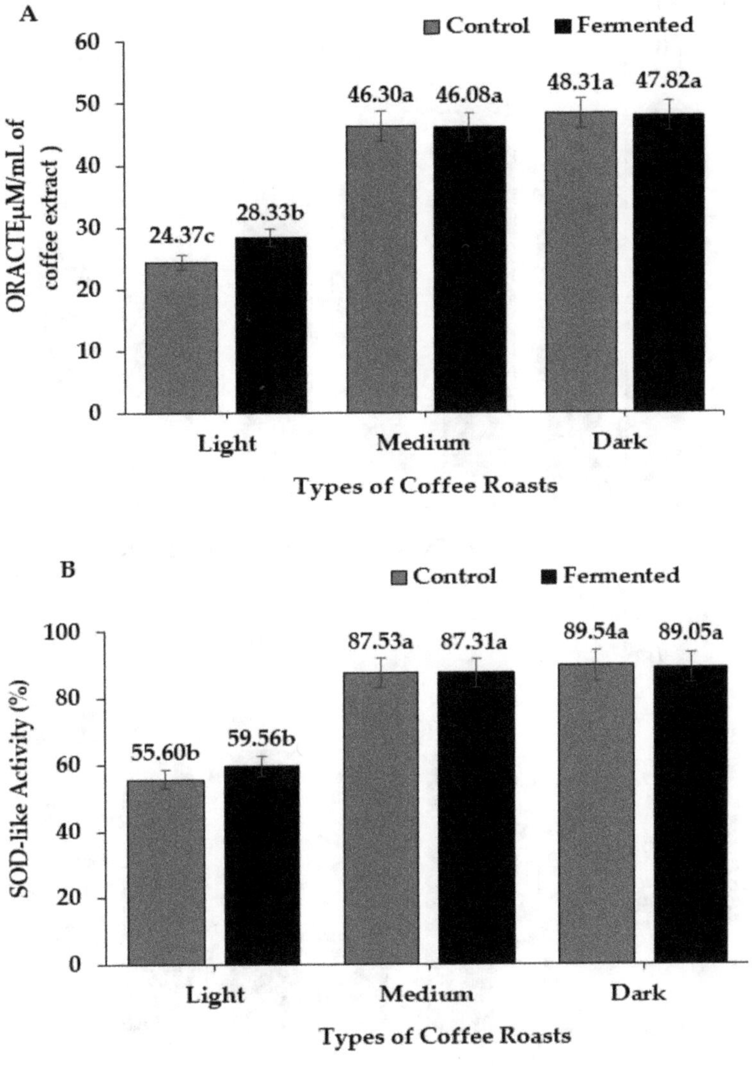

Figure 4. The oxygen radical absorbance capacity (ORAC) and superoxide dismutase-like (SOD)-like activity of fermented coffee at different degrees of roast (**A** and **B**, respectively). Bars with different letters indicate a significant difference ($p < 0.5$) among treatments.

3.3.2. DPPH and FRAP

DPPH and FRAP are among the various methods of evaluating the antioxidant properties of foods and drinks. The percentage of DPPH inhibition was higher in the fermented coffee than the non-fermented coffee in all roasting conditions (Figure 5A). Like the DPPH, the FRAP was improved in the fermented coffee in the light, medium, and dark roast types (Figure 5B). The maximum DPPH inhibition resulted from the fermented treatment in the light roasting condition (46.31%). The highest FRAP was found in the fermented coffee in the dark roasting condition (176.11 AAE μg/mL of coffee extract). In general, fermentation of green coffee beans using *W. anomalus* improved the DPPH and

FRAP. Adetuyi and Ibrahim [51] reported that fermentation enhanced the DPPH radical-scavenging ability and FRAP of okra seeds. The increase in DPPH capacity after the fermentation process shows that fermentation likely has great potential in generating some metabolites with superior radical scavenging activity [52]. The percentage of DPPH inhibition dropped as the roasting time increased (from the light to dark roasting condition). The result is in agreement with Jung et al. [53], who reported that lightly roasted coffee extract has the highest antioxidant activity with a low roasting temperature and that the DPPH decreases in the dark roasted extract. However, the FRAP improved as the roasting time increased. A supporting result published by Song et al. [54] showed that as the roasting time increased from 11 to 13 min (medium-light to medium-dark) the FRAP increased as well.

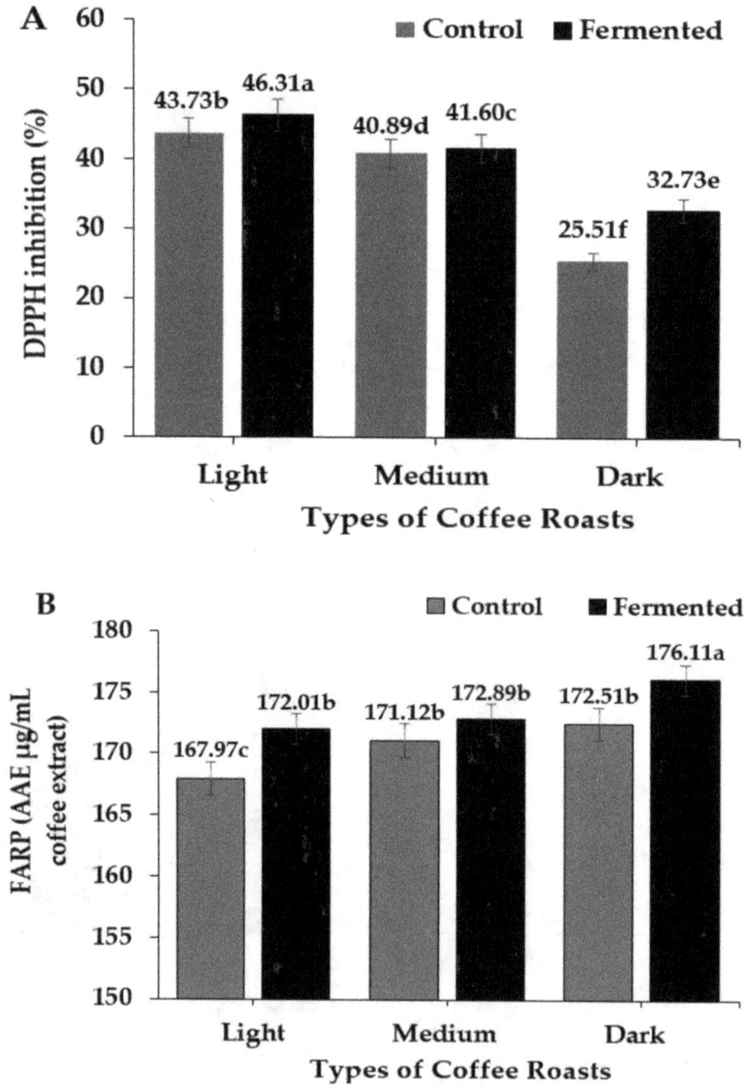

Figure 5. The DPPH inhibition (%) and Ferric reducing antioxidant power (AAE µg/mL of coffee extract) of fermented coffee. (**A** and **B**, respectively). Bars with different letters indicate a significant difference ($p < 0.5$) among treatments.

3.4. Total Phenol, Flavonoid and Tannin Content

The TPC of yeast-fermented coffee extracts had significant differences ($p < 0.05$) compared to the non-fermented coffee in all roasting types (Figure 6). During the medium roasting condition, the highest TPC difference was observed between the fermented and non-fermented coffee. The TPC in the control treatment showed a decreasing trend as the roasting time increased (light to medium to dark). The TPC in the fermented treatment was 0.98, 1.29 and 0.91 GAEmg/mL of coffee extract in the light, medium and dark roasting condition, respectively (Figure 6). The fermentation of green coffee beans with *W. anomalus* increased the TPC of the coffee extract by 0.09, 0.46, and 0.21 GAE mg/mL of coffee extract compared to non-fermented coffee at each degree of roasting (light, medium and dark roast, respectively) (Figure 6). Generally, fermentation improved the TPC compared to non-fermented coffee. These results are in agreement with several reports on fermented seeds, where fermentation induced an improvement in the phenolic content of seeds, such as legumes [55–59], okra seeds [51], soybeans [60], and coffee [21,22]. Similarly, fermentation elevated the phenolic and flavonoid contents and improved the antioxidant properties of the following plant seeds: chestnut, adlay, walnut, and lotus seed [61]. This might be associated with, proteolytic enzymes from the starter organism that hydrolyze the complexes of phenolics into simple, soluble-free phenols and biologically more active forms during fermentation, which are readily absorbed by organisms [56,62]. However, other researchers' findings indicate the proteolytic activities alone did not show a significant TPC increase; rather, when it was combined with a pectinase enzyme, the olive oil polyphenol content showed a significantly high increment [63]. These findings lead us to conclude *W. anomalus's* ability to produce pectinase enzymes linked with the increased TPC content in this experiment. In our experiment, the TPC was decreased as the roasting time increased, except the fermented coffee in the medium roast condition which significantly increased the TPC, regardless of the fermentation. A supporting result was published by Odžaković, B., et al. [64], the TPC decreased as the roasting temperature increased. During coffee roasting the degradation of polyphenols [36,44–46], which are sensitive to heat, are affected by roasting temperature. The reduction in the polyphenol content which is linked to the extended roasting is due to the heat-sensitive nature of such compounds and the lengthened roasting duration, as well as the high processing temperature. Phenol reductions or losses during the processing steps are undesirable because of their profound effects on human health.

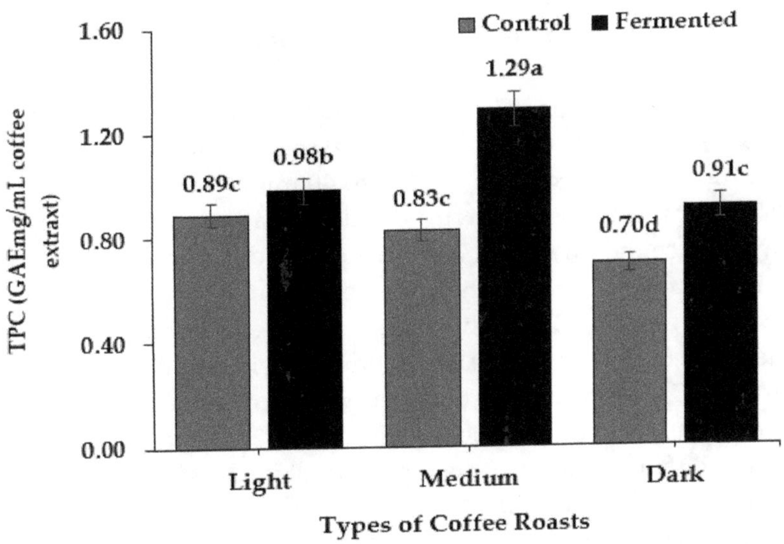

Figure 6. Total phenol contents (GAE mg/mL of coffee extract) of fermented coffee. Bars with different letters indicate a significant difference ($p < 0.5$) among treatments.

Fermentation of green coffee beans with *W. anomalous* significantly ($p < 0.05$) increased the TFC compared to non-fermented coffee in all roasting conditions (light, medium and dark) (Figure 7). TFC was approximately consistent in the control treatment in all roasting conditions. However, in the fermented coffee, the TFC was significantly higher in the light roasting compared to medium and dark roasting conditions. As shown in Figure 7, the TFC in the control treatment was 0.64, 0.67, and 0.64. The TFC in the fermented coffee at different roasting levels was 0.86 (light), 0.74 (medium), and 0.72 (dark) QE mg/mL of coffee extract. Fermentation improves the TFC content by 0.22, 0.07, and 0.08 QE mg/mL of coffee extract in the light, medium, and dark roasting conditions, respectively (Figure 7). These findings agree with several published papers on different fermented seeds, where fermentation significantly enhanced the TFC compared to the TFC in unfermented seeds, such as in soybeans [60], legumes [45,55,56], coffee beans [21,22], and okra seeds [51]. The increase in the flavonoid content might be linked to the rise in acidic values during fermentation, which releases bound flavonoid components and makes them more bioavailable [51].

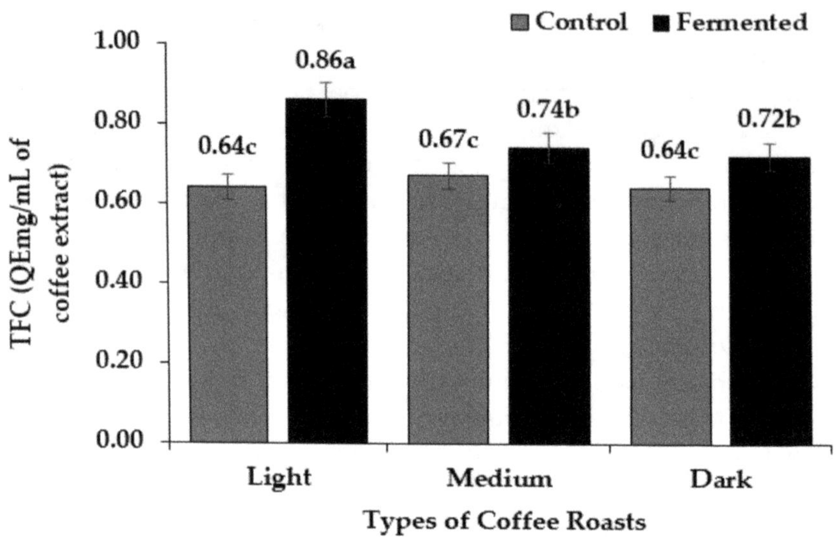

Figure 7. Total flavonoid contents (QE mg/mL of coffee extract) of fermented coffee. Bars with different letters indicate a significant difference ($p < 0.5$) among treatments.

Regardless of the fermentation, the average TFC from both treatments at each roasting condition was 0.75 ± 0.11, 0.71 ± 0.03, and 0.68 ± 0.04 QE mg/mL of coffee extract at the light, medium, and dark roasting levels, respectively. The average TFC value showed a decreasing trend as the roasting time increased, regardless of the fermentation. The result in our study coincides with Tiwari, B. K., et al. [65], who found that the TFC was reduced significantly as the roasting temperature increased from 80 °C to 120 °C and roasting duration proceeded from 20 min to 40 min. The cause for the reduction in the TFC at a higher temperature might be associated with the degradation of flavonoids. The preparation and processing of food may reduce the flavonoid contents depending on the techniques used [66].

The yeast-fermented coffee extracts had a significant difference ($p < 0.05$) compared to non-fermented coffee with regard to TTC in the light and dark roasting conditions (Figure 8). The TTC of non-fermented coffee was 0.53, 0.33, and 0.21 mg tannic acid/mL of coffee extract in light, medium and dark roasting conditions, respectively. In the fermented coffee, the TTC was 0.73, 0.28, and 0.15 mg tannic acid/mL coffee extract in the light, medium, and dark roasting conditions, respectively. In the light roasting condition, the tannin content was higher in fermented coffee than the control and it was lower than the control in the medium and dark roasting conditions (Figure 8). It has been reported that fermentation significantly decreased the tannin content of fermented sorghum cultivars [67], and in fermentation of *Xuan mugua* fruit with lactic acid bacteria, where it caused a significant decrease in the tannin content [68]. We have also found a supporting result in our previous experiment that indicates the fermentation of green coffee beans with *S. cerevisiae* and *S. fibuligera* strains reduced the TTC compared to non-fermented coffee [22]. Regardless of the fermentation, the TTC was reduced as the roasting time increased. The average TTC in each roasting condition was 0.63, 0.31 and 0.19 mg tannic acid mg/mL coffee extract in the light, medium, and dark roasting condition (Figure 8). Likewise, the tannin content was decreased in soya bean flour (0.01–0.30 g/100 g dry weight) [69], and 22% in maize [70]; because of soaking in water for 48 h and roasting. Tannin content has an astringent characteristic that contributes to the bitterness of coffee, so the reduction in tannins can be viewed as a positive aspect.

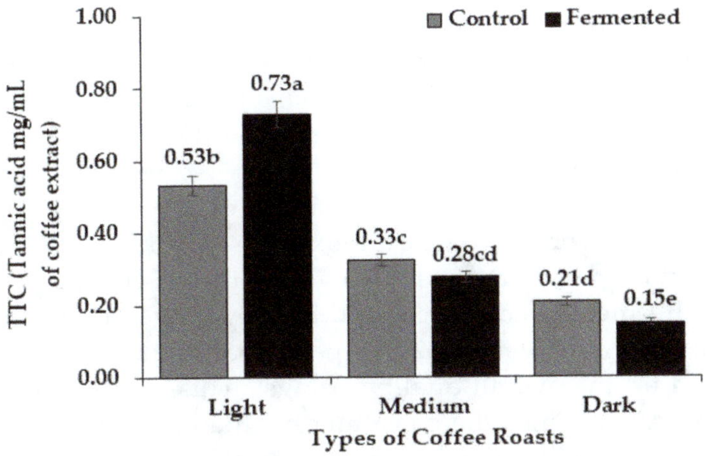

Figure 8. Total tannin contents (tannic acid mg/mL of coffee extract) of fermented coffee. Bars with different letters indicate a significant difference ($p < 0.5$) among treatments.

3.5. Sensory Evaluation

Based on the results of consumer acceptance ratings, the non-fermented coffee received a higher score than the fermented coffee in terms of its color, sourness, mouthfeel, acidity, and overall quality. However, the fermented coffee received the highest rating with respect to aroma, bitterness, and astringency. The average score for the overall quality of the coffee was 5.22 and 4.51 for non-fermented and fermented coffee, respectively (Figure 9). The average score for acidity was 3.75 for the fermented coffee and 4.15 for the non-fermented coffee. Kwak et al. [21] have evaluated the consumer acceptability of fermented green coffee beans with different yeasts, noting that one of their fermented coffee treatment acceptability ratings was lower but insignificant when compared to the control treatment. In addition, they mentioned that all the fermented coffee received lower ratings when compared to the control. Fermentation has both positive and negative influences on the flavor and aroma characteristics of coffee. Based on the overall quality rating, we have categorized the consumer responses into two groups: people who preferred the non-fermented coffee over the fermented coffee (55.48%) and people who preferred the fermented coffee over the non-fermented coffee (44.52%). A supporting result was published by Kwak et al. [21], who found that 39.4% of consumers preferred fermented coffee while 60.6% preferred non-fermented coffee by disliking the control (non-fermented) treatments based on the overall quality ratings. However, consumer acceptance evaluations are very subjective to the individual.

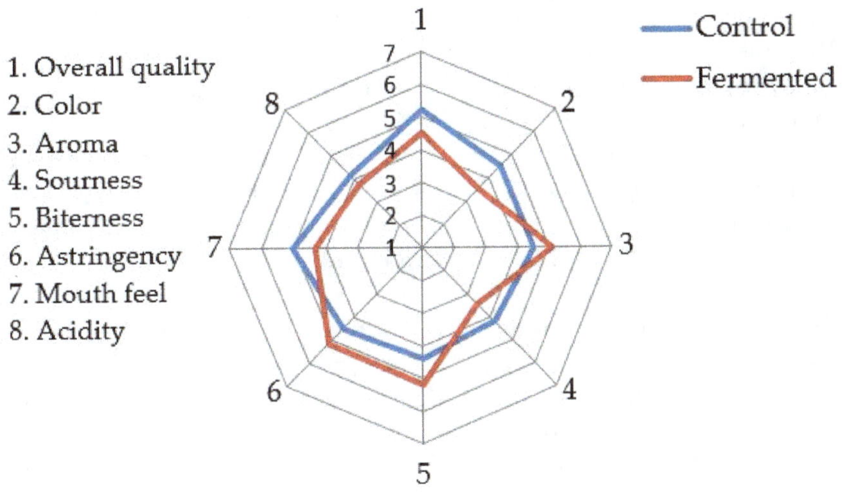

Figure 9. The consumer acceptability ratings of fermented and non-fermented coffee.

4. Conclusions

This study showed the responses of the coffee beans fermented with *W. anomalus* strain KNU18Y3. The fermentation of green coffee beans with *W. anomalus* for 24 h increased the DPPH, FRAP, TPC, and TFC when compared to non-fermented coffee, whereas the SOD-like and ORAC did not significantly differ between the fermented and non-fermented coffee beans. We have also found that the fermentation of coffee beans for 24 h is enough to modify the functionality of coffee beans. The extent of roasting had both negative and positive effects on the overall antioxidants, TPC, TFC, and TTC. The degradation of the TPC, TFC, and TTC was observed as the roasting time increased from 11.5 to 16 min, regardless of the fermentation. The fermented coffee received better ratings in terms of its aroma, bitterness, and astringency parameters. Moreover, 44.52% of participants chose the fermented coffee over the non-fermented one based on the overall quality ratings. This result shows that fermented coffee is also preferred by consumers. Since the consumer evaluation experiment was conducted blind (the participants were not informed of the coffee types), the increased antioxidants and phenolic and flavonoid compounds will probably attract more consumers when they are informed of these positive aspects of fermented coffee. While the *W. anomalus* yeast has the potential to be used for green coffee bean fermentation, further selection and evaluation of microorganisms should be continued to maximize the functionality of the coffee bean and its health benefits.

Author Contributions: Conceptualization, W.H.K., M.H.; methodology, M.H.; Data curation, M.H.; validation, W.H.K.; formal analysis, M.H.; software, M.H.; investigation, W.H.K.; resources, W.H.K.; funding acquisition, W.H.K.; writing—original draft preparation, M.H.; writing—review and editing, W.H.K., M.H.; supervision, W.H.K.; project administration, W.H.K. All authors have read and agreed to the published version of the manuscript.

References

1. Mesfin, H.; Kang, W.H. The role of microbes in coffee fermentation and their impact on coffee quality. *J. Food Qual.* **2019**. [CrossRef]
2. Martins, S.; Mussatto, S.I.; Martínez-Avila, G.; Montañez-Saenz, J.; Aguilar, C.N.; Teixeira, J.A. Bioactive Phenolic Compounds: Production and Extraction by Solid-State Fermentation. A Review. *Biotechnol. Adv.* **2011**, *29*, 365–373. [CrossRef] [PubMed]
3. Torino, M.I.; Limón, R.I.; Martínez-Villaluenga, C.; Mäkinen, S.; Pihlanto, A.; Vidal-Valverde, C.; Frias, J. Antioxidant and Antihypertensive Properties of Liquid and Solid State Fermented Lentils. *Food Chem.* **2013**, *136*, 1030–1037. [CrossRef] [PubMed]
4. Dulf, F.V.; Vodnar, D.C.; Socaciu, C. Effects of solid-state fermentation with two filamentous fungi on the total phenolic contents, flavonoids, antioxidant activities and lipid fractions of plum fruit (*Prunus domestica* L.) by-products. *Food Chem.* **2016**, *209*, 27–36. [CrossRef]
5. Zhang, X.Y.; Chen, J.; Li, X.L.; Yi, K.; Ye, Y.; Liu, G.; Wang, Z.G. Dynamic changes in antioxidant activity and biochemical composition of tartary buckwheat leaves during *Aspergillus niger* fermentation. *J. Funct. Foods* **2017**, *32*, 375–381. [CrossRef]
6. Belitz, H.D.; Grosh, W.; Schieberle, P. Coffee, Tea, Cocoa. In *Food Chemistry*; Springer-Verlag: Heidelberg/Berlin, Germany; New York, NY, USA, 2009.
7. Rawel, H.M.; Kulling, S.E. Nutritional contribution of coffee, cacao and tea phenolics to human health. *J. Verbraucherschutz Lebensmittelsicherheit* **2007**, *2*, 399–406. [CrossRef]
8. Richelle, M.; Tavazzi, I.; Offord, E. Comparison of the antioxidant activity of commonly consumed polyphenolic beverages (coffee, cocoa, and tea) prepared per cup serving. *J Agric. Food Chem.* **2001**, *49*, 3438–3442. [CrossRef]
9. Cheynier, V. Polyphenols in foods are more complex than often thought. *Am. J. Clin. Nutr.* **2005**, *81*, 223S–229S. [CrossRef]
10. Manach, C.; Scalbert, A.; Morand, C.; Rémésy, C.; Jimenez, L. Polyphenols: Food sources and bioavailability. *Am. J. Clin. Nutr.* **2004**, *79*, 727–747. [CrossRef]
11. Kumar, S.; Mishra, A.; Pandey, A.K. Antioxidant mediated protective effect of *Parthenium hysterophorus* against oxidative damage using in vitro models. *BMC Complement Altern. Med.* **2013**, *13*, 120. [CrossRef]

12. Kumar, S.; Pandey, A.K. Phenolic content, reducing power and membrane protective activities of *Solanum xanthocarpum* root extracts. *Vegetos* **2013**, *26*, 301–307.
13. Yao, L.H.; Jiang, Y.M.; Shi, J.; Tomás-Barberán, F.A.; Datta, N.; Singanusong, R.; Chen, S.S. Flavonoids in food and their health benefits. *Plant Foods Hum. Nutr.* **2004**, *59*, 113–122. [CrossRef] [PubMed]
14. Kurtzman, C.P.; Fell, J.W. *The Yeast, a Taxonomical Study*, 4th ed.; Elsevier Science: Amsterdam, The Netherlands, 1988; p. 1055.
15. De Hoog, G.S. Risk assessment of fungi reported from humans and animals. *Mycoses* **1996**, *39*, 407–417. [CrossRef]
16. Comitini, F.; De Ingenis, J.; Pepe, L.; Mannazu, I.; Ciani, M. *Pichia anomala* and *Kluyveromyces wikerhamii* killer toxins as new tool against *Dekkera/Brettanomyces* spoilage yeasts. *FEMS Microbiol. Lett.* **2004**, *238*, 235–240. [CrossRef] [PubMed]
17. Lee, Y.J.; Choi, Y.R.; Lee, S.Y.; Park, J.T.; Shim, J.H.; Park, K.H.; Kim, J.W. Screening wild yeast strains for alcohol fermentation from various fruits. *Mycobiology* **2011**, *39*, 33–39. [CrossRef] [PubMed]
18. Ye, M.; Yue, T.; Yuan, Y. Effects of sequential mixed cultures of *Wickerhamomyces anomalus* and *Saccharomyces cerevisiae* on apple cider fermentation. *FEMS Yeast Res.* **2014**, *14*, 873–882. [CrossRef] [PubMed]
19. Madrigal, T.; Maicas, S.; Tolosa, J.J.M. Glucose and ethanol tolerant enzymes produced by *Pichia* (*Wickerhamomyces*) isolates from enological ecosystems. *Am. J. Enol. Vitic.* **2013**, *64*, 126–133. [CrossRef]
20. Zha, M.; Sun, B.; Wu, Y.; Yin, S.; Wang, C. Improving flavor metabolism of Saccharomyces cerevisiae by mixed culture with *Wickerhamomyces anomalus* for Chinese Baijiu making. *J. Biosci. Bioeng.* **2018**, *126*, 189–195. [CrossRef]
21. Kwak, H.S.; Jeong, Y.; Kim, M. Effect of Yeast Fermentation of Green Coffee Beans on Antioxidant Activity and Consumer Acceptability. *J. Food Qual.* **2018**. [CrossRef]
22. Mesfin, H.; Kang, W.H. Antioxidant Activity, Total Polyphenol, Flavonoid and Tannin Contents of Fermented Green Coffee Beans with Selected Yeasts. *Fermentation* **2019**, *5*, 29.
23. Lee, L.W.; Cheong, M.W.; Curran, P.; Yu, B.; Liu, S.Q. Modulation of coffee aroma via the fermentation of green coffee beans with *Rhizopus oligosporus*: I. Green Coffee. *Food Chem.* **2016**, *211*, 916–924. [CrossRef] [PubMed]
24. Lee, L.W.; Tay, G.Y.; Cheong, M.W.; Curran, P.; Yu, B.; Liu, S.Q. Modulation of the volatile and non-volatile profiles of coffee fermented with *Yarrowia lipolytica*: I. Green coffee. *LWT- Food Sci. Technol.* **2017**, *77*, 225–232. [CrossRef]
25. Mesfin, H.; Kang, W.H. Isolation, Identification, and Characterization of Pectinolytic Yeasts for Starter Culture in Coffee Fermentation. *Microorganisms* **2019**, *7*, 401.
26. Ou, B.; Hampsch-Woodill, M.; Prior, R.L. Development and validation of an improved oxygen radical absorbance capacity assay using fluorescein as the fluorescent probe. *J. Agric. Food Chem.* **2001**, *49*, 4619–4626. [CrossRef] [PubMed]
27. Marklund, S.; Marklund, G. Involvement of the superoxide anion radical in the autoxidation of pyrogallol and a convenient assay for superoxide dismutase. *Eur. J. Biochem.* **1974**, *47*, 469–474. [CrossRef]
28. Pataro, G.; Sinik, M.; Capitoli, M.M.; Donsì, G.; Ferrari, G. The influence of post-harvest UV-C and pulsed light treatments on quality and antioxidant properties of tomato fruits during storage. *Innov. Food Sci. Emerg. Technol.* **2015**, *30*, 103–111. [CrossRef]
29. Oyaizu, M. Studies on products of browning reaction. *Jpn. J. Nutr. Diet.* **1986**, *44*, 307–315. [CrossRef]
30. Singleton, V.L.; Orthofer, R.; Lamuela-Raventós, R.M. Analysis of total phenols and other oxidation substrates and antioxidants by means of Folin-ciocalteu reagent. *Methods Enzymol.* **1999**, *299*, 152–178.
31. Dewanto, V.; Wu, X.; Liu, R.H. Processed sweet corn has higher antioxidant activity. *J. Agric. Food Chem.* **2002**, *50*, 4959–4964. [CrossRef]
32. CI, K.C.; Indira, G. Quantitative estimation of total phenolic, flavonoids, tannin and chlorophyll content of leaves of *Strobilanthes Kunthiana* (Neelakurinji). *J. Med. Plants* **2016**, *4*, 282–286.
33. Coote, N.; Kirsop, B.H. Factors responsible for the decrease in pH during beer fermentations. *J. Inst. Brew.* **1976**, *82*, 149–153. [CrossRef]
34. Rodrigues, C.I.; Marta, L.; Maia, R.; Miranda, M.; Ribeirinho, M.; Máguas, C. Application of solid-phase extraction to brewed coffee caffeine and organic acid determination by UV/HPLC. *J. Food Compos. Anal.* **2007**, *20*, 440–448. [CrossRef]

35. Afoakwa, E.O.; Paterson, A.; Fowler, M.; Ryan, A. Flavor formation and character in cocoa and chocolate: A critical review. *Food Sci. Nutr.* **2008**, *48*, 840–857. [CrossRef] [PubMed]
36. Kwon, D.Y.; Choi, K.H.; Kim, S.J.; Choi, D.W.; Kim, Y.S.; Kim, Y.C. Peroxyl radical-scavenging activity of coffee brews. *Eur. Food Res. Technol.* **2005**, *221*, 471–477.
37. Gómez-Ruiz, J.Á.; Leake, D.S.; Ames, J.M. In Vitro Antioxidant Activity of Coffee Compounds and Their Metabolites. *J. Agric. Food Chem.* **2007**, *55*, 6962–6969. [CrossRef] [PubMed]
38. Nicoli, M.C.; Anese, M.; Manzocco, L.; Lerici, C.R. Antioxidant properties of coffee brews in relation to the roasting degree. *Lebensm. Wiss. Technol.* **1997**, *30*, 292–297. [CrossRef]
39. Liu, Y.; Kitts, D.D. Confirmation that the Maillard reaction is the principle contributor to the antioxidant capacity of coffee brews. *Food Res. Int.* **2011**, *44*, 2418–2424. [CrossRef]
40. Vignoli, J.A.; Viegas, M.C.; Bassoli, D.G.; Benassi, M.T. Roasting process affects differently the bioactive compounds and the antioxidant activity of arabica and robusta coffees. *Food Res. Int.* **2014**, *61*, 279–285. [CrossRef]
41. Sánchez-González, I.; Jiménez-Escrig, A.; Saura-Calixto, F. In vitro antioxidant activity of coffees brewed using different procedures (Italian, espresso and filter). *Food Chem.* **2005**, *90*, 113–139. [CrossRef]
42. Chandrasekara, N.; Shahidi, F. Effect of roasting on phenolic content and antioxidant activities of whole cashew nuts, kernels, and testa. *J. Agric. Food Chem.* **2011**, *59*, 5006–5014. [CrossRef]
43. Del Castillo, M.D.; Ames, J.M.; Gordon, M.H. Effect of roasting on the antioxidant activity of coffee brews. *J. Agric. Food Chem.* **2002**, *50*, 3698–3703. [CrossRef] [PubMed]
44. Cämmerer, B.; Kroh, L.W. Antioxidant activity of coffee brews. *Eur. Food Res. Technol.* **2006**, *223*, 469–474. [CrossRef]
45. Perrone, D.; Farah, A.; Donangelo, C.M. Influence of coffee roasting on the incorporation of phenolic compounds into melanoidins and their relationship with antioxidant activity of the brew. *J. Agric. Food Chem.* **2012**, *60*, 4265–4275. [CrossRef] [PubMed]
46. Szymanowska, K.; Wołosiak, R. The effect of coffee roasting on selected parameters of its quality. *Res. Teach. Appar.* **2014**, *19*, 77–83.
47. Duarte, S.M.S.; Abreu, C.M.P.; Menezes, H.C.; Santos, M.H.; Gouvêa, C.M.C.P. Effect of processing and roasting on the antioxidant activity of coffee brews. *Ciênc. Tecnol. Aliment.* **2005**, *25*, 387–393. [CrossRef]
48. Hecimovic, I.; Belscak-Cvitanovic, A.; Horzic, D.; Komes, D. Comparative study of polyphenols and caffeine in different coffee varieties affected by the degree of roasting. *Food Chem.* **2011**, *129*, 991–1000. [CrossRef]
49. Sacchetti, G.; Mattia, C.D.; Pittia, P.; Mastrocola, D. Effect of roasting degree, equivalent thermal effect and coffee type on the radical scavenging activity of coffee brews and their phenolic fraction. *J. Food Eng.* **2009**, *90*, 74–80. [CrossRef]
50. Summa, C.A.; Calle, B.; Brohee, M.; Stadler, R.H.; Anklama, E. Impact of the roasting degree of coffee on the in vitro radical scavenging capacity and content of acrylamide. *LWT- Food Sci. Technol.* **2007**, *40*, 1849–1854. [CrossRef]
51. Adetuyi, F.O.; Ibrahim, T.A. Effect of fermentation time on the phenolic, flavonoid and vitamin C contents and antioxidant activities of okra (*Abelmoschus esculentus*) seeds. *Niger. Food J.* **2014**, *32*, 128–137. [CrossRef]
52. Abu, F.; Taib, M.; Norma, C.; Moklas, M.; Aris, M.; Mohd Akhir, S. Antioxidant Properties of Crude Extract, Partition Extract, and Fermented Medium of *Dendrobium sabin* Flower. *J. Evid.-Based Complement. Altern. Med.* **2017**. [CrossRef]
53. Jung, S.; Kim, M.H.; Park, J.H.; Jeong, Y.; Ko, K.S. Cellular antioxidant and anti-inflammatory effects of coffee extracts with different roasting levels. *J. Med. Food.* **2017**, *20*, 626–635. [CrossRef] [PubMed]
54. Song, J.L.; Asare, T.S.; Kang, M.Y.; Lee, S.C. Changes in Bioactive Compounds and Antioxidant Capacity of Coffee under Different Roasting Conditions. *Korean J. Plant Resour.* **2018**, *31*, 704–713.
55. Moktan, B.; Saha, J.; Sarkar, P.K. Antioxidant activities of soybean as affected by *Bacillus*-fermentation to kinema. *Food Res. Int.* **2008**, *41*, 586–593. [CrossRef]
56. Ademiluyi, A.O.; Oboh, G. Antioxidant properties of condiment produced from fermented bambara groundnut (*Vigna subterranea* L. Verdc). *J. Food Biochem.* **2011**, *35*, 1145–1160. [CrossRef]
57. Dajanta, K.; Janpum, P.; Leksing, W. Antioxidant capacities, total phenolics and flavonoids in black and yellow soybeans fermented by *Bacillus subtilis*: A comparative study of Thai fermented soybeans (thua nao). *Int. Food Res. J.* **2013**, *20*, 3125.

58. Guzmán-Uriarte, M.L.; Sánchez-Magaña, L.M.; Angulo-Meza, G.Y.; Cuevas-Rodríguez, E.O.; Gutiérrez-Dorado, R.; Mora-Rochín, S.; Reyes-Moreno, C. Solid state bioconversion for producing common bean (Phaseolus vulgaris L.) functional flour with high antioxidant activity and antihypertensive potential. *Food Nutr. Sci.* **2013**, *4*, 480.
59. Plaitho, Y.; Kangsadalampai, K.; Sukprasansap, M. The protective effect of Thai fermented pigmented rice on urethane induced somatic mutation and recombination in Drosophila melanogaster. *J. Med. Plant Resour.* **2013**, *7*, 91–98.
60. Amadou, I.; Yong-Hui, S.; Sun, J.; Guo-Wei, L. Fermented soybean products: Some methods, antioxidants compound extraction and their scavenging activity. *Asian J. Biochem.* **2009**, *4*, 68–76.
61. Wang, C.Y.; Sz-Jie, W.; Shyu, Y.T. Antioxidant Properties of Certain Cereals as Affected by Food-Grade Bacteria Fermentation. *J. Biosci. Bioeng.* **2014**, *117*, 449–456. [CrossRef]
62. Shrestha, A.K.; Dahal, N.R.; Ndungutse, V. Bacillus fermentation of soybean: A review. *J. Food Sci. Technol.* **2010**, *6*, 1–9. [CrossRef]
63. Moustakime, Y.; Hazzoumi, Z.; Joutei, K.A. Effect of proteolytic activities in combination with the pectolytic activities on extractability of the fat and phenolic compounds from olives. *Springer Plus* **2016**, *5*, 739. [CrossRef] [PubMed]
64. Odžaković, B.; Džinić, N.; Kukrić, Z.; Grujić, S. Effect of roasting degree on the antioxidant activity of different Arabica coffee quality classes. *Acta Sci. Pol. Technol. Aliment.* **2016**, *15*, 409–417. [CrossRef] [PubMed]
65. Tiwari, B.K.; Brunton, N.P.; Brennan, C. Handbook of plant food phytochemicals: Sources, stability and extraction. John Wiley and Sons: Hoboken, NJ, USA, 2013.
66. Kumar, S.; Pandey, A.K. Antioxidant, lipo-protective and antibacterial activities of phytoconstituents present in Solanum xanthocarpum root. *Int. Rev. Biophys. Chem.* **2012**, *3*, 42–47.
67. Hassan, I.A.; El Tinay, A.H. Effect of fermentation on tannin content and in-vitro protein and starch digestibilities of two sorghum cultivars. *Food Chem.* **1995**, *53*, 149–151. [CrossRef]
68. Shang, Y.F.; Cao, H.; Ma, Y.L.; Zhang, C.; Ma, F.; Wang, C.X.; Wei, Z.J. Effect of lactic acid bacteria fermentation on tannins removal in Xuan Mugua fruits. *Food Chem.* **2019**, *274*, 118–122. [CrossRef]
69. Agume, A.; Njintang, N.; Mbofung, C. Effect of soaking and roasting on the physicochemical and pasting properties of soybean flour. *Foods* **2017**, *6*, 12. [CrossRef]
70. Oboh, G.; Ademiluyi, O.A.; Akindahunsi, A.A. The effect of roasting on the nutritional and antioxidant properties of yellow and white maize varieties. *Int. J. Food Sci. Technol.* **2010**, *45*, 1236–1242. [CrossRef]

5

A *Pichia anomala* Strain (*P. anomala* M1) Isolated from Traditional Greek Sausage is an Effective Producer of Extracellular Lipolytic Enzyme in Submerged Fermentation

Maria Papagianni [1,*] and Emmanuel M. Papamichael [2]

1 Department of Hygiene and Technology of Food of Animal Origin, School of Veterinary Medicine, Aristotle University of Thessaloniki, Thessaloniki 54006, Greece
2 Department of Chemistry, University of Ionnina, Ioannina 45500, Greece; epapamic@cc.uoi.gr
* Correspondence: mp2000@vet.auth.gr

Abstract: Ayeast isolate, selected for its lipolytic activity from a meat product, was characterized as *Pichia anomala*. Lipolytic activity, determined on *p*-NPA as esterase, was maximum at 28 °C, pH 6.5, and induced by the short chain triglyceride tributyrin. Fermentations in 2 L and 10 L stirred tank bioreactors, with 20 and 60 g/L glucose respectively, showed that in the second case lipolytic activity increased 1.74-fold, while the biomass increased 1.57-fold. Under otherwise identical aeration conditions, improved mixing in the 10 L reactor maintained higher dissolved oxygen levels which, along with the elevated glucose concentration, resulted in significant increase of specific rates of lipolytic activity (51 vs. 7 U/g/L), while specific rates of growth and glucose consumption maintained lower. The Crabtree-negative yeast (glucose insensitive growth) exhibited a Pasteur effect at lower dissolved oxygen concentrations while elevated glucose prevented ethanol formation under oxygen saturation. The particular physiological traits can be exploited to obtain significant lipolytic activity in a scalable aerobic process.

Keywords: *Pichia anomala*; characterization; lipolytic activity; esterase; fermentation; metabolism

---

1. Introduction

The yeast *Pichia anomala* (syn. *Wickerhamomyces anomalus*, former *Hansenula*) has been isolated from very diverse habitats, including plants, animals, soil, water, food and hospitals [1,2]. In food and feed, *P. anomala* has either a beneficial role, as flavor enhancer, biopreservation agent, probiotic, enzyme producer, or a detrimental role, as spoilage yeast. Biopreservation is due to its unusual-for-a-yeast broad spectrum antimicrobial activity against several fungal, yeast, bacterial species and viruses. *P. anomala* produces several metabolites with potential biotechnological exploitation. Examples, apart from the antimicrobial agents that make it an attractive biocontrol organism for applications in the agri-food sector, include bioethanol, isobutanol, enzymes, sophorolipids, γ-aminobutyric acid, several volatile organic compounds, and beverage starter cultures [2]. Regarding enzyme production, *P. anomala* strains have been reported as producers of phytase, amylase, β-glucosidase, peptidase and lipase [3]. *P. anomala* is therefore a biotechnologically interesting organism which deserves in-depth study in view of future commercialization, as several authors have stressed in their reports [4–6].

A preliminary study with yeast isolates from colonies on the surface of traditional Greek sausages revealed a significant number of *Pichia* spp. *Pichia* isolates were tested for lipolytic activity and upon obtaining positive results, further studies were carried out which are reported in the present work. A particular *Pichia* isolate with the highest lipolytic activity was identified as *P. anomala* and designated as M1 strain.

Lipolytic enzyme production is common among yeasts. Only a few yeasts however (mainly *Candida, Yarrowia* and *Geotrichum* spp.) are capable of producing lipases and esterases with interesting characteristics and in sufficient amounts that allow characterization as industrial products [7,8]. Apart from the established producers, there are several promising lipase-producing yeasts as for example, *Issatchenkia orientalis, Kluyveromyces marxianus* and others [9]. To the best of our knowledge, *P. anomala* is reported as a lipolytic enzyme producer in only two cases, the works of Banerjee et al. [10] and Ionita et al. [11], both being screening works. Strains of *P. anomala* were cultivated in solid media and shake flask cultures and moderate lipolytic activity, optimum at pH 6.5–7.5, was detected. In the present work, the *P. anomala* M1 strain was cultivated in stirred tank bioreactors and produced significant lipolytic activity compared to known sources. The characteristics of lipolytic enzyme production were investigated and evaluated.

Lipolytic enzymes share some important characteristics that can be application specific, e.g., substrate specificity, regionspecificity and chiral selectivity [12]. Novel applications have been successfully established using lipases, as for example in the synthesis of biopolymers, flavor compounds, enantiopure pharmaceuticals, agrochemicals and biosensors [12,13]. Innovations need novel biocatalysts and a useful approach for obtaining them is the isolation of microorganisms from natural sources and subsequent screening for the desired enzymatic activity. *P. anomala* exhibits many interesting and potentially exploitable characteristics. It can grow under extreme environmental stress conditions with regard to oxygen, pH, a_W and sugar concentration [3,4,14,15]. As a Crabtree-negative yeast and therefore insensitive to elevated levels of glucose concentrations in its substrate, it can be grown easily in batch fermenters without the need of fed-batch cultivation. These characteristics show an organism of practical significance and therefore a potential industrial microorganism in the case of effective production of a valuable metabolite.

2. Materials and Methods

2.1. Yeast Isolation

Yeast colonies were transferred from the surface of sausages on agar plates containing YP-glucose medium (YP, yeast-peptone; 1% yeast extract, 2% bacteriological peptone, 2% glucose, 2% technical agar). The plates were incubated at 28 °C for 2 days. Selected colonies were sub-cultured on new plates and purified by repeated streaking. The isolated colonies were maintained on YP-glucose agar at 4 °C. The isolated yeast colonies were subjected to standard tests and classification procedures according to Kreger-van Rij [16].

2.2. Molecular Characterization of an Isolated Pichia Anomala Strain

Yeast identification was done by amplification and sequence analysis of the ribosomal DNA internal transcribed space region (ITS) [17,18]. The following primers were used: ITS1 (5′-CGGGATCCGTAGGTGAACCTGCGG-3′) and ITS4 (5′-CGGGATCCTCCGCTTATTGATATGC-3′). The amplification mixture was consisted of: 20 pmol of each primer, 300 ng genomic DNA template, 0.2 mM dNTP, 1.5 mM $MgCl_2$, and 1 U Taq polymerase (final volume of 20 μL). Reactions were run for 40 cycles at 94 °C for 1 min, annealing at 55 °C for 1 min and extension at 72 °C for 2 min, according to Tao et al. [18]. The polymerase chain reaction products were cloned into the plasmid vector pMD18T-Simple (2693 bp) and sequenced. PCR derived sequences were compared with the deposit in the GenBank ITS region (http://www.ncbi.nlm.nih.gov) and the similarity was evaluated using the BLAST program.

2.3. Yeast Cultivation

Cultures were grown in 250 mL Erlenmeyer flasks containing 50 mL of YP-glucose broth, pH 6.5, which contained in addition olive oil (1% *v/v*), tributyrin (1 g/L), or triolein (1 g/L), alternatively. The flasks containing sterile media were inoculated with a loopful of yeast directly from a plated culture. The inoculated flasks were incubated in a shaker incubator at 150 rpm and 28 °C for 48 h. 20 h flask cultures were used to inoculate the bioreactors in volumes giving a final optical density at 600 nm of 0.1 to 0.15.

Bioreactor cultivations were carried out in two stirred tank bioreactors of different capacities: a 2 L working volume (BioFlo, New Brunswick Scientific, NJ, USA) and a 10 L working volume (BIOFLO 410, New Brunswick Scientific, NJ, USA). The agitator of the 2 L reactor had two six-bladed Rushton-type impellers (52 mm). The reactor vessel was equipped with baffles. The agitation system of the 10 L reactor consisted of three disc turbine impellers, 8 cm in diameter, with six flat blades. The reactor vessel was also equipped with baffles. The operating stirrer speed in both reactors was 200 rpm. The airflow rate was adjusted at 1 L of air per liter of reactor working volume per minute (1 vvm). The temperature was maintained at 28 °C, while the pH at 6.5 throughout fermentations by automatic addition of titrants (1 M NaOH and HCl solutions). Dissolved oxygen tension (DOT) was monitored in both reactors during fermentations.

Fermentations in the 2 L bioreactor carried out using the above-described media and lasted 72 h. In the 10 L reactor fermentations, the medium was modified to include 60 g/L glucose and 1 g/L tributyrin and fermentations lasted 96 h. Samples were taken at regular intervals, cells were removed by centrifugation at 14,000 rpm for 10 min, and the supernatants were examined for lipolytic activity. Batch fermentations were repeated at least three times and each sample was analyzed in triplicate.

2.4. Analytical Methods

The reagents used throughout the study were analytical grade by Sigma. Growth was estimated by measuring the optical density at 600 nm and by measuring the cell dry weight (CDW). Broth samples were filtered through pre-weighed glass microfiber filters (Whatman, Maidstone, UK), washed twice and dried in a microwave oven at the defrost setting for 15 min. The filters were placed in a desiccator for the next 24 h and weighed. The OD_{600} was correlated with the CDW and 1 unit OD_{600} equals 0.42 ± 0.3 mg cells/mL.

Lipolytic activity was preliminary evaluated on YP-glucose agar plates to which 1% *v*/*v* olive oil had been added. Formation of a clearer halo around yeast colonies after 5 days of incubation at 28 °C was proof of lipolytic activity. The relative enzyme activity (REA) was calculated by dividing the diameter of the zone of clearance by the diameter of the respective colony (in millimeters). Experiments were carried out in triplicate and mean values are presented.

In liquid culture filtrates, lipolytic activity was assayed using the esters *p*-nitrophenyl palmitate (*p*-NPP) and *p*-nitrophenylacetate (*p*-NPA) as substrates, following the protocol of our earlier work with molds [19], which was based on the method of Winkler and Stuckmann [20]. The reaction mixture was consisted of 405 µL buffer and 45 µL of substrate. The buffer was as: 200 mg Triton X-100, 50 mg gum arabic in 50 mL of 50 mM phosphate buffer, pH 6.5. As substrate, 15 mg *p*-NPP or *p*-NPA in 10 mL 2-propanol was used. The reaction mixture was emulsified in an ultrasonic bath for 5 min. 50 µL of the sample solution containing the lipolytic enzyme was added to the mixture warmed at 40 °C and after a 5 min incubation the reaction was terminated by addition of 0.5 mL of 2% Trizma base. Centrifugation was followed at 12,000 rpm/4 min and the OD of the supernatant was measured at 410 nm. One unit (U) of lipase activity was defined as the amount of enzyme that released 1 µmol of *p*-nitrophenol/min under the specified conditions.

Neutral protease activity was determined in liquid culture filtrates using the casein assay according to Sigma's protocol [21]. Alkaline protease activity was determined using the azocasein assay according to Reichard et al. [22] with the pH of the 0.1 M Tris-HCl buffer adjusted at 9.0 instead of the suggested 8.0. One unit (U) of protease activity was defined as the absorbance variation in the assay conditions of the above methods.

Glucose concentrations were determined with the glucose oxidase/peroxidase method of Kunst et al. [23]. Ethanol, acetate and glycerol concentrations were determined calorimetrically using assay kits by Sigma-Aldrich (St. Louis, MO, USA) (MAK076, MAK086 and MAK117, respectively).

All values are given as mean ± standard deviation in triplicate for each point. Kinetic parameters of growth and production were calculated from raw data using numerical differentiation.

3. Results and Discussion

3.1. Strain Isolation and Identification

The yeast strain *P. anomala* M1 studied in this work was isolated from colonies developed on the surface of traditional Greek sausages. Although the yeast population in the sausages is significant, there are no reports in the literature either on its biodiversity or on its role in the development of product characteristics. A preliminary study with yeast isolates from colonies on the surface of the particular type of sausages revealed a significant number of *Debaryomyces* and *Pichia* spp. (data not shown). Studies on Spanish and Italian relevant products show the dominance of *Debaryomyces* and significant numbers of *Candida, Pichia, Rhodotorula* and *Torulopsis* spp. [24,25]. The yeasts ferment sugars and have proteolytic and lipolytic activity and therefore, produce a large array of metabolites that contribute to the development of flavor and aroma [26]. In addition, some yeasts metabolize the lactic acid produced by lactic acid bacteria and cause an increase in the pH toward neutrality which is translated in the product as milder taste [25].

Colonies of the M1 strain on agar plates containing YP-glucose medium were off-white in color, flat, smooth and opaque. Pseudohyphae were not observed, while 1 to 4 ascospores were detected per ascus (Figure 1). The strain was able to grow in media containing as carbon source glucose, sucrose, galactose, and maltose. It was not able to grow on lactose. It could assimilate soluble starch, raffinose, ethanol, glycerol, D-mannitol, as well as lactic and citric acids, but not L-rhamnose. The yeast was able to grow on substrates containing 1.5 M NaCl (osmotolerant) and in the absence of vitamins, while it produces ethanol under anaerobic conditions (negative Crabtree effect). The yeast was preliminary identified as *Pichia* while possession of traits such as high osmotolerance, fermentation of sucrose, assimilation of raffinose but not of L-rhamnose, puts it under the species classification of *P. anomala* according to Kurtzman [1].

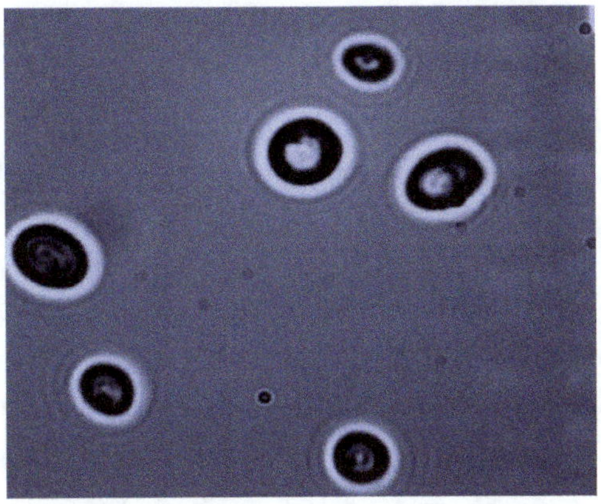

Figure 1. *Pichia anomala* M1 cell morphology (phase contrast microscopy, ×400).

Taxonomic identity of the M1 strain was confirmed through 18s rRNA ITS region cloning and sequencing. Following PCR amplification, a fragment of about 600 bp was obtained (Figure 2). Sequencing results of the ITS region excluding the primers were as: aaggatcattatagtattctattg ccagcgcttaattgcgcggcgataaaccttacacacattgtctagtttttttgaactttgctttgggtggtgagcctg gcttactgcccaaaggtctaaacacatttttaatgttaaacctttaaccaatagtcatgaaattttaacaaaattaaaatcttcaaaactttcaacaa cggatctcttggttctcgcaacgatgaagaacgcagcgaaatgcgatacgtattgtgaattgcagattttcgtgaatccgaatctttgaacgcacattg caccctctggtattccagagggtatgcctgtttgagcgtcatttctctctcaaaccttcgggtttggtattgagtgatactctgtcaagggttaacttgaa atattgacttagcaagagtgtactaataagcagtctttctgaaataatgtattaggttcttccaactcgttatatcagctaggcaggttagaagtattttagg ctcggcttaacaacaataaactaaaagtttgacctcacaggtaggactacccgctgaacttaa (570 bp).

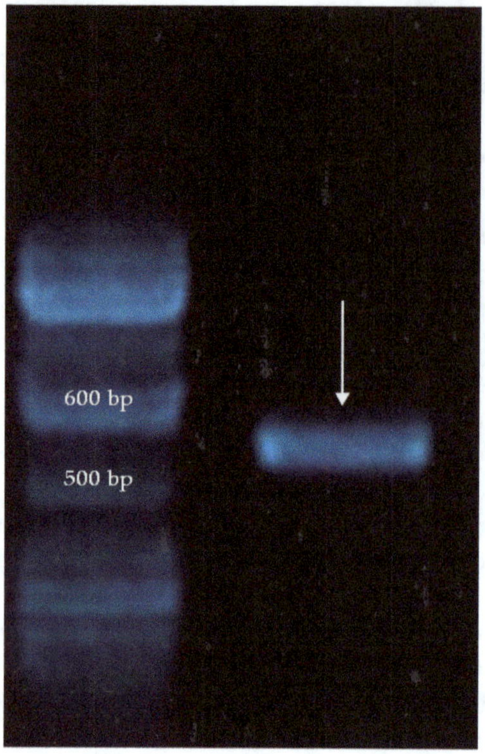

Figure 2. PCR product (pointed) of the internal transcribed spacer (ITS) region of *P. anomala* M1.

A BLAST search for the above sequence showed that it shared almost complete identity (99%) with a number of *Wickerhamomyces anomalus* strains-the name preferred by NCBI for *Pichia anomala*. The sequence was deposited to GenBank under the accession number BankIt2015357 (MF076893).

3.2. Relative Lipolytic Activity in Solid State Cultivation

Research on *P. anomala* growth conditions has shown that it is able to tolerate a relatively wide range of environmental growth conditions, e.g., temperatures ranging from 3 to 37 °C; pH values from 2.0 to 12.0; and osmotic conditions as low as a_w 0.92 for NaCl and 0.85 for glycerol [15]. Tao et al. [18] in their work with *P. anomala* Y1 reported that good growth was obtained in YPD medium with the pH in the range of 3.5–6.5 and temperature 25–30 °C, while optimum growth conditions were reached at 30 °C and pH 5.0. It would be necessary therefore to investigate for lipolytic activity by *P. anomala* M1 over a wide range of temperatures and pHs and this was feasible by evaluating first the relative lipolytic activity in solid media containing olive oil, over a temperature range at pH 6.5, since most of yeast and fungal lipases are reported to be most active and stable in neutral and alkaline pHs [19,27].

Figure 3 shows the relative lipolytic activity of *P. anomala* M1 on YP-glucose agar containing olive oil at various temperatures, ranging from 25 to 40 °C. The pH of the medium was adjusted at 6.5. The highest activity was obtained at 28 °C. The lipolytic activity was lower at 30 °C but significantly higher compared to values obtained at 25 and 35 °C. The relative lipolytic activity at 40 °C was close to zero. An experiment was also carried out keeping the temperature at 28 °C and changing the pH of the medium in a range from 3.0 to 7.5. As it is shown in Figure 4, the highest lipolytic activity was obtained at pH 6.5. *P. anomala* M1 was able to produce lipolytic activity within the pH range 5.0–7.0 but the relative lipolytic activity values obtained at pHs 6.5 and 6.0 were significantly higher. Lipolytic activity was not detected at pHs 3.0, 3.5, and 7.0, while it was marginally detected at pHs 4.0 and 4.5. Considering both Figures 3 and 4, it can be concluded that the optimum conditions of temperature and pH for lipolytic enzyme production by *P. anomala* M1 are 28 °C and pH 6.5, and these conditions were applied in submerged fermentations.

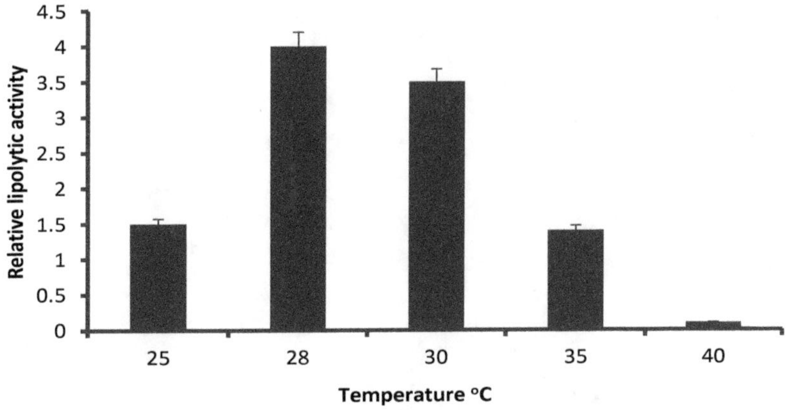

Figure 3. The effect of temperature on the relative lipolytic activity of *P. anomala* M1 grown on YP-glucose agar supplemented with olive oil at pH 6.5. Experiments were carried out in triplicate and the Y-error bars represent the standard deviation of the mean value.

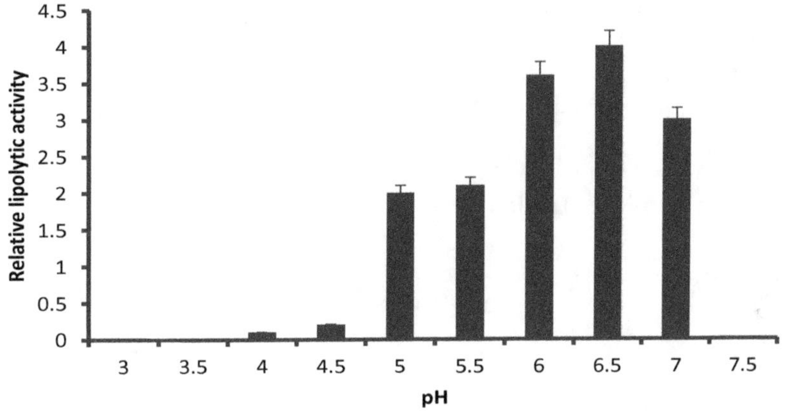

Figure 4. The effect of pH on the relative lipolytic activity of *P. anomala* M1 grown on YP-glucose agar supplemented with olive oil at 28 °C. Experiments were carried out in triplicate and the Y-error bars represent the standard deviation of the mean value.

3.3. Production of Extracellular Lipase in Submerged Fermentation

Lipases can be distinguished from carboxyl esterases by their substrate spectra, using *p*-NPP (cleaved by lipases) versus *p*-NPA (cleaved by esterases). The lipolytic activity assay produced results only on the *p*-NPA substrate and therefore, the activity is due to an esterase enzyme.

Figures 5 and 6 show the time-courses of biomass and lipolytic activity of *P. anomala* grown in shaken flasks with olive oil, triburyrin (short chain triglyceride, $C_{15}H_{26}O_6$) or triolein (long chain triglyceride $C_{57}H_{104}O_6$) added in the medium as enzyme inducer. There is an obvious enhancement of both growth but mainly lipolytic activity in the case of tributyrin. Comparing results, it appears that final lipolytic activity with tributyrin was 1.91 times higher than the obtained with olive oil while 2.43 times higher than the obtained with triolein. The effect on biomass was not as sound -although results of the same trend were obtained. Biomass yield with tributyrin was 1.16 times higher compared to the obtained with olive oil and 1.47 times higher than that of triolein. In all cases, lipolytic activity increased with increasing biomass and resumed in about 40 h when cultures were well into the stationary phase of growth. Although in shaken flasks cultivation important process parameters (e.g., the air supply or the culture pH) were not under control, the effect of the inclusion of tributyrin in the medium was clear.

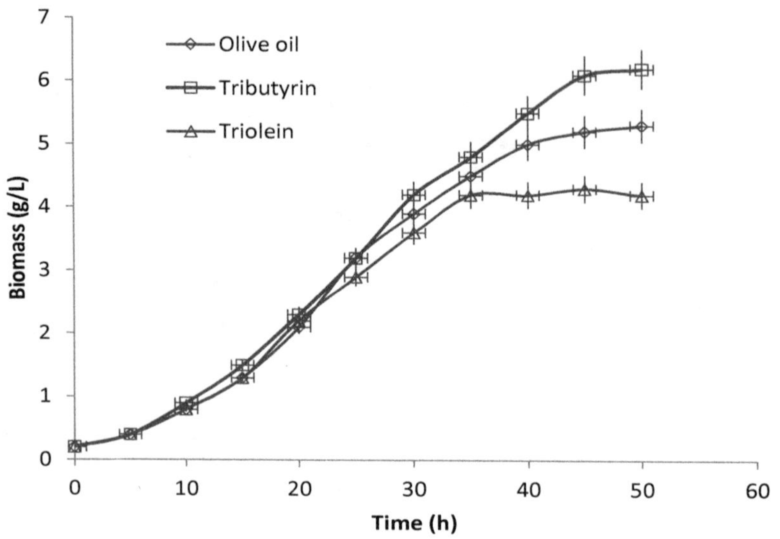

Figure 5. Time courses of biomass concentration of *P. anomala* M1 in shaken flasks cultures (200 rpm). The YP-glucose medium was supplemented alternatively with olive oil, tributyrin and triolein.

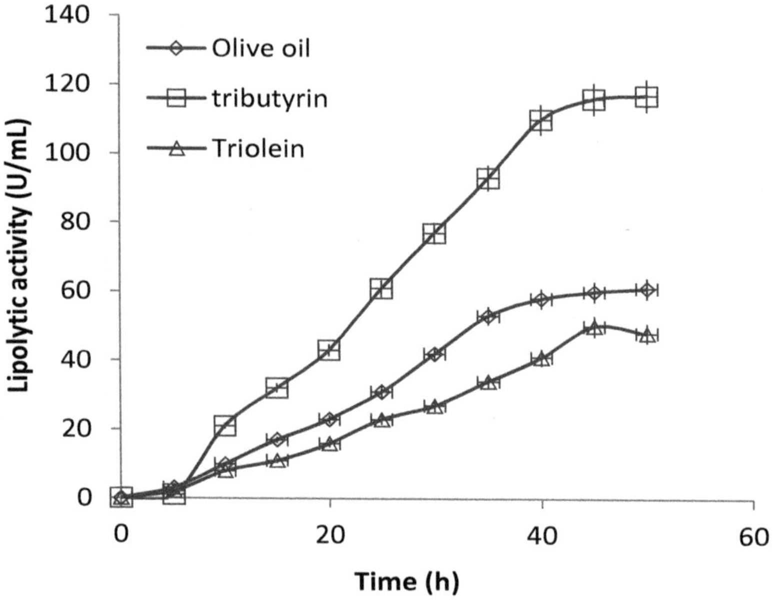

Figure 6. Time courses of extracellular lipolytic activity of *P. anomala* M1 in shaken flasks cultures (200 rpm). The YP-glucose medium was supplemented alternatively with olive oil, tributyrin and triolein.

The profile of extracellular lipolytic activity of *P. anomala* M1 was next evaluated in submerged bioreactor cultivations of two different scales under the conditions described in the previous section, using tributyrin as inducer. Concerning oxygen requirements of *P. anomala*, it is known that it is able to grow in both oxygen-replete and oxygen-limited conditions (facultative yeast). However, under aerobic conditions growth rates and biomass yields are significantly higher than those obtained under anaerobic conditions [14]. Since extracellular lipases are products of the primary metabolism with growth-associated characteristics, fermentations were carried out under aerobic conditions to ensure increased growth rates.

Figure 7 shows the time-courses of biomass, lipolytic activity and residual glucose concentrations in a run in the 2 L stirred tank bioreactor. *P. anomala* grew exponentially with a maximum specific growth rate μ_{max} of 0.17 h^{-1} and a biomass yield $Y_{x/s}$ of 0.43 g of CDW g of glucose^{-1} (Table 1).

Only low amounts of ethanol, glycerol and acetate were detected (results at 50 h, Table 1). Dissolved oxygen tension decreased during growth and the lowest levels monitored were 70% of saturation, which were high enough to consider oxygen limitation (Figure 7). Maximum biomass concentration (8.6 g/L) was obtained around 55 h while maximum lipolytic activity (159 U/L) around 50 h of fermentation. With depletion of glucose (50 h samples), biomass started decreasing and a sharp fall in lipolytic activity was observed. Running the process for another 24 h after glucose depletion resulted in a loss of lipolytic activity of the order of 19.2%. This was observed in several cases of lipase production in fermentation [27] and was attributed mainly to proteolytic degradation. Protease activity was determined in the present case and as shown in Figure 7, proteases appeared in the broth after 30 h of fermentation (1.2 U/mL) and proteolytic activity was increasing towards the end of the process to reach 11 U/mL. Ethanol, glycerol and acetate were not detected beyond 65 h, indicating their consumption in the absence of glucose.

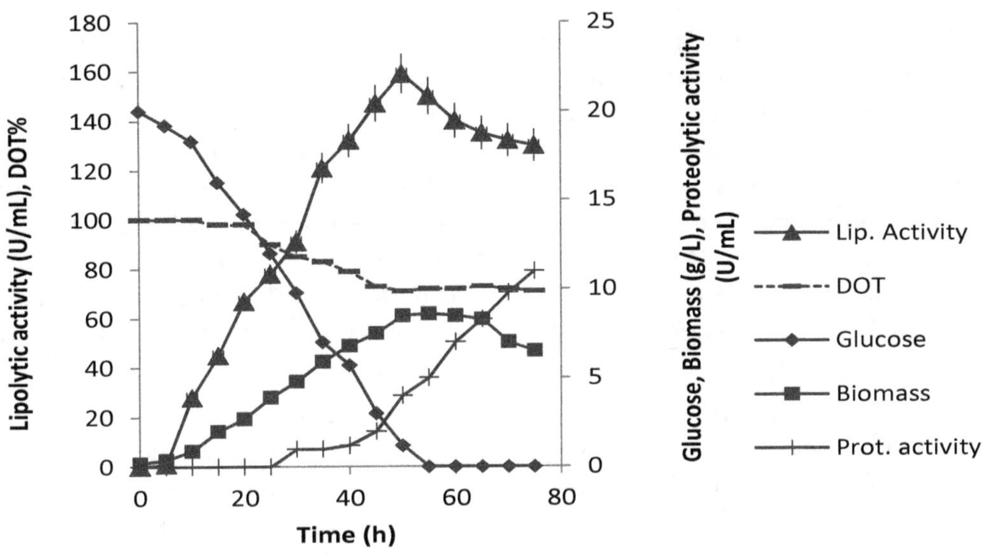

Figure 7. *P. anomala* fermentation in a 2 L working volume stirred tank reactor (200 rpm, 28 °C, pH 6.5, 1 vvm aeration). The YP-20 g/L glucose medium was supplemented with tributyrin.

Table 1. Kinetic parameters of *P. anomala* M1 cultures. Values are averages ± standard deviations based on the results of three independent cultivations.

Parameter	2 L Bioreactor 20 g/L Glucose	10 L Bioreactor 60 g/L Glucose
Maximum Specific Growth Rate (h^{-1})	0.17 ± 0.02	0.12 ± 0.03
Maximum Specific Consumption Rate of Glucose (mmol/g/h)	1.9 ± 0.2	1.3 ± 0.3
Specific Production Rate of		
Lipolytic activity (U/g/h)	7 ± 0.8	51 ± 1.3
Ethanol (mmol/g/h)	0.12 ± 0.02	0.09 ± 0.03
Acetate (mmol/g/h)	0.14 ± 0.04	0.11 ± 0.01
Glycerol (mmol/g/h)	0.11 ± 0.03	0.10 ± 0.02
Yield on Glucose (g/g)		
Biomass	0.43 ± 0.14	0.24 ± 0.02
Ethanol	0.03 ± 0.002	0.05 ± 0.01
Acetate	0.05 ± 0.01	0.04 ± 0.02
Glycerol	0.02 ± 0.009	0.02 ± 0.001

Figure 7 shows the *P. anomala* M1 fermentation process in the 10 L bioreactor with 60 g/L initial glucose concentration. Both biomass production and lipolytic activity were higher compared to those obtained in the 2 L fermentation. The maximum values obtained, were 14.9 g/L biomass and 213 U/mL lipolytic activity (75 h), which is 1.57 times higher biomass, and 1.74 times higher lipolytic activity compared to those obtained from the 2 L reactor fermentation. Comparing the kinetic parameters for growth in the two reactors it appears that the values for maximum specific growth rateand the yield of biomass on glucose were lower in the 10 L fermentation: 0.12 compared to 0.17 h^{-1} and 0.24 compared to 0.43 g of CDW g of glucose $^{-1}$ g, respectively (Table 1). The lower biomass levels however produced the lipolytic enzyme (in terms of activity) at a higher rate: 0.48 U/mL/h compared to 0.06 U/mL/h and in terms of specific production rate, 51 U/g/h biomass in the 10 L process compared to 7 U/g/h biomass in the 2 L process (timing: 50 h). Glucose concentration fell to zero in 72 h and the following 24 h of the process were characterized by declining biomass and enzyme activity levels. The loss of lipolytic activity was 20.1% in the case of the 10 L fermentation. Only low amounts of ethanol, glycerol and acetate were detected in this case again (results at 50 h, Table 1). Dissolved oxygen tension decreased during fermentation but higher values were monitored compared to those of the 2 L bioreactor fermentation: about 85% of saturation at maximum biomass levels.

Since the lipolytic activity is highly dependent on the experimental conditions and the applied methodology, direct comparisons of the results with those reported for various microbial lipases become rather difficult. However, the obtained values for kinetic parameters and product titers are either comparable or even higher. Kar and co-workers [28] obtained maximum specific rates of extracellular lipase production by *Y. lipolytica* at around 50 h of fermentation as in the present study, the values however were expressed per million of cells and not per g CDW and cannot be compared since it is a different yeast species. With fungal lipases, the production period is longer as for example with *Mucor hiemalis* lipase whose maximum activity was obtained at 192 h of fermentation and reached 110 U/mL [29], or a *Penicillium nalgiovense* lipase with a maximum activity of 82 U/mL at 168 h of fermentation [19].

The applied culture conditions in the 10 L bioreactor (Figure 8) resulted in a significant increase in the specific rate of lipolytic activity while the specific rates of growth and glucose consumption were decreased compared to those of the 2 L bioreactor. The main differences between the two systems were the concentrations of glucose and the dissolved oxygen tension. *P. anomala* is a respiratory yeast that lacks a Crabtree effect and therefore, there is no induction of alcoholic fermentation by high glucose concentrations. In addition, similarly to the other respiratory yeast *K. lactis*, *P. anomala* exhibits minor fermentation despite the elevated activities of the key fermentative enzymes PDC and ADH [14,30]. This metabolic profile was apparent in the present study. Although the same stirrer speed (200 rpm) and air supply rate (1 vvm) were applied in the two reactor systems, the dissolved oxygen tension was maintained at higher levels in the 10 L bioreactor despite the increased biomass concentration. This can be explained by a more efficient mixing regime in the 10 L reactor that facilitated a smaller mixing time of the broth, however, mixing studies were not undertaken at that stage to quantify the effects. The decrease in DOT in the 2 L bioreactor may have acted as the inducing stimulus for the activation of fermentative metabolism and the increased glucose consumption rate points to an ongoing Pasteur effect. Glucose availability however, did not induce ethanol formation.

Intensive studies of *P. anomala* metabolism and physiology in response to oxygen and glucose availability carried out by Fredlund and co-workers (2004) [14] showed that with the particular yeast, the signals that induce respiratory and fermentative metabolism are completely different from those known for *Saccharomyces cerevisiae*. Respiratory growth was obtained from aerobic batch cultures with a high glucose concentration and fermentation could be induced by shifting the cultures to oxygen limitation (shift from 50% to 0% DOT). Oxygen therefore is the main regulator of the central energy metabolism while glucose does not repress respiration but possibly repress the ethanol assimilation pathway. Having this metabolic and physiological background, lipolytic enzyme production by *P. anomala* appears to be advantageous compared to other sources as it can take place under oxygen-saturated conditions using substrates with elevated glucose concentrations.

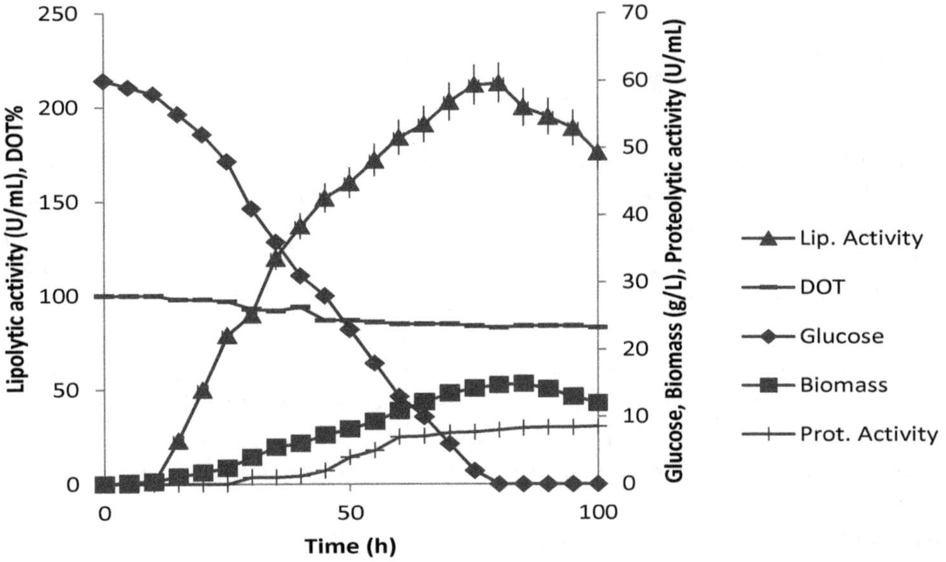

Figure 8. *P. anomala* fermentation in a 10 L working volume stirred tank reactor (200 rpm, 28 °C, pH 6.5, 1 vvm aeration). The YP-60 g/L glucose medium was supplemented with tributyrin.

From the applications point of view, the non-conventional yeast *P. anomala* has been extensively studied mainly as a biopreservation agent for various post-harvest systems. Although it was reported in the past as lipase-positive yeast, its production potential in fermentation systems was not evaluated. The current study shows that a wild strain isolated from a rich in fat meat product can produce significant lipolytic activity in a scalable aerobic process. Further studies will focus on the characteristics and properties of the produced esterase as well as in the process itself. Factors for example, identified in this study as controlling process parameters, e.g., the DOT, should be studied in a scaling-up perspective.

4. Conclusions

This study shows that a wild strain of *P. anomala* is capable of producing extracellular lipolytic (esterase) activity in a process with practical significance. Experiments carried out in agar plates, shaken flasks, a 2 L stirred tank bioreactor and a 10 L stirred tank bioreactor showed that by setting the appropriate conditions the physiological traits of the particular yeast can be exploited for effective enzyme production. The Crabtree-negative yeast *P. anomala* exhibits a Pasteur effect at lower dissolved oxygen concentrations but elevated glucose concentrations are not inhibitory and depress ethanol formation in an oxygen-saturated substrate. The main stimulus for the regulation of central carbon metabolism as it appears from its effects on the specific rates of growth and enzyme production is oxygen and, therefore, setting an aerobic process with a high glucose concentration in the substrate advances enzyme production.

Author Contributions: Both authors contributed equally in this study.

References

1. Kurtzman, C.P.; Fell, J.W. (Eds.) *The Yeasts, a Taxonomic Study*; Elsevier: Amsterdam, The Netherlands, 1998.
2. Passoth, V.; Fredlund, E.; Druvefors, U.Ä.; Schnürer, J. Biotechnology, physiology and genetics of the yeast *Pichia anomala*. *FEMS Yeast Res.* **2006**, *6*, 3–13. [CrossRef] [PubMed]
3. Walker, G.M. *Pichia anomala*: Cell physiology and biotechnology relative to other yeasts. *Antonie van Leeuwenhoek* **2011**, *99*, 25–34. [CrossRef] [PubMed]

4. Fletcher, E.; Feizi, A.; Kim, S.; Siewers, V.; Nielsen, J. RNA-Seq analysis of *Pichia anomala* reveals important mechanisms required for survival at low pH. *Microb. Cell Fact.* **2015**, *14*, 143. [CrossRef] [PubMed]
5. Passoth, V.; Olstorpe, M.; Schnürer, J. Past, present and future research directions with *Pichia anomala*. *Antonie Van Leeuwenhoek* **2011**, *99*, 121–125. [CrossRef] [PubMed]
6. Schneider, J.; Rupp, O.; Trost, E.; Jaenicke, S.; Passoth, V.; Goesmann, A.; Tauch, A.; Brinkrolf, K. Genome sequence of *Wickerhamomyces anomalus* DSM 6766 reveals genetic basis of biotechnologically important antimicrobial activities. *FEMS Yeast Res.* **2012**, *12*, 382–386. [CrossRef] [PubMed]
7. Goncalves, F.A.G.; Colen, G.; Takahasi, J.A. *Yarrowia lipolytica* and its multiple applications in the biotechnological industry. *Sci. World J.* **2014**. [CrossRef]
8. Satyanarayana, T.; Kunze, G. *Yeast Biotechnology: Diversity and Applications*; Springer: Berlin, Germany, 2009.
9. Stergiou, P.Y.; Foukis, A.; Sklivaniti, H.; Zacharaki, P.; Papagianni, M.; Papamichael, E.M. Experimental investigation and optimization of process variables affecting the production of extracellular lipase by *Kluyveromyces marxianus* IFO 0288. *Appl. Biochem. Biotechnol.* **2012**, *168*, 672–680. [CrossRef] [PubMed]
10. Banerjee, S.; Sengupta, I.; Majumdar, S.K. Lipase Production by *Hansenula anomala var.Schnegii*. *J. Food Sci. Technol.* **1985**, *22*, 137–139.
11. Ionita, A.; Moscovici, M.; Popa, C.; Vamanu, A.; Popa, O.; Dinu, L. Screening of yeast and fungal strains for lipolytic potential and determination of some biochemical properties of microbial lipases. *J. Mol. Catal. B. Enzym.* **1997**, *3*, 147–151. [CrossRef]
12. De Miranda, A.S.; Miranda, L.S.M.; de Suza, R.O.M.A. Lipases: Valuable catalysts for dynamic kinetic resolutions. *Biotechnol. Adv.* **2015**, *33*, 372–393. [CrossRef] [PubMed]
13. Faber, K. *Biotransformations in Organic Chemistry*, 6th ed.; Springer: Berlin, Germany, 2011.
14. Fredlund, E.; Blank, L.M.; Schnürer, J.; Sauer, U.; Passoth, V. Oxygen- and glucose-dependent regulation of central carbon metabolism in *Pichia anomala*. *Appl. Environ. Microbiol.* **2004**, *70*, 5905–5911. [CrossRef] [PubMed]
15. Fredlund, E.; Druvefors, U.; Ligsten, K.J.; Boysen, M.E.; Schnürer, J. Physiological characteristics of the biocontrol yeast *Pichia anomala* J121. *FEMS Yeast Res.* **2002**, *2*, 395–402. [PubMed]
16. Kreger-van Rij, N.J.W. *The Yeasts: A Taxonomic Study*, 3rd ed.; Elsevier: Amsterdam, The Netherlands, 1984.
17. Caggia, C.; Restuccia, C.; Pulvirenti, A.; Giudici, P. Identification of *Pichia anomala* isolated from yoghurt by RFLP of the ITS region. *Int. J. Food Microbiol.* **2001**, *71*, 71–73. [CrossRef]
18. Tao, N.; Gao, Y.; Liu, X. Isolation and characterization of a *Pichia anomala* strain: A promising candidate for bioethanol production. *Braz. J. Microbiol.* **2011**, *42*, 668–675. [CrossRef] [PubMed]
19. Papagianni, M. An evaluation of the proteolytic and lipolytic potential of *Penicillium* spp. isolated from traditional Greek sausages in submerged fermentation. *Appl. Biochem. Biotechnol.* **2014**, *172*, 767–775. [CrossRef] [PubMed]
20. Winkler, U.K.; Stuckmann, M. Glycogen, hyaluronate, and some other polysaccharides greatly enhance the formation of exolipase by *Serratia marcescens*. *J. Bacteriol.* **1979**, *138*, 663–670. [PubMed]
21. Sigma-Aldrich. Universal Protease Activity Assay: Casein as a Substrate. 2013. Available online: http://www.sigmaaldrich.com/life-science/learning-center/life-science-video/universal-protease.html. (accessed on 1 July 2017).
22. Reichard, U.; Buttner, S.; Eifferst, H.; Staib, F.; Ruchel, R. Purification and characterisation of an extracellular serine proteinase from *Aspergillus fumigatus* and its detection in tissue. *J. Med. Microbiol.* **1990**, *33*, 243–251. [CrossRef] [PubMed]
23. Kunst, A.; Draeger, B.; Ziegenhom, J. Colorimetric methods with glucose oxidase. *Methods Enzym. Anal.* **1986**, *6*, 178–185.
24. Gardini, F.; Suzzi, G.; Lombardi, A.; Galgano, F.; Crudele, M.A.; Andrighetto, C.; Schirone, M.; Totalo, R. A survey of yeast in traditional sausages of southern Italy. *FEMS Yeast Res.* **2001**, *1*, 161–167. [CrossRef] [PubMed]
25. Mendoça, R.C.S.; Gouvêa, D.M.; Hungaro, H.M.; Sodré, A.F.; Querol-Simon, A. Dynamics of the yeast flora in artisanal country style and industrial dry cured sausage (yeast in fermented sausage). *Food Control* **2013**, *29*, 143–148. [CrossRef]
26. Durá, M.A.; Flores, M.; Toldrá, F. Effect of growth phase and dry-cured sausage processing conditions on *Debaryomyces* spp. generation of volatile compounds from branched-chain amino acids. *Food Chem.* **2004**, *86*, 391–399. [CrossRef]

27. Bussamara, R.; Fuentefria, A.M.; de Oliveira, E.S.; Broetto, L.; Simcikowa, M.; Valente, P.; Schrank, A.; Vainstein Henning, M. Isolation of a lipase-secreting yeast for enzyme production in a pilot-plant scale batch fermentation. *Biores. Technol.* **2010**, *101*, 268–275. [CrossRef] [PubMed]
28. Kar, T.; Delvigne, F.; Masson, M.; Destain, J.; Thonart, P. Investigation of the effect of different extracellular factors on the lipase production by *Yarrowia lipolytica* on the basis of a scale-down approach. *J. Ind. Microbiol. Biotechnol.* **2008**, *35*, 1053–1059. [CrossRef] [PubMed]
29. Hiol, A.; Jonzo, M.D.; Bruet, D.; Comeu, L. Production, purification and characterization of an extracellular lipase from *Mucor hiemalis f. hiemalis*. *Enzyme Microb. Technol.* **1999**, *25*, 80–87. [CrossRef]
30. Kiers, J.; Zeeman, A.M.; Luttic, C.; Thiele, C.; Castrillo, J.I.; Steensma, H.Y.; van Dijken, J.P.; Pronk, J.T. Regulation of alcoholic fermentation in batch and chemostat cultures of *Kluyveromyces lactis* CBS 2359. *Yeast* **1998**, *14*, 459–469. [CrossRef]

6

Application of Non-*Saccharomyces* Yeasts to Wine-Making Process

José Juan Mateo * and Sergi Maicas

Departament de Microbiologia i Ecologia, Universitat de València, E-46100-Burjassot, Spain; Sergi.Maicas@uv.es
* Correspondence: Jose.J.Mateo@uv.es

Academic Editor: Ronnie G. Willaert

Abstract: Winemaking is a complex process involving the interaction of different microbes. The two main groups of microorganisms involved are yeasts and bacteria. Non-*Saccharomyces* yeasts are present on the grape surface and also on the cellar. Although these yeasts can produce spoilage, these microorganisms could also possess many interesting technological properties which could be exploited in food processing. It has been shown that some of the metabolites that these yeasts produce may be beneficial and contribute to the complexity of the wine and secrete enzymes providing interesting wine organoleptic characteristics. On the other hand, non-*Saccharomyces* yeasts are the key to obtain wines with reduced ethanol content. Among secreted enzymes, β-glucosidase activity is involved in the release of terpenes to wine, thus contributing to varietal aroma while β-xylosidase enzyme is also interesting in industry due to its involvement in the degradation of hemicellulose by hydrolyzing its main heteroglycan (xylan).

Keywords: non-*Saccharomyces* yeasts; wine; flavor; β-glucosidase; β-xylosidase

1. Introduction

Since Louis Pasteur elucidated the conversion of grape juice into wine, this process and the role of the yeast therein has been studied extensively [1]. More than 130 years later, there are many areas that are still not well understood [2]. This is especially the case for the roles of the numerous non-*Saccharomyces* yeasts normally associated with grape must and wine. These yeasts, present in all wine fermentations, are metabolically active and their metabolites can impact on wine quality. In the past, the influence of non-*Saccharomyces* yeasts in wine was restricted and even eliminated by inoculation with pure *S. cerevisiae* cultures because they have been regarded as spoilage yeasts [3,4]. However, in the past three decades, great interest has grown in the beneficial role of non-*Saccharomyces* yeasts in wine biotechnology [5,6]. It has been shown that some of the metabolites that these yeasts produce may be beneficial and contribute to the complexity of the wine when they are used in mixed fermentations with *S. cerevisiae* cultures [7,8]. Evidence supporting this fact has been published [9] and the role of the non-*Saccharomyces* yeasts in wine fermentation is receiving increasingly more attention by wine microbiologists in wine-producing countries [10].

Non-*Saccharomyces* yeasts are found on the grapes, but also in lesser numbers on the cellar equipment [11]. The initial environment that affects the microbial makeup of a wine fermentation is that of the vineyard. Although a drastically different environment than juice or wine, the types of microbes present on grapes will have an impact on the ensuing ecology in the wine fermentation, particularly in the early stages. Microorganisms appear to colonize around the grape stomata where small amounts of exudate are secreted [12,13]. The apiculate yeasts, *Hanseniaspora* and *Kloeckera*, its asexual anamorph, are the most prevalent vineyard yeasts and typically represent over half the yeast flora on grapes [14]. Other yeast genera present on berries include: *Metschnikowia*, *Candida*,

Pichia, Wickerhamomyces, Zygosaccharomyces, and *Torulaspora* [15]. Also present in the vineyard are numerous other yeasts, some of which have an impact on wine: *Sporidiobolus, Kluyveromyces,* and *Hansenula* [16]. *Saccharomyces* species are relatively scarce among healthy berries (Table 1) [17,18]. Before inoculation with *S. cerevisiae,* they are the yeasts present in the highest numbers in the grape must. During the fermentation there is a sequence of dominance by the various non-*Saccharomyces* yeasts, followed by *S. cerevisiae,* which then completes the fermentation [19]. This is especially evident in spontaneously fermenting grape must, which has a low initial *S. cerevisiae* concentration. Research has shown that non-*Saccharomyces* yeast strains can be detected throughout wine fermentation [20] and their dominance during the early part of fermentation can leave an imprint on the final composition of the wine [21].

Table 1. Main non-*Saccharomyces* yeasts isolated from grape musts and wines.

Aureobasidium pullulans	*Hansenula* sp
Brettanomyces sp	*Issatchenkia terricola*
B. anomalus	*Kluyveromyces thermotolerans*
Candida guilliermondii	*Lachancea thermotolerans*
C. molischiana	*Metschnikowia pulcherrima/C. pulcherrima*
C. stellata	*Pichia angusta*
C. utilis	*P. anomala*
C. zemplinina	*P. capsulata*
Debaryomyces castellii	*P. guilliermondii*
D.hansenii	*P. kluyvery*
D.polymorphus	*P. membranifaciens*
D.pseudopolymorphus	*Saccharomycodes ludwigii*
D. vanriji	*Schizosaccharomyces pombe*
Hanseniaspora sp. (*Kloeckera*)	*Sporidiobolus pararoseus*
H. guilliermondii	*Torulaspora delbrueckii*
H. osmophila	*Trichosporon asahii*
H. vineae	*Wickerhamomyces anomalus*
H. uvarum	*Zygosaccharomyces bailii*

In the past, the influence of non-*Saccharomyces* yeasts in wine was restricted and even eliminated by inoculation with pure *S. cerevisiae* cultures because they have been regarded as spoilage yeasts [17]. However, in the past three decades, great interest has grown in the beneficial role of non-*Saccharomyces* yeasts in wine biotechnology [18,19]. It has been shown that some of the metabolites that these yeasts produce may be beneficial and contribute to the complexity of the wine when they are used in mixed fermentations with *S. cerevisiae* cultures [20,21].

It is believed that when pure non-*Saccharomyces* yeasts are cultivated with *S. cerevisiae* strains, their negative metabolic activities may not be expressed or could be modified by the metabolic activities of the *S. cerevisiae* strains [22]. Several strains belonging to different non-*Saccharomyces* species have been extensively studied in relation to the formation of some metabolic compounds affecting the bouquet of the final product. Diverse studies on the growth and metabolic interactions between non-*Saccharomyces* and *Saccharomyces* yeasts in mixed cultures have shown that their impact on ethanol content, wine flavor, aromatic profile, and quality and control of spoilage yeasts depends on the strains and the inoculation strategies [23,24]. In addition, a great number of studies inform about enzyme activities in winemaking and fermentations [25,26]. However, there are no known reports that associate the production of enzymatic activities in mixed cultures of *Saccharomyces* and non-*Saccharomyces* during the fermentation with the final aromatic profile of wines.

2. Contribution of Non-*Saccharomyces* Yeast Reduction in the Ethanol Content of Wines

Consumer and market demand for wines containing lower ethanol has shaped research to develop and evaluate strategies to generate low-ethanol wines [27]. Numerous studies have reported lower ethanol yields when using non-*Saccharomyces* yeast [23,28]. Another alternative is to exploit the oxidative metabolism observed in some non-*Saccharomyces* species [29]. Nevertheless, only one study has reported the use of aerobic yeast for the production of reduced alcohol wine. Wines containing 3% *v/v* ethanol were obtained after fermentation of grape must by *Williopsis saturnus* and *Pichia subpelliculosa* under intensive aerobic conditions. These reduced alcohol wines were considered to be of an adequate quality [30].

Microbiological approaches for decreasing ethanol concentrations take advantage of the differences in energy metabolism among the wine yeast species. Several strategies that use genetically-modified yeasts have been proposed for the production of low-alcohol wine. Recently, Tilloy and co-workers [31] using evolution-based strategies together with breeding strategy showed that evolved or hybrid strains produced an ethanol reduction of 0.6%–1.3% (v/v). Another approach to reduce the production of ethanol could be the use of non-*Saccharomyces* wine yeasts, in combination with *S. cerevisiae*, to improve the quality and enhance the complexity of wine. Following numerous studies on the influence of non-*Saccharomyces* yeast in winemaking, there has been a re-evaluation of the role of these yeasts. Indeed, some non-*Saccharomyces* yeast can enhance the profile of the wine, and for this reason the use of controlled multi-starter fermentation using selected cultures of non-*Saccharomyces* and *S. cerevisiae* yeast strains has been encouraged [32]. Indeed, nowadays one of the most recent technological advances in winemaking is the practice of co-inoculation of grape juice with selected culture of a non-*Saccharomyces* coupled with a *S. cerevisiae* starter strain [25]. In this context, non-*Saccharomyces* wine yeasts in multi-starter fermentations could be an interesting way to reduce the ethanol content in wine. In addition, different respiro-fermentative regulatory mechanisms of some non-*Saccharomyces* yeasts compared to *S. cerevisiae* could be a modality to reduce the ethanol production through partial and controlled aeration of the grape juice. Indeed, in this way sugar is consumed via respiration rather than fermentation. Both of these approaches have indicated the promising use of non-*Saccharomyces* wine yeast to limit ethanol production [33,34].

3. Contribution to Wine by Non-*Saccharomyces* Yeast

According to Fleet [35], yeast influences wine aroma by different mechanisms; of these, novo biosynthesis of aroma compounds is probably the most important [36]. The variety of odor compounds produced by non-*Saccharomyces* yeasts is known [37]. The contribution of non-*Saccharomyces* yeasts to wine quality can take various forms. Production of glycerol by *Candida stellata* and esters by *C. pulcherrima* has been reported [15]. Other non-*Saccharomyces* yeasts are also widely recognized for producing glucosidase enzymes [38], which, by hydrolyzing such bonds, are capable of releasing volatile compounds linked to sugars, giving greater complexity to the wine's aromatic profile [39]. Conversely, others such as *Kloeckera apiculata* are associated with the production of acetic acid, which lowers wine quality [40]. Therefore, to determine the potential of non-*Saccharomyces* yeasts to be used in the wine industry, it is necessary to check that their activity in mixed culture does not affect the development of *Saccharomyces*, or produce compounds that may harm wine quality. These metabolic products include terpenoids, esters, higher alcohols, glycerol, acetaldehyde, acetic acid, and succinic acid [41].

Therefore, sensory differences were found [42]. Over 160 esters have been distinguished in wine [43]. These esters can have a helpful effect on wine quality, especially in wine from varieties with neutral flavors [44]. Non-*Saccharomyces* can be divided into two groups, neutral yeasts (producing little or no flavor compounds) and flavor-producing species. Flavor-producing yeasts included *P. anomala* (*Hansenula anomala*) and *K. apiculata*. *Candida pulcherrima* is also known to be a high producer of esters [41]. The accumulation of esters in wine is determined by the balance between the yeast's ester-synthesizing enzymes and esterases (responsible for cleavage and in some cases, formation of ester bonds) [45]. Although extracellular esterases are known to occur in *S. cerevisiae* [46], the situation for non-*Saccharomyces* needs further investigation.

Different non-*Saccharomyces* yeasts produce different levels of higher alcohols (n-propanol, isobutanol, isoamyl alcohol, and active amyl alcohol) [44]. This is important during wine production, as high concentrations of higher alcohols are generally not desired, whereas lower values can add to wine complexity.

Glycerol, the next major yeast metabolite produced during wine fermentation after ethanol, is important in yeast metabolism for regulating redox potential in the cell [47]. Glycerol contributes to smoothness, sweetness, and complexity in wines, but the grape variety and wine style will govern the extent to which glycerol impacts on these properties [48]. Several non-*Saccharomyces* yeasts, particularly *C. zemplinina*, can consistently produce high glycerol concentrations during wine fermentation [25]. Unfortunately, increased glycerol production is usually linked to increased acetic acid production [49], which can be detrimental to wine quality. Spontaneously-fermented wines have higher glycerol levels, indicating a possible contribution by non-*Saccharomyces* yeasts [14].

Other compounds that are known to play a role in the sensory quality of wine include volatile fatty acids, carbonyl, and sulfur compounds [44]. Volatile thiols greatly contribute to the varietal character of some grape varieties, particularly Sauvignon Blanc [50]. Some non-*Saccharomyces* strains, specifically isolates from *C. zemplinina* and *P. kluyveri*, can produce significant amounts of the volatile thiols 3-mercaptohexan-1-ol (3MH) and 3-mercaptohexan-1-ol acetate (3MHA), respectively, in Sauvignon Blanc wines [51]. Similarly, *T. delbrueckii*, *M. pulcherrima*, and *Lachancea thermotolerans* have also been described as able to release important quantities of 3MH from its precursor during Sauvignon Blanc fermentation [52].

However, the use of some non-*Saccharomyces* yeast in mixed fermentations with *S. cerevisiae* can generate wines with increased volatile acidity and acetic acid concentration [25]. Some non-*Saccharomyces* yeasts are able to form succinic acid [48]. This correlates with high ethanol production and ethanol tolerance. Succinic acid production could positively influence the analytical profile of wines by contributing to the total acidity in wines with insufficient acidity. Nevertheless, succinic acid has a "salt-bitter-acid" taste and excessive levels will negatively influence wine quality. Other non-*Saccharomyces* metabolites can act as intermediaries in aroma metabolic pathways. Acetoin is considered a relatively odorless compound in wine [53]. However, diacetyl and 2,3-butanediol (potentially off-flavors in wine) can be derived from acetoin by chemical oxidation and yeast-mediated reduction, respectively. This indicates that acetoin can play a role in off-flavor formation in wines. Definitely, high concentrations of acetoin produced by non-*Saccharomyces* yeasts can be utilized by *S. cerevisiae* in mixed and sequential culture fermentations [54].

Non-*Saccharomyces* yeasts have also been reported to affect the concentration of polysaccharides in wine [55]. Polysaccharides can positively influence wine taste and mouth-feel by increasing the perception of wine "viscosity" and "fullness" on the palate [56]. The early death of some non-*Saccharomyces* yeasts during fermentation can also be a source of specific nutrients for *S. cerevisiae* enabling it to ferment optimally. These nutrients include cellular constituents such as cell wall polysaccharides (mannoproteins). For this method of nutrient supply to be effective, any killer or other inhibitory effects by the non-*Saccharomyces* yeasts against *S. cerevisiae* should be known [57] so that the subsequent *S. cerevisiae* fermentation is not adversely affected.

Other non-*Saccharomyces* extracellular enzymatic activities, such as proteolytic and pectinolytic (polygalacturonase) enzymes, might also be beneficial to winemaking [58]. For example, proteolytic activity of some non-*Saccharomyces* yeast could lead to a reduction in protein levels with accompanying increase in protein stability of the end-product. Species found to produce the greatest number of extracellular enzymes are *C. stellata*, *H. uvarum* and *M. pulcherrima* [59].

Certain flavor and aroma compounds are present in grapes as glycosidic precursors with no sensory properties [60]. These compounds may be hydrolyzed by the enzyme D-glucosidase to form free volatiles that can improve the flavor and aroma of wine, but this enzyme is not encoded by the *S. cerevisiae* genome [61]. However, certain non-*Saccharomyces* yeasts belonging to the genera *Debaryomyces*, *Hansenula*, *Candida*, *Pichia*, and *Hanseniaspora* possess various degrees of D-glucosidase activity [21] and can play a role in releasing volatile compounds from non-volatile precursors. An intracellular D-glucosidase has also been isolated and purified from *Debaryomyces hansenii*. This enzyme, which is not inhibited by glucose and ethanol, was used during fermentation of Muscat grape juice, resulting in an increase in concentration of monoterpenols in the wine [62].

4. Non-*Saccharomyces* Strains as Glycosidase Producers for Vinification

4.1. β-Glucosidases

Glycosidically-bound volatiles are highly complex and diverse, especially regarding the aglycone moiety. The sugar parts consist of β-D-glucopyranosides and different diglycosides: 6-*O*-α-L-arabinofuranosyl-β-D-glucopyranosides, 6-*O*-α-L-arabinopyranosyl-β-D-glucopyranosides (vicianosides), 6-*O*-α-L-rhamnopyranosyl-β-D-glucopyranoside (rutinosides), 6-*O*-β-D-apiofuranosyl-β-D-glucopyranosides, 6-*O*-β-D-glucopyranosyl-β-D-glucopyranosides, and 6-*O*-β-D-xilopyranosyl-β-D-glucopyranosides (primeverosides). The aglycon part is often formed with terpenols, but other flavor precursors can occur, such as linear or cyclic alcohols, C-13 norisoprenoids, phenolic acids, and probably volatile phenols such as vanillin [63].

If we consider only glycosides with the most flavorant aglycons the most abundant in grape juice are apiosylglycosides (up to 50% according to grape variety) followed by rutinosides (6% to 13%) and, finally, glucosides (4% to 9%). All glycosides are not present in all cultivars and their proportions also differ according to grapes [64]. The glycoside flavor potential from grapes remains quite stable during winemaking and in young wines as well. These findings opened a new field of intensive research on the chemistry of glycoconjugated flavor compounds to exploit this important flavor source. Some aglycones are already odorous when released from glycosides and can contribute to the varietal aroma of wines [65].

Terpene glycosides can be hydrolysed by an enzymic way [66] to enrich wine flavor by release of free aromatic compounds from natural glycoside precursors. Enzymatic hydrolysis of glycosides is carried out with various enzymes which act sequentially according to two steps: firstly, α-L-rhamnosidase, α-L-arabinosidase, or β-D-apiosidase make the cleavage of the terminal sugar and rhamnose, arabinose, or apiose and the corresponding β-D-glucosides are released; subsequent liberation of monoterpenol takes place after action of a β-D-glucosidase (Figure 1) [67]. Nevertheless, one-step hydrolysis of disaccharide glycosides has also been described; enzymes catalysing this reaction have been isolated from grapes [67]. Enzymic hydrolysis of glycoside extracts from Muscat, Riesling, Semillon, Chardonnay, Sauvignon, and Sirah varieties have provoked the liberation not only of terpenes, but also C-13 norisoprenoids, such as 3-oxo-β-ionol and 3-hydroxy-β-damascenona [68].

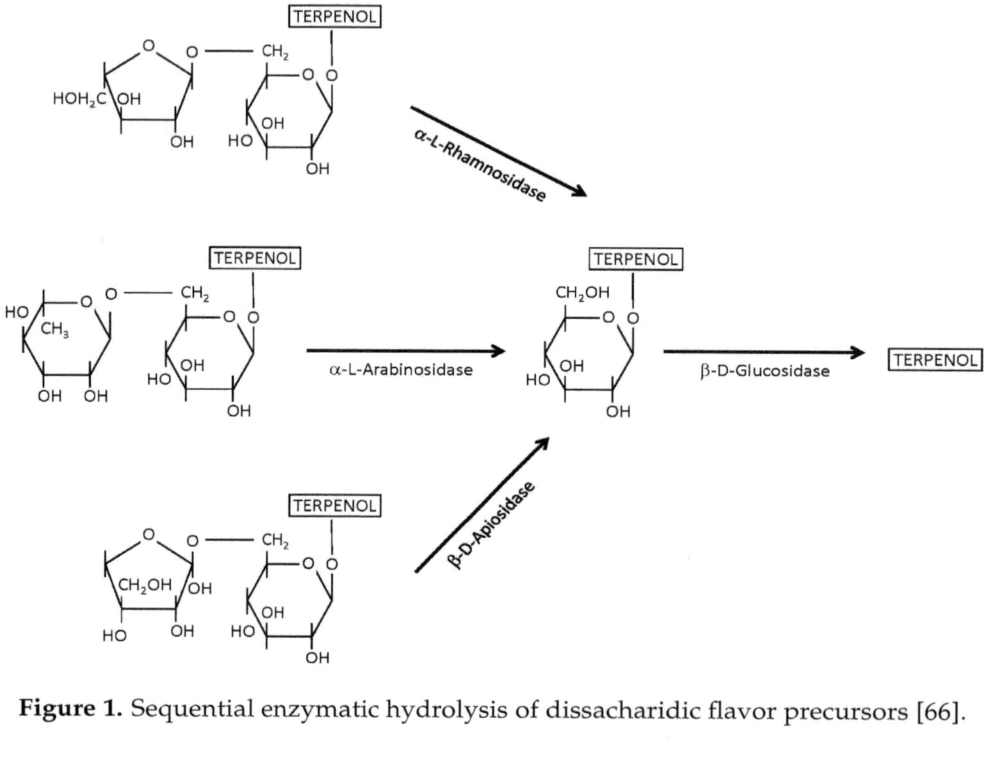

Figure 1. Sequential enzymatic hydrolysis of dissacharidic flavor precursors [66].

Few data are available regarding glycosidase activities of oenological yeast strains and the technological properties of the enzymes. Low α-rhamnosidase, α-arabinosidase, or β-apiosidase activities were detected in S. cerevisiae [69], but different efforts have been made to clone these genes obtained from differents microorganisms in S. cerevisiae [70]. Nevertheless, data on β-glucosidase activity on Saccharomyces are contradictory. First, results showed that these yeasts had a very low activity [71] but Delcroix et al. [69] found three enological strains showing high β-glucosidase activity. On the other hand, Darriet et al. [72] have shown that hydrolases located in the periplasmic space of a strain of S. cerevisiae were able to hydrolyse monoterpene glucosides of Muscat grapes; they also found that the activity of this β-glucosidase was glucose independent. Mateo and Di Stefano [73] detected β-glucosidase activity in different Saccharomyces strains on the basis of its hydrolytic activity on p-nitrophenyl-β-D-glucoside (pNPG) and terpene glucosides of Muscat juice. This enzymatic activity is induced by the presence of bound β-glucose as carbon source in the medium and seems to be a characteristic of the yeast strain. This β-glucosidase is associated with the yeast cell wall, is quite glucose independent but is inhibited by ethanol. This β-glucosidase is associated with the yeast cell wall is quite glucose independent but is inhibited by ethanol. These results could open new pathways regarding other glycosidase activities in S. cerevisiae; α-rhamnosidase, α-arabinosidase or β-apiosidase activities could be induced in wine yeast by changing the composition of the medium including inductive compounds, as well as in filamentous fungi [74].

Several flavor and aroma compounds in grapes are present as glycosylated flavorless precursors [2]. These compounds may be hydrolyzed by the enzyme β-glucosidase to form free volatiles that can increase the flavor and aroma of wine, but this enzyme is not encoded by the S. cerevisiae genome [46]. In contrast, non-Saccharomyces yeasts belonging to the genera Debaryomyces, Hansenula, Candida, Pichia, and Hanseniaspora possess various degrees of β-glucosidase activity and can play a role in releasing volatile compounds from non-volatile precursors [75]. Co-fermentation of Chardonnay grape juice with D. pseudopolymorphus and S. cerevisiae resulted in an increased concentration of the terpenols: citronellol, nerol, and geraniol in wine [76]. Similarly, cofermentation of Muscat grape juice with D. vanriji and S. cerevisiae produced wines with increased concentration of several terpenols [77]. Equally, mixed cultures of Sauvignon Blanc grape juice with *C. zemplinina / S. cerevisiae* and *T. delbrueckii / S. cerevisiae* produced wines with high concentrations of terpenols compared to wines only fermented with S. cerevisiae [23].

It has been reported that non-*Saccharomyces* yeasts can produce β-glucosidase [58]. The β-glucosidases from non-*Saccharomyces* species, such as *C. molischiana*, *C. wickerhamii*, and *P. anomala* were found to be more tolerant of winemaking conditions (for example, low pH values, low temperatures, high sugar, or ethanol levels) and tend to be more specific for glycosides than those from other yeast species [78]. Attempts have previously been made to enhance wine aroma using non-*Saccharomyces* yeasts and their glycosidases [79] because there is substantial yeast diversity in grapes and wines. Screening indigenous yeasts with glycosidases and their application in winemaking may allow wineries to make wines with more pleasant, typical, varietal aroma characteristics. Therefore, it is important to explore the potential of indigenous yeast biodiversity from specific enological ecosystems for specific and abundant β-glucosidases. β-Glucosidase-producing strains can be screened in Petri dishes with media containing cellulose-congo red, p-nitrophenyl-β-D-glucopyranoside (pNPG), or 4-methylumbelliferyl-β-dglucuronide (4-MUG) [80,81].

Yeasts of the *Hansenula* species isolated from fermenting must were reported to have an inducible β-glucosidase activity, but this enzyme was inhibited by glucose [82]. Other yeast strains, such as *C. molischiana* and *C. wickerhamii*, also possess activities towards various β-glucosides and they were little influenced by the nature of aglycon [83]. β-Glucosidase from *C. molischiana* was immobilized to Duolite A-568 resin, showing similar physicochemical properties to those of free enzyme. The immobilized enzyme was found to be very stable under wine conditions and could be used repeatedly for several hydrolyzes of bound aroma [84]. *Endomyces fibuliger* also produces extracellular β-glucosidase when grown in malt extract broth [85].

Screening 370 strains belonging to 20 species of yeasts, all of the strains of the species *D. castelli*, *D. hansenii*, *D. polymorphus*, *K. apiculata*, and *H. anomala* showed β-glucosidase activity [86]. A strain of *D. hansenii* exhibited the highest exocellular activity and some wall-bound and intracellular activity and its synthesis, occurred during exponential growth, was enhanced by aerobic conditions and repressed by high glucose concentration. The optimum condition for this enzyme was pH 4.0–5.0 and 40 °C. This enzyme was immobilized using a one-step procedure on hydroxyapatite. The immobilized enzyme exhibited a lower activity than the purified free enzyme, but was much more stable than the enzyme in cell-free supernatant [87]. Their studies have shown the ability of several wine yeasts to hydrolyse terpenoids, norisoprenoids and benzenoids glycosides; among wine yeasts *H. uvarum* was able to hydrolyze both glyco-conjugated forms of pyranic and furanic oxides of linalool [88]. Other authors have also shown the important role of non-*Saccharomyces* species in releasing the glycosidic-bound fraction of grape aroma components [89].

A total of 17 *Pichia* (*Wickerhamomyces*) isolates obtained from enological ecosystems in the Utiel-Requena Spanish region were characterized by physiological and molecular techniques (PCR-RFLP and sequencing) as belonging to the species *P. fermentans*, *P. membranifaciens*, and *W. anomalus*. In order to characterize their enzymatic abilities, xylanase, β-glucosidase, lipase, esterase, protease, and pectinase qualitative and quantitative assays were made. *W. anomalus* and *P. membranifaciens* showed to be the most interesting species to be used as sources of enzymes for the winemaking industry. Glycosidase enzymes show a high degree of tolerance to high levels of glucose and ethanol, making them of great interest to be used in enological procedure [90].

The sensorial characteristics of the wines produced with Muscat grapes are related to the level of terpene alcohols, so an improvement of such a level, as a result of hydrolytic processes conducted by *Hanseniaspora*, is expected. Isolates from *H. uvarum* and *H. vineae* have been proved to be good candidates to be used in commercial vinification processes to enhance wine properties. Wine inoculated with yeasts showed an increase in the level of aromatic compounds (Table 2) [91].

Table 2. Terpene compounds in Muscat wine. Concentration expressed as μg/L [a] [91].

Compound	Control [b]	*Hanseniaspora* Inoculated		
		H. uvarum H107	*H. vineae* G26	*H. vineae* P38
Oxide A [c]	29.7 (1.2)	30.4 (2.1)	33.7 (3.2)	26.9 (3.4)
Oxide B [d]	nd	nd	nd	nd
Linalool	20.0 (0.9)	40.4 * (3.9)	47.4 * (3.4)	38.2 * (5.3)
Ho-trienol	24.0 (3.2)	51.3 *(5.3)	35.1 * (4.2)	24.9 * (0.6)
2-Phenylethanol	1890.2 (43.4)	3057.5 * (39.8)	2747.8 * (26.8)	2568.5 * (45.6)
Oxide C [e]	nd	nd	nd	nd
Oxide D [f]	nd	nd	nd	nd
Terpineol	53.3 (3.4)	67.2 * (4.7)	65.1 *(1.2)	54.5 (3.9)
Nerol	24.6 (2.8)	25.8 (1.1)	23.4 (3.1)	26.3 (1.2)
Geraniol	59.8 (5.0)	61.3 (3.7)	56.9 (1.7)	62.8 (1.7)
Diol 1 [g]	43.2 (4.7)	87.9 * (2.1)	80.2 * (2.1)	81.2 * (3.2)
4-Vinylphenol	63.2 (1.2)	89.7 * (2.4)	75.7 * (5.8)	62.1 (0.9)
Endiol [h]	nd	58.8 * (2.1)	52.0 * (3.4)	34.1 * (4.2)
Diol 2 [i]	12.0 (0.6)	13.4 (0.9)	7.8 (2.6)	10.1 (0.9)
2-Phenylethyl acetate	28.0 (4.1)	56.2 * (7.2)	23.3 (1.2)	25.8 (4.7)
2-Methoxy-4-vinylphenol	89.0 (6.1)	103.0 * (5.3)	105.4 * (6.5)	94.1 (2.9)

[a] Values in brackets represent standard deviation ($n = 3$). ANOVA one factor, significant difference is indicated as * ($p < 0.05$); [b] Wine produced only with *Saccharomyces cerevisiae*; [c] *cis*-5-vinyltetrahydro-1,1,5-trimethyl-2-furanmethanol; [d] *trans*-5-vinyltetrahydro-1,1,5-trimethyl-2-furanmethanol; [e] *cis*-6-vinyltetrahydro-2,2,6-trimethyl-2H-pyran-3-ol; [f] *trans*-6-vinyltetrahydro-2,2,6-trimethyl-2H-pyran-3-ol; [g] 2,6-Dimethyl-3,7-octadien-2,6-diol; [h] 2,6-Dimethyl-7-octene-2,6-diol; [i] 2,6-Dimethyl-2,7-octadien-1,6-diol; nd: not detected

The potential applications of wild yeast strains with β-glucosidase activity have been investigated under simulated oenological conditions, coupled with the exploration of the potential applications of the β-glucosidases by studying the enzymatic activity and stability under similar oenological conditions [92]. The effects of different oenological factors on β-glucosidase production indicated that one isolate from the *T. asahii* strain had higher β-glucosidase production than the other strains under low pH conditions. However, isolates from *H. uvarum* and *S. cerevisiae* strain showed higher β-glucosidase production under high-sugar conditions. Furthermore, the influence of oenological factors on the activity and stability of the β-glucosidases revealed that the enzyme from the *T. asahii* strain had a stronger low-pH-value resistance than the other yeast β-glucosidases [92].

Hu et al. [93] applied a semiquantitative colorimetric assay to screen yeasts from three different regions of China. Among 493 non-*Saccharomyces* isolates belonging to eight genera, three isolates were selected for their high levels of β-glucosidase activity and were identified as *H. uvarum*, *P. membranifaciens*, and *Rhodotorula mucilaginosa*. β-Glucosidase from the *H. uvarum* strain showed the highest activity in winemaking conditions among the selected isolates. For aroma enhancement in winemaking, the glycosidase extract from *H. uvarum* exhibited catalytic specificity for aromatic glycosides of C13-norisoprenoids and some terpenes, enhancing fresh floral, sweet, berry, and nutty aroma characteristics in wine.

4.2. Xylanases

β-1,4-xylan is a heteroglycan with a backbone of β-(1→4)-linked D-xylopyranose residues that can be substituted with L-arabinofuranose, D-glucuronic acid, and/or 4-*O*-methyl-D-glucuronic acid [94]. It constitutes the major component of hemicelluloses found in the cell walls of monocots and hardwoods, and represents one of the most abundant biomass resources. Recently, xylanolytic enzymes of microbial origin have received great attention due to their possible industrial applications for sustainable fuel-ethanol production from xylan. Two key reactions proceed during hydrolysis of the xylan backbone; endo-1,4-β-xylanases (1,4-β-D-xylan xylanohydrolase) hydrolyze internal β-(1→4)-xylosidic linkages in the insoluble xylan backbone to yield soluble xylooligosaccharides, while 1,4-β-xylosidases are exoglycosidases that cleave terminal xylose monomers from the non-reducing end of short-chain xylooligosaccharides [95]. Additional enzyme activities, such as α-L-arabinofuranosidase, α-D-glucuronidase, and acetyl xylan esterase, remove side-chain substituents. 1,4-β-Xylosidase is important in xylan degradation, considering that xylans are not completely hydrolyzed by xylanases alone. The finding of new isolates of non-*Saccharomyces* yeasts, showing beneficial enzymes (such as β-glucosidase and β-xylosidase) can contribute to the production of quality wines [96]. In a selection and characterization program we have studied 114 isolates of non-*Saccharomyces* yeasts, four isolates were selected because of their both high β-glucosidase and β-xylosidase activities. The ribosomal D1/D2 regions were sequenced to identify them as *P. membranifaciens*, *H. vineae*, *H. uvarum*, and *W. anomalus*. The induction process was optimized to be carried on YNB-medium supplemented with 4% xylan, inoculated with 10^6 cfu/mL and incubated 48 h at 28 °C without agitation. Most of the strains had a pH optimum of 5.0 to 6.0 for both the β-glucosidase and β-xylosidase activities. The effect of sugars was different for each isolate and activity. Each isolate showed a characteristic set of inhibition, enhancement or null effect for β-glucosidase and β-xylosidase. The volatile compounds liberated from wine incubated with each of the four yeasts were also studied, showing an overall terpene increase when wines were treated with non-*Saccharomyces* isolates. In detail, terpineol, 4-vinyl-phenol, and 2-methoxy-4-vinylphenol increased after the addition of *Hanseniaspora* isolates. Wines treated with *Hanseniaspora*, *Wickerhamomyces*, or *Pichia* produced more 2-phenyl ethanol than those inoculated with other yeasts [97].

An ethanol-tolerant 1,4-β-xylosidase was purified from cultures of a strain of *P. membranifaciens* grown on xylan at 28 °C. The enzyme was purified by sequential chromatography on DEAE-cellulose and Sephadex G-100. The relative molecular mass of the enzyme was determined to be 50 kDa by SDS-PAGE. The activity of 1,4-β-xylosidase was optimum at pH 6.0 and 35 °C. The activity had a Km of 0.48 ± 0.06 $mmol \cdot L^{-1}$ and a Vmax of 7.4 ± 0.1 $\mu mol \cdot min^{-1} \cdot mg^{-1}$ protein for p-nitrophenyl-β-D-xylopyranoside. The enzyme characteristics (pH and thermal stability, low inhibition rate by glucose and ethanol tolerance) make this enzyme a good candidate to be used in enzymatic production of xylose and improvement of hemicellulose saccharification for production of bioethanol [98].

5. Conclusions

It is generally accepted that the wealth of yeast biodiversity with hidden potential, especially for oenology, is largely untapped. In order to exploit the potential benefits of non-*Saccharomyces* yeasts in wine production and to minimize potential spoilage, the yeast populations on grapes and in must, as well as the effect of wine-making practices on these yeasts, must be known, as must the metabolic characteristics of non-*Saccharomyces* yeasts. Strain selection will be very important, as not all strains within a species will necessarily show the same desirable characteristics. For example, significant variability is found in the formation of β-glucosidase amongst strains within some non-*Saccharomyces* yeast species.

Whatever the outcome of the search for non-*Saccharomyces* yeasts for use in wine production, the accepted list of desirable characteristics as pertaining to the wine yeast *S. cerevisiae* will not necessarily apply to non-*Saccharomyces* yeasts. High fermentation efficiency, sulfite tolerance and killer properties, for example, might not be needed in the new technology of wine production. The new non-*Saccharomyces* wine yeasts will necessarily have a different list of desired characteristics: efficient sugar utilization, enhanced production of desirable volatile esters, enhanced liberation of grape terpenoids and production of glycerol to improve wine flavor and other sensory properties can be met by selected non-*Saccharomyces* wine yeasts.

Acknowledgments: The Universitat de València kindly supported this work included in the project "Identification and biotechnological characterization of yeasts isolated from agrifood residues of the Valencian Community", grant INV-AE-336499.

Author Contributions: Sergi Maicas and José Juan Mateo wrote the paper.

References

1. Padilla, B.; Gil, J.V.; Manzanares, P. Past and future of non-*Saccharomyces* yeasts: From spoilage microorganisms to biotechnological tools for improving wine aroma complexity. *Front. Microbiol.* **2016**. [CrossRef] [PubMed]
2. Pretorius, I.S. Tailoring wine yeast for the new millennium: Novel Approaches to the ancient art of winemaking. *Yeast* **2000**, *16*, 675–729. [CrossRef]
3. Jackson, R.S. *Wine Science: Principles and Applications*, 1st ed.; Elsevier: San Diego, CA, USA, 1994.
4. Andorrà, I.; Berradre, M.; Rozès, N.; Mas, A.; Guillamón, J.M.; Esteve-Zarzoso, B. Effect of pure and mixed Cultures of the Main Wine yeast Species on Grape Must Fermentations. *Eur. Food Res. Technol.* **2010**, *231*, 215–224. [CrossRef]
5. Gil, J.V.; Mateo, J.J.; Jiménez, M.; Pastor, A.; Huerta, T. Aroma compounds in wine as influenced by apiculate yeasts. *J. Food Sci.* **1996**, *61*, 1247–1250. [CrossRef]
6. Wang, C.; Mas, A.; Esteve, B. The interaction between *Saccharomyces cerevisiae* and non-*Saccharomyces* yeast during alcoholic fermentation is species and strain specific. *Front. Microbiol.* **2016**. [CrossRef] [PubMed]
7. Mateo, J.J.; Jiménez, M.; Huerta, T.; Pastor, A. Contribution of different yeasts isolated from musts of Monastrell grapes to the aroma of wine. *Int. J. Food Microbiol.* **1991**, *14*, 153–160. [CrossRef]
8. Rodríguez-Gómez, F.; Arroyo-López, F.N.; López-López, A.; Bautista-Gallego, J.; Garrido-Fernández, A. Lipolytic activity of the yeast species associated with the fermentation/storage phase of ripe olive processing. *Food Microbiol.* **2010**, *27*, 604–612. [CrossRef] [PubMed]
9. Heard, G. Novel yeasts in winemaking—Looking to the future. *Food Aust.* **1999**, *51*, 347–352.
10. Barata, A.; Malfeito-Ferreira, M.; Loureiro, V. The microbial ecology of wine grape berries. *Int. J. Food Microbiol.* **2012**, *153*, 243–259. [CrossRef] [PubMed]
11. Martini, A. Origin and domestication of the wine yeast *Saccharomyces cerevisiae*. *J. Wine Res.* **1993**, *4*, 165–176. [CrossRef]

12. Ribéreau-Gayon, P.; Dubordieu, D.; Donèche, B.; Lonvaud, A. *Handbook of Enology Volume 1: The Microbiology of Wine and Vinifications*; John Wiley & Sons Ltd.: Chichester, UK, 2000.
13. Johnson, E.A. Biotechnology of non-*Saccharomyces* yeasts—The Ascomycetes. *Appl. Microbiol. Biotechnol.* **2013**, *97*, 503–517. [CrossRef] [PubMed]
14. Pretorius, I.S.; Van Der Westhuizen, T.J.; Augustyn, O.P.H. Yeast biodiversity in vineyards and wineries and its importance to the South African wine industry—A review. *S. Afr. J. Enol. Vitic.* **1999**, *20*, 61–75.
15. Kurtzman, C.P. *The Yeasts: A Taxonomic Study*, 5th ed.; Elsevier Science: Waltham, MA, USA, 2012.
16. Davenport, R.R. Microecology of yeast and yeastlike organisms associated with and English vineyard. *Vitis* **1974**, *13*, 123–130.
17. Jolly, N.P.; Vera, C.; Pretorius, I.S. Not your ordinary yeast: Non-Saccharomyces yeasts in wine production uncovered. *FEMS Yeast Res.* **2014**, *14*, 215–237. [CrossRef] [PubMed]
18. Boynton, P.J.; Duncan, G. The ecology and evolution of non-domesticated *Saccharomyces* species. *Yeast* **2014**, *12*, 449–462.
19. Henick-Kling, T.; Edinger, W.; Daniel, P.; Monk, P. Selective effects of sulfur dioxide and yeast starter culture addition on indigenous yeast populations and sensory characteristics of the wine. *J. Appl. Microbiol.* **1998**, *84*, 865–876. [CrossRef]
20. Jolly, N.P.; Augustyn, O.H.P.; Pretorius, I.S. The effect of non-*Saccharomyces* yeasts on fermentation and wine quality. *S. Afr. J. Enol. Vitic.* **2003**, *24*, 55–62.
21. Romano, P.; Suzzi, G.; Domizio, P.; Fatichenti, F. Secondary products formation as a tool for discriminating non-*Saccharomyces* wine strains. *Antonie Leeuwenhoek* **1997**, *71*, 239–242. [CrossRef] [PubMed]
22. Ciani, M.; Comitini, F. Non-*Saccharomyces* wine yeasts have a promising role in biotechnological approaches to winemaking. *Ann. Microbiol.* **2011**, *61*, 25–32. [CrossRef]
23. Sadoudi, M.; Tourdot-Maréchal, R.; Rousseaux, S.; Steyer, D.; Gallardo-Chacón, J.J.; Ballester, J.; Vichi, S.; Guérin-Schneider, R.; Caixach, J.; Alexandre, H. Yeast-yeast interactions revealed by aromatic profile analysis of Sauvignon Blanc wine fermented by single or co-culture of non-*Saccharomyces* and *Saccharomyces* yeasts. *Food Microbiol.* **2012**, *32*, 243–253. [CrossRef] [PubMed]
24. Sun, L.; Hebert, A.S.; Yan, X.; Zhao, Y.; Westphall, M.S.; Rush, M.J.P.; Zhu, G.; Champion, M.M.; Coon, J.J.; Dovichi, N.J. Over 10000 peptide identifications from the HeLa proteome by using single-shot capillary zone electrophoresis combined with tandem mass spectrometry. *Angew. Chem.* **2014**, *126*, 14151–14153. [CrossRef]
25. Comitini, F.; Gobbi, M.; Domizio, P.; Romani, C.; Lencioni, L.; Mannazzu, I.; Ciani, M. Selected non-*Saccharomyces* wine yeasts in controlled multistarter fermentations with *Saccharomyces cerevisiae*. *Food Microbiol.* **2011**, *28*, 873–882. [CrossRef] [PubMed]
26. Maturano, Y.P.; Rodríguez Assaf, L.A.; Toro, M.E.; Nally, M.C.; Vallejo, M.; Castellanos de Figueroa, L.I.; Combina, M.; Vazquez, F. Multi-enzyme production by pure and mixed cultures of *Saccharomyces* and non-*Saccharomyces* yeasts during wine fermentation. *Int. J. Food Microbiol.* **2012**, *155*, 43–50. [CrossRef] [PubMed]
27. Kutyna, D.; Varela, C.; Henschke, P.; Chambers, P.; Stanley, G. Microbiological approaches to lowering ethanol concentration in wine. *Trends Food Sci. Technol.* **2010**, *21*, 293–302. [CrossRef]
28. Di Maio, S.; Polizzotto, G.; Di Gangi, E.; Foresta, G.; Genna, G.; Verzera, A. Biodiversity of indigenous *Saccharomyces* populations from old wineries of South-Eastern Sicily (Italy): Preservation and economic potential. *PLoS ONE* **2012**. [CrossRef]
29. Gonzalez, R.; Quirós, M.; Morales, P. Yeast respiration of sugars by non-*Saccharomyces* yeast species: A promising and barely explored approach to lowering alcohol content of wines. *Trends Food Sci. Technol.* **2013**, *29*, 55–61. [CrossRef]
30. Erten, H.; Tanguler, H. Influence of *Williopsis saturnus* yeasts in combination with *Saccharomyces cerevisiae* on wine fermentation. *Lett. Appl. Microbiol.* **2010**, *50*, 474–479. [CrossRef] [PubMed]
31. Tilloy, V.; Ortiz-Julien, A.; Dequin, S. Reduction of ethanol yield and improvement of glycerol formation by adaptive evolution of the wine yeast *Saccharomyces* cerevisiae under hyperosmotic conditions. *Appl. Environ. Microbiol.* **2014**, *80*, 2623–2632. [CrossRef] [PubMed]
32. Ciani, M.; Comitini, F.; Mannazzu, I.; Domizio, P. Controlled mixed culture fermentation: A new perspective on the use of non-*Saccharomyces* yeasts in winemaking. *FEMS Yeast Res.* **2010**, *10*, 123–133. [CrossRef] [PubMed]

33. Quiros, M.; Rojas, V.; Gonzalez, R.; Morales, P. Selection of non-*Saccharomyces* yeast strains for reducing alcohol levels in wine by sugar respiration. *Int. J. Food Microbiol.* **2014**, *181*, 85–91. [CrossRef] [PubMed]
34. Gobbi, M.; De Vero, L.; Solieri, L.; Comitini, F.; Oro, L.; Giudici, P.; Ciani, M. Fermentative aptitude of non-*Saccharomyces* wine yeast for reduction in the ethanol content in wine. *Eur. Food Res. Technol.* **2014**, *239*, 41–48. [CrossRef]
35. Fleet, G.H. Wine yeasts for the future. *FEMS Yeast Res.* **2008**, *8*, 979–995. [CrossRef] [PubMed]
36. Styger, G.; Prior, B.; Bauer, F.F. Wine flavor and aroma. *J. Ind. Microbiol. Biotechnol.* **2011**, *38*, 1145–1159. [CrossRef] [PubMed]
37. Swiegers, J.H.; Bartowsky, E.J.; Henschke, P.A.; Pretorius, I.S. Yeast and bacterial modulation of wine aroma and flavour. *Aust. J. Grape Wine Res.* **2005**, *11*, 139–173. [CrossRef]
38. Zironi, R.; Romano, P.; Suzzi, G.; Battistutta, F.; Comi, G. Volatile metabolites produced in wine by mixed and sequentialcultures of *Hanseniaspora guilliermondii* or *Kloeckera apiculata* and *Saccharomyces cerevisiae*. *Biotechnol. Lett.* **1993**, *15*, 235–238. [CrossRef]
39. Arévalo-Villena, M.; Úbeda-Iranzo, J.F.; Gundllapalli, S.B.; Cordero-Otero, R.R.; Briones-Pérez, A.I. Characterization of an exocellular β-glucosidase from *Debaryomyces pseudopolymorphus*. *Enzyme Microb. Technol.* **2006**, *39*, 229–234. [CrossRef]
40. Ocón, E.; Gutiérrez, A.R.; Garijo, P.; López, R.; Santamaría, P. Presence of non-*Saccharomyces* yeasts in cellar equipment and grape juice during harvest time. *Food Microbiol.* **2010**, *27*, 1023–1027. [CrossRef] [PubMed]
41. Clemente-Jiménez, J.M.; Mingorance-Cazorla, L.; Martínez-Rodríguez, S.; Las Heras-Vázquez, F.J.; Rodríguez-Vico, F. Molecular characterization and oenological properties of wine yeasts isolated during spontaneous fermentation of six varieties of grape must. *Food Microbiol.* **2004**, *21*, 149–155. [CrossRef]
42. Arévalo-Villena, M.; Úbeda-Iranzo, J.F.; Briones-Pérez, A.I. β-Glucosidase activity in wine yeasts: Application in enology. *Enzyme Microb. Technol.* **2007**, *40*, 420–425. [CrossRef]
43. Jackson, R.S. *Wine Science: Principles, Practice, Perception*, 2nd ed.; Elsevier: San Diego, CA, USA, 2000.
44. Lambrechts, M.G.; Pretorius, I.S. Yeast and its importance to wine aroma–A review. *S. Afr. J. Enol. Vitic.* **2000**, *21*, 97–129.
45. Swiegers, J.H.; Pretorius, I.S. Yeast modulation of wine flavor. *Adv. Appl. Microbiol.* **2005**, *57*, 131–175. [PubMed]
46. Ubeda-Iranzo, J.F.; Briones-Perez, A.I.; Izquierdo-Cañas, P.M. Study of the oenological characteristics and enzymatic activities of wine yeasts. *Food Microbiol.* **1998**, *15*, 399–406. [CrossRef]
47. Scanes, K.T.; Hohmann, S.; Prior, B.A. Glycerol production by the yeast *Saccharomyces* cerevisiae and its relevance to wine: A review. *S. Afr. J. Enol. Vitic.* **1998**, *19*, 17–24.
48. Ciani, M.; Maccarelli, F. Oenological properties of non-*Saccharomyces* yeasts associated with wine-making. *World J. Microbiol. Biotechnol.* **1998**, *14*, 199–203. [CrossRef]
49. Prior, B.A.; Toh, T.H.; Jolly, N.; Baccari, C.L.; Mortimer, R.K. Impact of yeast breeding for elevated glycerol production on fermentation activity and metabolite formation in Chardonnay. *S. Afr. J. Enol. Vitic.* **2000**, *21*, 92–99.
50. Swiegers, J.H.; Kievit, R.L.; Siebert, T.; Lattey, K.A.; Bramley, B.R.; Francis, I.L. The influence of yeast on the aroma of Sauvignon Blanc wine. *Food Microbiol.* **2009**, *26*, 204–211. [CrossRef] [PubMed]
51. Anfang, N.; Brajkovich, M.; Goddard, M.R. Co-fermentation with *Pichia kluyveri* increases varietal thiol concentrations in Sauvignon Blanc. *Aust. J. Grape Wine Res.* **2009**, *15*, 1–8. [CrossRef]
52. Zott, K.; Thibon, C.; Bely, M.; Lonvaud-Funel, A.; Dubourdieu, D.; Masneuf-Pomarede, I. The grape must non-*Saccharomyces* microbial community: Impact on volatile thiol release. *Int. J. Food Microbiol.* **2011**, *151*, 210–215. [CrossRef] [PubMed]
53. Romano, P.; Suzzi, G. Higher alcohol and acetoin production by *Zygosaccharomyces* wine yeasts. *J. Appl. Bacteriol.* **1993**, *75*, 541–545. [CrossRef]
54. Mateo, J.J.; Peris, L.; Ibañez, C.; Maicas, S. Characterization of glycolytic activities from non-*Saccharomyces* yeasts isolated from Bobal musts. *J. Ind. Microbiol. Biotechnol.* **2011**, *38*, 347–354. [CrossRef] [PubMed]
55. Domizio, P.; Liu, Y.; Bisson, L.F.; Barile, D. Use of non-*Saccharomyces* wine yeasts as novel sources of mannoproteins in wine. *Food Microbiol.* **2014**, *43*, 5–15. [CrossRef] [PubMed]
56. Vidal, S.; Francis, L.; Noble, A.; Kwiatkowski, M.; Cheynier, V.; Waters, E. Taste and mouth-feel properties of different types of tannin-like polyphenolic compounds and anthocyanins in wine. *Anal. Chim. Acta* **2004**, *513*, 57–65. [CrossRef]

57. Fleet, G.H.; Prakitchaiwattana, C.; Beh, A.L.; Heard, G. The yeast ecology of wine grapes. In *Biodiversity and Biotechnology of Wine Yeasts*; Ciani, M., Ed.; Research Signpost: Kerala, India, 2002; pp. 1–17.
58. Strauss, M.L.; Jolly, N.P.; Lambrechts, M.G.; Rensburg, P.V. Screening for the production of extracellular hydrolytic enzymes by non-*Saccharomyces* wine yeasts. *J. Appl. Microbiol.* **2001**, *91*, 182–190. [CrossRef] [PubMed]
59. Mateo, J.J.; Maicas, S.; Thieβen, C. Biotechnological characterisation of exocellular proteases produced by enological *Hanseniaspora* isolates. *Int. J. Food Sci. Technol.* **2015**, *50*, 218–225. [CrossRef]
60. Pretorius, I.S. The genetic analysis and tailoring of wine yeasts. In *Functional Genetics of Industrial Yeasts*; De Winde, J.H., Ed.; Springer-Verlag: Berlin, Germany, 2003; Volume 2, pp. 99–142.
61. Van Rensburg, P.; Stidwell, T.; Lambrechts, M.G.; Cordero-Otero, R.R.; Pretorius, I.S. Development and assessment of a recombinant *Saccharomyces cerevisiae* wine yeast producing two aroma-enhancing β-glucosidases encoded by the *Saccharomycopsis fibuligera* BGL_1 and BGL_2 genes. *Anal. Microbiol.* **2005**, *55*, 33–42.
62. Yanai, T.; Sato, M. Isolation and properties of β-glucosidase produced by *Debaryomyces hansenii* and its application in winemakig. *Am. J. Enol. Vitic.* **1999**, *50*, 231–235.
63. Sarry, J.E.; Gunata, Z. Plant and microbial glycoside hydrolases: Volatile release from glycosidic aroma precursors. *Food Chem.* **2004**, *87*, 509–521. [CrossRef]
64. Bayonove, C.; Gunata, Y.; Sapis, J.C.; Baumes, R.L.; Dugelay, I.; Grassin, C. L'aumento degli aromi nel vino mediante l'uso degli enzimi. *Vignevini* **1993**, *9*, 33–36. (In Italian)
65. Mateo, J.J.; Jiménez, M. Monoterpenes in grape juice and wines. *J. Chromatogr. A* **2000**, *881*, 557–567. [CrossRef]
66. Maicas, S.; Mateo, J.J. Hydrolysis of terpenyl glycosides in grape juice and other fruit juices: A review. *Appl. Microbiol. Biotechnol.* **2005**, *67*, 322–335. [CrossRef] [PubMed]
67. Gunata, Z.; Bitteur, S.; Brillouet, J.M.; Bayonove, C.; Cordonnier, R. Sequential enzymatic hydrolysis of potentially aromatic glycosides form grapes. *Carbohydr. Res.* **1988**, *184*, 139–149. [CrossRef]
68. Gunata, Z.; Bayonove, C.; Tapiero, C.; Cordonnier, R. Hydrolysis of grape monoterpenyl β-D-glucosides by various β-glucosidases. *J. Agric. Food Chem.* **1990**, *38*, 1232–1236. [CrossRef]
69. Delcroix, A.; Gunata, Z.; Sapis, J.C.; Salmon, J.M.; Bayonove, C. Glycosidase activities of three enological yeast strains during wine making. Effect on the terpenol content of Muscat wine. *Am. J. Enol. Vitic.* **1994**, *45*, 291–296.
70. Bisson, L.F.; Karpel, J.E. Genetics of yeast impacting wine quality. *Ann. Rev. Food Sci. Technol.* **2010**, *1*, 139–162. [CrossRef] [PubMed]
71. Gunata, Z.; Dugelay, I.; Sapis, J.C.; Baumes, R.; Bayonove, C. Action des glycosidases exogènes au cours de la vinification: Liberation de l'arôme à partir des précurseurs glycosidiques. *J. Int. Sci. Vigne Vin* **1990**, *24*, 133–144. (In French)
72. Darriet, P.; Boidron, J.N.; Dubourdieu, D. L'hydrolyse des hétérosides terpéniques du Muscat a Petit Grains par les enzymes périplasmiques de *Saccharomyces cerevisiae. Connaiss. Vigne Vin* **1988**, *22*, 189–195. (In French)
73. Mateo, J.J.; Di Stefano, R. Enological properties of β-glucosidase in wine yeasts. *Food Microbiol.* **1998**, *14*, 583–591. [CrossRef]
74. Dupin, I.; Gunata, Z.; Sapis, J.C.; Bayonove, C.; M'Bairaroua, O.; Tapiero, C. Production of a β-apiosidase by *Aspergillus niger*. Partial purification, properties and effect on terpenyl apiosylglucosides from grape. *J. Agric. Food Chem.* **1992**, *40*, 1886–1891. [CrossRef]
75. Spagna, G.; Barbagallo, R.N.; Palmeri, R.; Restuccia, C.; Giudici, P. Properties of endogenous β-glucosidase of a *Saccharomyces cerevisiae* strain isolated from Sicilian musts and wines. *Enzyme Microb. Technol.* **2002**, *31*, 1030–1035. [CrossRef]
76. Cordero Otero, R.R.; Ubeda Iranzo, J.F.; Briones-Perez, A.I.; Potgieter, N.; Villena, M.A.; Pretorius, I.S.; van Rensburg, P. Characterization of the β-glucosidase activity produced by enological strains of non-*Saccharomyces* yeasts. *J. Food Sci.* **2003**, *68*, 2564–2569. [CrossRef]
77. García, A.; Carcel, C.; Dulau, L.; Samson, A.; Aguera, E.; Agosin, E. Influence of a mixed culture with *Debaryomyces vanriji* and *Saccharomyces cerevisiae* on the volatiles of a Muscat wine. *J. Food Sci.* **2002**, *67*, 1138–1143. [CrossRef]
78. Jutaporn, S.; Sukanda, V.; Christian, E.B.; Kanit, V. The characterisation of a novel *Pichia anomala* β-glucosidase with potentially aroma-enhancing capabilities in wine. *Ann. Microbiol.* **2009**, *59*, 335–343.

79. Carrascosa, A.V.; Martinez-Rodriguez, A.; Cebollero, E.; Gonzalez, R. Molecular Wine Microbiology. In *Saccharomyces Yeast II: Secondary Fermentation*; Carrascosa, A.V., Muñoz, R., Gonzalez, R., Eds.; Elsevier: London, UK, 2011; pp. 33–49.
80. Hernandez, L.F.; Espinosa, J.C.; Fernandez, G.M.; Briones, A. β-glucosidase activity in a *Saccharomyces cerevisiae* wine strain. *Int. J. Food Microbiol.* **2003**, *80*, 171–176. [CrossRef]
81. Wang, Y.X.; Zhang, C.; Li, J.M.; Xu, Y. Different influences of β-glucosidases on volatile compounds and anthocyanins of Cabernet Gernischt and possible reason. *Food Chem.* **2013**, *140*, 245–254. [CrossRef] [PubMed]
82. Grossmann, M.; Rapp, A.; Rieth, W. Enzymatische Freisetzung gebundener Aromastoffe in Wein. *Dtsch. Lebensm. Rundsch.* **1987**, *83*, 7–12. (In German)
83. Gunata, Z.; Brillouet, J.M.; Voirin, S.; Baumes, B.; Cordonnier, R. Purification and some properties of an α-L-arabinofuranosidase from *Aspergillus niger*. Action on grape monoterpenyl arabinofuranosyl glucosidases. *J. Agric. Food Chem.* **1990**, *38*, 772–776. [CrossRef]
84. Gueguen, Y.; Chemardin, P.; Pien, S.; Arnaud, A.; Galzy, P. Enhancement of aromatic quality of Muscat wine by the use of immobilized β-glucosidase. *J. Biotechnol.* **1997**, *55*, 151–156. [CrossRef]
85. Brimer, L.; Nout, M.J.R.; Tuncel, G. Glycosidase (amygdalase and linamarase) from *Endomyces fibuliger* (LU677): Formation and crude enzyme properties. *Appl. Microbiol. Biotechnol.* **1998**, *49*, 182–188. [CrossRef] [PubMed]
86. Rosi, I.; Vinella, M.; Domizio, P. Characterization of β-glucosidase activity in yeasts of oenological origin. *J. Appl. Bacteriol.* **1994**, *77*, 519–527. [CrossRef] [PubMed]
87. Riccio, P.; Rossano, R.; Vinella, M.; Domizio, P.; Zito, F.; Sanseverino, F.; D'Elia, A.; Rosi, I. Extraction and immobilization in one step of two β-glucosidases released from a yeast strain of *Debaryomyces hansenii*. *Enzyme Microb. Technol.* **1999**, *24*, 123–129. [CrossRef]
88. Fernández-González, M.; Di Stefano, R.; Briones, A.I. Hydrolysis and transformation of terpene glycosides from Muscat must by different yeast species. *Food Microbiol.* **2003**, *20*, 35–41. [CrossRef]
89. Mendes, A.; Climaco, M.C.; Mendes, A. The role of non-*Saccharomyces* species in releasing glycosidic bound fraction of grape aroma components–A preliminary study. *J. Appl. Microbiol.* **2001**, *91*, 67–71. [CrossRef]
90. Madrigal, T.; Maicas, S.; Mateo, J.J. Glucose and ethanol tolerant enzymes produced by *Pichia* (*Wickerhamomyces*) isolates from enological ecosystems. *Am. J. Enol. Vitic.* **2013**, *64*, 126–133. [CrossRef]
91. Lopez, S.; Mateo, J.J.; Maicas, S. Characterization of *Hanseniaspora* isolates with potential aroma enhancing properties in Muscat wines. *S. Afr. J. Enol. Vitic.* **2014**, *35*, 292–303.
92. Wang, Y.; Xu, Y.; Li, J. A Novel extracellular β-glucosidase from *Trichosporon asahii*: Yield prediction, evaluation and application for aroma enhancement of cabernet sauvignon. *J. Food Sci.* **2012**, *77*, 505–515. [CrossRef] [PubMed]
93. Hu, K.; Qin, Y.; Tao, Y.S.; Zhu, X.L.; Peng, C.T.; Ullah, N. Potential of glycosidase from non-*Saccharomyces* isolates for enhancement of wine aroma. *J. Food Sci.* **2016**, *81*, 935–943. [CrossRef] [PubMed]
94. Bhat, M.K.; Hazlewood, G.P. Enzymology and other characteristics of cellulases and xylanases. In *Enzymes in Farm Animal Nutrition*; Bedford, M., Partridge, G., Eds.; CABI Publishing: Oxon, UK, 2001.
95. Polizeli, M.L.; Rizzati, A.C.; Monti, R.H.F.; Terenzi, C.G.; Jorge, J.A.; Amorin, D.S. Xylanases from fungi: Properties and industrial applications. *Appl. Microbiol. Biotechnol.* **2005**, *67*, 577–591. [CrossRef] [PubMed]
96. Linden, T.; Hahn-Hagerdal, B. Fermentation of lignocellulose hydrolysates with yeasts and xylose isomerase. *Enzyme Microb. Technol.* **1989**, *11*, 583589. [CrossRef]
97. López, M.C.; Mateo, J.J.; Maicas, S. Screening of β-glucosidase and β-xylosidase activities in four non-*Saccharomyces* yeast isolates. *J. Food Sci.* **2015**, *80*, 1696–1704. [CrossRef] [PubMed]
98. Romero, A.M.; Mateo, J.J.; Maicas, S. Characterization of an ethanol-tolerant 1,4-β-xylosidase produced by *Pichia membranifaciens*. *Lett. Appl. Microbiol.* **2012**, *55*, 354–361. [CrossRef] [PubMed]

7

Co-Existence of Inoculated Yeast and Lactic Acid Bacteria and their Impact on the Aroma Profile and Sensory Traits of Tempranillo Red Wine

Pedro Miguel Izquierdo-Cañas [1,2], María Ríos-Carrasco [3], Esteban García-Romero [1], Adela Mena-Morales [1], José María Heras-Manso [4] and Gustavo Cordero-Bueso [3,*]

1 Instituto Regional de Investigación y Desarrollo Agroalimentario y Forestal de Castilla-La Mancha (IRIAF), IVICAM, Ctra. Toledo-Albacete s/n., 13700 Tomelloso, Ciudad Real, Spain; pmizquierdo@jccm.es (P.M.I.-C.); estebang@jccm.es (E.G.-R.); amenam@jccm.es (A.M.-M.)
2 Parque Científico y Tecnológico de Castilla-La Mancha, Paseo de la Innovación 1, 02006 Albacete, Spain
3 Department of Biomedicine, Biotechnology and Public Health, University of Cádiz, Puerto Real, 11003 Cádiz, Spain; maria.rioscarrasco@alum.uca.es
4 Lallemand Península Ibérica, Tomás Edison 4, 28521 Madrid; jmheras@lallemand.com
* Correspondence: gustavo.cordero@uca.es

Abstract: This study investigates the effects of simultaneous inoculation of a selected *Saccharomyces cerevisiae* yeast strain with two different commercial strains of wine bacteria *Oenococcus oeni* at the beginning of the alcoholic fermentation on the kinetics of malolactic fermentation (MLF), wine chemical composition, and organoleptic characteristics in comparison with spontaneous MLF in Tempranillo grape must from Castilla-La Mancha (Spain). Evolution of MLF was assessed by the periodic analysis of L-malic acid through the enzymatic method, and most common physiochemical parameters and sensory traits were evaluated using a standardized sensory analysis. The samples were analyzed by GC/MS in SCAN mode using a Trace GC gas chromatograph and a DSQII quadrupole mass analyzer. Co-inoculation reduced the overall fermentation time by up to 2 weeks leading to a lower increase in volatile acidity. The fermentation-derived wine volatiles profile was distinct between the co-inoculated wines and spontaneous MLF and was influenced by the selected wine bacteria used in co-inoculation. Co-inoculation allows MLF to develop under reductive conditions and results in wines with very few lactic and buttery flavors, which is related to the impact of specific compounds like 2,3-butanedione. This compound has been also confirmed as being dependent on the wine bacteria used.

Keywords: Simultaneous inoculation; Alcoholic fermentation; Malolactic fermentation; *Sacccharomyces cerevisiae; Oenococcus oeni*; PN4^{TM}; Omega^{TM}; Aroma profile

---

1. Introduction

Malolactic fermentation (MLF) occurs in wine as the result of the metabolic activity of wine lactic acid bacteria. It can take place after the alcoholic fermentation, normally carried out by yeasts, but it can occur earlier. MLF reduces wine acidity by the conversion of dicarboxylic L-malic acid (malate) into monocarboxylic L-lactic acid (lactate) and carbon dioxide, which implies a modification in wine flavor and its microbiological stability, both of which are considered to be beneficial effects for wine quality [1,2]. Sensory studies carried out in the 1980s by Davis et al. [3] showed that malolactic activity of bacteria imparts recognizable changes to the flavor characteristics of wine. These modifications come from the biotransformation and enzymatic processing of grape nutrients into flavor-active compounds [4].

Several studies show that the combination of selected wine yeasts and bacteria strains has different sensory effects in wines due to the production of certain impact metabolites. They are synthesized in different amounts by these organisms according to the grape variety and bacterial strain, giving us the opportunity to determine if their contribution to aroma and flavor depends on the species or strains [5–7]. Research performed by Pozo-Bayón et al. revealed different malolactic behavior for *Oenococcus oeni* and *Lactobacillus plantarum*, concluding that MLF by both species may contribute to wine quality by modifying the concentration of some of the amino acids and aroma compounds of wine because of metabolic differences [8]. However, according to other previous studies, it was proved that the influence of lactic acid bacteria on wine sensory attributes is strain specific. Experiments carried out using different strains of *O. oeni* showed significant changes in aroma and flavor compounds, such as the modifications in fruity characters in red wines or the greater complexity acquired in Sauvignon Blanc wines [9–11].

Lactic acid bacteria can also produce undesirable aroma and flavor compounds, such as volatile phenols or biogenic amines, both related with quality loss and health problems for humans [12,13]. This problem could be solved with a custom selection of bacterial strains, according to the type of wine and its organoleptic properties, which would be modulated depending on the searched sensory profile and consumer preferences [10,14]. Additionally, the use of selected strains of wine bacteria allows for better control of the timeframe of L-malic acid degradation [15,16].

Once the wine bacterial strains are selected, another relevant parameter to study is the timing of the bacteria starter addition and the number of cells and viability in the wine after inoculation, since these can influence the sensory profile of wine. Three different strategies are being investigated: (i) inoculation at any time during alcoholic fermentation often performed simultaneously with yeast (co-inoculation); (ii) inoculation at pressing; and (iii) inoculation after the alcoholic fermentation (sequential inoculation) [17]. Co-inoculation of yeast and bacteria during fermentation shows several benefits when compared to the other strategies, as it shortens vinification times, enhances the wine organoleptic characteristics and reduces the probability of microbial spoilage [2]. Co-inoculation also influences the volatile chemical composition and the potential sensory attributes of the wine, often giving fruitier notes, as opposed to butter or nuts when the MLF begins after the end of the alcoholic fermentation [18].

The aim of the present study is to determine the effect of the coexistence of inoculated commercial yeast and novel selected lactic acid bacteria strains and its impact on the aroma profile and sensory traits of Tempranillo red wine, by analyzing chemical, physical, and sensory parameters.

2. Materials and Methods

2.1. Microorganisms

The yeast strain selected to perform alcoholic fermentation was Viacell C-58® (Lallemand Inc., Montreal, Quebec, Canada). This *Saccharomyces cerevisiae* yeast strain was isolated in Germany and used according to the manufacturer's recommendations. It has the ability to release polysaccharides to the wine, stabilizing color in red wines, and decreasing astringency of green tannins. Furthermore, it stimulates the development of fruity aromas that persist after barrel aging. This yeast strain also shows a high adaptation to low temperatures, which makes it suitable for cold pre-fermentative maceration.

The commercial wine lactic acid bacteria (LAB) used were *O. oeni* PN4™ (Lallemand Inc., Montreal, Quebec, Canada) and *O. oeni* Omega™ (Lallemand Inc., Montreal, Quebec, Canada). LAB strain PN4™ was isolated and selected at the Edmund Mach Institute (San Michele all'Adige, Trentino, Italy). This LAB stands out for being a robust strain, which showed the ability to complete malolactic fermentation in red and white wines under adverse conditions of pH, alcohol, SO_2, and temperature. In red wines, PN4™ seems to contribute to spicy notes. LAB strain *O. oeni* Omega™ was isolated and selected in the Languedoc region in France by the Institut Française de la Vigne et du Vin (IFV). According to the previous studies made by the manufacturer, it stands out for its ability to finish rapidly malolactic fermentation in a wide range of applications. This LAB endures low pH and high alcohol levels. It is effective in red, rose, and white wines. It complements a fresh and direct fruit style and helps to stabilize color due to its slow degradation of acetaldehyde. Both LAB had been described by the manufacturer (Lallemand Inc., Montreal, Quebec, Canada) to be highly suitable for the use as components of a mixed yeasts/bacteria co-inoculum.

2.2. *Vinifications*

Tempranillo grape-berries were harvested at the optimum technological maturity (23–25 °Brix) during the 2018 vintage in the Castilla-La Mancha winemaking area. Once grapes were received in the cellar, they were crushed and treated with 4 g/hL of SO_2, and then a dose of 2 g/hL of IOC enzyme (Lallemand Inc., Montreal, Quebec, Canada) was added to favor the color extraction. The initial chemical composition of must was: °Brix 23.50; total acidity 5.35 g/L; pH 3.85, and L-malic acid 1.95 g/L. Must with crushed grapes was homogeneously divided after crushing into six stainless steel vessels with a capacity of 100 L. Two of the vessels were used as control (AF1), in which AF means Alcoholic Fermentation and no lactic acid bacteria strains were added, and after alcoholic fermentation with the selected *S. cerevisiae*, wines carried out malolactic fermentation spontaneously. The four remaining deposits were inoculated with commercial *O. oeni* lactic acid bacteria strains 24 h after yeast inoculation, two of them with PN4^{TM} (AF2) and the other two with Omega^{TM} (AF3). The obtained wines were cold-stabilized at –3 °C for 25 days, and afterwards they were filtered and bottled. Once they were bottled, wines were sensorily and chemically analyzed.

2.3. *Microbiological Control*

Determination of microbial population was carried out as stated by Izquierdo-Cañas et al. [16]. The viable count of yeasts was performed by plating samples in malt extract agar media (Cultimed, Barcelona, Spain), and incubated at 28 °C for 48 h. In order to quantify lactic acid bacteria, samples were plated onto MLO Agar (MLOA, *O. oeni* Medium, Scharlab, Barcelona, Spain) supplemented with 10% *v/v* tomato juice, 50 mg/L sodium azide, and 100 mg/mL of cycloheximide. Plates were incubated under anaerobic conditions (Gas Pack System, Oxoid Ltd, Basingstoke, UK) at 30 °C for 5 days. The results were expressed in colony forming units per mL (CFU/mL) of wine.

The implantation of the inoculated lactic acid bacteria was assessed by the random selection of isolated colonies from MLOA plates at different fermentation times after inoculation, when L-malic acid concentration had decreased to 50% and when L-malic acid content was below 0.2 g/L.

2.4. *FML Progress and Physicochemical and Spectrophotometric Analysis*

FML evolution was determined by the periodic analysis of L-malic acid content, using the enzymatic method established by the International Organization of Vine and Wine (OIV) [19]. Most common physiochemical parameters, such as alcoholic strength, pH, total acidity, total and free SO_2, glucose and fructose content, and Folin–Ciocalteu index to determine the total phenolic content of wines were also measured following the recommendations of the official methods of the OIV [20]. Colorimetric parameters were measured at pH 3.6 by a CIE-Lab* color system following the protocol proposed by Ayala et al. [21]. Total anthocyanin content was measured by the sulfur dioxide bleaching method according to Riberèau-Gayon and Stonestree [22]. Concentration of different carboxylic acids (malic, lactic and citric acid) and glycerol content were determined by an isocratic HPLC system that was set up with a column block heater and refractive index detector. The mobile phase was 8 mmol/L H_2SO_4, 0.6 mL/min, and set at 75 °C in an Aminex HPX-87H 300 × 7.8 mm column. Finally, tannin content in wines was quantified by precipitation with methylcellulose following the protocol proposed by Sarneckis et al. [23].

2.5. Volatile Compound Analysis

Volatile compound extraction was performed according to the protocol described by Ibarz et al. [24]. SPE cartridges (Li-Chrolut EN, Merck, 0.2 g of phase, Darmstadt, Germany) and 4-nonanol (0.1 g/L) were used as internal standards. The extracts obtained were concentrated to a final volume of 150 μL by distillation in a Vigreux column and then under a nitrogen stream. Samples were kept at −20 °C until further analysis. A Focus-GC system coupled to a mass spectrometer ISQ with electron impact-ionization source and quadrupole analyzer with an auto-sampler Tri-Plus (Thermo-Quest, Waltham, MA, USA) was used to assess the free volatile composition of wines. We used a BP21 column (SGE, Ringwood, Australia) with a 50 m × 0.32 mm internal diameter and 0.25 μm thick of Free Fatty Acid Phase (FFAP). The chromatographic conditions were as follows: carrier helium gas (1.4 mL min^{-1}, split 1/57); injector temperature, 220 °C; and oven temperature, 40 °C for 15 min, 2 °C min^{-1} to 100 °C, 1 °C/min to 150 °C, 4 °C/min to 210 °C, and 55 min at 210 °C. The detector conditions were as follows: impact energy, 70 eV; electron multiplier voltage, 1250 V; ion source temperature, 250 °C; and mass scanning range, 40–250 amu. Volatile compounds were detected by chromatographic retention times and mass spectra, using commercial products as standards. Quantification of volatile compounds was carried out by analyzing the characteristic m/z fragment for each compound using the internal standard method. Concentrations of non-available volatile compounds were expressed at μg/L or mg/L as 4-nonanol equivalents obtained by normalizing the peak area to that of the internal standard and multiplying by the concentration of the internal standard.

2.6. Sensory Analysis

Descriptive and triangular tests were first performed to determine the differences among treatments. Fifteen tasters were selected to carry out a descriptive sensory analysis according to the Sensory Profile method, based on a scale from 0 (absence of the descriptor) to 5 (maximum intensity), and established by the ISO standard [25]. Selected attributes were color intensity, purplish-red color, aroma of red fruits, aroma of ripe red fruits, aroma of raisins, floral bouquet, aroma of spices, milky aromas, fresh aroma, astringent flavor, noble mature tannins in the mouth, and lengthy finish in the mouth.

2.7. Statistical Analysis

Significant differences among samples were determined for each chemical compound by analysis of variance (ANOVA) and paired Student's t-test ($p = 0.05$) for each chemical compound. A cobweb graph was performed to determine significant differences of the attributes after sensory analysis. Statistical analysis was performed with 20.0 SPSS (IBM Inc. Chicago, IL, USA) for Windows statistical package.

3. Results

3.1. Evolution of Microbial Populations and Fermentation Kinetics

The progress of alcoholic fermentation of the commercial *S. cerevisiae* Viacell C-58® strain was evaluated for each vessel by daily density measurement under the established experimental conditions: spontaneous malolactic fermentation, co-inoculation with *O. oeni* PN4™, and co-inoculation with *O. oeni* Omega™. After seven days, since density was always below 993 g/L, alcoholic fermentation was considered to be finished with similar results in all wines (Figure 1).

Although the presence of native LABs could have an adverse effect in non-sterile must fermentation, our results showed the favorable effect on the start-up period of inoculating pure LAB cultures. The presence of native lactic acid bacteria in the first stages of alcoholic fermentation did not affect the progressive decrease of density in wine (Figure 1). In fact, in the spontaneous malolactic fermentation (control), it was observed that population of *S. cerevisiae* was affected by the presence of native lactic acid bacteria, maintaining an average concentration of 10^6 cells/mL when must density decreased, until LAB took part in malolactic fermentation (Figure 1). In the co-inoculated vinification with the

O. oeni PN4™ strain, the development of the bacterial population showed a good adaptation to fermentation conditions, following fermentative kinetics similar to the corresponding kinetics of Viacell C-58® yeast. Conversely, in the co-inoculated fermentation with Omega™, a different behavior was observed in presence of Viacell C-58® during the first days of the alcoholic fermentation, reaching concentrations between 10^7 and 10^8 cells/mL, while it reached similar populations to PN4™ during the rest of the alcoholic fermentation (Figure 1).

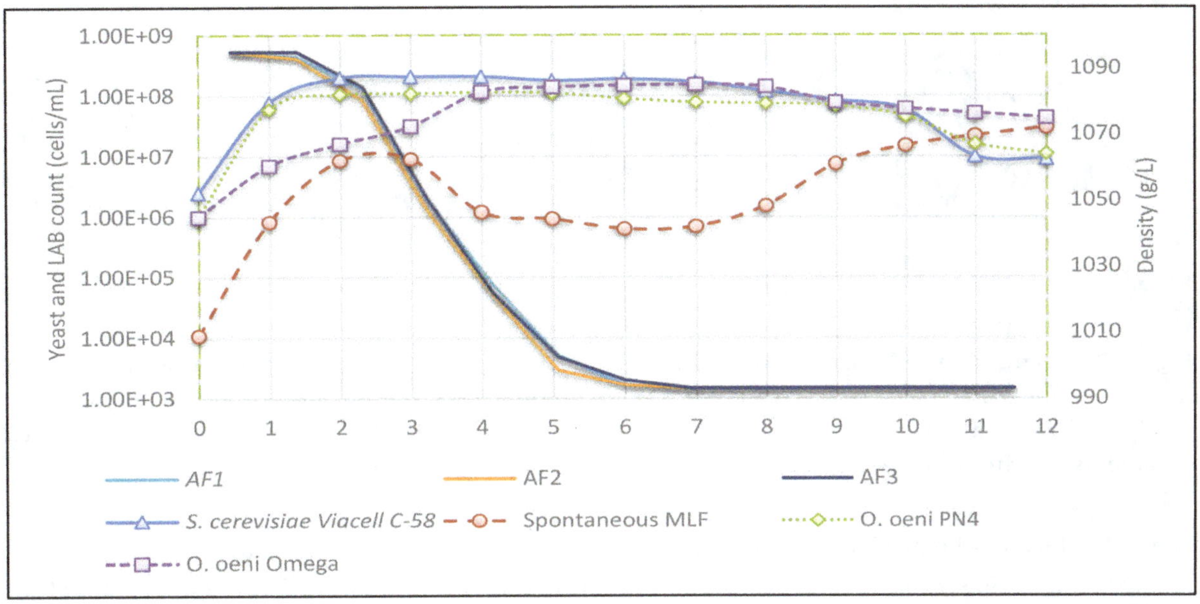

Figure 1. Progress of the yeast Viacell C-58® (**Δ**) and lactic acid bacteria population (dashed lines) and must density (continuous lines) during the three wine alcoholic fermentations denoted as (AF1, AF2, and AF3) and performed with Tempranillo musts using the following inoculation strategies: no lactic acid bacteria (LAB) inoculation (•); inoculation with LAB PN4™ (♦); and inoculation with LAB Omega™ (■). Data represented correspond to mean values obtained from duplicate fermentations; in all cases standard deviation was <1.5. MLF means malolactic fermentation.

3.2. Malolactic Fermentation

Malolactic fermentation dynamics were controlled by the daily determination of L-malic and L-lactic acid content. In co-inoculated fermentation with PN4™ and Omega™ strains, the behavior was similar during the first days, but variations in L-malic acid consumption appeared in the final stages of the alcoholic fermentation (Figure 2). The strain PN4™ could completely consume L-malic acid in 10 days in the presence of *S. cerevisiae* starters, while the Omega™ strain did the same two days later. Regarding the spontaneous malolactic fermentation, the consumption of L-malic acid was started from day 12, finishing MLF in 26 days (Figure 2). Again, this fact confirms that lactic acid bacteria co-existed with *S. cerevisiae* Viacell C-58® until L-malic acid was almost finished. It is important to highlight that the L-malic acid consumption was reduced between 14 and 16 days when *O. oeni* PN4™ and Omega™ strains were inoculated compared to the spontaneous MLF.

3.3. Determination of Physic-Chemical Parameters of Fermentations and Color

Despite observing significant differences when comparing residual sugars in both co-inoculated and control vinifications, in all cases must sugars were completely consumed, and dry wines were obtained. Regarding acetic acid levels, they decreased to 0.28 g/L for PN4™ and 0.26 g/L for Omega™ values, which were lower than those obtained for spontaneous MLF wine, 0.35 g/L (Table 1).

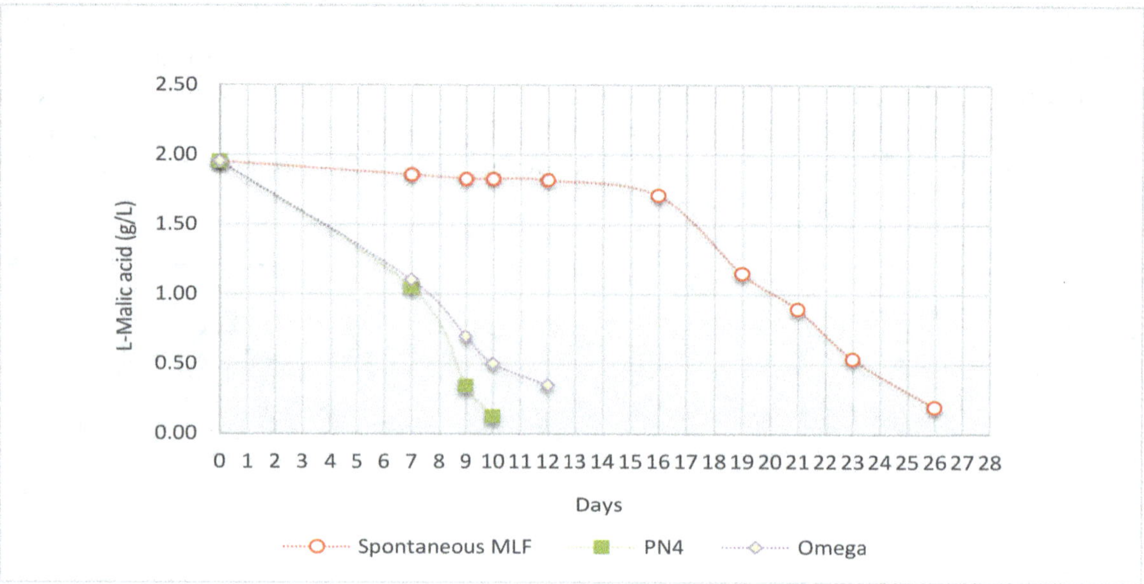

Figure 2. Progression of L-malic acid consumption (g/L) during vinification of Tempranillo must in the two samples inoculated with LAB PN4^{TM} (♦) and with LAB Omega^{TM} (■) at the beginning of the alcoholic fermentation and the uninoculated one (•). Data represented correspond to mean values obtained from duplicate fermentations; in all cases standard deviation was <1.5.

Table 1. Physicochemical parameters, organic acid concentrations, and color values of wines. Data represented correspond to mean and standard deviation values obtained from duplicate vinifications. Characters a, b, and c mean significant differences at $p < 0.05$.

	Spontaneous MLF	**PN4^{TM}**	**Omega^{TM}**
Alcoholic strength (% v/v)	13.34 ± 0.66 ^{a}	13.45 ± 0.15 ^{a}	13.52 ± 0.23 ^{a}
Total acidity (g/L)	3.49 ± 0.03 ^{a}	3.62 ± 0.18 ^{a}	3.63 ± 0.18 ^{a}
pH	4.07 ± 0.08 ^{a}	4.04 ± 0.07 ^{a}	4.01 ± 0.03 ^{a}
Volatile acidity (g/L)	0.35 ± 0.01 ^{b}	0.28 ± 0.03 ^{a}	0.26 ± 0.00 ^{a}
L-malic acid (g/L)	0.07 ± 0.04 ^{a}	0.05 ± 0.01 ^{a}	0.08 ± 0.02 ^{a}
L-lactic acid (g/L)	1.27 ± 0.00 ^{c}	1.22 ± 0.02 ^{b}	1.12 ± 0.06 ^{a}
Citric acid (g/L)	0.07 ± 0.03 ^{a}	0.20 ± 0.04 ^{b}	0.19 ± 0.04 ^{b}
Glycerol (g/L)	9.69 ± 0.93 ^{a}	9.91 ± 0.26 ^{a}	9.85 ± 0.28 ^{a}
L*	17.13 ± 3.29 ^{a}	15.34 ± 0.36 ^{a}	19.26 ± 2.23 ^{a}
a*	43.39 ± 1.31 ^{a}	43.46 ± 0.68 ^{a}	44.38 ± 0.09 ^{a}
b*	22.13 ± 0.84 ^{a}	21.28 ± 0.18 ^{a}	20.99 ± 0.82 ^{a}
Color intensity	6.44 ± 0.83 ^{a}	6.79 ± 0.32 ^{a}	5.89 ± 0.55 ^{a}
Color tone	0.83 ± 0.00 ^{b}	0.76 ± 0.04 ^{a}	0.79 ± 0.04 ^{ab}
Total anthocyanins (mg/L)	446.00 ± 39.60 ^{a}	432.00 ± 33.94 ^{a}	426.50 ± 27.58 ^{a}
Catechins (mg/L)	694.05 ± 67.39 ^{a}	730.10 ± 107.48 ^{a}	614.60 ± 55.58 ^{a}
Tannins (mg/L)	0.61 ± 0.12 ^{a}	0.64 ± 0.15 ^{a}	0.51 ± 0.30 ^{a}
Folin–Ciocalteu	33.00 ± 4.24 ^{a}	34.50 ± 3.54 ^{a}	31.00 ± 1.41 ^{a}

There were no significant differences in volatile acidity concentrations between the two co-inoculated wines, while the spontaneous MLF showed a higher volatile acidity (Table 1). Total acidity, glycerol, and pH were similar in all cases, which indicates that co-inoculation with commercial lactic acid bacteria did not affect in a negative way the chemical properties of wine (Table 1). L-lactic acid values differed for the three experimental conditions, and co-inoculated wine with the LAB Omega^{TM} showed the lowest values and the spontaneous MLF the highest. On the other hand, a decrease in citric acid content was observed in the spontaneous malolactic fermentation (Table 1). This fact indicates that citric acid degradation depends on the LAB strain used. In this study, the commercial strains degraded less citric acid and, as a consequence, the volatile acidity produced was lower than that of the non-co-inoculated one (Table 1). It is well known that during MLF, wine discoloration can take place. However, in this work, it was proved that there were no important differences in the spectral characteristics of wine color, and intensity was similar in all cases. Co-inoculation did not decrease wine color. No significant differences were observed in the rest of analyzed parameters (Table 1).

3.4. *Volatile Compound Analysis*

Aromatic profiles from the three different wines are shown in Table 2. Metabolism from Viacell C-58® and the native and commercial lactic acid bacteria decisively influenced the volatile composition of wine, including secondary aromas related to MLF. It is appreciated that esters and alcohols were significantly higher, and fatty acid content was lower in wines made by co-inoculation with commercial LAB strains compares to that made by spontaneous MLF, used as a control.

Ethyl acetate content was significantly different between the three fermentation strategies (Table 2). Fermentations carried out with the *O. oeni* Omega^{TM} strain showed a higher isoamyl acetate and hexyl acetate content, which are remarkable compounds in Tempranillo wines when compared to the other two wines. The lower content of 2-phenyl acetate was lower in the wine co-inoculated with the *O. oeni* PN4^{TM} strain (Table 2). Regarding the ethyl lactate, the LAB strain Omega^{TM} contributed to lower levels of lactic aromas, while the isoamyl acetate concentration was higher (Table 2).

Ethyl hexanoate is responsible for the fruity notes in red wines, and it was observed that its content was significantly higher in those wines made with *O. oeni* PN4^{TM} and Omega^{TM}, if compared to spontaneous MLF (Table 2). The same happened with 1-hexanol, which contributed to herbaceous notes (Table 2). It should also be emphasized that ethyl butyrate was significantly higher in wines made with Omega^{TM} (Table 2). No significant differences in the ethyl octanoate and ethyl decanoate contents were found between the three wine trials (Table 2).

Carboxylic acids are responsible for fatty, rancid, and buttery notes (Table 2). It was shown that co-inoculated wines had lower levels of hexanoic acid, octanoic acid, and decanoic acid with respect to the spontaneous MLF, highlighting that *O. oeni* PN4^{TM} notably reduced butyric acid and hexanoic acid content.

Taking into account that acetaldehyde content decreased during MLF, a decline was observed in wines with spontaneous MLF, in contrast with LAB co-inoculated wines, where PN4^{TM} showed the highest level of the compound. The 2,3-butanodione is a decisive compound in MLF, since it is mainly derived from lactic acid bacteria metabolism. High concentrations of this metabolite give heavy milky notes. Therefore, it is interesting to obtain it in low concentrations, as it will contribute to wine smoothness. High quantities of this compound were produced in spontaneous MLF wine in contrast to the co-inoculated ones. This fact can be very promising for wines produced by co-inoculation of commercial LAB since they have less lactic notes (octanoic and decanoic acid, 2,3-butanodione), which avoids the masking of fruity notes that give personality to Tempranillo wines made with the co-inoculation of *O. oeni* PN4^{TM} and Omega^{TM} strains.

3.5. *Sensory Analysis*

Regarding the sensory profile of wines and according to the tasters, differences in organoleptic properties were observed (Figure 3). Wines made by spontaneous MLF showed higher aromas of raisins, milky aromas, and color intensity. In contrast, wines developed by the inoculation of commercial lactic acid bacteria highlighted the fresh aroma and lengthy finish flavor and ripe red fruit character and less milky aromas (Figure 3).

Table 2. Data (Mean ± S.D.) of volatile composition related to the uninoculated fermentations (S) and inoculated fermentations with two commercial LAB strains (PN4^{TM} and Omega^{TM}). Characters a, b, and c mean significant differences at $p \leq 0.05$.

	Spontaneous MLF	**PN4^{TM}**	**Omega^{TM}**
Ethyl acetate (mg/L)	80.00 ± 15.89 ^{a}	95.18 ± 0.13 ^{b}	108.65 ± 2.93 ^{c}
Isoamyl acetate (mg/L)	5.85 ± 0.76 ^{a}	6.13 ± 0.10 ^{a}	8.85 ± 0.89 ^{b}
2-Phenylethyl acetate (μg/L)	182.61 ± 7.24 ^{b}	141.89 ± 7.89 ^{a}	170.63 ± 31.80 ^{b}
Hexyl acetate (μg/L)	13.54 ± 1.77 ^{a}	13.75 ± 0.84 ^{a}	21.21 ± 7.18 ^{b}
Ethyl lactate (mg/L)	145.86 ± 19.96 ^{b}	137.78 ± 6.11 ^{b}	117.54 ± 1.24 ^{a}
Ethyl butyrate (μg/L)	632.50 ± 106.77 ^{a}	687.50 ± 30.41 ^{a}	896.00 ± 94.75 ^{b}
Ethyl hexanoate (μg/L)	739.82 ± 26.96 ^{a}	790.88 ± 57.99 ^{ab}	866.78 ± 85.64 ^{b}
Ethyl octanoate (μg/L)	716.64 ± 49.17 ^{a}	677.01 ± 90.65 ^{a}	785.66 ± 85.41 ^{a}
Ethyl decanoate (μg/L)	97.26 ± 8.89 ^{a}	90.29 ± 14.61 ^{a}	100.97 ± 10.01 ^{a}
1-propanol (mg/L)	62.83 ± 8.25 ^{a}	54.74 ± 5.25 ^{a}	54.55 ± 0.51 ^{a}
Isobutanol (mg/L)	30.48 ± 3.71 ^{a}	28.96 ± 0.51 ^{a}	30.65 ± 1.20 ^{a}
1-butanol (mg/L)	1.46 ± 0.30 ^{a}	1.65 ± 0.12 ^{a}	1.46 ± 0.23 ^{a}
Isoamylic alcohols (mg/L)	342.50 ± 36.63 ^{a}	332.19 ± 1.25 ^{a}	327.92 ± 10.87 ^{a}
1-octanol (μg/L)	13.82 ± 2.45 ^{a}	16.14 ± 2.27 ^{a}	17.33 ± 1.77 ^{a}
3-methyl-1-pentanol (μg/L)	167.16 ± 42.26 ^{a}	212.66 ± 6.96 ^{c}	181.06 ± 1.16 ^{ab}
3-ethoxy-1-propanol (μg/L)	53.47 ± 13.10 ^{a}	48.15 ± 3.02 ^{a}	45.46 ± 2.95 ^{a}
1-hexanol (mg/L)	3.24 ± 0.12 ^{a}	3.46 ± 0.16 ^{b}	3.53 ± 0.08 ^{b}
t-3-hexenol (μg/L)	74.70 ± 0.47 ^{a}	75.24 ± 9.86 ^{a}	74.74 ± 0.86 ^{a}
c-3-hexen-1-ol (μg/L)	513.62 ± 88.94	516.16 ± 21.29	560.66 ± 4.26
2-phenylethanol (mg/L)	23.68 ± 2.79 ^{a}	26.07 ± 1.58 ^{a}	22.54 ± 1.49 ^{a}
Isobutiric acid (μg/L)	460.53 ± 49.70 ^{a}	428.63 ± 51.80 ^{a}	426.93 ± 10.52 ^{a}
Butyric acid (μg/L)	570.86 ± 1.69 ^{b}	436.89 ± 2.96 ^{a}	513.28 ± 91.90 ^{b}
Isovaleric acid (mg/L)	1.43 ± 0.05 ^{c}	1.31 ± 0.08 ^{b}	1.22 ± 0.06 ^{a}
Valeric acid (μg/L)	13.68 ± 3.89 ^{b}	7.71 ± 2.71 ^{a}	6.68 ± 4.72 ^{a}
Hexanoic acid (mg/L)	8.29 ± 0.46 ^{b}	7.28 ± 0.57 ^{a}	7.80 ± 0.34 ^{ab}
Octanoic acid (mg/L)	7.25 ± 1.65 ^{b}	5.02 ± 0.08 ^{a}	5.41 ± 0.16 ^{a}
Decanoic acid (mg/L)	3.78 ± 0.94 ^{b}	2.08 ± 0.10 ^{a}	1.92 ± 0.17 ^{a}
γ-nonalactone (μg/L)	6.35 ± 0.53 ^{a}	5.39 ± 0.56 ^{a}	6.14 ± 0.55 ^{a}
δ-dodecalactone (μg/L)	9.58 ± 0.67 ^{a}	6.89 ± 1.05 ^{a}	7.56 ± 3.70 ^{a}
δ-ethoxycarbonyl-δ-butyrolactone (μg/L)	321.46 ± 39.16 ^{ab}	275.65 ± 26.25 ^{a}	355.95 ± 31.85 ^{b}
Damascenone (μg/L)	6.03 ± 0.81 ^{a}	6.27 ± 0.23 ^{a}	5.74 ± 0.97 ^{a}
3-oxo-α-ionol (μg/L)	55.82 ± 0.60 ^{a}	49.64 ± 7.01 ^{a}	53.82 ± 4.76 ^{a}
Acetaldehyde (mg/L)	9.28 ± 0.05 ^{a}	10.87 ± 0.96 ^{b}	14.00 ± 1.16 ^{c}
2.3 Butanodione (mg/L)	18.35 ± 1.67 ^{c}	14.77 ± 1.61 ^{b}	8.55 ± 0.48 ^{a}
1-octen-3-ona (μg/L)	14.25 ± 0.59 ^{a}	14.02 ± 1.84 ^{a}	14.19 ± 2.08 ^{a}
3-methyl-thio-propanol (μg/L)	277.33 ± 85.07 ^{a}	311.44 ± 66.18 ^{a}	211.11 ± 10.75 ^{a}
Furaneol (μg/L)	12.32 ± 1.26 ^{b}	5.94 ± 3.73 ^{a}	10.39 ± 2.52 ^{b}
Guaiacol (μg/L)	12.39 ± 1.65 ^{a}	18.44 ± 0.37 ^{b}	19.88 ± 5.18 ^{b}

Table 2. *Cont.*

	Spontaneous MLF	PN4™	Omega™
Eugenol (µg/L)	3.97 ± 0.36 b	2.52 ± 0.79 a	2.71 ± 0.13 a
Phenol (µg/L)	8.33 ± 0.98 a	9.41 ± 0.66 a	10.24 ± 3.49 a
4-metil-2, 6-ditercbutil-fenol (µg/L)	95.91 ± 29.03 b	66.28 ± 8.35 a	63.06 ± 0.07 a

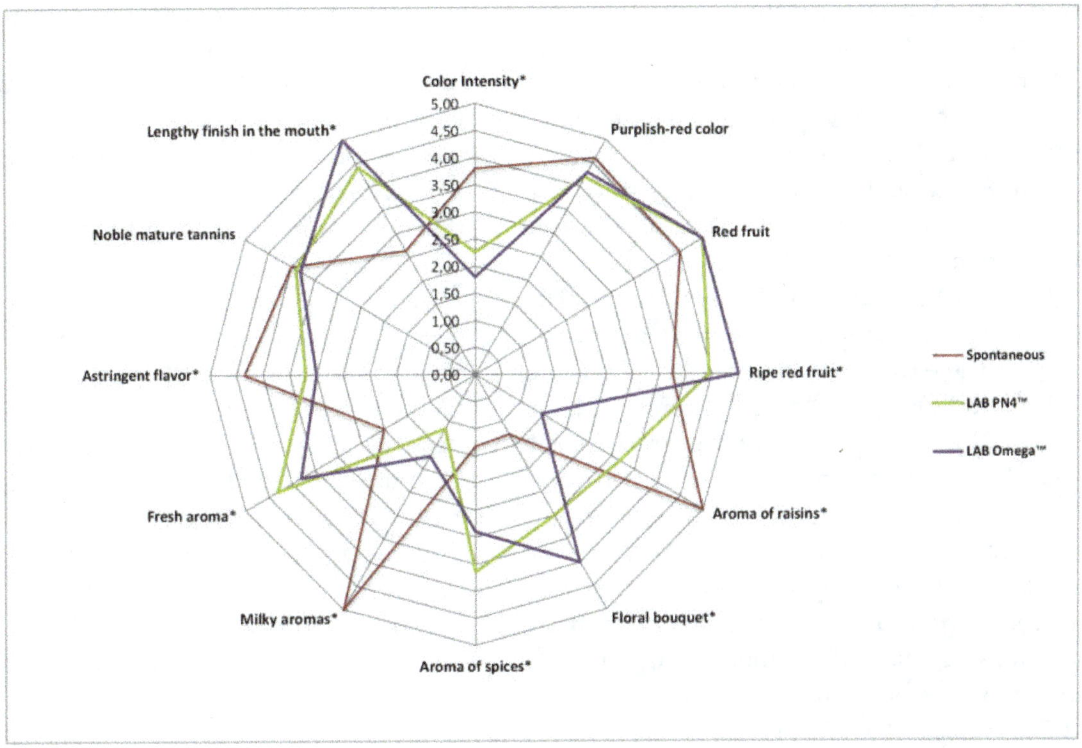

Figure 3. This figure shows the polar coordinate (cobweb) graph of mean sensory score ratings of "color intensity", "purplish-red color", "red fruit", "ripe red fruit", "aroma of raisins", "floral bouquet", "aroma of spices", "milky aroma", "fresh aroma", "astringent flavor", "noble mature tannins", and "lengthy finish in the mouth" for the spontaneous Tempranillo wine and the PN4™ and Omega™ LAB strains. In sensorial variables indicated with an asterisk (*) a difference between trials was verified for $p \leq 0.05$.

Panelists concluded that wines fermented with LAB reached more fruity and floral flavors, and spontaneous wines were less fruity and had stronger astringency and milky, raisin flavors. LAB Omega™ had strongly fruity and floral aromas, and PN4 in spices and fresh aromas in the Tempranillo wines. Spontaneous LAB produced the least ripe red fruit, floral, spices, and fresh aromas wines in this comparison. These wines tended to have milky, raisin, and astringent flavors. Statistical analysis revealed that the descriptors that the descriptors differed significantly in non-inoculated Tempranillo wines from those ones, which were inoculated,) suggesting that the different LAB strains deeply affected the flavor compounds of the wines (Figure 3).

4. Discussion

Under normal conditions, spontaneous malolactic fermentation (MLF) takes place once alcoholic fermentation has finished. This fact is due to a competition for must nutrients between yeast and lactic acid bacteria (LAB) naturally occurring in grapes. *S. cerevisiae* develops rapidly and consumes the essential nutrients for lactic acid bacteria growth, in addition to the production of ethanol and other byproducts that avoid their proliferation in the first stages of fermentation [26,27]. However, as alcoholic fermentation finishes, yeast suffer what is called lysis or autolysis, where they free intracellular nutrients that will be used as nitrogen and carbon sources to those lactic acid bacteria that resisted the ethanol concentration of the final wine [27]. Then, lactic acid bacteria population will exponentially proliferate to carry out MLF. It is possible to reproduce this natural phenomenon in controlled conditions. When winemakers use selected or commercial yeast starters to perform alcoholic fermentation, generally they inoculate lactic acid bacteria in sequential steps reproducing the succession that would take place naturally.

Other winemakers decide to co-inoculate LAB with fermentative yeast at the beginning of fermentation or just before alcoholic fermentation finishes [27–29]. This technique is gaining popularity because not only MLF is assured, but also carries recognized advantages by enologists and winemakers. One of the most important benefits of simultaneous inoculation of yeast/LAB is the reduction of total time of fermentation [18]. Jussier et al. [30] observed a significant reduction of MLF time in Chardonnay wines at pH of 3.53 and 13% ethanol (*v*/*v*) in co-inoculation when compared to sequential AF/MLF. This study corroborates this assertion at a pilot scale and is consistent with previous investigations performed at the laboratory scale [29,31–34].

Reduction of MLF times can be explained by specific interactions that occur between S. *cerevisiae* and O. *oeni* during AF and MLF, when the chosen strategy is the co-inoculation of both starter cultures [35]. In the experiments of this study, Viacell C-58® was used as fermentative yeast and two *O. oeni* (PN4 ™ and Omega™) as commercial LAB as a mixed co-inoculum of yeast/bacteria. Yeast and LAB strains had been described by the producers to be highly suitable for the use as component of a mixed yeast/bacteria co-inoculum. It can be observed that interactions between the yeast and the commercial co-inoculated bacteria are different from those that take place in spontaneous fermentation. The concentration of the inoculated yeast was about 10^6 CFU/mL in all cases for the commercial LAB, while the LAB initial population of spontaneous elaborations was 10^4 CFU/mL, a little higher than that reported in red wine musts [36]. MLF also occurred when AF was finished, but the duration of the total process was substantially superior than in the case of the two starter culture inoculations, due to the adaptation of bacterial starter to the must environment from the beginning of AF. Furthermore, viability of S. *cerevisiae* starter did not get influenced during the simultaneous progress of AF and MLF, as the yeast exponential growth phase did not decrease before reaching the stationary phase [32–37]. These evidences are consistent with those obtained in previous studies in Cabernet Franc, Tempranillo, and Merlot wines with other LAB and yeast strains [16,38] and Neroamaro wines [29].

This study also confirmed that MLF can take place in the presence of fermentable sugars without a significant increasing of acetic acid, in contrast with data obtained in other studies where the content of acetic acid increased in yeast and LAB co-inoculations [12,39,40]. This variability in the reduction of volatile acidity in co-inoculated wines with PN4™ and Omega™ strains could depend on the yeast strain used, on the LAB starter used, or both. This result shows consistency with those obtained by Tristezza et al. [29] in Neroamaro wines and du Plessis et al. [41] in Shiraz wines. In fact, in the vinification experiments, approximately 50% of the malic acid was consumed at the 7th day of the AF, and the total consumption of malic acid took place after the AF (days 7–12). This result reconfirmed a previous study of this research group in which an improvement of organoleptic properties of Cabernet Franc, Merlot, and Tempranillo wines was observed [16,38].

Regarding the color, no significant differences were observed in the spectral properties of color in co-inoculated MLF when compared to the control, but in contrast, lower shades were found in LAB co-inoculated wines. According to Burns and Osborne [42], MLF can affect red wine color independently of pH change. The species *O. oeni* could change the concentration of phenolic and non-phenolic compounds involved in red wine color development independently of pH change. As stated before, there were no strain specific differences in the color, but the tonality could be related to the *O. oeni* strain used.

The obtained results not only confirm that the correct selection of yeast and LAB strains for the performing of MLF by co-inoculation has a positive influence in the reduction of fermentation times and volatile acidity production, but also in the aromatic composition of wine. In fact, it was clearly proved the considerable effect of co-inoculation of the commercial yeast with two different LAB strains on the organoleptic properties of Tempranillo wines obtained, when compared with those obtained by spontaneous MLF. Recent investigations have highlighted the variation in aromatic profile in wine produced by different LAB inoculation processes [16,18,38,39]. Our data suggest that coexistence of yeast *Saccharomyces* with the lactic acid bacterial strain Omega^{TM} increases some ester levels, improving the fruity aroma according to the sensory analysis, as showed by Tristezza et al [29]. Nine esters were identified and quantified, and wines produced by co-inoculation contained higher concentrations of diethyl and monoethylsuccinate, ethyl lactate, 2-phenyl acetate, and ethyl fatty acids esters [2,29].

In general, the co-inoculation of both LAB strains gave as a result a significant change in the ester profile of wine, those of ethyl fatty acid being the most representative. This suggests that coexistence of yeast and lactic acid bacteria stimulates the formation of medium chain fatty acids and, consequently, the concentration of fatty acid esters varies in wine [29,38,43]. On the other hand, the presence of 2,3-butanodiol shows that, in the case of co-inoculation, bacteria were able to degrade diacethyl, a derived-MLF compound with high impact on wine organoleptic properties [18,39]. When this compound appears in high concentrations in wine, it can affect in a negative way its aroma, conferring buttery notes that interfere with fruity aromas [44]. Consequently, applying the co-inoculation technique, it is possible to reduce unwanted buttery and lactic notes, allowing the fruity ones to prevail, bringing out the characteristic varietal aroma of Tempranillo wines. Moreover, after sensory analysis, the wines from non-inoculated LAB fermentations had the majority of high scores for these descriptors, while those fermented with LAB had the lowest scores.

It is well known that the period from the end of the AF to the beginning of the FML is specifically conducive to the development of *Brettanomyces/Dekkera*. Early wine inoculation with LAB, immediately after AF or co-inoculation, has been proved a simple and effective method to prevent *Brettanomyces* development and the production of off-flavors due to high concentrations of ethylphenols [45]. This study showed that phenol content in wine elaborated by co-inoculation of LAB starters was significantly lower than those ones by spontaneous MLF. Co-inoculation reduced the overall fermentation time by up to 2 weeks leading to a lower increase in volatile acidity and an increase of the aromatic quality of wines with more varietal and fruity notes.

Co-inoculation allows MLF to develop under reductive conditions and results in wine with very few milky and buttery flavors, related to the impact of specific compounds like 2,3-butanedione. This compound has also confirmed as being dependent on the wine bacteria used.

Author Contributions: P.M.I.-C., E.G.-R., and A.M.-M. contributed to the conception and design of experiments; P.M.I.-C., M.R.-C., G.C.-B, and J.M.H.-M. performed the experiments; P.M.I.-C., G.C.-B., E.G.-R., and A.M.-M. analyzed the data; P.M.I.-C-, M.R.-C. and G.C.-B. contributed to writing the paper. All authors critically revised the manuscript before submission. All authors have read and agreed to the published version of the manuscript.

Acknowledgments: The authors thank J.J. Armario for critically revising the manuscript and the English grammar.

References

1. Davis, C.R.; Wibowo, D.; Eschenbruch, R.; Lee, T.H.; Fleet, G.H. Practical implications of malolactic fermentation: A review. *Am. J. Enol. Viticult.* **1985**, *36*, 290–301.
2. Versari, A.; Patrizi, C.; Parpinello, G.P.; Mattioli, A.U.; Pasini, L.; Meglioli, M.; Longhini, G. Effect of co-inoculation with yeast and bacteria on chemical and sensory characteristics of commercial Cabernet Franc red wine from Switzerland. *J. Chem. Technol. Biot.* **2016**. [CrossRef]
3. Davis, C.R.; Wibowo, D.; Fleet, G.H.; Lee, T.H. Properties of wine lactic acid bacteria: Their potential enological significance. *Am. Soc. Enol. Viticult.* **1988**, *39*, 137–142.
4. Styger, G.; Prior, B.; Bauer, F.F. Wine flavor and aroma. *J. Ind. Microbiol. Biot.* **2011**. [CrossRef] [PubMed]
5. Bokulich, N.A.; Collins, T.S.; Masarweh, C.; Allen, G.; Heymann, H.; Ebeler, S.E.; Mills, D.A. Associations among wine grape microbiome, metabolome, and fermentation behavior suggest microbial contribution to regional wine characteristics. *mBio* **2016**, *7*, e00631-16. [CrossRef] [PubMed]

6. Belda, I.; Ruiz, J.; Esteban-Fernández, A.; Navascués, E.; Marquina, D.; Santos, A.; Moreno-Arribas, M.V. Microbial contribution to Wine aroma and its intended use for Wine quality improvement. *Molecules* **2017**, 189. [CrossRef] [PubMed]
7. Bartowsky, E.J.; Borneman, A.R. Genomic variations of *Oenococcus oeni* strains and the potential to impact on malolactic fermentation and aroma compounds in wine. *Appl. Microbiol. Biot.* **2011**, *92*, 441–447. [CrossRef]
8. Pozo-Bayón, M.; G.-Alegría, E.; Polo, M.C.; Tenorio, C.; Martín-Álvarez, P.J.; Calvo De La Banda, M.T.; Ruiz-Larrea, F.; Moreno-Arribas, M.V. Wine volatile and amino acid composition after malolactic fermentation: Effect of *Oenococcus oeni* and *Lactobacillus plantarum* starter cultures. *J. Agr. Food Chem.* **2005**, *53*, 8729–8735.
9. Cappello, M.S.; Zapparoli, G.; Logrieco, A.; Bartowsky, E.J. Linking wine lactic acid bacteria diversity with wine aroma and flavour. *Int. J. Food Microbiol.* **2017**. [CrossRef]
10. De Revel, G.; Martin, N.; Pripis-Nicolau, L.; Lonvaud-Funel, A.; Bertrand, A. Contribution to the knowledge of malolactic fermentation influence on wine aroma. *J. Agr. Food Chem.* **1999**, *47*, 4003–4008. [CrossRef]
11. Malherbe, S.; Menichelli, E.; du Toit, M.; Tredoux, A.; Muller, N.; Næs, T.; Nieuwoudt, H. The relationships between consumer liking, sensory and chemical attributes of *Vitis vinifera* L. cv. Pinotage wines elaborated with different Oenococcus oeni starter cultures. *J. Sci. Food Agr.* **2013**. [CrossRef] [PubMed]
12. Liu, S. Malolactic fermentation in wine—beyond deacidification. *J. Appl. Microbiol.* **2002**, *92*, 589–601. [CrossRef]
13. Guo, Y.-Y.; Yang, Y.-P.; Peng, Q.; & Han, Y. Biogenic amines in wine: A review. *Int. J. Food Sci. Tech.* **2015**, *50*, 1523–1532. [CrossRef]
14. Inês, A.; Falco, V. Lactic Acid Bacteria Contribution to Wine Quality and Safety. In *Generation of Aromas and Flavours*; Vilela, A., Ed.; InTechOpen: London, UK, 2018. [CrossRef]
15. Gao, C.; Fleet, G.H. The degradation of malic acid by high-density cell suspensions of *Leuconostoc oenos*. *J. Appl. Bacteriol.* **1994**. [CrossRef]
16. Izquierdo-Cañas, P.M.I.; Pérez-Martín, F.; Romero, E.G.; Prieto, S.S.; de los Herreros, M.L.P. Influence of inoculation time of an autochthonous selected malolactic bacterium on volatile and sensory profile of Tempranillo and Merlot wines. *Int. J. Food Microbiol.* **2012**. [CrossRef]
17. Fugelsang, K.C.; Edwards, C.G. *Wine microbiology. Practical Applications and Procedures, in Lactic Acid Bacteria*, 2nd ed.; Springer: New York, NY, USA, 2007; pp. 29–41.
18. Abrahamse, C.E.; Bartowsky, E.J. Timing of malolactic fermentation inoculation in Shiraz grape must and wine: Influence on chemical composition. *W J. Microb Biot.* **1007**. [CrossRef] [PubMed]
19. OIV. *Compendium of international methods of analysis–OIV Chromatic Characteristics Method OIV-MA-AS2-07B: R2009*; OIV: Paris, France, 2009; Available online: http://www.oiv.int/oiv/files/6-Domainesscientifiques/6-4Methodesdanalyses/6-4-1/EN/OIV-MA-AS2-07B.pdf.
20. OIV. *Compendium of international methods of wine and must analysis*; International Organization of Vine and Wine: Paris, France.
21. Ayala, F.; Echávarri, J.F.; Negueruela, A.I. A new simplified method for measuring the color of wines. I. Red and Rosé wines. *Am. J. Enol. Viticult.* **1997**, *48*, 357–363.
22. Ribéreau-Gayon, P.; Stonestreet, E. Le dosage des anthocyanes dans les vins rouges. *Bull. Soc. Chim.* **1965**, *9*, 2649–2652.
23. Sarneckis, C.J.; Dambergs, R.G.; Jones, P.; Mercurio, M.; Herderich, M.J.; Smith, P.A. Quantification of condensed tannins by precipitation with methyl-cellulose: Development and validation of an optimized tool for grape and wine analysis. *Aust. J. Grape Wine Res.* **2006**, *12*, 39–42. [CrossRef]
24. Ibarz, M.; Ferreira, V.; Hernández-Orte, P.; Loscos, N.; Cacho, J. Optimization and evaluation of a procedure for the gas chromatographic-mass spectrometric analysis of the aromas generated by fast acid hydrolysis of flavors precursors extracted from grapes. *J. Chromatogr. A.* **2006**, *1116*, 217–229. [CrossRef]
25. ISO Standard 11035. In *Sensory analysis. Identification and selection of descriptors for stablishing a sensory profile*; ISO: Geneva, Switzerland, 1994.
26. Lucio, O.; Pardo, I.; Krieger-Weber, S.; Heras, J.M.; Ferrer, S. Selection of *Lactobacillus* strains to induce biological acidification in low acidity wines. *LWT Food Sci Tech.* **2016**. [CrossRef]
27. Pardo, I.; Ferrer, S. Yeast-Bacteria Co-inoculation. *Red Wine Technol.* **2019**. [CrossRef]
28. Lasik, M. The application of malolactic fermentation process to create good-quality grape wine produced in cool-climate countries: A review. *Eur. Food Res. Technol.* **2013**. [CrossRef]

29. Tristezza, M.; di Feo, L.; Tufariello, M.; Grieco, F.; Capozzi, V.; Spano, G.; Mita, G.; Grieco, F. Simultaneous inoculation of yeasts and lactic acid bacteria: Effects on fermentation dynamics and chemical composition of Negroamaro wine. *LWT Food Sci. Tech.* **2016**. [CrossRef]
30. Jussier, D.; Morneau, A.D.; De Orduña, R.M. Effect of simultaneous inoculation with yeast and bacteria on fermentation kinetics and key wine parameters of cool-climate Chardonnay. *Appl. Environ. Microb.* **2006**. [CrossRef]
31. Rosi, I.; Fia, G.; Canuti, V. Influence of different pH values and inoculation time on the growth and malolactic activity of a strain of *Oenococcus oeni*. *Aust. J. Grape. Wine Res.* **2003**. [CrossRef]
32. Massera, A.; Soria, A.; Catania, C.; Krieger, S.; Combina, M. Simultaneous inoculation of Malbec (*Vitis Vinifera*) musts with yeast and bacteria: Effects on fermentation performance, sensory and sanitary attributes of wines. *Food Technol. Biotech.* **2009**.
33. Zapparoli, G.; Tosi, E.; Azzolini, M.; Vagnoli, P.; Krieger, S. Bacterial inoculation strategies for the achievement of Malolactic fermentation in high-alcohol wines. *S. Afr J. Enol. Vitic.* **2009**. [CrossRef]
34. Antalick, G.; Perello, M.C.; De Revel, G. Co-inoculation with yeast and LAB under winery conditions: Modification of the aromatic profile of merlot wines. *S. Afr. J. Enol. Vitic.* **2013**. [CrossRef]
35. Alexandre, H.; Costello, P.J.; Remize, F.; Guzzo, J.; Guilloux-Benatier, M. *Saccharomyces cerevisiae-Oenococcus oeni* interactions in wine: Current knowledge and perspectives. *Int. J. Food Microbiol.* **2004**. [CrossRef]
36. Ribéreau-Gayon, P.; Glories, Y.; Maujean, A.; Dubourdieu, D. *Handbook of Enology, The Chemistry of Wine: Stabilization and Treatments: Second Edition*, 2nd ed.; Wiley: Hoboken, NJ, USA, 2006. [CrossRef]
37. Tristezza, M.; Fantastico, L.; Vetrano, C.; Bleve, G.; Corallo, D.; Mita, G.; Grieco, F. Molecular and technological characterization of *Saccharomyces cerevisiae* strains isolated from natural fermentation of susumaniello grape must in Apulia, Southern Italy. *Int. J. Microbiol.* **2014**. [CrossRef] [PubMed]
38. Izquierdo-Cañas, P.M.; Romero, E.G.; Pérez-Martín, F.; Seseña, S.; Palop, M.L. Sequential inoculation versus co-inoculation in Cabernet Franc wine fermentation. *Food Sci. Technol. Int.* **2015**. [CrossRef]
39. Knoll, C.; Fritsch, S.; Schnell, S.; Grossmann, M.; Krieger-Weber, S.; du Toit, M.; Rauhut, D. Impact of different malolactic fermentation inoculation scenarios on Riesling wine aroma. *World J. Microb. Biot.* **2012**. [CrossRef] [PubMed]
40. Garofalo, C.; El Khoury, M.; Lucas, P.; Bely, M.; Russo, P.; Spano, G.; Capozzi, V. Autochthonous starter cultures and indigenous grape variety for regional wine production. *J. Appl. Microbiol.* **2015**. [CrossRef]
41. Du Plessis, H.; Du Toit, M.; Nieuwoudt, H.; Van Der Rijst, M.; Kidd, M.; Jolly, N. Effect of *Saccharomyces*, non-*Saccharomyces* yeasts and malolactic fermentation strategies on fermentation kinetics and flavor of Shiraz wines. *Fermentation* **2017**, 64. [CrossRef]
42. Burns, T.R.; Osborne, J.P. Impact of malolactic fermentation on the color and color stability of pinot noir and Merlot wine. *Am. J. Enol. Viticult.* **2013**. [CrossRef]
43. Fang, Y.; Qian, M. Aroma compounds in Oregon Pinot Noir wine determined by aroma extract dilution analysis (AEDA). *Flavour Frag. J.* **2005**. [CrossRef]
44. Bartowsky, E.J.; Henschke, P.A. The "buttery" attribute of wine - Diacetyl - Desirability, spoilage and beyond. *Int J. Food Microbiol.* **2004**. [CrossRef]
45. Gerbaux, V.; Briffox, C.; Dumont, A.; Krieger, S. Infuence of inoculation with Malolactic bacteria on volatile phenols in wines. *Am. J. Enol. Viticult.* **2009**, *60*, 233–235.

8

Micro- and Nanoscale Approaches in Antifungal Drug Discovery

Ronnie G. Willaert [1,2]

[1] Alliance Research Group VUB-UGent NanoMicrobiology (NAMI), IJRG VUB-EPFL NanoBiotechnology & NanoMedicine (NANO), Research Group Structural Biology Brussels, Vrije Universiteit Brussel, 1050 Brussels, Belgium; Ronnie.Willaert@vub.be

[2] Department Bioscience Engineering, University Antwerp, 2020 Antwerp, Belgium

Abstract: Clinical needs for novel antifungal agents have increased due to the increase of people with a compromised immune system, the appearance of resistant fungi, and infections by unusual yeasts. The search for new molecular targets for antifungals has generated considerable research, especially using modern omics methods (genomics, genome-wide collections of mutants, and proteomics) and bioinformatics approaches. Recently, micro- and nanoscale approaches have been introduced in antifungal drug discovery. Microfluidic platforms have been developed, since they have a number of advantages compared to traditional multiwell-plate screening, such as low reagent consumption, the manipulation of a large number of cells simultaneously and independently, and ease of integrating numerous analytical standard operations and large-scale integration. Automated high-throughput antifungal drug screening is achievable by massive parallel processing. Various microfluidic antimicrobial susceptibility testing (AST) methods have been developed, since they can provide the result in a short time-frame, which is necessary for personalized medicine in the clinic. New nanosensors, based on detecting the nanomotions of cells, have been developed to further decrease the time to test antifungal susceptibility to a few minutes. Finally, nanoparticles (especially, silver nanoparticles) that demonstrated antifungal activity are reviewed.

Keywords: antifungal drug discovery; microfluidics; nanobiotechnology; omics-based approaches; antifungal susceptibility testing; nanomotion detection; nanoparticles

---

1. Introduction

Fungal infections are an extremely important health problem. Fungi infect about 1.2 billion people every year, yet their contribution to the global burden of disease is largely unrecognised [1,2]. Over 600 different fungi have been reported to infect humans. Most are "relatively" minor infections, but millions contract diseases that kill at least as many people as tuberculosis or malaria. More than 300 million people are affected by serious fungal infections worldwide. Invasive fungal infections are responsible for about 1.5 million deaths per year. Fungi are present everywhere in our environment and are, usually, harmless for people with a normal immune system. Fungal infections can be topical and local, such as infections on the skin or in the vaginal tract. Systemic infections arise when the fungi enter and proliferate in the bloodstream. Systemic fungal infections affect people with an altered immune system due to medical interventions (such as cancer therapy, organ transplantation, and immune-modulatory medications), immunosuppressive diseases (such as Acquired Immune Deficiency Syndrome (AIDS)) [3], or malnutrition (under- and overnutrition) [4–6].

The choice of available antifungal drugs to treat invasive fungal infections is limited, since only three structural classes of compounds are available, i.e., polyenes, azoles, and echinocandins. Antifungal therapy has become progressively more effective since the 1990s, however, no new antifungal classes have been reported since 2006 [7]. Current antifungal drugs show some limitations:

Amphothericin B (a polyene antibiotic) displays a considerable toxicity and undesirable side effects [8,9], issues with pharmacokinetic properties (such as a short half-life of echinocandins) and activity spectrum, a small number of targets [10,11], and they can interact with other drugs, such as chemotherapy agents and immunosuppressants [12,13]. The last approved antifungal (i.e., anidulafungin) by the European Medicines Agency and the Food and Drug Administration (FDA) dates back to 2006 [14]. There is an urgent need for safer and more effective antifungal drugs. Multiple types of antifungal compounds are in clinical development and these new agents have been recently reviewed [15–17].

Over the last 20 years, several approaches to antifungal discovery have been explored. The traditional approach seeks, first, to identify active compounds from large compound libraries using a panel of fungal pathogens in standardised assays when possible. In the genetic, genomic, or bioinformatics approach, the objective is, initially, to identify broadly represented targets in fungal pathogens and non-pathogens [18]. In this "target-centric" genomic approach, up-front genetic, bioinformatics, and biochemical target prioritisation is performed and, subsequently, an in vitro-based screening of individual targets is achieved. However, an exceedingly high rate of failure has been observed when applying this approach [19]. Additionally, not all bioactive compounds act through a target-specific mechanism of action (MoA). Various compounds act non-selectively as alkylating agents, intercalators, detergents, etc. Therefore, an integrated approach has been recently proposed for antimicrobial lead discovery that is rooted in empirical whole-cell screening for small molecules with intrinsic bioactivity whose MoA may be determined using a variety of forward or reverse genomic platforms to identify and, subsequently, validate their target [20]. In this "compound-centric" strategy, it is proposed to use existing genetic methods to evaluate the therapeutic effect of inactivating the target (both in vitro and in vivo). These methods often use the principles of genetic interaction, relying on the idea of genetic modifiers (enhancers or suppressors) to generate target hypotheses. Direct biochemical methods or computational inference have also been used for target identification. In many cases, however, a combination of approaches may be required to fully characterise on-target and off-target effects, and understand the mechanisms of small-molecule actions [21].

Recently, micro- and nanoscale approaches have been introduced in the field of antifungal drug discovery. In this paper, microfluidic approaches for antifungal screening and antifungal susceptibility testing (AFST), nanosensor development for AFST, and antifungal nanoparticles are reviewed.

2. Emerging Fungal Diseases and Antifungal Drugs

The epidemiology of invasive fungal infection is evolving [22–25]. A growing population of immunosuppressed patients have been diagnosed with invasive fungal infections. The epidemiology of candida infections has shifted over the last decade [26]. Non-*albicans Candida* species, with a reduced susceptibility to antifungals, are becoming more and more responsible for invasive candidiasis [27]. Several species, such as *Candida glabrata* and *C. krusei*, which are less susceptible to azole antifungal therapy, are increasing in some settings. *C. auris* has emerged in hospitals as a global concern, since the strains demonstrated multidrug resistance. *Trichosporon* species infections are the second most common cause of fungaemia in patients with haematological malignant disease and they were the third most commonly isolated non-*candida* species from patients in the ARTEMIS Global Antifungal Surveillance Program [28]. These species show resistance to amphotericin and echinocandins. *Rhodotorula* species are emerging opportunistic yeasts that are responsible for catheter-related fungaemia and sepsis, and invasive infections, particularly in immunosuppressed or -compromised patients [24]. The incidence of infections with non-*neoformans* crytococci has increased over the past 40 years, especially in patients with advanced HIV infection or cancer who are undergoing transplant [29]. Since the 1990s, there have been a growing number of reports about *Saccharomyces cerevisiae* invasive infections and novel strains continue to be identified [24,30,31].

Aspergillus species have, historically, been one of the most common causative organisms associated with invasive mould diseases [32]. *Aspergillus fumigatus* species were responsible for most of infections.

Recently, the epidemiology has changed towards non-*A. fumigatus* species and other moulds, such as *Zygomycetes* [33–36], *Fusarium* [37,38], and *Scedosporium* [38,39] species. This is due to the occurrence of azole resistant *Aspergillus* species [40,41].

Antifungal development is challenging, since, apart from the fungal cell wall, fungi are metabolically similar to mammalian cells and, therefore, offer few pathogen-specific targets [42]. As explained above, only three molecular classes that target three different fungal metabolic pathways are commonly used in clinical practice to treat, essentially, systemic fungal infections: Polyenes, azoles, and echinocandins (Table 1) [43]. The fluoropyrimidine analog 5-fluorocystosine is also used, but only in combination with amphotericin B.

Table 1. Overview of the mostly used antifungal agents, mechanism of action (MoA), and spectrum of activity [43–50].

Class Compound	MoA	Spectrum of Activity	Comment
Polyenes			
Amphotericin B	Selective binding to ergosterol cause the formation of pores in the membrane.	Treatment of deep mycoses, candidiasis, cryptococcosis, histoplasmosis, blastomycosis, paracoccodioidomycosis, cocciodioidomycosis, aspergillosis, extracutaneous sporotrichichosis, and some cases of mucormycosis, hyalohyphomycosis, and phaeohyphomycosis, *S. cerevisiae.*	Fungicidal, broad spectrum, intravenous, little resistance observed, significant nephrotoxicity, indirectly affects action of many drugs.
Nystatin/Nyotran	Selective binding to ergosterol cause the formation of pores in the membrane.	Candidiasis	Nyotran is a liposomal formulation of nystatin with lowered toxicity.
Natamycin	Binds to ergosterol in the plasma membrane, preventing ergosterol-dependent fusion of vacuoles, membrane fusion, and fission.	Keratinophilic fungi, corneal infections	
Azoles			
Fluconazole	Selective inhibition of fungal cytochrome P450-dependent lanosterol-14-α-demethylase.	*Candida immitis, C. neoformans, Paracoccidioides brasiliensis;* lower activity against *Aspergillus, Fusarium, Scedosporium, Penicillium* species and other filamentous fungi.	Fluconazole resistant *C. albicans* and non-*albicans* strains increasing.
Itraconazole	Selective inhibition of fungal cytochrome P450-dependent lanosterol-14-α-demethylase.	Most *Candida* species, *P. brasiliensis, H. capsulatum, Blastomyces dermatitidis, Aspergillus fumigatus, A. niger, Penicillium marneffei.*	Better than fluconazole in the treatment of cocciodioido-mycosis, not reaching the central nervous system; numerous drug interactions due to inhibition of CYP 3A4.
Voriconazole	Selective inhibition of fungal cytochrome P450-dependent lanosterol-14-α-demethylase.	*Candida* species, including *C. krusei, S. cerevisiae.*	More potent than fluconazole; very rapid metabolism in children; numerous drug interactions due to inhibition of CYP 3A4.
Econazole	Selective inhibition of fungal cytochrome P450-dependent lanosterol-14-α-demethylase.	*Trichophyton rubrum, T. Mentagrophytes, Epidermophyton floccosum.*	Is an immidazole antifungal for the treatment of tinea pedis and crusis, pityriasis versicolor.
Tioconazole	Selective inhibition of fungal cytochrome P450-dependent lanosterol-14-α-demethylase.	*C. albicans, Trichophyton* sp., *Epidermophyton* sp.	Is an immidazole antifungal for topical treatment of superficial mycoses (ringworm, jock itch, athlete's foot, tinea versicolor.

Table 1. *Cont.*

Class Compound	MoA	Spectrum of Activity	Comment
Echinocandins			
Caspofungin	Fungal β-1,3-glucan synthase inhibitors.	*Candida* species, *Aspergillus* species.	Fungicidal for *Candida*, fungistatic for *Aspergillus*; modest efficacy as first-line agent for invasive aspergillosis; intravenous formulation only; interacts with ciclosporin and rifampicin.
Anidulafungin	Fungal β-1,3-glucan synthase inhibitors.	*Candida* species, *Aspergillus* species.	Fungicidal for *Candida*, fungistatic for *Aspergillus*; licensed for the treatment of invasive and esophageal candidiasis.
Micafungin	Fungal β-1,3-glucan synthase inhibitors.	*Candida* species, *Aspergillus* species.	Fungicidal for *Candida*, fungistatic for *Aspergillus*; licensed invasive and esophageal candidiasis.
5-Fluoropyrimidine			
5-Fluorocystosine	Selective conversion of toxic intermediate (5-fluorouridine).	Cryptococcosis, candidiasis, chromoblastomycosis; high MIC for some strains of *Aspergillus*, *Penicillium* and several Zygomycetes, except for chromoblastomycosis.	5-Fluorocystosine is always used in combination with amphotericin B.

Recently, a lot of effort has also been made in analysing the antifungal action of natural compounds or natural bioactive compounds. The discovered natural antifungal derivatives or compounds derived from natural origins have been recently reviewed [17,51]. Some examples of these compounds include chitosan [52]; herbal compounds, such as thymol, carvacrol, eugenol, and menthol [53]; and extracts from plants, such as *Hypericum carinatum* [54], *Stenachaenium megapotamicum* [55], and *Acca sellowiana* [56].

3. Omics-Based Antifungal Drug Discovery Approaches

The traditional approach seeks, first, to identify active compounds (generally from large libraries of synthetic small molecules or natural products) that inhibit the growth of the fungus. The most commonly used assay to identify antifungal leads is liquid growth inhibition assay in which microbial growth is measured by optical density (OD) of the culture. These assays have some limitations, such as, sometimes, a poor correlation between growth and OD for fungi that grow as filaments, and these assays are unable to distinguish between molecules that inhibit growth (fungistatic) from those that kill the organism (fungicidal), a feature that is very important for the treatment of some fungi (e.g., *Cryptococcus*) [57]. Recently, cell viability screening assays based on other readouts have been developed. The most widely adopted approach is the use of the dyes, Alamar Blue and tetrazolium salt (XTT), as reporters for metabolic activity. These dyes are converted to fluorescent molecules when metabolised by viable molecules and can be used for high-throughput multiwell screening [58,59]. The dye, resazurin, was used to develop an assay for high-throughput screening of *A. fumigatus* [60]. An Alamar-Blue-based high-throughput protocol was devised to identify molecules with fungicidal activity against *C. neoformans* [61]. A second type of viability assay that was recently developed is based on the detection of extracellular adenylate kinase (AK) as a reporter of cell lysis [62]. AK is a conserved enzyme that has been used as a cytotoxicity reporter in mammalian cell culture assays and as a reporter of brewing yeast autolysis in the beverage industry [63].

Another approach is referred to as genetic, genomic, or bioinformatics in which the objective is, initially, to identify broadly represented targets in fungal pathogens and non-pathogens. The operational code for this strategy consists of three basic rules for identifying potential targets: (i) A bioinformatics-driven approach is used to identify pathogen-specific genes with the desired conservation and spectrum, and the absence of a human ortholog; (ii) genetic analyses are performed to confirm that loss-of-function mutations result in a nonviable growth phenotype and/or a nonvirulent phenotype under standard laboratory conditions and/or a nonvirulent phenotype in a relevant animal model of infection; and (iii) there should be some evidence, derived from sequence, structure, or biochemical information that the target is druggable [64]. This operational code was used to guide genomic and target-based antimicrobial discovery. Forward and reverse chemical-genetic methods have been used to identify the target of drug-like bioactive compounds and MoA. Target-based approaches have been pursued using high-throughput screening for inhibitor of enzymes and/or proteins that perform essential functions in microbes. Molecular target-based screening has provided mediocre to poor results in the field of anti-infective drug discovery, suggesting that cell-based screening for antimicrobials remains the most effective strategy [19]. Gene knockout microorganisms and small molecules with well-defined mechanisms can each be used to alter the functions of putative targets, uncovering dependencies on activity. Reverse chemical screens that exploit genome-wide collections of mutants have been employed to map susceptibility phenotypes to specific genes by systematically screening antimicrobial agents against a defined (ideally) comprehensive mutant collection [20]. Mating of laboratory and wild yeast strains can reveal patterns of small-molecule sensitivity with specific loci [65]. In another method, molecularly barcoded libraries of open reading frames have been used to detect small-molecule-resistant clones that are then identified by microarray analysis [66].

A proteomic profiling approach, based on two-dimensional difference gel electrophoresis and mass spectrometric identification of the proteins, has been used to predict the target protein of small molecules of interest [67]. It was demonstrated that this proteomic profiling system could discriminate small molecules by their mechanism of action. Omics methods have also been used successfully to identify the MoA of compounds that were discovered using traditional screening approaches. A key illustrative case is the discovery of the orotomide F901318 compound [68]. A combination of genetic and biochemical approaches revealed the target of F901318 in *A. fumigatus*, which was initially discovered in a high-throughput screen of compound libraries. It was found that F901318 acts via the inhibition of the pyrimidine biosynthesis enzyme, dihydroorotate dehydrogenase, in a fungal-specific manner.

Although only around 1100 of the 6000 genes of *S. cerevisiae* are essential under nutrient-rich growth conditions [69], almost all genes become essential in specific genetic backgrounds in which another non-essential gene has been deleted or otherwise attenuated, an effect termed synthetic lethality [70]. Genome-scale surveys suggest that over 200,000 binary synthetic lethal gene combinations dominate the yeast genetic landscape [71]. The genetic buffering phenomenon is also manifested as a phalanx of differential chemical-genetic interactions in the presence of sublethal doses of bioactive compounds [72]. These observations highlight the inherent redundancy of genetic networks and frame the problem of interdicting network functions with single agent therapeutics [73]. This genetic network organisation suggests that judicious combinations of small molecule inhibitors of both essential and non-essential targets may elicit additive or synergistic effects on cell growth [74,75]. Compounds that enhance the activity of known agents in yeasts have been identified by small molecule library screens [76–79]. Direct tests of synergistic compounds have successfully yielded combinations that are active against pathogenic fungi, such as the combination of fluconazole with chemical inhibitors (such as 17-AAG and 17-DMAG) of Hsp90 [80], calcineurin (such as miconazole, ketoconazole), or adenosine diphosphate (ADP) ribosylation factor (ARF) (such as brefeldin A [81], the antibiotic polymyxin B [78]), and compounds selected from an off-patent drugs library [79]. Combinatorial antifungal therapies have many advantages, including a decrease in the rate of selection of resistant strains, a lower required dosage of individual drugs, a decrease in host toxicity, and enhanced antifungal activity [82]. Syncretic combinations of drugs with improved antifungal properties can be readily identified in both model fungal species and highly pathogenic clinical isolates [83]. Additionally, it was shown that synergistic combinations usually yield enhanced selectivity without adverse side effects [84].

These, recently developed, omics-based approaches require expensive, automated, robotic screening platforms when applied in high throughput. Additionally, these platforms are usually based on multiwell screening, which has some limitations. As discussed in the next section, further improvements can be achieved by integrating these methods in microfluidic screening platforms.

4. Micro- and Nanoscale Approaches

4.1. Microfluidic High-Throughput Antifungal Drug Discovery

Growing cells in 96-, 384-, or 1536-well plates has miniaturised cell assays for drug discovery. These experiments with multiwell plates are typically integrated in a robotic analysis platform. Major drawbacks of robotic platforms are the expense of the instrumentation, the cost of experimental consumables, the systems are closed (no flow through of reagents or cell culture medium), and still a relatively high consumption of reagents compared to recent developed microfluidic chips. Additionally, as cell collections are growing, there is a need to further miniaturise the assays to increase the parallelism of analyses.

In microfluidic lab-on-a-chips, fluids are manipulated at the micrometer length scale [85]. Reducing the scale evidently reduces the reagent consumption and, consequently, the cost to perform assays, which becomes significant for high-throughput drug screening. The physics at micro/nano scale are exploited in microfluidic chip designs for drug discovery. Other physical phenomena dominate at this small length scale compared to the macroscale [86]. Fluid flow in microchannels is characterized by a low Reynolds number (a dimensional criterion that determines the relative importance of inertial and viscous effects). The dominant role of viscous forces results in laminar flow behavior and mass transport by diffusion between two adjacent fluids, resulting in the generation of stable concentration gradients [87]. Also, fast media and temperature changes can be obtained using the laminar flow [88]. Mass transport by diffusion becomes a fast mixing method when the length scale is reduced in small channels or wells. It has been demonstrated that, at these small length scales, time-resolved reactions have been conducted with millisecond resolution due to the extremely rapid diffusive mixing [89,90]. Surface tension plays an important role in the formation of small droplet emulsions in immiscible fluids in microfluidic channels [86]. Surface tension and viscous stresses destabilize the interface and create droplets when water is injected into a stream of oil at a T-junction or two perpendicular crossing microchannels [91]. Based on this principle, droplet-based microfluidic platforms have been developed to encapsulate and screen single cells [92].

Due to the ease of controlling the cellular environment, microfluidic technology has been used to perturb cellular physiology in screening arrays. Additionally, the cell number and density of a given area or volume can be controlled, which allows monitoring of a high spatial and temporal resolution and observation of the dynamic behavior of many cells [93,94]. Parallellisation of experimental conditions and automatization in microfluidic chips has resulted in designs for high-throughput cell screening [95], such as living cell microarrays [96–101]. Microfluidic parallelization and single-cell monitoring allows the measurement and averaging of parameters on hundreds of individual cells, compared to measuring parameters of a whole cell population. New insights can be obtained by observing single cells, such as the monitoring of certain classes of proteins, using fluorescently tagged proteins [102,103] and give information about the cell-to-cell variation in a heterogeneous microbial population [104–106]. Automation of microfluidic cell culture systems also leads to standardized manipulation and monitoring, which allows the perfect timing of protocols to characterize dynamic processes at high temporal resolution to be performed [94].

Cellular microarray platforms have been developed for high-throughput antifungal drug discovery based on the screening of nanoliter biofilms that are created in hydrogels (Table 2). Robotic printing is used to fabricate the cellular arrays. The formation of biofilms complicates antifungal therapy, since the ability of fungal cells to form biofilms is an important reason for the emergence of severe resistance to most clinically available antifungal agents [107–110]. A better understanding of fungal biofilms provides new opportunities for the development of urgently needed novel antifungal agents and strategies.

Table 2. Examples of microfluidic platforms for high-throughput antifungal drug discovery.

High-Throughput Technology	Application	Microorganism	Characteristics	References
Cellular microarray	Antifungal biofilm screening	*Candida albicans*	Cells robotically printed, 768 (48 × 16 array) cultures of 50 nL biofilms in collagen.	[111,112]
Cellular microarray	Antifungal biofilm screening	*C. albicans*	Cells robotically printed, 1200 (60 × 20 array) cultures of 30 nL biofilms in alginate.	[113]
Cellular microarray	Antibiotic and antifungal biofilm screening	*C. albicans, Staphylococcus aureus, Pseudomonas aeruginosa*	Robotically printing of mono- and polymicrobial biofilms, 576 (48 × 12 array) cultures of 30 nL biofilms in alginate.	[114]
Droplet microfluidics	Antifungal drug screening	*Phytophthora sojae*	The plant pathogen spores and the drug were encapsulated in liquid droplets. Phenotypic responses to the drug at different concentrations were microscopically quantified.	[115]

4.2. Microfluidics for Antifungal Susceptibility Testing

Resistant microbial infections are becoming a major threat to public health, causing increasing mortality worldwide. The fast emergence of multiresistant pathogens is caused by extensive, and sometimes unnecessary, use of antimicrobials and the lack of interest in developing new variants. The cost of antimicrobial resistance (AMR) is projected to increase significantly as some models predict a rise in global casualties from the present figure of one million to 10 million in 2050 [116]. A survey of antimicrobial susceptibility test (AST) methods available demonstrated a need for new approaches that would enable rapid, inexpensive, and sensitive tests that can quickly provide physicians with antibiotic profiles [117]. A quick diagnostic method is necessary to prescribe a patient that is infected by a life-threatening microbe with an effective antimicrobial [118]. AST is widely applied to determine antimicrobial resistance profiles of the microbial isolates and to help in the selection of an antimicrobial treatment option [117,119–121]. Currently, AST is usually not performed in the clinic, but in a clinical microbiology lab, which necessitates transportation of the patient samples [122]. Current susceptibility tests, that are based on cell culturing of the pathogen, can take several days (especially for slow growing microbes) and extends the time to make the correct diagnosis and decisions for appropriate and effective antimicrobial therapy. This leads to increased patient mortality and the use of broad-spectrum antimicrobials that promote resistance [120]. To survive this evolutionary war against microbial pathogens, we must pursue technologies that can rapidly perform AST to enable personalized therapies (narrow-spectrum antimicrobial administration) at the earliest possible treatment stage [122].

Standard antifungal susceptibility (AFST) methods rely on measuring fungal growth in the presence of antifungals over a few days. The broth microdilution (BMD) is the standard method for the evaluation of susceptibility to antifungal agents in *Candida* species [123–126]. Standardized micro-dilution-based procedures by the Clinical and Laboratory Standards Institute (CLSI) and the European Committee on Antibiotic Susceptibility Testing (EUCAST) are universally accepted for performing AFST [127]. However, these procedures are complex, time-consuming, and not intended for routine use.

Commercially available tests, such as Sensititre YeastOne (broth microdilution method), Etest (agar-based disk diffusion method), and the fully automated Vitek 2 (broth microdilution method) yeast susceptibility system, all easy-to-use modifications from CLSI/EUCAST reference methods, are widely used for testing antifungal susceptibility of relevant *Candida* and *Aspergillus* species [10]. New diagnostic approaches based on emerging technologies, such as flow cytometry, MALDI-TOF mass spectroscopy, and isothermal microcalorimetry, have been developed to expand, and potentially improve, the capability of the clinical microbiology laboratory to yield AFST results [126]. Flow cytometry is used to determine the effect of an antifungal compound by measuring changes in the viability of fungal cells that are fluorescently labeled [128,129]. A simple and rapid AFST assay based on MALDI-TOF was developed [130]. This approach facilitated the discrimination of the susceptible and resistant isolates of *C. albicans* after a 3-h incubation in the presence of "breakpoint" level drug concentrations of the echinocandin caspofungin (CSF). A microcalorimetry-based AFST assay has been developed for real-time susceptibility testing of *Aspergillus* spp. [131]. The method is based on measuring changes in growth-related heat production in the presence of the antifungal compound.

Semi-automated susceptibility systems (such as VITEK and MicroScan) decrease the turn-around-time and operator touch-time (culture and colony isolation are still required) compared to traditional culture-based methods, but can add significant cost to the tests [132]. The fastest test still requires 9 h and, therefore, these semi-automated systems do not provide information in time to influence initial treatment decisions [133,134].

Since microfluidics promises several advantages over existing macro-scale methods, several microfluidic platforms that can perform rapid antimicrobial susceptibility tests have been developed during the last years (Table 3) [132,135]. The recent development of microfluidic platforms was mostly focused on devices for antibiotics susceptibility testing and much less on antifungals susceptibility testing, although the same designs could, usually, be used for both (e.g., see [136,137]). Most systems rely on microscopic transmission or fluorescence observation of cells to quantify the effect of the antimicrobial on the cell growth or viability. The microorganisms are confined to a small volume (wells, channels, chambers, or droplets). In many systems, the cells are immobilized in a hydrogel, such as agarose or alginate (Table 3) (Figure 1).

Table 3. Examples of microfluidic platforms for antimicrobial susceptibility testing.

Microorganism	Measurement Principle	Description	Reference
Fungi			
Candida strains	Fluorescence-based distinction between living and dead cells.	Cell Chip kit [1] used as cell sorter and the determination of fluorescence histograms of previously labelled cells.	[138]
Candida albicans	Immunosorbent ATP-bioluminescence assay.	The microfluidic device employs a fiberglass membrane sandwiched between two polypropylene components, with capture antibodies immobilised on the membrane. Cells immobilised in alginate hydrogel.	[136]
Saccharomyces cerevisiae	Fluorescence staining and imaging after incubation.	Cell seeding and diffusive medium supply is provided by phase-guide technology, enabling operation of continuous culturing.	[137]
Bacteria			
Escherichia coli	Optical imaging of single cell growth (number of cells).	Growth of cells in channels, with a large surface-to-volume ratio.	[139]
E. coli	Magnetic bead rotation, which is inversely proportional to bacterial mass.	Droplet microfluidics where single cells are entrapped in liquid drops.	[140]
E. coli	Optical imaging of single cell growth (number of cells).	Individual cells grow in gas permeable (PDMS) microchannels, with dimensions comparable to a single cell.	[141]
E. coli	Fluorescence imaging of cells that express green fluorescent protein.	Growth of cells in 12 sets of quadruplicate microfluidic chambers. Quantification of the effect of four antibiotics and their combinations.	[142]
E. coli	Reflectometric interference spectroscopy of pH-sensitive chitosan hydrogel measured the accumulation of metabolic products.	Growth of cells in microfluidic channels.	[143]
E. coli	Fluorescence imaging of immunomicrobeads attached to the cells.	Growth in microfluidic chambers.	[144]

Table 3. *Cont.*

Microorganism	Measurement Principle	Description	Reference
Bacteria			
E. coli	Optical imaging.	The standard broth microdilution method was miniaturised in a microfluidic chip that generates an antibiotic concentration gradient and delivers antibiotic-containing culture media to eight 30-nL chambers for cell culture.	[145]
E. coli	Fluorescence imaging.	The microfluidic chip allows the carrying out of commonly executed antibiotic susceptibility assays in an array of nanoliter droplets.	[146]
E. coli	Optical phase-shift reflectometric interference spectroscopic measurements.	The use of biofunctionalised silicon micropillar arrays to provide both a preferable solid-liquid interface for bacteria networking and a simultaneous transducing element that monitors the response of bacteria when exposed to chosen antibiotics in real time.	[147]
E. coli	Spectral absorbance of cell suspensions.	An automated linear gradient generator based on centrifugal microfluidics.	[148]
E. coli, Nitrosomas europaea	Optical imaging of single cell growth (number of cells).	The cells grow in a layer of agarose upon which a gradient of the antibiotic is applied.	[149]
Enterobacter cloacae, E. coli, Klebsiella pneumoniae, P. aeruginosa, Acinetobacter baumannii	Optical imaging of bacterial replication.	A solid-phase microwell growth surface in a 384-well plate format was used, with inkjet printing–based application of both antimicrobials and bacteria at any desired concentrations.	[150]
Enterococcus faecalis, E. coli	Fluorescence staining and imaging after incubation.	Cell seeding and diffusive medium supply is provided by phase-guide technology, enabling operation of continuous culturing.	[137]
Mycobacterium tuberculosis	Optical imaging of single cell growth (number of cells).	Cells were immobilized in an agarose matrix, which was molded in a microfluidic chip.	[151]
Pseudomonas aeruginosa	Fluorescence imaging of GFP-expressing cells.	The microfluidic chip generates a logarithmic concentration gradient through semidirect dilution in a zero-flow condition and cells grow in nanoliter reactors.	[135]
Staphylococcus aureus	Fluorescence imaging of viability indicator.	Stochastic confinement of individual cells into liquid plugs (droplet microfluidics); distinction between sensitive and resistant bacteria.	[152]
S. aureus	Optical imaging of single cell growth (number of cells).	Antibiotics diffuse into a microfluidic channel containing the cells.	[153]
S. aureus	Fluorescence imaging for dead cells (rates of killing).	Cells are covalently bound to the bottom of the channels and fluid flow shear stress activation of pathways that are targets of antibiotics.	[154]
S. aureus	Optical imaging of single cell growth (number of cells).	Chip with 32 individual fluidic channels.	[155]
S. aureus, E. coli, K. pneumoniae, P. aeroginosa	Optical imaging of single cell growth (number of cells).	Antibiotics diffuse into a microfluidic agarose channels in 96-well format.	[156]
S. aureus, S. epidermitis, S. saprophyticus, E. coli, K. pneumonia, P. aeruginosa, P. mirabilis, Streptococcus pyrogenes, S. viridans	Immunosorbent ATP-bioluminescence assay.	The microfluidic device employs a fiberglass membrane sandwiched between two polypropylene components, with capture antibodies immobilised on the membrane. Cells immobilised in alginate hydrogel.	[136]
Salmonella thyphimurium, E. coli, S. aureus	Optical imaging of single cell growth (number of cells).	Cells immobilised in agarose with a gradient of the antibiotic in the gel slab.	[157]
P. aeruginosa	Optical imaging of single cell growth (number of cells).	Growth of cells in four parallel channels.	[158]

[1] Agilent Technologies.

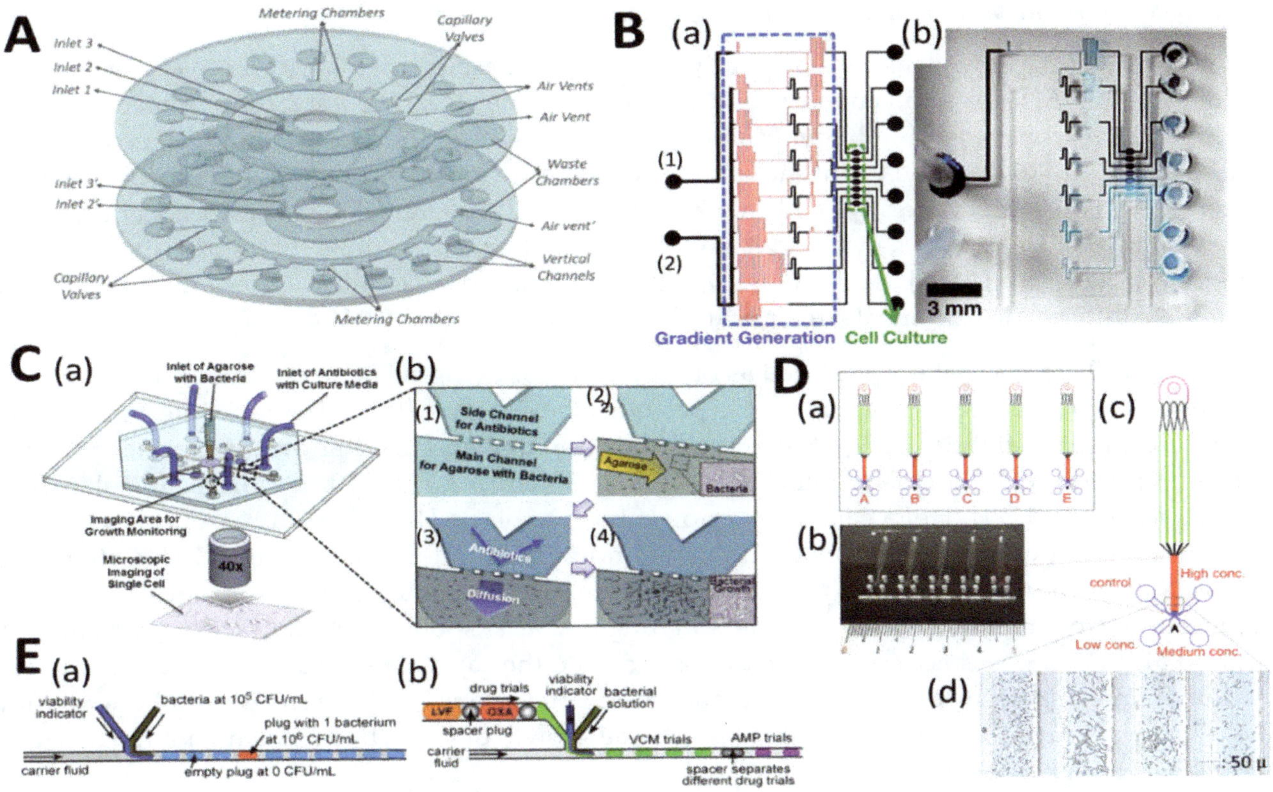

Figure 1. (**A**) An automated linear concentration gradient generator based on centrifugal microfluidics for antibiotic susceptibility testing (AST). The operation is based on the use of multi-layered microfluidics in which individual fluidic samples to be mixed together are stored and metered in their respective layers before, finally, being transferred to a mixing chamber. Reprinted with permission from [148]; (**B**) (**a**) The microfluidic microchip design: Black and red channels represent 149-μm and 4-μm tall features, respectively; (**b**) A photograph of the fabricated PDMS chip. Blue food dye solution and water were injected to upper (1) and lower (2) inlet ports. (**a**) Reprinted with permission from [145]; (**C**) Schematic diagram of the AST process for the microfluidic agarose channel (MAC) system. (**a**) The MAC chip was fabricated with PDMS on glass. An agarose–bacteria mixture solution was injected into the center of the chip, which flowed synchronously into the six main channels. Each interface between the agarose with the bacteria and antibiotic solutions was monitored microscopically to analyse bacterial cell growth. (**b**) (1) The empty channel before AST. (2) The bacteria were mixed with agarose and then injected into the main channels. (3) A sharp interface was generated due to the anchors and then liquid medium with different concentrations of antibiotic. (4) Time-lapse microscopic observation of a bacterial cell. Reprinted with permission from [153]; (**D**) The structure of the AST microfluidic device for MIC determination of five drugs. (**a**) Design. (**b**) Actual image. (**c**) Precise structure of one set of fluids. (**d**) Microscopic image of *Pseudomonas aeruginosa* grown in the presence of piperacillin. Reprinted with permission from [158]; (**E**) Stochastic confinement of bacteria into plugs. (**a**) Schematic drawing illustrates the increase in cell density, resulting from the stochastic confinement of an individual bacterium in a nanoliter-sized plug. (**b**) Screening of many antibiotics against the same bacterial sample. Antibiotics: Ampicillin (AMP), levofloxacin (LVF), vancomycin (VCN), and oxacillin (OXA). Reprinted with permission from [152].

4.3. Nanomotion Analysis for Evaluating Antifungal Suscessibility

Recently, very new sensitive sensor technologies based on microcantilevers have been developed [159–161]. Nanomechanical oscillators are being used for the detection of very small masses [162,163], for measuring buoyant mass, and determining the "instantaneous" growth rates of individual cells [164] for the quantitative time-resolved membrane protein (vancomycin)—ligand (mucopeptides) on cantilever arrays with 10 nM sensitivity and at clinically relevant concentrations in blood serum [165]. Cantilever nanosensors have been used to measure mass differences in the pico- to femtogram ranges in air [166] and liquid [167,168]. Many of the available systems are limited by the need to perform the measurements in air or in a humid environment [161].

Microcantilevers have been explored as nanosensors for living cell studies, since they offer many advantages, such as being highly sensitive, selective, label-free, performed in real time, and provide in situ detection capabilities [169]. Single cell detection and monitoring on the cantilever sensor has been reported for *S. cerevisiae* cells [164,170,171], *E. coli* [164,171,172], *Bacillus subtilis* [164,172], *Enterococcus faecalis* [171], HeLa cells [173], mouse lymphoblasts [164], and human lung carcinoma and mouse lymphocytic leukemia cells [171,174], and mouse and human T cells [171]. Cell growth detection has been demonstrated by monitoring resonance frequency changes of cantilevers as the mass increases from immobilized *S. cerevisiae* and fungal *A. niger* spores on the surface of the cantilevers in humid air [175]. *S. cerevisiae* cells were deposited onto the cantilever surface and its bending as a function of time and corresponded to the yeast growth behaviour [169]. Recently, serial microfluidic mass resonator arrays were used to measure single-cell growth of yeast, bacterial and mammalian cells in liquid and a higher throughput [171].

Recently, a nanomechanical detector, that can be used to assess the effects of chemicals on living organisms in a timeframe of minutes, was developed. In this technique, the living microorganisms are adsorbed to the surface of a nanomechanical sensor, i.e., an AFM cantilever, and its fluctuations are measured as a function of time (Figure 2). This approach requires only a few minutes without cell growth, since the decrease in cell activity in the presence of the antimicrobial is measured. Moreover, it is a label-free technique. These are major advantages compared to current methods. Antimicrobials can be introduced in the measuring chamber, allowing a rapid identification of the antimicrobial to which the microorganism is susceptible. The sensor position detection is similar to the one used in Atomic Force Microscopy (AFM): A laser beam is focused onto the sensor (cantilever) and ends its path onto a multisegment photodiode. Numerous proof-of-concept experiments have been performed involving several species of bacteria and fungi. By comparing this method with the traditional techniques in double-blind experiments, a success rate close to 90% could be achieved [176]. The variation of the nanovibration signal is directly proportional to the metabolic activity of the microbial cell. It allowed estimating with unpreceded speed (15 min per ampicillin dose), the minimum inhibitory (MIC) and bactericidal concentrations (MBC) for the effect of the antibiotic ampicillin on *E. coli* [177]. The high sensitivity of the nanosensor allowed differentiation between bacteriostatic (kanamycin) and bactericidal (ampicillin) effects when exposed to a resistant *E. coli*. The high sensitivity of the nanosensor made it possible to differentiate between bacteriostatic and bactericidal effects. Recently, this method was successfully applied to blood culture pellets to determine the antibiotic susceptibility against agents of bloodstream infection [178]. AFM-nanomotion detection (NMD) has also been applied for AFST, where the effect of a low (10 μg/mL) and high (40 μg/mL) caspofungin concentration on *C. albicans* was evaluated [179]. It was recently demonstrated that AFM-cantilever NMD is so sensitive that single-cell activities of *S. cerevisiae* cells could be detected [179], and even cell cycle progression could be observed [180].

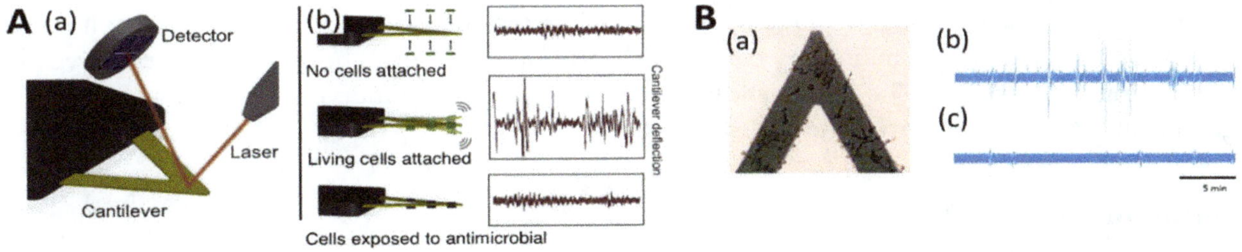

Figure 2. (**A**) Outline of experimental setup and description of experiments. (**a**) Schematics of nanomotion detector setup with a cantilever sensor; laser beam is focused on the surface of the sensor and reflection is used to monitor movements of the cantilever. (**b**) Representation of a typical nanomotion susceptibility test. When microorganisms are not attached to the sensor, fluctuations are driven only by thermal motion and are relatively low. After attachment of living cells, fluctuations are linked to their metabolic activity and are high. Finally, after exposure to an antimicrobial drug, the cells are nonviable and fluctuations return to low levels. Reprinted with permission from [178]. (**B**) (**a**) *C. albicans* deposited onto a cantilever. (**b**) Reprinted with permission from [161].

4.4. Antifungal Nanoparticles

Due to the development of antimicrobial resistance, pharmaceutical companies and researchers are searching for new antimicrobial agents. Nanoscale materials have emerged as novel antimicrobial agents, owing to their high surface area to volume ratio and their unique chemical and physical properties [181,182]. By definition, nanoparticles are structures that have dimensions in the 1–100 nm [183]. Silver nanoparticles (Ag-NPs) are one of the most commonly used nanomaterials in consumer and medical products because of their antimicrobial activity [184,185]. It is increasingly used in a variety of both medical and consumer products, resulting in an increase in human exposure [183,186]. A large number of in vitro studies indicate that Ag-NPs are also toxic to mammalian cells derived from the skin, liver, lung, brain, vascular system, reproductive organs, and the immune system. Although significant progress has been achieved on the elucidation of the antimicrobial mechanism of silver nanoparticles, the exact mechanism of action is still not completely known [187]. It has been demonstrated that non-cytotoxic doses of Ag-NPs could induce genes that are associated with cell cycle progression [188,189] and apoptosis in human hepatoma cells [189]. DNA damage by Ag-NPs in mammalian cells has also been reported [188,190]. The mechanisms for Ag-NP induced toxicity include the effects of this particle on cell membranes, mitochondria, and genetic material. It has been recently demonstrated that bacteria (*E. coli, P. aeruginosa*) can also develop resistance to silver nanoparticles after repeated exposure [191]. The biosynthesis of silver nanoparticles is now considered to be the most environmentally friendly and cost-effective method [185]. It can be achieved using a variety of organisms, such as by the fungi, *A. flavus* [185], *A. terreus* [192], *Arthroderma fulvum* [193], *Penicillium fellutanum* [194], and *P. expansum* [192].

Other nanostructures have been discovered that also show antifungal activity. It was demonstrated that cationic terephthalamide-bisurea molecules showed excellent microbial selectivity, with minimal host toxicity [195]. The terephthalamide-bisurea recognition motif facilitated spontaneous supramolecular self-assembly, with the formation of fibres in water. Antifungal activity against clinically isolated, drug-sensitive, and drug-resistant *Cryptococcus neoformans* strains was observed. These antifungal agents showed effectively dispersed *C. albicans* biofilms and excellent in vivo biocompatibility. Homogeneously dispersed copper nanoparticles (originating from n-Cu sepiolite fibres) in soda-lime glass powder showed high antibacterial properties against gram-positive (*Micrococcus luteus*) and gram-negative bacteria (*E. coli*), and antifungal (*Issatchenkia orientalis*) activity [196]. The observed high activity of the n-Cu glass powder was explained by the inhibitory synergistic effect of the Ca^{2+} lixiviated from the glass on the growth of the cells, since n-Cu sepiolite had no significant antifungal activity. Another approach of obtaining nanoparticles with antifungal activity is to immobilize, covalently, the antifungal amphotericin into nanomaterials, such as silica nanoparticles [197]. These antifungal nanoparticle conjugates demonstrated fungicidal activity against several strains of *Candida* sp., mainly by contact. In addition, they could be reused for up to five cycles without losing their activity. The results showed that the antifungal nanoparticle conjugates were more fungistatic and fungicidal than 10 nm colloidal silver.

5. Conclusions

Fungal infections continue to appear as the population of people with an altered immune system increases. An impaired immune system arises due to medical intervention, immunosuppressive diseases, or malnutrition. The infections become life-threatening for systemic infections. The therapeutic options for invasive fungal infections are limited, since a very limited number of structural classes of drug compounds are available and some demonstrate significant limitations. The discovery of new antifungals is mostly achieved by the screening of natural or synthetic/semisynthetic chemical compounds [16,17]. Using the genomics approach, substantial progress has been made in antifungal drug development for a multitude of potential drug targets [198] and inhibitors [199,200]. Recent advances that could support and refine the antifungal pipeline were focused on elucidating fungal pathways, targets, and mechanisms of action that could lead to new antifungal therapies; antifungal compounds and immune strategies currently in development that could become new antifungal therapies; improved formulations of existing compounds; and the repurposing of drugs approved for other indications and could show potential antifungal activity [16]

Surely, there is a need for novel antifungal discovery approaches. The current antifungal tools that are available to tackle the invasive fungal epidemic occurring in clinics and hospitals have improved, but are still inadequate for use in all patient groups [16]. Recently, micro- and nanoscale approaches have been introduced in antifungal drug discovery. It has become increasingly clear in trends in antimicrobial drug discovery that microfluidic approaches will have an increasing role. The capability of manipulating fluids, flexibility on geometries and materials, manipulation of a large number of cells simultaneously and independently, and ease of integrating numerous analytical standard operations and large-scale integration makes microfluidic devices a versatile tool for antifungal drugs. One of the future challenges lies in the construction of extended cell microarrays, or single-cell droplet arrays, and integration in the microfluidic chip [100], since the recently developed bioinformatics approach for antifungal discovery could be further improved by integrating these methods in a microfluidic screening platform. Living cell arraying methods, based on closed microchambers arrays in microfluidic bioreactors, could increase throughput significantly compared to classical multiwell-plate cell assays, with significantly reduced amounts spent on expensive test reagents, cells, and chemical compounds, and without the need for expensive robotic multiwell-plate screening facilities.

Significant progress has been made in the development of microfluidic platforms for antimicrobial susceptibility testing (AST). Recent efforts are focused on the development of AFST methods that are independent of cell growth and can provide susceptibility results in a very short time frame. These methods rely on measuring the cell death in the presence of the antifungal. A promising technique is the AFM-based nanomotion detection, which can give an answer about the susceptibility in a few minutes. Further parallelization of this method is required to introduce it in the clinic.

Nanoscale materials have emerged as novel antimicrobial agents. One of the most extensively studied nanoparticles are silver nanoparticles and this is for their antifungal, but also antibacterial and antiviral, activities. In minute concentrations, it was found not to be toxic to humans and microorganisms [185]. Further research should focus on a better understanding the toxicity and determining the exact mechanism of the interaction between nanoparticles and cells.

Acknowledgments: The Belgian Federal Science Policy Office (Belspo) and the European Space Agency (ESA) PRODEX program supported this work. The Research Council of the Vrije Universiteit Brussel (Belgium) and the University of Ghent (Belgium) are acknowledged to support the Alliance Research Group VUB-UGent

NanoMicrobiology (NAMI), and the International Joint Research Group (IJRG) VUB-EPFL BioNanotechnology & NanoMedicine (NANO).

References

1. Vos, T.; Flaxman, A.D.; Naghavi, M.; Lozano, R.; Michaud, C.; Ezzati, M.; Shibuya, K.; Salomon, J.A.; Abdalla, S.; Aboyans, V.; et al. Years lived with disability (YLDs) for 1160 sequelae of 289 diseases and injuries 1990–2010: A systematic analysis for the Global Burden of Disease Study 2010. *Lancet* **2012**, *380*, 2163–2196. [CrossRef]
2. Brown, G.D.; Denning, D.W.; Levitz, S.M. Tackling human fungal infections. *Science* **2012**, *336*, 647. [CrossRef] [PubMed]
3. Richardson, M.D. Changing patterns and trends in systemic fungal infections. *J. Antimicrob. Chemother.* **2005**, *56* (Suppl. 1), i5–i11. [CrossRef] [PubMed]
4. Huttunen, R.; Syrjänen, J. Obesity and the risk and outcome of infection. *Int. J. Obes.* **2013**, *37*, 333–340. [CrossRef] [PubMed]
5. Rytter, M.J.; Kolte, L.; Briend, A.; Friis, H.; Christensen, V.B. The immune system in children with malnutrition—A systematic review. *PLoS ONE* **2014**, *9*, e105017. [CrossRef] [PubMed]
6. Ibrahim, M.K.; Zambruni, M.; Melby, C.L.; Melby, P.C. Impact of childhood malnutrition on host defense and infection. *Clin. Microbiol. Rev.* **2017**, *30*, 919–971. [CrossRef] [PubMed]
7. Denning, D.W.; Bromley, M.J. Infectious Disease. How to bolster the antifungal pipeline. *Science* **2015**, *347*, 1414–1416. [CrossRef] [PubMed]
8. Lemke, A.; Kiderlen, A.F.; Kayser, O. Amphotericin B. *Appl. Microbiol. Biotechnol.* **2005**, *68*, 151–162. [CrossRef] [PubMed]
9. Bellmann, R. Pharmacodynamics and pharmacokinetics of antifungals for treatment of invasive aspergillosis. *Curr. Pharm. Des.* **2013**, *19*, 3629–3647. [CrossRef] [PubMed]
10. Pfaller, M.A. Antifungal drug resistance: Mechanisms, epidemiology, and consequences for treatment. *Am. J. Med.* **2012**, *125* (Suppl. 1), S3–S13. [CrossRef] [PubMed]
11. Pianalto, K.M.; Alspaugh, J.A. New Horizons in Antifungal Therapy. *J. Fungi.* **2016**, *2*, 26. [CrossRef] [PubMed]
12. Nivoix, Y.; Ubeaud-Sequier, G.; Engel, P.; Levêque, D.; Herbrecht, R. Drug-drug interactions of triazole antifungal agents in multimorbid patients and implications for patient care. *Curr. Drug. Metab.* **2009**, *10*, 395–409. [CrossRef] [PubMed]
13. Lewis, R.E. Current concepts in antifungal pharmacology. *Mayo Clin. Proc.* **2011**, *86*, 805–817. [CrossRef] [PubMed]
14. Campoy, S.; Adrio, J.L. Antifungals. *Mayo Clin. Proc.* **2017**, *133*, 86–96. [CrossRef] [PubMed]
15. McCarthy, M.W.; Walsh, T.J. Drugs currently under investigation for the treatment of invasive candidiasis. *Expert. Opin. Investig. Drugs* **2017**, *26*, 825–831. [CrossRef] [PubMed]
16. Perfect, J.R. The antifungal pipeline: A reality check. *Nat. Rev. Drug Discov.* **2017**, *16*, 603–616. [CrossRef] [PubMed]
17. Fuentefria, A.M.; Pippi, B.; Dalla Lana, D.F.; Donato, K.K.; de Andrade, S.F. Antifungals discovery: An insight into new strategies to combat antifungal resistance. *Lett. Appl. Microbiol.* **2018**, *66*, 2–13. [CrossRef] [PubMed]
18. Weig, M.; Brown, A.J. Genomics and the development of new diagnostics and anti-Candida drugs. *Trends Microbiol.* **2007**, *15*, 310–317. [CrossRef] [PubMed]
19. Payne, D.J.; Gwynn, M.N.; Holmes, D.J.; Pompliano, D.L. Drugs for bad bugs: Confronting the challenges of antibacterial discovery. *Nat. Rev. Drug Discov.* **2007**, *6*, 29–40. [CrossRef] [PubMed]
20. Roemer, T.; Boone, C. Systems-level antimicrobial drug and drug synergy discovery. *Nat. Chem. Biol.* **2013**, *9*, 222–231. [CrossRef] [PubMed]
21. Schenone, M.; Dančík, V.; Wagner, B.K.; Clemons, P.A. Target identification and mechanism of action in chemical biology and drug discovery. *Nat. Chem. Biol.* **2013**, *9*, 232–240. [CrossRef] [PubMed]
22. Nucci, M.; Marr, K.A. Emerging fungal diseases. *Clin. Infect. Dis.* **2005**, *41*, 521–526. [CrossRef] [PubMed]
23. Castón-Osorio, J.J.; Rivero, A.; Torre-Cisneros, J. Epidemiology of invasive fungal infection. *Int. J. Antimicrob. Agents* **2008**, *32* (Suppl. 2), S103–S109. [CrossRef]

24. Miceli, M.H.; Díaz, J.A.; Lee, S.A. Emerging opportunistic yeast infections. *Lancet Infect. Dis.* **2011**, *11*, 142–151. [CrossRef]
25. Galimberti, R.; Torre, A.C.; Baztán, M.C.; Rodriguez-Chiappetta, F. Emerging systemic fungal infections. *Clin. Dermatol.* **2012**, *30*, 633–650. [CrossRef] [PubMed]
26. Enoch, D.A.; Yang, H.; Aliyu, S.H.; Micallef, C. The Changing Epidemiology of Invasive Fungal Infections. In *Human Fungal Pathogen Identification. Methods in Molecular Biology*; Lion, T., Ed.; Humana Press: New York, NY, USA, 2017; Volume 1508, pp. 17–65. ISBN 978-1-4939-6513-7.
27. Schwartz, I.S.; Patterson, T.F. The Emerging Threat of antifungal resistance in transplant infectious diseases. *Curr. Infect. Dis. Rep.* **2018**, *20*, 2. [CrossRef] [PubMed]
28. Pfaller, M.A.; Diekema, D.J.; Gibbs, D.L.; Newell, V.A.; Ellis, D.; Tullio, V.; Rodloff, A.; Fu, W.; Ling, T.A. Global Antifungal Surveillance Group. Results from the ARTEMIS DIS Global Antifungal Surveillance Study, 1997 to 2007: A 10.5-year analysis of susceptibilities of Candida Species to fluconazole and voriconazole as determined by CLSI standardized disk diffusion. *J. Clin. Microbiol.* **2010**, *48*, 1366–1377. [CrossRef] [PubMed]
29. Khawcharoenporn, T.; Apisarnthanarak, A.; Mundy, L.M. Non-neoformans cryptococcal infections: A systematic review. *Infection* **2007**, *35*, 51–58. [CrossRef] [PubMed]
30. Enache-Angoulvant, A.; Hennequin, C. Invasive *Saccharomyces* infection: A comprehensive review. *Clin. Infect. Dis.* **2005**, *41*, 1559–1568. [CrossRef] [PubMed]
31. Anoop, V.; Rotaru, S.; Shwed, P.S.; Tayabali, A.F.; Arvanitakis, G. Review of current methods for characterizing virulence and pathogenicity potential of industrial Saccharomyces cerevisiae strains towards humans. *FEMS Yeast Res.* **2015**, *15*, fov057-2. [CrossRef] [PubMed]
32. Lass-Flörl, C.; Cuenca-Estrella, M. Changes in the epidemiological landscape of invasive mould infections and disease. *J. Antimicrob. Chemother.* **2017**, *72* (Suppl. 1), i5–i11. [CrossRef] [PubMed]
33. Ribes, J.A.; Vanover-Sams, C.L.; Baker, D.J. Zygomycetes in human disease. *Clin. Microbiol. Rev.* **2000**, *13*, 236–301. [CrossRef] [PubMed]
34. Kontoyiannis, D.P.; Marr, K.A.; Park, B.J.; Alexander, B.D.; Anaissie, E.J.; Walsh, T.J.; Ito, J.; Andes, D.R.; Baddley, J.W.; Brown, J.M.; et al. Prospective surveillance for invasive fungal infections in hematopoietic stem cell transplant recipients, 2001–2006: Overview of the Transplant-Associated Infection Surveillance Network (TRANSNET) Database. *Clin. Infect. Dis.* **2010**, *50*, 1091–1100. [CrossRef] [PubMed]
35. Roilides, E.; Zaoutis, T.E.; Walsh, T.J. Invasive zygomycosis in neonates and children. *Clin. Microbiol. Infect.* **2009**, *15* (Suppl. 5), 50–54. [CrossRef] [PubMed]
36. Lanternier, F.; Lortholary, O. Zygomycosis and diabetes mellitus. *Clin. Microbiol. Infect.* **2009**, *15* (Suppl. 5), 21–25. [CrossRef] [PubMed]
37. Nucci, M.; Anaissie, E. *Fusarium* infections in immunocompromised patients. *Clin. Microbiol. Rev.* **2007**, *20*, 695–704. [CrossRef] [PubMed]
38. Tortorano, A.M.; Richardson, M.; Roilides, E.; Diepeningen, A.V.; Caira, M.; Munoz, P.; Johnson, E.; Meletiadis, J.; Pana, Z.D.; Lackner, M.; et al. ESCMID & ECMM Joint Guidelines on Diagnosis and Management of Hyalohyphomycosis: *Fusarium* spp., *Scedosporium* spp., and others. *Clin. Microbiol. Infect.* **2014**, *20* (Suppl. 3), 27–46. [PubMed]
39. Guarro, J.; Kantarcioglu, A.S.; Horré, R.; Luis Rodriguez-Tudela, J.; Cuenca Estrella, M.; Berenguer, J.; Sybren De Hoog, G. *Scedosporium apiospermum*: Changing clinical spectrum of a therapy-refractory opportunist. *Med. Mycol.* **2006**, *44*, 295–327. [CrossRef] [PubMed]
40. Montesinos, I.; Dodemont, M.; Lagrou, K.; Jacobs, F.; Etienne, I.; Denis, O. New case of azole-resistant *Aspergillus fumigatus* due to TR46/Y121F/T289A mutation in Belgium. *J. Antimicrob. Chemother.* **2014**, *69*, 3439–3440. [CrossRef] [PubMed]
41. Buil, J.B.; Meis, J.F.; Melchers, W.J.; Verweij, P.E. Are the TR46/Y121F/T289A Mutations in Azole-Resistant Aspergillosis Patient Acquired or Environmental? *Antimicrob. Agents Chemother.* **2016**, *60*, 3259–3260. [CrossRef] [PubMed]
42. Ostrosky-Zeichner, L.; Casadevall, A.; Galgiani, J.N.; Odds, F.C.; Rex, J.H. An insight into the antifungal pipeline: Selected new molecules and beyond. *Nat. Rev. Drug Discov.* **2010**, *9*, 719–727. [CrossRef] [PubMed]
43. Vandeputte, P.; Ferrari, S.; Coste, A.T. Antifungal resistance and new strategies to control fungal infections. *Int. J. Microbiol.* **2012**, *2012*, 713687. [CrossRef] [PubMed]

44. Ballard, S.A.; Lodola, A.; Tarbit, M.H. A comparative study of 1-substituted imidazole and 1,2,4-triazole antifungal compounds as inhibitors of testosterone hydroxylations catalysed by mouse hepatic microsomal cytochromes P-450. *Biochem. Pharmacol.* **1988**, *37*, 4643–4651. [CrossRef]
45. DiDomenico, B. Novel antifungal drugs. *Curr. Opin. Microbiol.* **1999**, *2*, 509–515. [CrossRef]
46. Arévalo, M.P.; Carrillo-Muñoz, A.J.; Salgado, J.; Cardenes, D.; Brió, S.; Quindós, G.; Espinel-Ingroff, A. Antifungal activity of the echinocandin anidulafungin (VER002, LY-303366) against yeast pathogens: A comparative study with M27-A microdilution method. *J. Antimicrob. Chemother.* **2003**, *51*, 163–166. [CrossRef] [PubMed]
47. Mukherjee, P.K.; Sheehan, D.; Puzniak, L.; Schlamm, H.; Ghannoum, M.A. Echinocandins: Are they all the same? *J. Chemother.* **2011**, *23*, 319–325. [CrossRef] [PubMed]
48. Firooz, A.; Nafisi, S.; Maibach, H.I. Novel drug delivery strategies for improving econazole antifungal action. *Int. J. Pharm.* **2015**, *495*, 599–607. [CrossRef] [PubMed]
49. Denning, D.W.; Hope, W.W. Therapy for fungal diseases: Opportunities and priorities. *Trends Microbiol.* **2010**, *18*, 195–204. [CrossRef] [PubMed]
50. Terra, L.; Abreu, P.A.; Teixeira, V.L.; Paixão, I.C.; Pereira, R.; Leal, B.; Lourenço, A.L.; Rampelotto, P.H.; Castro, H.C. Mycoses and Antifungals: Reviewing the basis of a current problem that still is a biotechnological target for marine products. *Front. Mar. Sci.* **2014**, *1*, 12. [CrossRef]
51. Vengurlekar, S.; Sharma, R.; Trivedi, P. Efficacy of some natural compounds as antifungal agents. *Pharmacogn. Rev.* **2012**, *6*, 91–99. [CrossRef] [PubMed]
52. Lopez-Moya, F.; Colom-Valiente, M.F.; Martinez-Peinado, P.; Martinez-Lopez, J.E.; Puelles, E.; Sempere-Ortells, J.M.; Lopez-Llorca, L.V. Carbon and nitrogen limitation increase chitosan antifungal activity in Neurospora crassa and fungal human pathogens. *Fungal. Biol.* **2015**, *119*, 154–169. [CrossRef] [PubMed]
53. Abbaszadeh, S.; Sharifzadeh, A.; Shokri, H.; Khosravi, A.R.; Abbaszadeh, A. Antifungal efficacy of thymol, carvacrol, eugenol and menthol as alternative agents to control the growth of food-relevant fungi. *J. Mycol. Med.* **2014**, *24*, e51–e56. [CrossRef] [PubMed]
54. Barros, F.M.C.; Pippi, B.; Dresch, R.R.; Dauber, B.; Luciano, S.C.; Apel, M.A.; Fuentefria, A.M.; Von Poser, G.L. Antifungal and antichemotactic activities and quantification of phenolic compounds in lipophilic extracts of Hypericum spp. native to South Brazil. *Ind. Crops Prod.* **2013**, *44*, 294–299. [CrossRef]
55. Danielli, L.J.; Dos Reis, M.; Bianchini, M.; Camargo, G.S.; Bordignon, S.A.L.; Guerreiro, I.K.; Fuentefria, A.; Apel, M.A. Antidermatophytic activity of volatile oil and nanoemulsion of *Stenachaenium megapotamicum* (Spreng.) Baker. *Ind. Crops Prod.* **2013**, *50*, 23–28. [CrossRef]
56. Machado, G.R.; Pippi, B.; Dalla Lana, D.F.; Amaral, A.P.S.; Teixeira, M.L.; De Souza, K.C.B.; Fuentefria, A.M. Reversal of fluconazole resistance induced by a synergistic effect with *Acca sellowiana* in Candida glabrata strains. *Pharm. Biol.* **2016**, *54*, 2410–2419. [CrossRef] [PubMed]
57. Bicanic, T.; Muzoora, C.; Brouwer, A.E.; Meintjes, G.; Longley, N.; Taseera, K.; Rebe, K.; Loyse, A.; Jarvis, J.; Bekker, L.G.; et al. Independent association between rate of clearance of infection and clinical outcome of HIV-associated cryptococcal meningitis: Analysis of a combined cohort of 262 patients. *Clin. Infect. Dis.* **2009**, *49*, 702–709. [CrossRef] [PubMed]
58. Pierce, C.G.; Uppuluri, P.; Tristan, A.R.; Wormley, F.L., Jr.; Mowat, E.; Ramage, G.; Lopez-Ribot, J.L. A simple and reproducible 96-well plate-based method for the formation of fungal biofilms and its application to antifungal susceptibility testing. *Nat. Protoc.* **2008**, *3*, 1494–1500. [CrossRef] [PubMed]
59. LaFleur, M.D.; Lucumi, E.; Napper, A.D.; Diamond, S.L.; Lewis, K.N. Novel high-throughput screen against *Candida albicans* identifies antifungal potentiators and agents effective against biofilms. *J. Antimicrob. Chemother.* **2011**, *66*, 820–826. [CrossRef] [PubMed]
60. Monteiro, M.C.; de la Cruz, M.; Cantizani, J.; Moreno, C.; Tormo, J.R.; Mellado, E.; De Lucas, J.R.; Asensio, F.; Valiante, V.; Brakhage, A.A.; et al. A new approach to drug discovery: High-throughput screening of microbial natural extracts against *Aspergillus fumigatus* using resazurin. *J. Biomol. Screen* **2012**, *17*, 542–549. [CrossRef] [PubMed]
61. Rabjohns, J.L.A.; Park, Y.D.; Dehdashti, J.; Henderson, C.; Zelazny, A.; Metallo, S.J.; Zheng, W.; Williamson, P.R. A high-throughput screening assay for fungicidal compounds against *Cryptococcus neoformans*. *J. Biomol. Screen* **2014**, *19*, 270–277. [CrossRef] [PubMed]

62. DiDone, L.; Scrimale, T.; Baxter, B.K.; Krysan, D.J. A high-throughput assay of yeast cell lysis for drug discovery and genetic analysis. *Nat. Protoc.* **2010**, *5*, 1107–1114. [CrossRef] [PubMed]
63. Cameron-Clarke, A.; Hulse, G.A.; Clifton, L.; Cantrell, I.C. The Use of Adenylate Kinase Measurement to Determine Causes of Lysis in Lager Yeast. *J Am. Soc. Brew. Chem.* **2003**, *61*, 152–156. [CrossRef]
64. Roemer, T.; Krysan, D.J. Antifungal drug development: Challenges, unmet clinical needs, and new approaches. *Cold Spring Harb. Perspect. Med.* **2014**, *4*, A019703. [CrossRef] [PubMed]
65. Perlstein, E.O.; Ruderfer, D.M.; Roberts, D.C.; Schreiber, S.L.; Kruglyak, L. Genetic basis of individual differences in the response to small-molecule drugs in yeast. *Nat. Genet.* **2007**, *39*, 496–502. [CrossRef] [PubMed]
66. Pierce, S.E.; Fung, E.L.; Jaramillo, D.F.; Chu, A.M.; Davis, R.W.; Nislow, C.; Giaever, G. A unique and universal molecular barcode array. *Nat. Methods* **2006**, *3*, 601–603. [CrossRef] [PubMed]
67. Muroi, M.; Kazami, S.; Noda, K.; Kondo, H.; Takayama, H.; Kawatani, M.; Usui, T.; Osada, H. Application of proteomic profiling based on 2D-DIGE for classification of compounds according to the mechanism of action. *Chem. Biol.* **2010**, *17*, 460–470. [CrossRef] [PubMed]
68. Oliver, J.D.; Sibley, G.E.; Beckmann, N.; Dobb, K.S.; Slater, M.J.; McEntee, L.; du Pré, S.; Livermore, J.; Bromley, M.J.; Wiederhold, N.P.; et al. F901318 represents a novel class of antifungal drug that inhibits dihydroorotate dehydrogenase. *Proc. Natl. Acad. Sci. USA* **2016**, *133*, 12809–12814. [CrossRef] [PubMed]
69. Winzeler, E.A.; Shoemaker, D.D.; Astromoff, A.; Liang, H.; Anderson, K.; Andre, B.; Bangham, R.; Benito, R.; Boeke, J.D.; Bussey, H.; et al. Functional characterization of the S. cerevisiae genome by gene deletion and parallel analysis. *Science* **1999**, *285*, 901–906. [CrossRef] [PubMed]
70. Tong, A.H.; Evangelista, M.; Parsons, A.B.; Xu, H.; Bader, G.D.; Pagé, N.; Robinson, M.; Raghibizadeh, S.; Hogue, C.W.; Bussey, H.; et al. Systematic genetic analysis with ordered arrays of yeast deletion mutants. *Science* **2001**, *294*, 2364–2368. [CrossRef] [PubMed]
71. Costanzo, M.; Baryshnikova, A.; Bellay, J.; Kim, Y.; Spear, E.D.; Sevier, C.S.; Ding, H.; Koh, J.L.; Toufighi, K.; Mostafavi, S.; et al. The genetic landscape of a cell. *Science* **2010**, *327*, 425–431. [CrossRef] [PubMed]
72. Hillenmeyer, M.E.; Fung, E.; Wildenhain, J.; Pierce, S.E.; Hoon, S.; Lee, W.; Proctor, M.; St Onge, R.P.; Tyers, M.; Koller, D.; et al. The chemical genomic portrait of yeast: Uncovering a phenotype for all genes. *Science* **2008**, *320*, 362–365. [CrossRef] [PubMed]
73. Hopkins, A.L. Network pharmacology: The next paradigm in drug discovery. *Nat. Chem. Biol.* **2008**, *4*, 682–690. [CrossRef] [PubMed]
74. Fitzgerald, J.B.; Schoeberl, B.; Nielsen, U.B.; Sorger, P.K. Systems biology and combination therapy in the quest for clinical efficacy. *Nat. Chem. Biol.* **2006**, *2*, 458–466. [CrossRef] [PubMed]
75. Lehár, J.; Stockwell, B.R.; Giaever, G.; Nislow, C. Combination chemical genetics. *Nat. Chem. Biol.* **2008**, *4*, 674–681. [CrossRef] [PubMed]
76. Borisy, A.A.; Elliott, P.J.; Hurst, N.W.; Lee, M.S.; Lehar, J.; Price, E.R.; Serbedzija, G.; Zimmermann, G.R.; Foley, M.A.; Stockwell, B.R.; et al. Systematic discovery of multicomponent therapeutics. *Proc. Natl. Acad. Sci. USA* **2003**, *100*, 7977–7982. [CrossRef] [PubMed]
77. Zhang, L.; Yan, K.; Zhang, Y.; Huang, R.; Bian, J.; Zheng, C.; Sun, H.; Chen, Z.; Sun, N.; An, R.; et al. High-throughput synergy screening identifies microbial metabolites as combination agents for the treatment of fungal infections. *Proc. Natl. Acad. Sci. USA* **2007**, *104*, 4606–4611. [CrossRef] [PubMed]
78. Zhai, B.; Zhou, H.; Yang, L.; Zhang, J.; Jung, K.; Giam, C.Z.; Xiang, X.; Lin, X. Polymyxin B, in combination with fluconazole, exerts a potent fungicidal effect. *J. Antimicrob. Chemother.* **2010**, *65*, 931–938. [CrossRef] [PubMed]
79. Spitzer, M.; Griffiths, E.; Blakely, K.M.; Wildenhain, J.; Ejim, L.; Rossi, L.; De Pascale, G.; Curak, J.; Brown, E.; Tyers, M.; et al. Cross-species discovery of syncretic drug combinations that potentiate the antifungal fluconazole. *Mol. Syst. Biol.* **2011**, *7*, 499. [CrossRef] [PubMed]
80. Cowen, L.E.; Singh, S.D.; Köhler, J.R.; Collins, C.; Zaas, A.K.; Schell, W.A.; Aziz, H.; Mylonakis, E.; Perfect, J.R.; Whitesell, L.; Lindquist, S. Harnessing Hsp90 function as a powerful, broadly effective therapeutic strategy for fungal infectious disease. *Proc. Natl. Acad. Sci. USA* **2009**, *106*, 2818–2823. [CrossRef] [PubMed]
81. Epp, E.; Vanier, G.; Harcus, D.; Lee, A.Y.; Jansen, G.; Hallett, M.; Sheppard, D.C.; Thomas, D.Y.; Munro, C.A.; Mullick, A.; et al. Reverse genetics in *Candida albicans* predicts ARF cycling is essential for drug resistance and virulence. *PLoS Pathog.* **2010**, *6*, e1000753. [CrossRef] [PubMed]

82. Sharom, J.R.; Bellows, D.S.; Tyers, M. From large networks to small molecules. *Curr. Opin. Chem. Biol.* **2004**, *8*, 81–90. [CrossRef] [PubMed]
83. Jansen, G.; Lee, A.Y.; Epp, E.; Fredette, A.; Surprenant, J.; Harcus, D.; Scott, M.; Tan, E.; Nishimura, T.; Whiteway, M.; et al. Chemogenomic profiling predicts antifungal synergies. *Mol. Syst. Biol.* **2009**, *5*, 338. [CrossRef] [PubMed]
84. Lehár, J.; Krueger, A.S.; Avery, W.; Heilbut, A.M.; Johansen, L.M.; Price, E.R.; Rickles, R.J.; Short, G.F., 3rd; Staunton, J.E.; Jin, X.; et al. Synergistic drug combinations tend to improve therapeutically relevant selectivity. *Nat. Biotechnol.* **2009**, *27*, 659–666. [CrossRef]
85. Whitesides, G.M. The origins and the future of microfluidics. *Nature* **2006**, *442*, 368–373. [CrossRef] [PubMed]
86. Squires, T.M.; Quake, S.R. Microfluidics: Fluid physics at the nanoliter scale. *Rev. Mod. Phys.* **2005**, *77*, 977–1026. [CrossRef]
87. Dertinger, S.K.W.; Chiu, D.T.; Jeon, N.L.; Whitesides, G.M. Generation of Gradients Having Complex Shapes Using Microfluidic Networks. *Anal. Chem.* **2001**, *73*, 1240–1246. [CrossRef]
88. Sackmann, E.K.; Fulton, A.L.; Beebe, D.J. The present and future role of microfluidics in biomedical research. *Nature* **2014**, *507*, 181–189. [CrossRef] [PubMed]
89. Knight, J.B.; Vishwanath, A.; Brody, J.P.; Austin, R.H. Hydrodynamic focusing on a silicon chip: Mixing nanoliters in microseconds. *Phys. Rev. Lett.* **1998**, *80*, 3863–3866. [CrossRef]
90. Pollack, L.; Tate, M.W.; Finnefrock, A.C.; Kalidas, C.; Trotter, S.; Darnton, N.C.; Lurio, L.; Austin, R.H.; Batt, C.A.; Gruner, S.M.; et al. Time resolved collapse of a folding protein observed with small angle X-ray scattering. *Phys. Rev. Lett.* **2001**, *86*, 4962–4965. [CrossRef] [PubMed]
91. Wen, N.; Zhao, Z.; Fan, B.; Chen, D.; Men, D.; Wang, J.; Chen, J. Development of Droplet Microfluidics Enabling High-Throughput Single-Cell Analysis. *Molecules* **2016**, *21*, 881. [CrossRef] [PubMed]
92. Clausell-Tormos, J.; Lieber, D.; Baret, J.C.; El-Harrak, A.; Miller, O.J.; Frenz, L.; Blouwolff, J.; Humphry, K.J.; Köster, S.; Duan, H.; et al. Droplet-based microfluidic platforms for the encapsulation and screening of Mammalian cells and multicellular organisms. *Chem. Biol.* **2008**, *15*, 427–437. [CrossRef] [PubMed]
93. Kim, D.; Wu, X.; Young, A.T.; Haynes, C.L. Microfluidics-based in vivo mimetic systems for the study of cellular biology. *Acc. Chem. Res.* **2014**, *47*, 1165–1173. [CrossRef] [PubMed]
94. Mehling, M.; Tay, S. Microfluidic cell culture. *Curr. Opin. Biotechnol.* **2014**, *25*, 95–102. [CrossRef] [PubMed]
95. Velve-Casquillas, G.; le Berre, M.; Piel, M.; Tran, P.T. Microfluidic tools for cell biological research. *Nano Today* **2010**, *5*, 28–47. [CrossRef] [PubMed]
96. Castel, D.; Pitaval, A.; Debily, M.A.; Gidrol, X. Cell microarrays in drug discovery. *Drug Discov. Today* **2006**, *11*, 616–622. [CrossRef] [PubMed]
97. Chen, D.S.; Davis, M.M. Molecular and functional analysis using live cell microarrays. *Curr. Opin. Chem. Biol.* **2006**, *10*, 28–34. [CrossRef] [PubMed]
98. Yarmush, M.L.; King, K.R. Living-cell microarrays. *Annu. Rev. Biomed. Eng.* **2009**, *11*, 235–257. [CrossRef] [PubMed]
99. Willaert, R.; Sahli, H. On-chip living-cell microarrays for network biology. In *Bioinformatics—Trends and Methodologies*; Mahdavi, M.A., Ed.; InTech-Open Access: PublisherRejeka, Croatia, 2011; pp. 609–630. ISBN 978-953-307-282-1.
100. Willaert, R.G.; Goossens, K. Microfluidic bioreactors for cellular microarrays. *Fermentation* **2015**, *1*, 38–78. [CrossRef]
101. Jonczyk, R.; Kurth, T.; Lavrentieva, A.; Walter, J.G.; Scheper, T.; Stahl, F. Living cell microarrays: An overview of concepts. *Microarrays* **2016**, *5*, 11. [CrossRef] [PubMed]
102. Charvin, G.; Cross, F.R.; Siggia, E.D. Forced periodic expression of G1 cyclins phase-locks the budding yeast cell cycle. *Proc. Natl. Acad. Sci. USA* **2009**, *106*, 6632–6637. [CrossRef] [PubMed]
103. Bean, J.M.; Siggia, E.D.; Cross, F.R. Coherence and timing of cell cycle start examined at single-cell resolution. *Mol. Cell* **2006**, *21*, 3–14. [CrossRef] [PubMed]
104. Lidstrom, M.E.; Konopka, M.C. The role of physiological heterogeneity in microbial population behavior. *Nat. Chem. Biol.* **2010**, *6*, 705–712. [CrossRef] [PubMed]
105. Martins, B.M.; Locke, J.C. Microbial individuality: How single-cell heterogeneity enables population level strategies. *Curr. Opin. Microbiol.* **2015**, *24*, 104–112. [CrossRef] [PubMed]
106. Rosenthal, K.; Oehling, V.; Dusny, C.; Schmid, A. Beyond the bulk: Disclosing the life of single microbial cells. *FEMS Microbiol. Rev.* **2017**, *41*, 751–780. [CrossRef] [PubMed]

107. Pierce, C.G.; Srinivasan, A.; Uppuluri, P.; Ramasubramanian, A.K.; López-Ribot, J.L. Antifungal therapy with an emphasis on biofilms. *Curr. Opin. Pharmacol.* **2013**, *13*, 726–730. [CrossRef] [PubMed]
108. Borghi, E.; Borgo, F.; Morace, G. Fungal Biofilms: Update on Resistance. *Adv. Exp. Med. Biol.* **2016**, *931*, 37–47. [CrossRef] [PubMed]
109. Koo, H.; Allan, R.N.; Howlin, R.P.; Stoodley, P.; Hall-Stoodley, L. Targeting microbial biofilms: Current and prospective therapeutic strategies. *Nat. Rev. Microbiol.* **2017**, *15*, 740–755. [CrossRef] [PubMed]
110. Wu, S.; Wang, Y.; Liu, N.; Dong, G.; Sheng, C. Tackling Fungal Resistance by Biofilm Inhibitors. *J. Med. Chem.* **2017**, *60*, 2193–2211. [CrossRef] [PubMed]
111. Srinivasan, A.; Uppuluri, P.; Lopez-Ribot, J.; Ramasubramanian, A.K. Development of a high-throughput *Candida albicans* biofilm chip. *PLoS ONE* **2011**, *6*, e19036. [CrossRef] [PubMed]
112. Srinivasan, A.; Lopez-Ribot, J.L.; Ramasubramanian, A.K. *Candida albicans* biofilm chip (CaBChip) for high-throughput antifungal drug screening. *J. Vis. Exp.* **2012**, *18*, e3845. [CrossRef]
113. Srinivasan, A.; Leung, K.P.; Lopez-Ribot, J.L.; Ramasubramanian, A.K. High-throughput nano-biofilm microarray for antifungal drug discovery. *MBio* **2013**, *4*, E00331-13. [CrossRef] [PubMed]
114. Srinivasan, A.; Torres, N.S.; Leung, K.P.; Lopez-Ribot, J.L.; Ramasubramanian, A.K. nBioChip, a Lab-on-a-Chip Platform of Mono- and Polymicrobial Biofilms for High-Throughput Downstream Applications. *mSphere* **2017**, *2*. [CrossRef] [PubMed]
115. Yang, H.; Bharracharyya, M.K.; Dong, L. Plant Pathogen Spores Grow in Microfluidic Droplets: A High-Throughput Approach to Antifungal Drug Screening. In Proceedings of the IEEE Conference paper Transducers' 11, Beijing, China, 5–9 June 2011.
116. Michael, C.A.; Dominey-Howes, D.; Labbate, M. The antimicrobial resistance crisis: Causes, consequences, and management. *Front. Public Health* **2014**, *2*, 145. [CrossRef] [PubMed]
117. Jorgensen, J.H.; Ferraro, M.J. Antimicrobial susceptibility testing: A review of general principles and contemporary practices. *Clin. Infect. Dis.* **2009**, *49*, 1749–1755. [CrossRef] [PubMed]
118. Barenfanger, J.; Drake, C.; Kacich, G. Clinical and financial benefits of rapid bacterial identification and antimicrobial susceptibility testing. *J. Clin. Microbiol.* **1999**, *37*, 1415–1418. [PubMed]
119. Arikan, S. Current status of antifungal susceptibility testing methods. *Med. Mycol.* **2007**, *45*, 569–587. [CrossRef] [PubMed]
120. Daniels, R. Surviving the first hours in sepsis: Getting the basics right (an intensivist's perspective). *J. Antimicrob. Chemother.* **2011**, *66*, 11–23. [CrossRef] [PubMed]
121. Alcazar-Fuoli, L.; Mellado, E. Current status of antifungal resistance and its impact on clinical practice. *Br. J. Haematol.* **2014**, *166*, 471–484. [CrossRef] [PubMed]
122. Syal, K.; Mo, M.; Yu, H.; Iriya, R.; Jing, W.; Guodong, S.; Wang, S.; Grys, T.E.; Haydel, S.E.; Tao, N. Current and emerging techniques for antibiotic susceptibility tests. *Theranostics* **2017**, *7*, 1795–1805. [CrossRef] [PubMed]
123. Clinical and Laboratory Standards Institute. *Reference Method for Broth Dilution Antifungal Susceptibility Testing of Yeasts. Approved Standard M27-A3*, 3rd ed.; CLSI: Wayne, PA, USA, 2008.
124. Arendrup, M.C.; Cuenca-Estrella, M.; Lass-Florl, C.; Hope, W.; Eucast, A. EUCAST technical note on the EUCAST definitive document EDef 7.2: Method for the determination of broth dilution minimum inhibitory concentrations of antifungal agents for yeasts EDef 7.2 (EUCAST-AFST). *Clin. Microbiol. Infect.* **2012**, *18*, E246–E247. [CrossRef] [PubMed]
125. Pulido, M.R.; García-Quintanilla, M.; Martín-Peña, R.; Cisneros, J.M.; McConnell, M.J. Progress on the development of rapid methods for antimicrobial susceptibility testing. *J. Antimicrob. Chemother.* **2013**, *68*, 2710–2717. [CrossRef] [PubMed]
126. Posteraro, B.; Torelli, R.; De Carolis, E.; Posteraro, P.; Sanguinetti, M. Antifungal susceptibility testing: Current role from the clinical laboratory perspective. *Mediterr. J. Hematol. Infect. Dis.* **2014**, *6*, e2014030. [CrossRef] [PubMed]
127. Subcommittee on Antifungal Susceptibility Testing of the ESCMID European Committee for Antimicrobial Susceptibility Testing. EUCAST Technical Note on the method for the determination of broth dilution minimum inhibitory concentrations of antifungal agents for conidia-forming moulds. *Clin. Microbiol. Infect.* **2008**, *14*, 982–984.
128. Vale-Silva, L.A.; Buchta, V. Antifungal susceptibility testing by flow cytometry: Is it the future? *Mycoses* **2006**, *49*, 261–273. [CrossRef] [PubMed]

129. Vale-Silva, L.A.; Pinto, P.; Lopes, V.; Ramos, H.; Pinto, E. Comparison of the Etest and a rapid flow cytometry-based method with the reference CLSI broth microdilution protocol M27-A3 for the echinocandin susceptibility testing of Candida spp. *Eur. J. Clin. Microbiol. Infect. Dis.* **2012**, *31*, 941–946. [CrossRef] [PubMed]
130. Vella, A.; De Carolis, E.; Vaccaro, L.; Posteraro, P.; Perlin, D.S.; Kostrzewa, M.; Posteraro, B.; Sanguinetti, M. Rapid antifungal susceptibility testing by matrix-assisted laser desorption ionization time of flight mass spectrometry analysis. *J. Clin. Microbiol.* **2013**, *51*, 2964–2969. [CrossRef] [PubMed]
131. Furustrand Tafin, U.; Clauss, M.; Hauser, P.M.; Bille, J.; Meis, J.F.; Trampuz, A. Isothermal microcalorimetry: A novel method for real- time determination of antifungal susceptibility of *Aspergillus* species. *Clin. Microbiol. Infect.* **2012**, *18*, E241–E245. [CrossRef] [PubMed]
132. Campbell, J.; McBeth, C.; Kalashnikov, M.; Boardman, A.K.; Sharon, A.; Sauer-Budge, A.F. Microfluidic advances in phenotypic antibiotic susceptibility testing. *Biomed. Microdevices* **2016**, *18*, 103. [CrossRef] [PubMed]
133. Mittman, S.A.; Huard, R.C.; Della-Latta, P.; Whittier, S. Comparison of BD phoenix to vitek 2, microscan MICroSTREP, and Etest for antimicrobial susceptibility testing of Streptococcus pneumoniae. *J. Clin. Microbiol.* **2009**, *47*, 3557–3561. [CrossRef] [PubMed]
134. Chatzigeorgiou, K.S.; Sergentanis, T.N.; Tsiodras, S.; Hamodrakas, S.J.; Bagos, P.G. Phoenix 100 versus Vitek 2 in the identification of gram-positive and gram-negative bacteria: A comprehensive meta-analysis. *J. Clin. Microbiol.* **2011**, *49*, 3284–3291. [CrossRef] [PubMed]
135. Dai, J.; Suh, S.J.; Hamon, M.; Hong, J.W. Determination of antibiotic EC50 using a zero-flow microfluidic chip based growth phenotype assay. *Biotechnol. J.* **2015**, *10*, 1783–1791. [CrossRef] [PubMed]
136. Dong, T.; Zhao, X. Rapid identification and susceptibility testing of uropathogenic microbes via immunosorbent ATP-bioluminescence assay on a microfluidic simulator for antibiotic therapy. *Anal. Chem.* **2015**, *87*, 2410–2418. [CrossRef] [PubMed]
137. Puchberger-Enengl, D.; van den Driesche, S.; Krutzler, C.; Keplinger, F.; Vellekoop, M.J. Hydrogel-based microfluidic incubator for microorganism cultivation and analyses. *Biomicrofluidics* **2015**, *9*, 014127. [CrossRef] [PubMed]
138. Bouquet, O.; Kocsis, B.; Kilár, F.; Lóránd, T.; Kustos, I. Amphotericin B and fluconazole susceptibility of *Candida* species determined by cell-chip technology. *Mycoses* **2012**, *55*, e90–e96. [CrossRef] [PubMed]
139. Chen, C.H.; Lu, Y.; Sin, M.L.; Mach, K.E.; Zhang, D.D.; Gau, V.; Liao, J.C.; Wong, P.K. Antimicrobial susceptibility testing using high surface-to-volume ratio microchannels. *Anal. Chem.* **2010**, *82*, 1012–1019. [CrossRef] [PubMed]
140. Sinn, I.; Kinnunen, P.; Albertson, T.; McNaughton, B.H.; Newton, D.W.; Burns, M.A.; Kopelman, R. Asynchronous magnetic bead rotation (AMBR) biosensor in microfluidic droplets for rapid bacterial growth and susceptibility measurements. *Lab Chip* **2011**, *11*, 2604–2611. [CrossRef] [PubMed]
141. Lu, Y.; Gao, J.; Zhang, D.D.; Gau, V.; Liao, J.C.; Wong, P.K. Single cell antimicrobial susceptibility testing by confined microchannels and electrokinetic loading. *Anal. Chem.* **2013**, *85*, 3971–3976. [CrossRef] [PubMed]
142. Mohan, R.; Mukherjee, A.; Sevgen, S.E.; Sanpitakseree, C.; Lee, J.; Schroeder, C.M.; Kenis, P.J. A multiplexed microfluidic platform for rapid antibiotic susceptibility testing. *Biosens. Bioelectron.* **2013**, *49*, 118–125. [CrossRef] [PubMed]
143. Tang, Y.; Zhen, L.; Liu, J.; Wu, J. Rapid antibiotic susceptibility testing in a microfluidic pH sensor. *Anal. Chem.* **2013**, *85*, 2787–2794. [CrossRef] [PubMed]
144. He, J.; Mu, X.; Guo, Z.; Hao, H.; Zhang, C.; Zhao, Z.; Wang, Q. A novel microbead-based microfluidic device for rapid bacterial identification and antibiotic susceptibility testing. *Eur. J. Clin. Microbiol. Infect. Dis.* **2014**, *33*, 2223–2230. [CrossRef] [PubMed]
145. Kim, S.C.; Cestellos-Blanco, S.; Inoue, K.; Zare, R.N. Miniaturized Antimicrobial Susceptibility Test by Combining Concentration Gradient Generation and Rapid Cell Culturing. *Antibiotics* **2015**, *4*, 455–466. [CrossRef] [PubMed]
146. Derzsi, L.; Kaminski, T.S.; Garstecki, P. Antibiograms in five pipetting steps: Precise dilution assays in sub-microliter volumes with a conventional pipette. *Lab Chip* **2016**, *16*, 893–901. [CrossRef] [PubMed]
147. Leonard, H.; Halachmi, S.; Ben-Dov, N.; Nativ, O.; Segal, E. Unraveling Antimicrobial Susceptibility of Bacterial Networks on Micropillar Architectures Using Intrinsic Phase-Shift Spectroscopy. *ACS Nano* **2017**, *11*, 6167–6177. [CrossRef] [PubMed]

148. Tang, M.; Huang, X.; Chu, Q.; Ning, X.; Wang, Y.; Kong, S.K.; Zhang, X.; Wang, G.; Ho, H.P. A linear concentration gradient generator based on multi-layered centrifugal microfluidics and its application in antimicrobial susceptibility testing. *Lab Chip* **2018**, *17*. [CrossRef] [PubMed]
149. Li, B.; Qiu, Y.; Glidle, A.; McIlvenna, D.; Luo, Q.; Cooper, J.; Shi, H.C.; Yin, H. Gradient microfluidics enables rapid bacterial growth inhibition testing. *Anal. Chem.* **2014**, *86*, 3131–3137. [CrossRef] [PubMed]
150. Smith, K.P.; Richmond, D.L.; Brennan-Krohn, T.; Elliott, H.L.; Kirby, J.E. Development of MAST: A Microscopy-Based Antimicrobial Susceptibility Testing Platform. *SLAS Technol.* **2017**, *22*, 662–674. [CrossRef] [PubMed]
151. Choi, J.; Yoo, J.; Kim, K.J.; Kim, E.G.; Park, K.O.; Kim, H.; Kim, H.; Jung, H.; Kim, T.; Choi, M.; et al. Rapid drug susceptibility test of *Mycobacterium tuberculosis* using microscopic time-lapse imaging in an agarose matrix. *Appl. Microbiol. Biotechnol.* **2016**, *100*, 2355–2365. [CrossRef] [PubMed]
152. Boedicker, J.Q.; Li, L.; Kline, T.R.; Ismagilov, R.F. Detecting bacteria and determining their susceptibility to antibiotics by stochastic confinement in nanoliter droplets using plug-based microfluidics. *Lab Chip* **2008**, *8*, 1265–1272. [CrossRef] [PubMed]
153. Choi, J.; Jung, Y.G.; Kim, J.; Kim, S.; Jung, Y.; Na, H.; Kwon, S. Rapid antibiotic susceptibility testing by tracking single cell growth in a microfluidic agarose channel system. *Lab Chip* **2013**, *13*, 280–287. [CrossRef] [PubMed]
154. Kalashnikov, M.; Lee, J.C.; Campbell, J.; Sharon, A.; Sauer-Budge, A.F. A microfluidic platform for rapid, stress-induced antibiotic susceptibility testing of *Staphylococcus aureus*. *Lab Chip* **2012**, *12*, 4523–4532. [CrossRef] [PubMed]
155. Price, C.S.; Kon, S.E.; Metzger, S. Rapid antibiotic susceptibility phenotypic characterization of *Staphylococcus aureus* using automated microscopy of small numbers of cells. *J. Microbiol. Methods* **2014**, *98*, 50–58. [CrossRef] [PubMed]
156. Choi, J.; Yoo, J.; Lee, M.; Kim, E.G.; Lee, J.S.; Lee, S.; Joo, S.; Song, S.H.; Kim, E.C.; Lee, J.C.; et al. A rapid antimicrobial susceptibility test based on single-cell morphological analysis. *Sci. Transl. Med.* **2014**, *6*, 267ra174. [CrossRef] [PubMed]
157. Hou, Z.; An, Y.; Hjort, K.; Hjort, K.; Sandegren, L.; Wu, Z. Time lapse investigation of antibiotic susceptibility using a microfluidic linear gradient 3D culture device. *Lab Chip* **2014**, *14*, 3409–3418. [CrossRef] [PubMed]
158. Matsumoto, Y.; Sakakihara, S.; Grushnikov, A.; Kikuchi, K.; Noji, H.; Yamaguchi, A.; Iino, R.; Yagi, Y.; Nishino, K. A Microfluidic Channel Method for Rapid Drug-Susceptibility Testing of Pseudomonas aeruginosa. *PLoS ONE* **2016**, *11*, e0148797. [CrossRef] [PubMed]
159. Hansen, K.M.; Thundat, T. Microcantilever biosensors. *Methods* **2005**, *37*, 57–64. [CrossRef] [PubMed]
160. Fritz, J. Cantilever biosensors. *Analyst* **2008**, *133*, 855–863. [CrossRef] [PubMed]
161. Willaert, R.; Kasas, S.; Devreese, B.; Dietler, G. Yeast nanobiotechnology. *Fermentation* **2016**, *2*, 18. [CrossRef]
162. Braun, T.; Ghatkesar, M.K.; Backmann, N.; Grange, W.; Boulanger, P.; Letellier, L.; Lang, H.P.; Bietsch, A.; Gerber, C.; Hegner, M. Quantitative time-resolved measurement of membrane protein-ligand interactions using microcantilever array sensors. *Nat. Nanotechnol.* **2009**, *4*, 179–185. [CrossRef] [PubMed]
163. Godin, M.; Tabard-Cossa, V.; Miyahara, Y.; Monga, T.; Williams, P.J.; Beaulieu, L.Y.; Bruce Lennox, R.; Grutter, P. Cantilever-based sensing: The origin of surface stress and optimization strategies. *Nanotechnology* **2010**, *21*, 75501. [CrossRef] [PubMed]
164. Godin, M.; Delgado, F.F.; Son, S.; Grover, W.H.; Bryan, A.K.; Tzur, A.; Jorgensen, P.; Payer, K.; Grossman, A.D.; Kirschner, M.W.; et al. Using buoyant mass to measure the growth of single cells. *Nat. Methods* **2010**, *7*, 387–390. [CrossRef] [PubMed]
165. Ndieyira, J.W.; Watari, M.; Barrera, A.D.; Zhou, D.; Vögtli, M.; Batchelor, M.; Cooper, M.A.; Strunz, T.; Horton, M.A.; Abell, C.; et al. Nanomechanical detection of antibiotic-mucopeptide binding in a model for superbug drug resistance. *Nat. Nanotechnol.* **2008**, *3*, 691–696. [CrossRef] [PubMed]
166. Lang, H.P.; Baller, M.K.; Berger, R.; Gerber, C.; Gimzewski, J.K.; Battiston, F.M.; Fornaro, P.; Ramseyer, J.P.; Meyer, E.; Guntherodt, H.J. An artificial nose based a micromechanical cantilever array. *Anal. Chim. Acta* **1999**, *393*, 59–65. [CrossRef]
167. Braun, T.; Barwich, V.; Ghatkesar, M.K.; Bredekamp, A.H.; Gerber, C.; Hegner, M.; Lang, H.P. Micromechanical mass sensors for biomolecular detection in a physiological environment. *Phys. Rev. E Stat. Nonlin. Soft Matter Phys.* **2005**, *72*, 031907. [CrossRef] [PubMed]

168. Hosaka, S.; Chiyoma, T.; Ikeuchi, A.; Okano, H.; Sone, H.; Izumi, T. Possibility of a femtogram mass biosensor using a self-sensing cantilever. *Curr. Appl. Phys.* **2006**, *6*, 384–388. [CrossRef]
169. Liu, Y.; Schweizerb, L.M.; Wanga, W.; Reubena, R.L.; Schweizer, M.; Shu, W. Label-free and real-time monitoring of yeast cell growth by the bending of polymer microcantilever biosensors. *Sens. Actuator B-Chem.* **2013**, *178*, 621–626. [CrossRef]
170. Bryan, A.K.; Goranov, A.; Amon, A.; Manalis, S.R. Measurement of mass, density, and volume during the cell cycle of yeast. *Proc. Natl. Acad. Sci. USA* **2010**, *107*, 999–1004. [CrossRef] [PubMed]
171. Cermak, N.; Olcum, S.; Delgado, F.F.; Wasserman, S.C.; Payer, K.R.; Murakami, M.A.; Knudsen, S.M.; Kimmerling, R.J.; Stevens, M.M.; Kikuchi, Y.; et al. High-throughput measurement of single-cell growth rates using serial microfluidic mass sensor arrays. *Nat. Biotechnol.* **2016**, *34*, 1052–1059. [CrossRef] [PubMed]
172. Burg, T.P.; Godin, M.; Knudsen, S.M.; Shen, W.; Carlson, G.; Foster, J.S.; Babcock, K.; Manalis, S.R. Weighing of biomolecules, single cells and single nanoparticles in fluid. *Nature* **2007**, *446*, 1066–1069. [CrossRef] [PubMed]
173. Park, K.; Jang, J.; Irimia, D.; Sturgis, J.; Lee, J.; Robinson, J.P.; Toner, M.; Bashir, R. 'Living cantilever arrays' for characterization of mass of single live cells in fluids. *Lab Chip* **2008**, *8*, 1034–1041. [CrossRef] [PubMed]
174. Bryan, A.K.; Hecht, V.C.; Shen, W.; Payer, K.; Grover, W.H.; Manalis, S.R. Measuring single cell mass, volume, and density with dual suspended microchannel resonators. *Lab Chip* **2014**, *14*, 569–576. [CrossRef] [PubMed]
175. Nugaeva, N.; Gfeller, K.Y.; Backmann, N.; Lang, H.P.; Düggelin, M.; Hegner, M. Micromechanical cantilever array sensors for selective fungal immobilization and fast growth detection. *Biosens. Bioelectron.* **2005**, *21*, 849–856. [CrossRef] [PubMed]
176. Stupar, P. Atomic Force Microscopy of Biological Systems: Quantitative Imaging and Nanomotion Detection. Ph.D. Thesis, EPFL Scientific Publications, Lausanne, Switzerland, 2018.
177. Longo, G.; Alonso-Sarduy, L.; Rio, L.M.; Bizzini, A.; Trampuz, A.; Notz, J.; Dietler, G.; Kasas, S. Rapid detection of bacterial resistance to antibiotics using AFM cantilevers as nanomechanical sensors. *Nat. Nanotechnol.* **2013**, *8*, 522–526. [CrossRef] [PubMed]
178. Stupar, P.; Opota, O.; Longo, G.; Prod'hom, G.; Dietler, G.; Greub, G.; Kasas, S. Nanomechanical sensor applied to blood culture pellets: A fast approach to determine the antibiotic susceptibility against agents of bloodstream infections. *Clin. Microbiol. Infect.* **2017**, *23*, 400–405. [CrossRef] [PubMed]
179. Stupar, P.; Yvanoff, C.; Chomicki, W.; Dietler, G.; Kasas, S.; Willaert, R. Exploring Nanoscale Motions of Yeast Cells. In Proceedings of the XIX Annual Linz Winter Workshop, Linz, Austria, 3–6 February 2017.
180. Vanden Boer, P.; Stupar, P.; Chomicki, W.; Dietler, G.; Kasas, S.; Willaert, R. Nanomotion Detection of Single Yeast Cell Growth. In Proceedings of the XX Annual Linz Winter Workshop, Linz, Austria, 2–5 February 2018.
181. Morones, J.R.; Elechiguerra, J.L.; Camacho, A.; Ramirez, J.T. The bactericidal effect of silver nanoparticles. *Nanotechnology* **2005**, *16*, 2346–2353. [CrossRef] [PubMed]
182. Kim, J.S.; Kuk, E.; Yu, K.N.; Kim, J.H.; Park, S.J.; Lee, H.J.; Kim, S.H.; Park, Y.K.; Park, Y.H.; Hwang, C.Y.; et al. Antimicrobial effects of silver nanoparticles. *Nanomed. Nanotechnol. Biol. Med.* **2007**, *3*, 95–101. [CrossRef] [PubMed]
183. Ahamed, M.; Alsalhi, M.S.; Siddiqui, M.K. Silver nanoparticle applications and human health. *Clin. Chim. Acta* **2010**, *411*, 1841–1848. [CrossRef] [PubMed]
184. Chen, X.; Schleusener, H.J. Nanosilver: A nanoproduct in medical application. *Toxicol. Lett.* **2008**, *176*, 1e12. [CrossRef] [PubMed]
185. Rai, M.; Yadav, A.; Gade, A. Silver nanoparticles as a new generation of antimicrobials. *Biotechnol. Adv.* **2009**, *27*, 76–83. [CrossRef] [PubMed]
186. De Jong, W.H.; Van Der Ven, L.T.; Sleijffers, A.; Park, M.V.; Jansen, E.H.; Van Loveren, H.; Vandebriel, R.J. Systemic and immunotoxicity of silver nanoparticles in an intravenous 28 days repeated dose toxicity study in rats. *Biomaterials* **2013**, *34*, 8333–8343. [CrossRef] [PubMed]
187. Durán, N.; Durán, M.; de Jesus, M.B.; Seabra, A.B.; Fávaro, W.J.; Nakazato, G. Silver nanoparticles: A new view on mechanistic aspects on antimicrobial activity. *Nanomedicine* **2016**, *12*, 789–799. [CrossRef] [PubMed]
188. Ahamed, M.; Karns, M.; Goodson, M.; Rowe, J.; Hussain, S.M.; Schlager, J.J.; Hong, Y. DNA damage response to different surface chemistry of silver nanoparticles in mammalian cells. *Toxicol. Appl. Pharmacol.* **2008**, *233*, 404–410. [CrossRef] [PubMed]
189. Kawata, K.; Osawa, M.; Okabe, S. In vitro toxicity of silver nanoparticles at noncytotoxic doses to HepG2 human hepatoma cells. *Environ. Sci. Technol.* **2009**, *43*, 6046–6051. [CrossRef] [PubMed]

190. AshaRani, P.V.; Low Kah Mun, G.; Hande, M.P.; Valiyaveettil, S. Cytotoxicity and genotoxicity of silver nanoparticles in human cells. *ACS Nano* **2009**, *3*, 279–290. [CrossRef] [PubMed]
191. Panáček, A.; Kvítek, L.; Smékalová, M.; Večeřová, R.; Kolář, M.; Röderová, M.; Dyčka, F.; Šebela, M.; Prucek, R.; Tomanec, O.; et al. Bacterial resistance to silver nanoparticles and how to overcome it. *Nat. Nanotechnol.* **2018**, *13*, 65–71. [CrossRef] [PubMed]
192. Ammar, H.A.; El-Desouky, T.A. Green synthesis of nanosilver particles by *Aspergillus terreus* HA1N and Penicillium expansum HA2N and its antifungal activity against mycotoxigenic fungi. *J. Appl. Microbiol.* **2016**, *121*, 89–100. [CrossRef] [PubMed]
193. Xue, B.; He, D.; Gao, S.; Wang, D.; Yokoyama, K.; Wang, L. Biosynthesis of silver nanoparticles by the fungus Arthroderma fulvum and its antifungal activity against genera of Candida, *Aspergillus* and *Fusarium*. *Int. J. Nanomedicine* **2016**, *11*, 1899–1906. [CrossRef] [PubMed]
194. Khan, N.T.; Jameel, N. Antifungal Activity of Silver Nanoparticles Produced from Fungus, *Penicillium fellutanum* at Different pH. *J. Microb. Biochem. Technol.* **2016**, *8*, 5. [CrossRef]
195. Fukushima, K.; Liu, S.; Wu, H.; Engler, A.C.; Coady, D.J.; Maune, H.; Pitera, J.; Nelson, A.; Wiradharma, N.; Venkataraman, S.; et al. Supramolecular high-aspect ratio assemblies with strong antifungal activity. *Nat. Commun.* **2013**, *4*, 2861. [CrossRef] [PubMed]
196. Esteban-Tejeda, L.; Malpartida, F.; Esteban-Cubillo, A.; Pecharromán, C.; Moya, J.S. Antibacterial and antifungal activity of a soda-lime glass containing copper nanoparticles. *Nanotechnology* **2009**, *20*, 505701. [CrossRef] [PubMed]
197. Paulo, C.S.; Vidal, M.; Ferreira, L.S. Antifungal nanoparticles and surfaces. *Biomacromolecules* **2010**, *11*, 2810–2817. [CrossRef] [PubMed]
198. Perfect, J.R. Fungal virulence genes as targets for antifungal chemotherapy. *Antimicrob. Agents Chemoth.* **1996**, *40*, 1577–1583.
199. Kitamura, A.; Someya, K.; Hata, M.; Nakajima, R.; Takemura, M. Discovery of a small-molecule inhibitor of β-1,6-glucan synthesis. *Antimicrob. Agents Chemoth.* **2009**, *53*, 670–677. [CrossRef] [PubMed]
200. Baxter, B.K.; DiDone, L.; Ogu, D.; Schor, S.; Krysan, D.J. Identification, in vitro activity and mode of action of phosphoinositide-dependent-1 kinase inhibitors as antifungal molecules. *ACS Chem. Biol.* **2011**, *6*, 502–510. [CrossRef] [PubMed]

9

Improvement of Malvar Wine Quality by use of Locally-Selected *Saccharomyces cerevisiae* Strains

Gustavo Cordero-Bueso [1,2,*], Braulio Esteve-Zarzoso [3], Mar Gil-Díaz [1], Margarita García [1], Juan Mariano Cabellos [1] and Teresa Arroyo [1]

1 Departamento de Agroalimentación, Instituto Madrileño de Investigación y Desarrollo Rural Agrario y Alimentario, Autovía A2 km 38.2, Alcalá de Henares, Madrid 28800, Spain; mar.gil.diaz@madrid.org (M.G.-D.); margarita_garcia_garcia@madrid.org (M.G.); juanmariano.cabellos@madrid.org (J.M.C.); teresa.arroyo@madrid.org (T.A.)

2 Departamento de Biomedicina, Biotecnología y Salud Pública, Universidad de Cádiz, Avenida de la República Saharaui s/n, Puerto Real, Cádiz 11510, Spain

3 Biotecnologia Enològica, Departament de Bioquimica i Biotecnologia, Facultat d'Enologia, Universitat Rovira i Virgili, Marcel li Domingo s/n, Tarragona 43007, Spain; Braulio.esteve@urv.cat

* Correspondence: gustavo.cordero@uca.es

Academic Editor: Ronnie G. Willaert

Abstract: Malvar grape juice offers relatively little in the way of a sensory experience. Our interest lies in the use of locally-selected yeast strains in experimental fermentations to improve the sensory characteristics of Malvar wines. Two locally-selected strains of *Saccharomyces cerevisiae* were used as starter cultures in vinifications and compared with spontaneous fermentations of the same cultivar musts. Wine quality was investigated by their principal oenological parameters, analysis of the volatile aroma components, and corroborated by an experienced taster panel. The most salient chemical attributes were its high concentrations of isoamyl acetate and hexyl acetate and the high acidity, which have been detected to be key constituents in setting the fruity and fresh character of Malvar wines. Winemakers of winegrowing areas where this grape variety is cultivated will have improved options to elaborate new white wines styles, using selected yeast strains that enhance its aromatic properties.

Keywords: sensory analysis; inoculation; spontaneous fermentation; yeast selection

---

1. Introduction

One of the most cultivated varieties in the winegrowing region of Madrid, with a total extension of 11,758 ha, is the white grape variety Malvar (*Vitis vinifera* L.). Part of the economic development of this area is based on wine production using grapes from this cultivar. However, Malvar grape juice offers relatively little in the way of a sensory experience, but after wine fermentation, it has a multitude of attributes [1]. Hence, the intervention of the yeasts in the fermentation process can be noteworthy. In a previous work, we characterized the enological relevance of 12 selected non-*Saccharomyces* wine yeast species isolated from spontaneous fermentations of this variety to propose their use in mixed starter cultures to improve the organoleptic properties of the Malvar wines [1]. The yeast strain *Torulaspora delbrueckii* CLI 918 was defined as a yeast strain with potential interest for its contribution to the aromatic Malvar wine profile with flowery and fruity aromas. This yeast species could be used in mixed starter cultures with *S. cerevisiae*. It is well known that ethanol-tolerant yeasts, such as *S. cerevisiae*, play an essential role in the evolution of yeast-derived and grape-derived flavor molecules [2–6]. However, not all wine yeasts strains are equal; juice from the same grapes will deliver quite different wines, depending on the choice of yeast [7] and the management of the fermentation processes [8]. Thus, revealing new data relevant to Malvar wines is necessary.

Autochthonous yeast strains selected for their use as starter cultures is a profitable approach. Thus, lately, in a challenge to enrich distinctive aromatic properties, some researchers have addressed their attention to the selection of yeasts from restricted areas [8–12]. Ecological studies of wine yeasts are essential for finding novel strains with new molecular and enological attributes. Some winemakers proclaim that wines with geographical characteristics can be obtained only with selected yeast starters originated from the same area where wines are elaborated and the use of autochthonous wine yeasts selected from each vine-growing area is widely spread [8,9].

Uninoculated fermentation is a complex microbial process accomplished by the sequential action of non-*Saccharomyces* and the different *Saccharomyces* yeast strain populations present on the skin of the grapes, winery equipment, in the musts, and in the wine [13]. On the other hand, inoculation with indigenous yeasts as a starter culture is engaged to set a dominant population of a selected strain from the beginning to the end of fermentation [14,15]. The yeast activity during a spontaneous fermentation is capricious and could contribute less desirable attributes to the wine. Moreover, risks related with natural fermentations include both sluggish or arrested fermentations and the propagation of contaminant yeasts [13,14]. To keep away of these problems, commercially available wine yeast exhibits great diversity in degrees of robustness to dryness but, unfortunately, the most resilient strains do not necessarily deliver optimal sensory characteristics to the wine [7,15]. In addition, the use of commercial starters could disguise the distinctive properties that characterize some local wines [7,15]. Thus, understanding how yeasts influence the principal properties of wine aroma, flavor, and color provided the basic steps for selection of autochthonous yeast strains for use as starter cultures and control of the alcoholic fermentation as a new commercial option for wine makers. Furthermore, the use of selected native yeast strains in starter cultures is rather preferred since they are better acclimatized to the environmental circumstances and may ensure the maintenance of the typical sensory characteristics of the wines of a certain region [11].

The goal of this study is the evaluation of Malvar wines made with indigenous and selected yeast strains of *S. cerevisiae* in order to evaluate the effect, to devise the use of these yeast strains to make wine, and to determine the most remarkable chemical and sensory characteristics of such wines belonging to the Appellation of Origin "Vinos de Madrid" (Madrid, Spain). Moreover, no investigations have been carried out to improve its enological characteristics using selected or autochthonous strains of *S. cerevisiae* selected in this area.

2. Experimental Section

2.1. *Yeast Strains and Vinification Procedure*

The yeast strains utilized in this study were two *S. cerevisiae* yeast strains (coded as CLI 889 and CLI 892) previously isolated among 18 different genetic profiles obtained by PFGE with different occurrence in the Madrid winegrowing region. These were selected and characterized in our laboratories based on some established and desirable enological criteria, such as high fermentation performance, resistance to ethanol, low production of hydrogen sulfide and sulfur dioxide, and volatile-derived, among others [16] (see Supplementary Material). Fermentations were conducted at the IMIDRA's experimental cellar. Must obtained showed 240 g/L of reducing sugars, 170 mg/L of yeast absorbable nitrogen (YAN), 900 mg/L of total amino acids, the pH value was 3.61, and the titratable acidity (expressed as g/L of tartaric acid) was 5.78. Musts were carefully racked, homogenized, and dislodged statically (at 4 °C) adding 0.01 g/L of pectolytic enzymes (Enozym Altair, Agrovin, Spain) and 50 mg/L of sulfur dioxide (SO_2). This must was divided into nine stainless steel vats of 100 L coded as A1, A2, and A3 (those inoculated with CLI 889); B1, B2, and B3 (those inoculated with CLI 892), and the spontaneous fermentations as S1, S2, and S3. Triplicate fermentations were carried out at a controlled temperature of 18 °C. Musts were inoculated with a final concentration of 10^6 cells/ml of pure selected yeast. Fermentation kinetic was controlled by monitoring daily the density. When its value was the same during two consecutive days, residual sugars were analyzed by enzymatic methodology (Roche diagnostics, Darmstadt, Germany). Fermentation was considered to be completed when residual sugars concentration was less than 2 g/L. After fermentation, the wines were clarified by cold settling using 3 mL/L of colloidal silica (Silisol, Agrovin, Alcázar de San Juan, Ciudad Real, Spain) and 2 mg/L of gelatin fining (Vinigel, Agrovin, Alcázar de San Juan, Ciudad Real, Spain). After three months of cold stabilization at 4 °C, wines were filtered (0.6 μm) and bottled. Then, they were subjected to chemical analysis.

2.2. Yeast Isolation, Identification, and Typification

Samples were taken from every vat during the vinification process at different density values, initial (D1) = 1090 g/L, D2 = 1085–1070 g/L, D3 = 1060–1050 g/L, D4 = 1030–1025 g/L, D5 = 1010–1000 g/L and D6 = 990 g/L (<2 g/L of residual sugars of the fermentations). Thus, fifty-four 100 mL sterile plastic flasks were filled with the Malvar must/wine from different parts of the vessels, kept under refrigeration (4 °C), and transported to the laboratory. Aliquots of tenfold dilution of the samples were spread onto YGC agar plates (Laboratorios Conda, Madrid, Spain). The plates were incubated for 3–4 days at 26 °C. After yeast colony counting, 30 colonies were randomly selected from each fermentation sample for their identification and subsequent monitoring of the implantation rate.

DNA extraction and quantification from isolates was performed as stated by Cordero-Bueso *et al.* [17]. Identification was carried out by the amplification of the ITS1-5.8S-ITS2 r DNA region and subsequent RFLP analysis using endonucleases *Cfo*I, *Hinf*I, *Hae*II, and *Dde*I [18]. Those isolated and identified different to *S. cerevisiae* were not analyzed in this study. In order to monitor the yeast strain dominance during the fermentation processes, two methodologies were used, karyotyping by pulsed field gel electrophoresis (PFGE) and amplification by PCR of microsatellite regions. The molecular karyotype was obtained following the protocol proposed by Rodríguez *et al.* [14]. Microsatellite reaction mix and amplification protocols were identical as those used by Vaudano and García-Moruno [19]. Amplified products were scattered on an agarose gel (2.5% *w/v*) with a final concentration of 5 μL/mL of ethidium bromide, in 1× TBE buffer at 100 V for 90 min. DNA fragment sizes were resolved by comparison with a molecular ladder marker of 100 bp (Promega, Madison, WI, USA). Fragment differentiation and allele size determination was carried out by single capillary automatic electrophoresis (CE) in ABI 3130 Genetic Analyzer (Applied Biosystems, Foster City, CA, USA).

2.3. Enological Parameters of the Fermentation Assays

Enological parameters measurements were performed as stated in Cordero-Bueso *et al.* [1]. Alcoholic titer was measured by near-infrared reflectance, the fermentative capacity was calculated as the difference between the initial and final sugar content. Fermentation velocity (V_F) (or alcohol production expressed as grams of sugar consumed daily) was measured checking daily the sugar percentage lost during the fermentation. In addition, V_{50} amount of sugar daily transformed by the yeasts when 50% of the sugar content had been used up was evaluated. Free and total sulfur dioxide, pH, titratable acidity, and volatile acidity were performed according to standard methods in the enological sector (OIV methods, Official Methods established by the European Union). Glycerol and 2,3-butanodiol compounds were determined using the Feuilles verts 588 (FV) method also suggested by the OIV.

2.4. Determination of Carboxylic Acids of the Malvar Wines

Carboxylic acids were quantified by ionic chromatography using Dionex DX 500 (Salt Lake City, UT, USA) equipment with a CD20 conductivity detector. Standard stock solutions of the organic acids (Panreac, Barcelona, Spain) were prepared by dissolution of the acids or the salts with deionized water. After filtering (0.22 μm) and dilution (1:20) with sterilized water, duplicate wine samples were injected into the chromatograph equipped with an IonPac ICE-AS6 capillary column. A concentration of 0.4 mM heptafluorobutyric acid (HFBA) (FlukaChemie AG, Buchs, Switzerland) was used as eluent at flow rate of 1.0 mL/min in isocratic mode. Anion-Ice micro-membrane was used as suppressor column and tetrabutylamonium hydroxide (Riedel-de Haën, Seelze, Germany) as a regenerator with a flow rate of 5 mL/min. The working conditions were as follow; temperature at 25 °C, 25 μL of injection volume, and 10 μs FS of the detector conductivity.

2.5. *Determination of the Volatile Fraction of the Fermentations*

Six major volatile compounds of the Malvar wines obtained were settled by gas chromatography coupled to a flame ionization detector (GC-FID) and 19 minor volatiles by gas chromatography combined with mass spectrometry (GC-MS). Major volatiles (acetaldehyde, acetoin, ethyl acetate, and the higher major alcohols, 1-propanol, isobutanol, and the isoamylic alcohols (2-methyl-1-butanol and 3-methyl-1-butanol)) were determined after steam distillation using a Hewlett Packard Series II (Palo Alto, CA, USA) gas chromatograph with a flame ionization detector. At the same time, minor volatiles were extracted after a liquid–liquid process using a mixture of ether hexane (1:1 v/v) as extractant. The organic phases were concentrated under a stream of N_2 and injected into a Hewlett-Packard 6890 gas chromatograph (Agilent, Avondale, PA, USA) fitted to a mass spectrometer detector HP Mass Selective 5973 (Agilent, Avondale, PA, USA). The reference Gil *et al.* [20] shows a complete description of both methods.

2.6. *Sensory Analysis*

Sensory evaluations were done under ISO standards [21–23] related to methodology and sensory analysis vocabulary (ISO 8586-1:1993), selection and formation of tasters (ISO 11035:1994), and tasting room (ISO 8589:2007). Two sensory analyses were performed in only one session by eleven skilled judges. Malvar wine sample positions were randomized every time and the sensory profile was defined using 13 descriptors previously described by Lozano *et al.* [24] and chosen by the taster panel in a previous session as stated in the ISO 11035:1994 rules and according to their importance in Malvar wines.

The first wine tasting was carried out by filling in a blank official tasting scorecard used in the Appellation of Origin "Vinos de Madrid". Penalizing scores were used; thus, the better quality wines obtained a lower score. Six variables (appearance, aroma quality, aroma intensity, taste intensity, taste quality, and harmony) were selected for estimation of wine quality, and a scale of seven categories (excellent: 0–7, very good: 8–23, good: 24–44, correct: 45–52, ordinary: 53–78, defective: 79–90, eliminated > 90) like those proposed by Vilanova *et al.* [25]. The total scores given by the eleven tasters for each parameter corresponding to the sensorial characteristics of wine were then statistically analyzed. After the first wine tasting, all judges were also asked to evaluate the sensory profile of the samples on a 0–5 point scale of intensity filling in a second official tasting scorecard containing the aroma descriptors mentioned above. Scale zero (0) implied that the descriptor was not perceived, while a score of five (5) was equal to the highest perception.

2.7. *Statistical Analysis*

One-way ANOVA and Tukey's test were accomplished to emphasize the effects of yeast strains on sensorial descriptors. Discriminant analysis was performed to point up any differences due to yeast strain. Principal component analysis (PCA) was carried out with 20.0 SPSS (Inc. Chicago, IL, USA) for Windows statistical package (significance level $p = 0.05$).

3. Results and Discussion

3.1. Fermentation Kinetics

Figure 1 shows the average of the fermentative kinetics and yeast population evolution of Malvar musts. A lag phase of two days was observed in all fermentations and proceeding to dryness between eight and nine days in the case of the inoculated fermenters, and between 10 and 11 days in the spontaneous fermentations. Although different kinetic behaviors were found, all strains were able to finish the fermentation consuming over 98.5% of initial sugar. Differences on sugar consumption in the middle stages of the fermentations have been observed. In spite of the similar profile obtained for the inoculated fermentations during the first days (until the third day) and at the completion of fermentation (from the sixth day), the strain CLI 892 showed higher fermentation rate than the CLI 889 strain and spontaneous fermentations, which showed the same rate during this period of time.

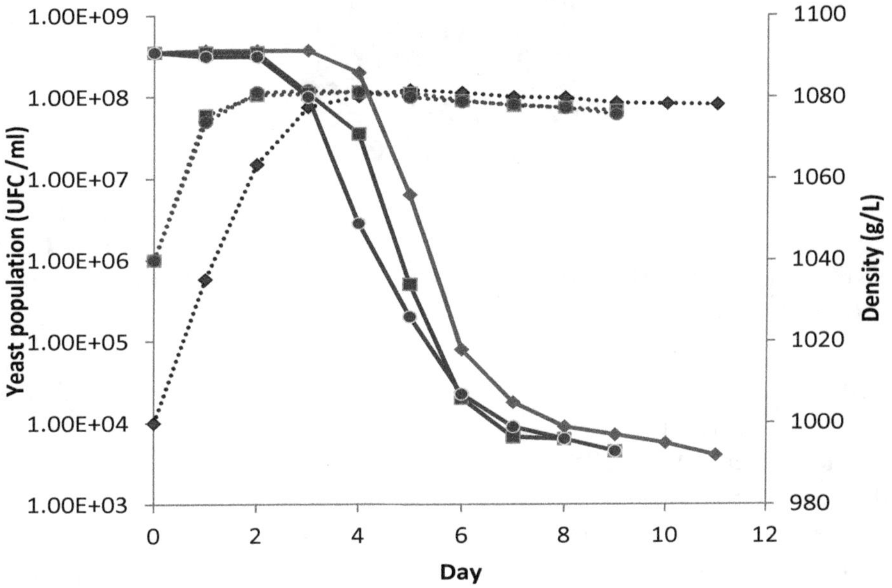

Figure 1. Evolution of the total yeast population (dashed lines) and must density (continuous lines) during three wine fermentations performed with Malvar musts using the following inoculation procedures: no inoculation (♦); inoculation with yeast strain CLI 889 (■); and inoculation with yeast strain CLI 892 (●). Data correspond to mean values obtained from triplicate experiments.

3.2. Implantation Rate

A total of 1620 colonies were isolated from the different stages of the fermentations of Malvar among the nine fermenters. Molecular identification using ITS-5.8S amplification and restriction analysis, and comparing the restriction profiles with those obtained by Esteve-Zarzoso *et al.* [18], showed that from first day of inoculated fermentations all isolates belonged to *S. cerevisiae*, while in the uninoculated musts, 131 of 540 colonies were identified as non-*Saccharomyces* in the initial and middle stages of the spontaneous fermentations.

All colonies identified as *S. cerevisiae* were characterized by PFGE, showing five different karyotypes. Isolates from inoculated vessels with the yeast strain CLI 889 showed two different karyotype patterns during the early and middle phases of the fermentations, but at the end of the fermentation process only the karyotype (A) corresponding to the inoculated yeast strain was found (Figure 2), while colonies of *S. cerevisiae* from the three vessels inoculated with the yeast strain CLI 892 showed one unique profile (B), throughout the entire fermentation process and corresponding to the inoculated yeast (Figure 2). The implantation rate at the end of the fermentation 100% fit the profile of the inoculated strain in both tanks. While in the uninoculated vessels, 90% of the population was represented by a single karyotype (A) as showed Figure 2. Dominance or competitiveness of a starter yeast strain could have an impact on the sensorial quality of wine by dominating its aromatic profile or eliminating the collaborative role of natural *S. cerevisiae* populations [13].

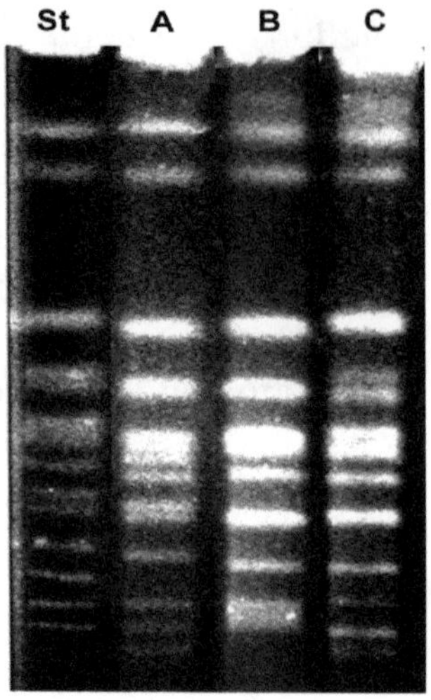

Figure 2. Electrophoretic karyotypes of the majority at the end of the fermentation *Saccharomyces cerevisiae* yeast strain isolated from the spontaneous fermentations (**A**) and selected inoculated yeast strains CLI 889 (**B**) and CLI 892 (**C**) The chromosomes of the standard *S cerevisiae* YNN 295 (Bio-Rad) were used as a reference (St).

The microsatellite PCR analysis of the *S. cerevisiae* strains, CLI 892 and CLI 889 showed that both strains are homozygous for the three alleles fingerprinted. The sizes of the alleles were as follows: SCPTSY7 (235 bp and 268 bp), SC8132X (215 bp and 209 bp), and YOR267C (416 bp and 451 bp), respectively. In the case of the strain isolated from spontaneous fermentations (S) allele sizes were as follows: SCPTSY7 (292 bp and 292 bp), SC8132X (212 bp and 310 bp), and YOR267C (308 bp and 389 bp). The comparison of these profiles with the five obtained by PFGE during the last phases of the fermentation of Malvar musts enabled that the number of patterns obtained and implantation rates during all fermentation processes in the different tanks were identical.

3.3. Principal Enological Parameters of the Fermentation Assays

Table 1 shows the means and standard deviations of the principal enological parameters, including organic acids of Malvar must fermentations carried out with the selected strains and uninoculated fermentations. Fermentative velocity (V_F) was higher in spontaneous fermentations, although the V_{50} was higher in the must inoculated with the yeast strain CLI 892. This finding is in agreement with the fact that those *S. cerevisiae* naturally present in must are better adapted to the fermentation conditions and environment than allochthonous inoculated yeasts [10]. It is interesting to point out that volatile acidity and acetic acid content were significantly lower in the uninoculated musts than the inoculated ones.

Regarding carboxylic acids, these compounds may contribute favorably to the organoleptic properties of young white wines. Our results showed that the total amount of these compounds was similar in wines from the CLI 892 strain and spontaneous fermentation, and higher than those obtained in wines from the CLI 889 strain. The excessive production of glycerol during wine fermentations for its positive sensory attributes gives rise to an increase in acetic acid concentration [26]. According to Erasmus *et al.* [27] it is, therefore, conceivable that different yeast strains experiencing the same fermentation conditions will respond by producing different concentrations of glycerol and acetic acid. The wine yeast strain isolated from the Malvar spontaneous fermentation (S), which produces low concentrations of acetic acid and conducts fermentations efficiently, seems to be a great candidate for the future production of high-quality Malvar wines in the Madrid winegrowing region, alone, or using mixed starter cultures.

Table 1. Enological parameters and organic acid concentrations of fermented wines (average and standard deviation (SD), Fisher's test (F, and significance (Sig) factors according to one-way ANOVA, test of comparison of means (Tukey): 2 and 11 degrees of freedom; the characters a, b, and c mean significant differences at $p \leqslant 0.05$. V_{50} = fermentation velocity consumption of 50% of the sugar content; V_F = fermentation velocity (% of daily sugar consumption).

Parameters	S	CLI 889	CLI 892	F	Sig
Alcoholic degree % (v/v)	12.50 ± 0.08 ^{a}	12.54 ± 0.06 ^{a}	12.70 ± 0.02 ^{b}	9.69	0.0132
Fermentative capacity	12.00 ± 0.07 ^{a}	12.61 ± 0.01 ^{c}	12.50 ± 0.01 ^{b}	186.53	0.0000
V_{50}	18.00 ± 0.04 ^{b}	15.45 ± 0.64 ^{a}	20.30 ± 0.2 ^{c}	128.70	0.0000
V_F	6.7 ± 0.3 ^{b}	5.40 ± 0.42 ^{a}	4.9 ± 0.7 ^{a}	10.27	0.0115
pH	3.81 ± 0.01 ^{b}	3.77 ± 0.01 ^{a}	3.79 ± 0.02 ^{b}	7.94	0.0206
Free SO_2 (mg/L)	6.0 ± 0.3 ^{a}	9.5 ± 0.71 ^{b}	10.0 ± 1.0 ^{b}	26.82	0.0010
Total SO_2 (mg/L)	15.0 ± 2.7	16.1 ± 1.56	14.0 ± 2.0	0.72	0.5230
Volatile acidity (g/L; acetic acid)	0.14 ± 0.01 ^{a}	0.23 ± 0.03 ^{b}	0.31 ± 0.04 ^{c}	25.04	0.0012
Titratable acidity (g/L; tartaric acid)	4.90 ± 0.07 ^{a}	5.14 ± 0.06 ^{b}	5.20 ± 0.07 ^{b}	16.93	0.0034
Citric acid (g/L)	0.31 ± 0.01 ^{b}	0.30 ± 0.00 ^{a,b}	0.29 ± 0.01 ^{a}	4.00	0.0787
Malic acid (g/L)	2.82 ± 0.05 ^{b}	2.58 ± 0.01 ^{a}	2.72 ± 0.09 ^{b}	27.44	0.0010
Lactic acid (g/L)	0.59 ± 0.06 ^{b}	0.55 ± 0.02 ^{b}	0.45 ± 0.04 ^{a}	8.36	0.0184
Acetic acid (g/L)	0.17 ± 0.01 ^{a}	0.32 ± 0.02 ^{b}	0.39 ± 0.03 ^{c}	69.14	0.0001
Succinic acid (g/L)	0.29 ± 0.01 ^{a,b}	0.28 ± 0.00 ^{a}	0.31 ± 0.01 ^{b}	5.55	0.0433
Reducing sugar (g/L)	1.5 ± 0.3	1.55 ± 0.07	1.3 ± 0.2	1.17	0.3730
Glycerol (g/L)	7.1 ± 1.6 ^{a,b}	7.85 ± 0.92 ^{b}	4.40 ± 1.5 ^{a}	5.13	0.0503
2,3-butanodiol (mg/L)	373.5 ± 67.5 ^{a}	386.8 ± 16.3 ^{a}	518.4 ± 91.2 ^{b}	7.86	0.0211

3.4. Aromatic Profile of Malvar Wines Fermented with Different Yeast Strains

Table 2 shows the average and standard deviations of the volatile compounds detected in the different fermentations. From all the volatile compounds identified, those presented at concentrations higher than their OTH (OAV higher than 1) are mainly considered as aroma-contributing compounds, and indicated in bold in Table 2.

Table 2. Data (Mean ± S.D.) of volatile composition related to the uninoculated fermentations (S) and inoculated fermentations with two locally-selected *S. cerevisiae* yeast strains (CLI 889 and CLI 892). Odor descriptors (ODE) and odor thresholds (OTH) described in the literature are included. Thresholds were calculated in a 10%–12% water/ethanol mixture. Odor activity values were also calculated. F = Fisher's test; Sig, one-way ANOVA analysis, test of comparison of means (Tukey): 2 and 11 degrees of freedom; the characters a, b, and c mean significant differences at $p \leqslant 0.05$.

Compound (mg/L)	S	CLI 889	CLI 892	F	Sig	ODE	OTH (mg/L)	OAV *		
								S	CLI 889	CLI 892
Acetaldehyde	50.75 ± 8.85 ^{a}	69.61 ± 8.19 ^{b}	60.35 ± 10.30 ^{ab}	3.18	0.1142	Pleasant, fruity	0.0025 [2]	**20.30**	**27.84**	**24.14**
Acetoin	tr ^{a}	2.4±1.03 ^{b}	1.27±0.73 ^{ab}	8.14	0.0195	Flowery, wet	150.0 [1]	<0.1	<0.1	<0.1
Ethyl acetate	68.71 ± 1.01 ^{a}	69.28 ± 4.41 ^{a}	64.65 ± 10.66 ^{a}	0.43	0.6700	Fruit, solvent	12.26 [1]	**5.60**	**5.65**	**5.27**
1-Propanol	35.02 ± 1.83 ^{b}	36.63 ± 1.41 ^{b}	25.80 ± 2.58 ^{a}	25.63	0.0012	Alcohol, ripe fruit	306.0 [1]	0.11	0.11	<0.1
Isobutanol	37.39 ± 0.58 ^{b}	39.50 ± 1.53 ^{b}	23.52 ± 2.63 ^{a}	70.62	0.0001	Fusel, alcohol	40.00 [1]	0.93	0.98	0.58
Isoamylic alcohols	175.23 ± 1.74 ^{b}	182.20 ± 2.99 ^{c}	156.38 ± 3.25 ^{a}	71.47	0.0001	Bitter, harsh	30.00 [1]	**5.84**	**6.07**	**5.21**
Σ Higher major alcohols	247.65 ± 0.66 ^{b}	258.13 ± 5.63 ^{b}	205.71 ± 8.45 ^{a}	66.86	0.0001					
1-Hexanol	0.81 ± 0.01 ^{a}	0.95 ± 0.00 ^{a}	1.00 ± 0.06 ^{a}	0.42	0.6754	Green grass	8.00 [1]	0.1	0.11	0.12
2-Phenylethanol	10.79 ± 0.29 ^{b}	9.29 ± 0.85 ^{a}	9.99 ± 0.82 ^{ab}	3.46	0.100	Roses	14.00 [1]	0.77	0.66	0.71
Σ Higher minor alcohols	11.59 ± 0.27 ^{b}	10.22 ± 0.83 ^{a}	10.99 ± 0.80 ^{ab}	3.03	0.1233					

Table 2. *Cont.*

Compound (mg/L)	S	CLI 889	CLI 892	F	Sig	ODE	OTH (mg/L)	OAV *		
								S	CLI 889	CLI 892
Isobutyl acetate	0.16 ± 0.00^{b}	0.18 ± 0.00^{c}	0.13 ± 0.01^{a}	37.75	0.0004	Sweet fruit	1.60 [1]	0.1	0.11	<0.1
Isoamyl acetate	7.66 ± 0.18^{b}	9.37 ± 0.54^{c}	6.19 ± 0.08^{a}	67.73	0.0001	Banana	0.030 [1]	**255.33**	**312.33**	**206.33**
Hexyl acetate	0.14 ± 0.01^{b}	0.19 ± 0.02^{b}	0.06 ± 0.05^{a}	11.07	0.0097	Fruity, green, pear	0.020 [3]	**7.00**	**9.50**	**3.00**
Phenylethyl acetate	0.56 ± 0.03^{b}	0.56 ± 0.07^{b}	0.38 ± 0.04^{a}	13.14	0.0064	Pleasant, flowery	0.250 [1]	**2.24**	**2.24**	**1.52**
Σ Higher alcohol acetates	8.38 ± 0.20^{b}	10.10 ± 0.62^{c}	6.70 ± 0.04^{a}	61.06	0.0001					
Ethyl butyrate	0.33 ± 0.01^{ab}	0.39 ± 0.00^{b}	0.27 ± 0.06^{a}	10.24	0.0116	Acid fruit	0.020 [1]	**16.5**	**19.50**	**13.5**
Ethyl hexanoate	0.60 ± 0.05^{b}	0.52 ± 0.07^{ab}	0.39 ± 0.13^{a}	4.16	0.0735	Green apple	0.014 [1]	**42.86**	**37.14**	**27.86**
Ethyl octanoate	1.01 ± 0.25^{a}	1.10 ± 0.11^{a}	0.73 ± 0.42^{a}	1.34	0.3314	Sweet, soap	0.005 [1]	**202.00**	**220.00**	**146.00**
Ethyl decanoate	0.08 ± 0.06^{b}	0.23 ± 0.02^{c}	tr ^{a}	28.09	0.0009	Pleasant, soap	0.200 [1]	0.40	**1.15**	<0.1
Σ Fatty acid esters	2.01 ± 0.37^{a}	2.22 ± 0.19^{a}	$1.40 \pm 0.0.58^{a}$	3.21	0.1127					
Ethyl lactate	tr ^{a}	2.24 ± 1.01^{b}	tr ^{a}	14.57	0.005	Lactic	0.157 [2]	<0.1	**14.26**	<0.1
Diethyl succinate	0.03 ± 0.01^{a}	0.09 ± 0.04^{b}	0.05 ± 0.01^{ab}	4.67	0.0599	Apple, fruity	0.20 [2]	0.15	0.45	0.25
Isobutyric acid	3.01 ± 0.02^{a}	3.02 ± 0.03^{a}	2.91 ± 0.01^{a}	0.94	0.4429	acid, fatty	0.230 [1]	**13.09**	**13.13**	**12.65**
Butyric acid	1.90 ± 0.04^{a}	1.93 ± 0.21^{a}	1.31 ± 1.14^{a}	0.81	0.4876	Cheese	0.173 [1]	**10.98**	**11.15**	**7.57**
Isovaleric acid	0.08 ± 0.06^{a}	0.26 ± 0.07^{b}	0.02 ± 0.01^{a}	14.60	0.005	Blue cheese	0.033 [1]	**2.42**	**7.87**	0.60
Σ SCFA	6.67 ± 0.07^{a}	6.65 ± 0.15^{a}	6.08 ± 1.06^{a}	0.88	0.4640					
Hexanoic acid	4.97 ± 0.23^{a}	4.29 ± 0.36^{a}	4.54 ± 0.11^{ab}	5.27	0.0477	Cheese	0.420 [1]	**11.83**	**10.21**	**10.80**
Octanoic acid	6.11 ± 0.20^{c}	5.45 ± 0.50^{b}	5.07 ± 0.20^{a}	17.91	0.0030	Rancid, harsh	0.500 [1]	**12.22**	**10.90**	**10.14**
Decanoic acid	3.70 ± 0.77^{a}	3.68 ± 0.32^{a}	2.85 ± 0.21^{a}	2.89	0.1319	Fatty	1.00 [1]	**3.70**	**3.68**	**2.85**
Σ MCFA	14.78 ± 1.20^{b}	13.39 ± 1.15^{ab}	12.46 ± 0.35^{a}	4.23	0.0713					
4-Vinylguaiacol	0.47 ± 0.18^{ab}	0.53 ± 0.04^{b}	0.29 ± 0.09^{a}	3.33	0.1062	Pleasant, phenolic	1.10 [1]	0.42	0.48	0.26

tr = traces; 1 = thresholds from Gil *et al.* [20]; 2 = thresholds from Duarte *et al.* [28]; 3 = threshold from Falqué *et al.* [29]; SCFA Small Chain Fatty Acids; MCFA Medium Chain Fatty Acids; * in bold, compounds with OAV > 1.

In all cases, achieved acetaldehyde levels were higher than the threshold proposed by Duarte *et al.* [28]. Among these values the highest values corresponds to the inoculated musts, while in the spontaneous fermentations the values were slightly lower (Table 2). Only free acetaldehyde has flavor relevance; at low levels it provides fruity flavors, while high concentrations (>200 mg/L) contribute "flatness" in wines [20]. All of the studied Malvar wines can be treated as correct due to their mean content being within the range previously studied in other non-oxidized white wines [30]. High concentration of acetoin is mostly related to non-*Saccharomyces* fermentations [1,8] but in our study, high concentrations were only detected in the inoculated fermentations. On the other hand, ethyl acetate concentrations showed higher values than the threshold proposed (OAVs, from 5.2 to 5.6) over all Malvar wines (Table 2). Ethyl acetate may confer with pleasant and fruity fragrances to the global wine aroma at concentrations lower than 150 mg/L.

Quantitatively, the largest group of volatile compounds present in wines were higher alcohols. Many of these compounds are strongly correlated with dislikable aromas in wines, but even though significant differences were found among the uninoculated and inoculated wines, the values obtained in wines fermented by the CLI 892 strain were significantly lower than the other two wines (Table 2). The amount of higher alcohols ranged from 205.71 ± 8.45 to 258.13 ± 5.63 mg/L, this variation principally is due to the isoamyl alcohol produced, which represent in all cases close to 60% of the total higher alcohols (Table 2). Discrete concentrations of fusel alcohols contribute to the wine's aromatic complexity. With respect to the higher minor alcohols, this group of volatiles had no odor activity contribution.

In terms of the number of components analyzed, esters represent the largest group (Table 2). Higher alcohol acetates are an important group of fermentative aromas, which are normally linked to fruity descriptors, but in a wine can significantly modify the global aroma. The fruity character associated to the aroma of Madrid white wines is mainly related to fruit notes such as banana, green apple, or pear [20,31,32], produced by the acetates of higher alcohols and fatty acids.

Significant differences were detected among the different fermentations for isoamyl acetate, hexyl acetate, and phenyl ethyl acetate (Table 2). The major ester in the Malvar wines analyzed was isoamyl acetate which was present in high concentrations and above of the OTH and OAV in all fermentations. Ethyl esters of fatty acids showed important variations in the concentrations, but in the majority of the cases, they showed higher values than the OTH and OAVs values proposed (Table 2). The real contribution of ethyl lactate and diethyl succinate to the white wines of the "D.O. Vinos de Madrid" has been previously described as insignificant [20,32]. Interestingly, Malvar wines fermented by the yeast strain CLI 889 showed a high concentration of ethyl lactate. It is now widely understood that ester concentration is conditioned by several factors, such as yeast strain, fermentation temperature, aeration degree and sugar content [33–35]. In our experimental conditions, in spite of the use of the same must under the same conditions, the different amounts of esters found can be attributed to the yeast strain used.

Fatty acids have been described with cheese, harsh, fatty, and rancid notes [20]. All compounds in the different fermentations reached the OTH and OAVS values proposed (Table 2), with the exception of the isovaleric acid in the fermentations carried out by CLI 892. Supposing that the presence of fatty acids is frequently associated to off flavors, they play an important role in the aromatic equilibrium in wines because they are antagonistic to the hydrolysis of the analogous esters [20].

High concentrations of vinyl phenols can be responsible for strong pharmaceutical odors in white wines [18,36], but at low or moderate concentrations they could be linked with grassy, herbaceous, or pleasant spicy aromas. Accordingly, Grando et al. [37] pointed out that 4-vinylguaiacol was the principal responsible compound for the spicy aroma of Gewürztraminer's wines. In our study, 4-vinylguaiacol was detected among the uninoculated and inoculated white wines. Statistically significant differences were observed when using Malvar juices fermented spontaneously or inoculated with the selected yeast strains.

PCA was performed to disclose the compounds that differentiated best among the uninoculated and inoculated Malvar wines (Figure 3). Hence, only yeast volatile fermentation constituents enabled a suitable discrimination between inoculated and spontaneous fermentation wines.

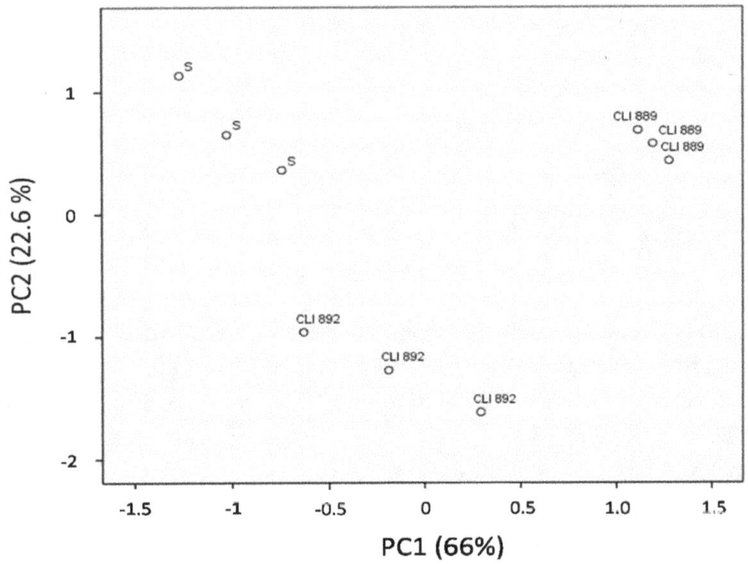

Figure 3. Principal components analysis of volatile composition data. Projection of the Malvar wines fermented spontaneously (S), and with CLI 892 and CLI 889 *S. cerevisiae* strains in the dials formed by the PC1 (22.6%) and PC2 (66%).

The PCA explained the 88.63% of the total variance. Musts fermented with the strain CLI 889 shaped a clear group, which was associated with the esters ethyl octanoate, isoamyl acetate, and ethyl lactate, as long as fermentation inoculated with CLI 892 was located in a different dial of the PCA plot and most associated with those compounds contained in the PC2 (ethyl hexanoate, 4-vinylguaiacol, ethyl butyrate, and ethyl acetate). The uninoculated wines showed a cluster mostly associated with those compounds of the PC1, but none was nearly correlated with any of the inoculated fermentation wines (data not shown). Thus, taking into account that fermentations were conducted in the same must, if the yeast starter dominates on native yeast population, the wine will exhibit singular aroma and sensory profiles of the each yeast starter involved [14]. Confirming this, PCA analysis and data of Table 2 showed that by comparison with inoculated wines with the yeast strains CLI 889 and CLI 892, uninoculated fermentation wines showed a high variability in the composition of volatile compounds that also contributes to wine aroma.

3.5. Sensory Analysis of Wines

After the first sensory analysis using a penalizing system on all of the wines, attributes related to the appearance, taste intensity, taste quality, and harmony were not statistically different. However, differences ($p < 0.05$) were found within the different Malvar wines in terms of their aroma intensity, and quality. The ratings obtained were; uninoculated 15.73 $\pm$ 3.20, inoculated with the yeast strain CLI 889 13.30 $\pm$ 1.92, and 16.40 $\pm$ 4.55 using the strain CLI 892. Thus, judges considered that wine made by inoculation of CLI 889 seem to have the better quality. On the other hand, the sensory analysis using the different selected descriptors reached that some descriptors were statistically influenced by the yeast strain (Figure 4).

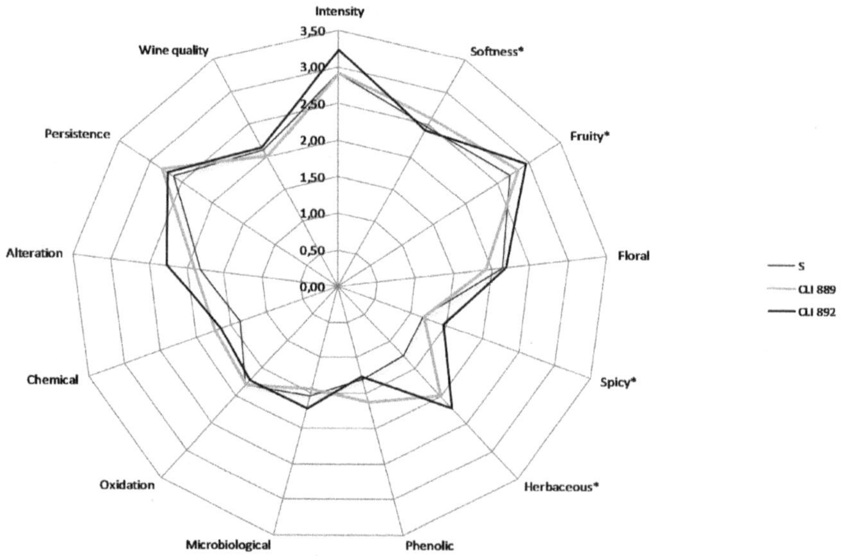

Figure 4. Polar coordinate (cobweb) graph of mean sensory scores rating of "alteration", "chemical", "herbaceous", "intensity", "floral", "fruity", "microbiological", "oxidation", "persistence", "phenolic", "softness", "spicy", and "wine quality" for wines made with the uninoculated *S. cerevisiae* (S) and locally-selected CLI 889 and CLI 892 yeast strains. In sensorial variables indicated with an asterisk (*) a difference between some trials is verified for $p \leqslant 0.05$.

The greatest softness of the spontaneously fermented wines was deserved to a lower acidity, and apparently to a large production of glycerol and as reported by Feuillat [38]. Glycerol is sweet, however its contribution to the sweet taste or palatableness of a wine remains unclear. The threshold concentration proposed by Noble and Bursick [39] is considered high (5.2 g/L) as quoted by Ugliano and Henschke [40], and the high acidity in wines is wont to interact with the remaining sweetness, so the limit to which glycerol contributes to sweetness or mouthfeel characters is still unrevealed. Nevertheless, is in agreement with Scacco *et al.* [41], at the concentrations obtained in this work (Table 1), glycerol contributes positively to softness and viscosity of the wines. The differences concerning fruitiness cannot be considered to be related to the amount of wine flavor compounds. Substantially, wines fermented by the strain CLI 892 was characterized by a considerable content in higher alcohol acetates (10.10 ± 0.62 mg/L) with banana, pear, and herbaceous notes, followed by spontaneous fermentation with 8.38 ± 0.62 mg/L (Table 2). These compounds are usually linked to fruity descriptors; hence, it is possible to assume that the fruity descriptor in wines made with CLI 889 and CLI 892 were swayed by the lower amount of substances with a masking effect on fruity, than in the spontaneous fermentations, as reported by Campo *et al.* [42]. The spiciness was lower in wines fermented by strain CLI 892, apparently because this strain produced a lower quantity of some compounds able to influence this descriptor than the other strains, such as 4-vinylguaiacol. Sensory analysis showed that the best wines were those fermented by CLI 889, which had great gustatory and aromatic characteristics, usual in high-quality white wines. These results could be associated to a reasonable production of several fruity esters, such as isoamyl acetate, and hexyl acetate, which highlight the organoleptic properties and regional characteristic of Malvar white wines.

4. Conclusions

In the context of what is covered in this work, winemakers of the winegrowing regions where Malvar is cultivated are set to benefit from this recent development, because they will have improved options to elaborate wine. Uninoculated and locally-selected yeast strains clearly transform the fermentation and have influence over the volatile profile of Malvar white wines. In spite of wines fermented spontaneously and yeast strains CLI 889 and 892 being considered accurate by judges, the wine fermented with the strain CLI 889 was qualified as the best. The most salient chemical attributes were its high concentrations of isoamyl acetate and hexyl acetate and the high acidity, which have been detected to be key constituents in setting the fruity and fresh character of Malvar wines. Thus, this autochthonous yeast strain has been deposited at the Spanish Type Collection Culture (CECT) with the accession number CECT 13145. In addition, a new low acetic acid-producing *S. cerevisiae* yeast strain isolated from the spontaneous fermentations was characterized. Other features of its contribution to the general characteristics of these white wines are now being investigated in order to enrich the Malvar wine quality.

Acknowledgments: We are grateful to Matthew Tyndale-Tozer for revising the manuscript. This work was financed by the Instituto Nacional de Investigación y Tecnología Agraria y Alimentaria (INIA- RM 2006-00012-00-00), by the Instituto Madrileño de Investigación y Desarrollo Rural Agrario y Alimentario (IMIDRA-FP07-AL2-Lev), and by the European Union's Seventh Framework Programme via the Marie Curie Action, "Co-funding of Regional, National and International Programs" to stimulate research activities without mobility restrictions, co-financed by the "Junta de Andalucía" and the European Commission under grant agreement no 291780.

Author Contributions: All authors contributed to the conception and design of experiments; Mar Gil-Díaz and Margarita García performed the experiments, Gustavo Cordero-Bueso performed the sensory analysis; all authors analyzed the data; Gustavo Cordero-Bueso and Teresa Arroyo contributed to writing the paper.

References

1. Cordero-Bueso, G.; Esteve-Zarzoso, B.; Cabellos, J.M.; Gil-Díaz, M.; Arroyo, T. Biotechnological potential of non-*Saccharomyces* yeasts isolated during spontaneous fermentations of Malvar (*Vitis vinifera*cv L). *Eur. Food Res. Technol.* **2013**, *236*, 193–207. [CrossRef]
2. Laffort, J.F.; Romat, H.; Darriet, P. Les levuresetl'expressionaromatique des vinsblancs. *Rev. Oenol.* **1989**, *53*, 9–12.
3. Günata, Y.Z.; Dugelay, I.; Sapis, J.C.; Baumes, R.; Bayonove, C. Role of enzymes in the use of the flavor potential from grape glycosides in winemaking. In *Progress in Flavor Precursor Studies*; Schreier, P., Winter-Halter, P., Eds.; Allured Publishing Corporation: Carol Stream, IL, USA, 1993; Volume 3, pp. 219–234.

4. Swiegers, J.H.; Bartowsky, E.J.; Henschke, P.A.; Pretorius, I.S. Yeast and bacteria modulation of wine aroma and flavor. *Aust. J. Grape Wine Res.* **2005**, *11*, 139–173. [CrossRef]
5. Swiegers, J.H.; Pretorius, I.S. Yeast modulation of wine flavor. *Adv. Appl. Microbiol.* **2007**, *57*, 131–175.
6. Carrascosa, A.V.; Bartolomé, B.; Robredo, A.; León, A.; Cebollero, E.; Juega, M.; Nuñez, Y.P.; Martínez, M.C.; Martínez-Rodríguez, A.J. Influence of locally-selected yeast on the chemical and sensorial properties of Albariño White wines. *LWT Food Sci. Technol.* **2012**, *46*, 319–325. [CrossRef]
7. Varela, C.; Siebert, T.; Cozzolino, D.; Rose, L.; McLean, H.; Henschke, P. Discovering a chemical basis for differentiating wines made by fermentation with wild indigineous and inoculated yeasts; role of yeast volatile compounds. *Aust. J. Grape Wine Res.* **2009**, *15*, 238–248. [CrossRef]
8. Vigentini, I.; Fabrizio, V.; Faccincani, M.; Picozzi, C.; Comasio, A.; Foschino, R. Dynamics of *Saccharomyces cerevisiae* populations in controlled and spontaneous fermentations for Franciacorta D.O.C.G. base wine production. *Ann. Microbiol.* **2013**, *64*, 639–651. [CrossRef]
9. Capece, A.; Pietrafesa, R.; Romano, P. Experimental approach for target selection of wild wine yeasts from spontaneous fermentation of "Inzolia" grapes. *World J. Microbiol. Biotechnol.* **2011**, *27*, 2775–2783. [CrossRef]
10. Csoma, H.; Zakany, N.; Capece, A.; Romano, P.; Sipiczki, M. Biological diversity of *Saccharomyces* yeasts of spontaneously fermenting wines in four wine regions: Comparative genotypic and phenotypic analysis. *Int. J. Food Microbiol.* **2010**, *140*, 239–248. [CrossRef] [PubMed]
11. Esteve-Zarzoso, B.; Gostincar, A.; Bobet, R.; Uruburu, F.; Querol, A. Selection and molecular characterization of wine yeasts isolated from the "El Penedes" area Spain. *Food Microbiol.* **2000**, *17*, 553–562. [CrossRef]
12. Orlic, S.; Redzepovic, S.; Jeromel, A.; Herjavec, S.; Iacumin, L. Influence of indigenous *Saccharomyces paradoxus* strains on Chardonnay wine fermentation aroma. *Int. J. Food Sci. Technol.* **2007**, *42*, 95–101. [CrossRef]
13. Fleet, G.H. Wine yeast for the future. *FEMS Yeast Res.* **2008**, *8*, 979–995. [CrossRef] [PubMed]
14. Rodríguez, M.E.; Infante, J.J.; Molina, M.; Domínguez, M.; Rebordinos, L.; Cantoral, J.M. Genomic characterization and selection of wine yeast to conduct industrial fermentations of a white wine produced in a SW Spain winery. *J. Appl. Microbiol.* **2010**, *108*, 1292–1302. [CrossRef] [PubMed]
15. Suzzi, G.; Arfelli, G.; Schirone, M.; Corsetti, A.; Perpetuini, G.; Tofalo, R. Effect of grape indigenous *Saccharomyces cerevisiae* strains on Montepulciano d'Abruzzo red wine quality. *Food Res. Int.* **2012**, *46*, 22–29. [CrossRef]
16. Arroyo, T. Estudio de la influencia de diferentes tratamientos enológicos en la evolución de la microbiota y en la calidad de los vinos elaborados con la variedad "Airén", en la D.O. "Vinos de Madrid". Ph.D. Thesis, University of Alcalá, Alcalá de Henares, Madrid, Spain, 2000.
17. Cordero-Bueso, G.; Arroyo, T.; Serrano, A.; Tello, J.; Aporta, I.; Vélez, M.D.; Valero, E. Influence of the farming system and vine variety on yeast communities associated with grape-berries. *Int. J. Food Microbiol.* **2011**, *45*, 132–139. [CrossRef] [PubMed]
18. Esteve-Zarzoso, B.; Belloch, C.; Uruburu, F.; Querol, A. Identification of yeast by RFLP analysis of the 5.8 rRNA and the two ribosomal internal transcribed spacers. *Int. J. Syst. Bacteriol.* **1999**, *49*, 329–337. [CrossRef] [PubMed]
19. Vaudano, E.; Garcia-Moruno, E. Discrimination of *Saccharomyces cerevisiae* wine strains using microsatellite multiplex PCR and band pattern analysis. *Food Microbiol.* **2008**, *25*, 56–64. [CrossRef] [PubMed]
20. Gil, M.; Cabellos, J.M.; Arroyo, T.; Prodanov, M. Characterization of the volatile fraction of young wines from the Denomination of Origin "Vinos de Madrid" Spain. *Anal. Chim. Acta* **2006**, *563*, 145–153. [CrossRef]
21. International Organization for Standardization. *ISO 11035. Sensory Analysis—Identification and Selection of Descriptors for Establishing a Sensory Profile by Multidimensional Approach*; ISO: Geneva, Switzerland, 1994.
22. International Organization for Standardization. *ISO 8589. Sensory Analysis—General Guidance for the Design of Test Room*; ISO: Geneva, Switzerland, 2007.
23. International Organization for Standardization. *ISO 8586-1. Sensory Analysis—General Guidance for the Selection, Training, and Monitoring of Assessors. Part 1: Selected Assessors*; ISO: Geneva, Switzerland, 1993.
24. Lozano, J.; Santos, J.P.; Arroyo, T.; Aznar, M.; Cabellos, J.M.; Gil, M.; Horrillo, M.C. Correlating e-nose responses to wine sensorial descriptors and gas Chromatography—Mass spectrometry profiles using partial least squares regression analysis. *Sens. Actuators B* **2007**, *127*, 267–276. [CrossRef]
25. Vilanova, M.; Masneuf-Pomarède, I.; Dubourdieu, D. Influence of *Saccharomyces cerevisiae* strains on general composition and sensorial properties of white wines made from Vitis vinifera cv. Albariño. *Food Technol. Biotechnol.* **2005**, *43*, 79–83.

26. Remize, F.; Roustan, J.L.; Sablayrolles, J.M.; Barre, P.; Dequin, S. Glycerol overproduction by engineered *Saccharomyces cerevisiae* wine yeast strains leads to substantial changes in byproduct formation and to a stimulation of fermentation rate in stationary phase. *Appl. Environ. Microbiol.* **1999**, *65*, 143–149. [PubMed]
27. Erasmus, D.J.; Cliff, M.; van Vuuren, H.J.J. Impact of yeast strain on the production of acetic acid, glycerol, and the sensory attributes of icewine. *Am. J. Enol. Viticult.* **2004**, *55*, 371–378.
28. Duarte, F.W.; Dias, R.D.; Oliveira, J.M.; Vilanova, M.; Teixeira, J.A.; Almeida e Silva, J.B.; Schwan, R.F. Raspberry *Rubusidaeus* L. wine: Yeast selection, sensory evaluation and instrumental analysis of volatile and other compounds. *Food Res. Int.* **2010**, *43*, 2303–2314. [CrossRef]
29. Falqué, E.; Fernández, E.; Dubourdieu, D. Differentation of white wines by their aromatic index. *Talanta* **2001**, *54*, 271–281. [CrossRef]
30. Escudero, A.; Asensio, E.; Cacho, J.; Ferreira, V. Sensory and chemical changes of young white wines stored under oxygen. An assessment of the role played by aldehydes and some other important odorants. *Food Chem.* **2002**, *77*, 325–331. [CrossRef]
31. Arroyo, T.; Lozano, J.; Cabellos, J.M.; Gil-Díaz, M.; Santos, J.P.; Horrillos, M.C. Evaluation of wine aromatic compounds by a sensory human panel and an electronic nose. *J. Agric. Food Chem.* **2009**, *57*, 11449–11582. [CrossRef] [PubMed]
32. Santos, J.P.; Arroyo, T.; Aleixandre, M.; Lozano, J.; Sayago, I.; García, M.; Fernández, M.J.; Arés, L.; Gutiérrez, J.; Cabellos, J.M.; *et al.* A comparative study of sensor array and GC-MS: application to Madrid wines characterization. *Sens. Actuators B* **2004**, *102*, 299–307. [CrossRef]
33. Hernández-Orte, P.; Cersosimo, M.; Loscos, N.; Cacho, J.; García-Moruno, E.; Ferreira, V. The development of varietal aroma from non-floral grapes by yeast of different genera. *Food Chem.* **2008**, *107*, 1064–1077. [CrossRef]
34. Rojas, V.; Gil, J.V.; Piñaga, F.; Manzanares, P. Acetate ester formation in wine by mixed cultures in laboratory fermentations. *Int. J. Food Microbiol.* **2003**, *86*, 181–188. [CrossRef]
35. Valero, E.; Moyano, L.; Millán, M.C.; Medina, M.; Ortega, J.M. Higher alcohols and esters production by *Saccharomyces cerevisiae*. Influence of the initial oxygenation of the grape must. *Food Chem.* **2002**, *78*, 57–61. [CrossRef]
36. Gómez-Míguez, M.J.; Cacho, J.F.; Ferreira, V.; Vicario, I.M.; Heredia, F.J. Volatile components of Zalema white wines. *Food Chem.* **2007**, *100*, 1464–1473. [CrossRef]
37. Grando, M.S.; Versini, G.; Nicolini, G.; Mattivi, F. Selective use of wine yeast strains having different volatile phenols production. *Vitis* **1993**, *32*, 43–50.
38. Feuillat, M. Yeast macromolecules: Origin, composition, and enological interest. *Am. J. Enol. Vitic.* **2003**, *54*, 211–213.
39. Noble, A.C.; Bursick, G.F. The contribution of glycerol to perceived viscosity and sweetness in white wine. *Am. J. Enol. Vitic.* **1984**, *35*, 110–112.
40. Ugliano, M.; Henschke, P.A. Yeasts and wine flavour. In *Wine Chemistry and Biochemistry*; Moreno-Arribas, M.V., Polo, M.C., Eds.; Springer Verlag: New York, NY, USA, 2009; pp. 313–392.
41. Scacco, A.; Oliva, D.; Di Maio, S.; Polizzotto, G.; Genna, G.; Tripodi, G.; Lanza, C.M.; Verzera, A. Indigenous *Saccharomyces cerevisiae* strains and their influence on the quality of Cataratto, Inzolia and Grillo white wines. *Food Res. Int.* **2012**, *46*, 1–9. [CrossRef]
42. Campo, E.; Ferreira, V.; Escudero, A.; Marqués, J.C.; Cacho, J. Quantitative gas chromatography-olfactometry and chemical quantitative study of the aroma of four Madeira wines. *Anal. Chim. Acta* **2006**, *563*, 180–187. [CrossRef]

10

Ustilago Rabenhorstiana—An Alternative Natural Itaconic Acid Producer

Susan Krull, Malin Lünsmann, Ulf Prüße and Anja Kuenz *

Thünen-Institute of Agricultural Technology, Bundesallee 47, 38116 Braunschweig, Germany; susan.krull@thuenen.de (S.K.); malinluensmann@ymail.com (M.L.); ulf.pruesse@thuenen.de (U.P.)
* Correspondence: anja.kuenz@thuenen.de

Abstract: Itaconic acid is an industrial produced chemical by the sensitive filamentous fungus *Aspergillus terreus* and can replace petrochemical-based monomers for polymer industry. To produce itaconic acid with alternative renewable substrates, such as lignocellulosic based hydrolysates, a robust microorganism is needed due to varying compositions and impurities. Itaconic acid producing basidiomycetous yeasts of the family *Ustilaginaceae* provide this required characteristic and the species *Ustilago rabenhorstiana* was examined in this study. By an optimization of media components, process parameters, and a fed-batch mode with glucose the final titer increased from maximum 33.3 g·L^{-1} in shake flasks to 50.3 g·L^{-1} in a bioreactor. Moreover, itaconic acid was produced from different sugar monomers based on renewable feedstocks by *U. rabenhorstiana* and the robustness against weak acids as sugar degradation products was confirmed. Based on these findings, *U. rabenhorstiana* has a high potential as alternative natural itaconic acid producer besides the well-known *U. maydis* and *A. terreus*.

Keywords: *Ustilago*; itaconic acid; process improvement; lignocellulosic feedstock

---

1. Introduction

Itaconic acid is an interesting chemical for the polymer industry, which is produced in a biotechnological process based on renewable substrates [1]. Petrochemical-based substances, like methacrylic or acrylic acid, can be replaced by this single unsaturated dicarbonic acid and its derivatives. Therefore, the field of products and applications is widespread, e.g., synthetic latex, styrene-butadiene rubber, superabsorbent polymers, or unsaturated polyester resins [2–6].

Since the 1960s, the filamentous fungus *Aspergillus terreus* is industrially used with a titer of 85–100 g·L^{-1}, whereas in laboratory scale, final titers of 160 g·L^{-1} itaconic acid are described [4,7–9]. *A. terreus* achieves a productivity up to 1.15 g (L·h)$^{-1}$ and a yield of 0.64 (*w/w*), whereby the theoretical yield with glucose is 0.72 (*w/w*) [7,8]. Besides pure glucose, itaconic acid was successfully produced by *A. terreus* with glycerol, starch hydrolysates, molasses, and different monosaccharides, like xylose, arabinose, galactose, and rhamnose [10]. A great cultivation challenge is caused by sugar degradation products or other impurities in lignocellulosic hydrolysates, which influence the morphology and itaconic acid production of the fungus. Due to the sensitivity of the fungus, complex purification processes are used for such hydrolysates or more resistant strains are generated by mutagenesis [11–13]. Another alternative is itaconic acid producing yeasts of the species *Candida*, *Pseudozyma*, or *Ustilago*, which are more robust and not as sensitive to metal ions as *A. terreus* [14–18]. For wildtype strains of the species *Ustilago*, low final titers of 44.5 g·L^{-1} itaconic acid, low yields up to 0.24 (*w/w*), and a low productivity of maximum 0.31 g (L·h)$^{-1}$ are disadvantageous [19]. This is due to a variety of by-products like other organic acids, glycolipids, and intracellular triacylglycerols, which are produced in parallel to itaconic acid [16,18,20,21]. Nevertheless, in addition to the robustness of the yeasts, the formation of haploid yeast-like cells is an advantage compared to the filamentous growth or formation of pellets of *A. terreus* with a decreased oxygen supply or increased viscosity [5,22].

In recent years, the research interest in itaconic acid production with the phytopathogenic basidiomycete *Ustilago maydis* increased. It was found, that an ammonium limitation triggers the itaconic acid overproduction in *U. maydis* [5,19] and itaconic acid is synthesized in the cytosol via the intermediate cis-aconitate and trans-aconitate and can be further converted to 2-hydroxyparaconic acid [23–25]. The itaconic acid gene cluster was also characterized and relevant enzymes, transporters, and promoters were found [23,25,26], whereby a summary of metabolic aspects is given by Wierckx et al. [27]. Based on these findings, metabolic engineering strategies and process optimization of *U. maydis* resulted in a reduction of by-product concentrations of malic acid and 2-hydroxyparaconic acid with a significant increased itaconic acid titer of 63.2 $g{\cdot}L^{-1}$ and a yield of 0.48 (*w/w*) [23]. All in all, detailed examinations are available for itaconic acid production of *U. maydis*, but also other wildtype strains of the family *Ustilaginaceae* could offer advantages of less sensibility or a yeast-like morphology for using second-generation feedstocks. This family is well-known for organic acid production [16,18,28,29], but the level of knowledge about alternative itaconic acid producer, like *U. cynodontis* or *U. rabenhorstiana*, are low.

This study considers the cultivation of *Ustilago rabenhorstiana* for itaconic acid production and its potential as alternative natural producer. Although the used strain is known as natural itaconic acid producer [18,30], the microorganism was not examined in literature more precisely. Only the growth of the organism with glycerol as substrate was described, whereby non-formation of organic acids was detected [28]. Concerning itaconic acid production based on renewable resources, usability of different sugar monomers and robustness towards influence of sugar degradation products were examined in this study. Moreover, the influences of media and fermentation parameters on the production of itaconic acid and by-products as well as the morphology of the yeast were investigated.

2. Materials and Methods

2.1. Microorganism

The basidiomycete *Ustilago rabenhorstiana* NBRC 8995 was purchased from the National Institute of Technology and Evaluation (Tokyo, Japan) and was stored at −80 °C as 50% (*v/v*) glycerol stock culture.

2.2. Media Compositions

YEPS-medium was used for the preparation of agarplates and preculture (20 $g{\cdot}L^{-1}$ sucrose, 10 $g{\cdot}L^{-1}$ yeast extract, 20 $g{\cdot}L^{-1}$ peptone, optional 20 $g{\cdot}L^{-1}$ agar-agar).

If not mentioned otherwise, the production media was a Tabuchi-medium [18] containing 120 $g{\cdot}L^{-1}$ glucose, 0.5 $g{\cdot}L^{-1}$ KH_2PO_4, 1.6 $g{\cdot}L^{-1}$ NH_4Cl, 0.2 $g{\cdot}L^{-1}$ $MgSO_4{\cdot}7\,H_2O$, 10 $mg{\cdot}L^{-1}$ $FeSO_4{\cdot}7\,H_2O$, 1 $g{\cdot}L^{-1}$ yeast extract, and 30 $g{\cdot}L^{-1}$ $CaCO_3$. All components were prepared separately in stock solutions; the pH-value was adjusted to pH 6.0 for all solutions with 0.5 M H_2SO_4 or 1 M NaOH and autoclaved. The pH-value of the iron-solution was not corrected, and the solution was sterile filtered. $CaCO_3$ was weighed in the glassware and autoclaved.

All media components were p.a. quality and purchased from Merck (Darmstadt, Germany), Sigma-Aldrich (St. Louis, MO, USA) or Roth (Karlsruhe, Germany). In fed-batch cultivation the glucose concentration was monitored during the cultivation. If necessary, glucose was added in solid form without previous sterilization to prevent a glucose limitation.

2.3. Cultivation

Precultures were conducted in a 250 mL shake flask with three baffles, a filling volume of 50 mL, and inoculated with a single colony from a YEPS-agar plate (30 °C, 3 days). The preculture was cultivated at 30 °C and 120 rpm (50 mm shaking diameter) for 24 h until an optical density of 10 at 605 nm was achieved. All experiments were inoculated with 1% (*v/v*) of the preculture.

The main cultures in shake flasks were carried out at 30 °C and 120 rpm in 250 mL shake flasks with three baffles and a filling volume of 100 mL Tabuchi-medium.

To identify potential impurities and the utilization of monosaccharides based on lignocellulosic feedstock, test tubes with Kapsenberg caps were used and a working volume of 2 mL (ø 16 mm × 100 mm). The test tubes were incubated for 4 days at 30 °C and 120 rpm in an inclined test tube holder with an inclination angle of 30°. Shake flasks and test tubes were continuously rotated by hand while sampling avoiding inhomogeneity.

The cultivation in bioreactors were conducted in four parallel 1 L-bioreactors, equipped with a Rushton impeller and an L-sparger (model SR0700ODLS, DASGIP GmbH, Jülich, Germany). DASGIP Control software (DASGIP GmbH, Jülich, Germany) was used for the regulation of gassing, temperature, pH-value, and stirring rate, as well as recording the data of dissolved oxygen (DO) and pH. The pH regulation to pH 6.0 was carried out with 4 M NaOH, if not otherwise mentioned. At the beginning of the cultivation, 0.5 mL antifoam solution (Ucolup N-115, Brenntag, Mühlheim/Ruhr, Germany) was added to the broth. The experiments were carried out at 30 °C, 500 rpm, a filling volume of 500 mL, and an aeration of 0.1 vvm, unless otherwise mentioned. All cultivations were carried out in minimum duplicates, whereby the deviation from the mean value was <5%. All results are presented as mean values without error bars on account of readability.

2.4. *Analytical Methods*

The samples were centrifuged at 21,000 *g* for 20 min at 20 °C and the supernatant was used for further analysis. A Shimadzu HPLC (Shimadzu Corp., Kyoto, Japan) with a HPX-87H column (BioRad, Munich, Germany) with a refractive index detector (RI) and UV detector at 210 nm was used to analyze the concentrations of sugars and organic acids. The column was tempered at 40 °C and as mobile phase a 5 M H_2SO_4 solution at a flow rate of 0.6 $mL \cdot min^{-1}$ was used. The concentration of an unknown product was estimated by the peak area of the RI-signal compared to a calibration of succinic acid. For samples of bioreactor experiments, the pellet was washed twice with deionized water and dried to a constant weight at 105 °C for at least 48 h to determine the cell dry weight (CDW).

The composition of fatty acids was analyzed by transesterification of the fatty acids as described by Lewis et al. [31]. The biomass of reactor cultivation was separated from the broth by centrifugation (21,000 *g* for 20 min at 20 °C). The supernatant was discarded, and the pellet was washed twice with 0.9% (*v/v*) NaCl-solution and suspended in 0.9% (*v/v*) NaCl-solution. The cells were disrupted by an ultrasonic-homogenisator on ice (4 cycles: 15 s at 65%, break 30 s; Sonopuls HD2200 with sonotrodetype UW2200, Bandelin electronic, Berlin, Germany). The suspension was stored at −80 °C and freeze-dried (Alpha 1-2 LD, Christ, Osterode, Germany). The fatty acids were derivatized to fatty acid methyl esters (FAME) [31] and analyzed by GC-MS on a GC-17A (Shimadzu Corp., Kyoto, Japan) with a Zebron™ ZB-WAX plus column (60 m × 0.25 mm × 0.25 μm), using 1.4 $mL \cdot min^{-1}$ helium as carrier gas. The temperature gradient of 60 °C was increased to 150 °C at a rate of 30 $°C \cdot min^{-1}$, and afterwards increased up to 240 °C at a rate of 13 $°C \cdot min^{-1}$. The temperature of 240 °C was kept for 30 min and raised to 255 °C for 5 min. The FAMEs were identified with the software LabSolutions (Shimadzu Corp., Kyoto, Japan) and the mass spectral data were compared with the database of the National Institute of Standards and Technology (Gaithersburg, MD, USA).

2.5. *Microscopy*

The cells were examined using a phase-contrast microscope (Axioplan, Carl Zeiss AG, Jena, Germany) with the software analysis pro (Analysis 5.1, Olympus Soft Imaging Solutions GmbH, Münster, Germany). Intracellular lipids were visualized after coloring with nil-red by fluorescence microscopy [32].

3. Results

3.1. Standard Cultivation in Shake Flasks

A standard cultivation of *U. rabenhorstiana* with pure glucose as substrate was performed in shake flasks (Figure 1). After one day, the itaconic acid production started, and additionally, succinic acid and malic acid were produced. Furthermore, an unknown product accumulated after 48 h approximately in a concentration range of <1 g·L^{-1}. Malic acid was consumed in further course of cultivation, and after four days, α-ketoglutaric acid was formed and increased parallel with the itaconic acid concentration. Glucose was completely consumed after 9.7 days, resulting in 31.3 g·L^{-1} itaconic acid, 13.6 g·L^{-1} α-ketoglutaric acid, 2.3 g·L^{-1} malic acid, and traces of an unknown metabolite, followed by a further production of α-ketoglutaric acid. The overall productivity was 0.13 g (L·h)$^{-1}$ with a yield of 0.26 (*w/w*) after 9.8 days. Despite the use of $CaCO_3$ as buffer, the pH-value constantly decreased from pH 6.8 to 4.9 throughout the cultivation. The morphology of *U. rabenhorstiana* changed from yeast-like single cells (0–2 days; Figure 1A) via a development of pseudomycel (2–7 days, Figure 1B) to filamentous growth like long branched mycel (7–11 days, Figure 1C). Moreover, intracellular lipids deposits were visible under the microscope, which became smaller in size after the glucose limitation at day 9.7.

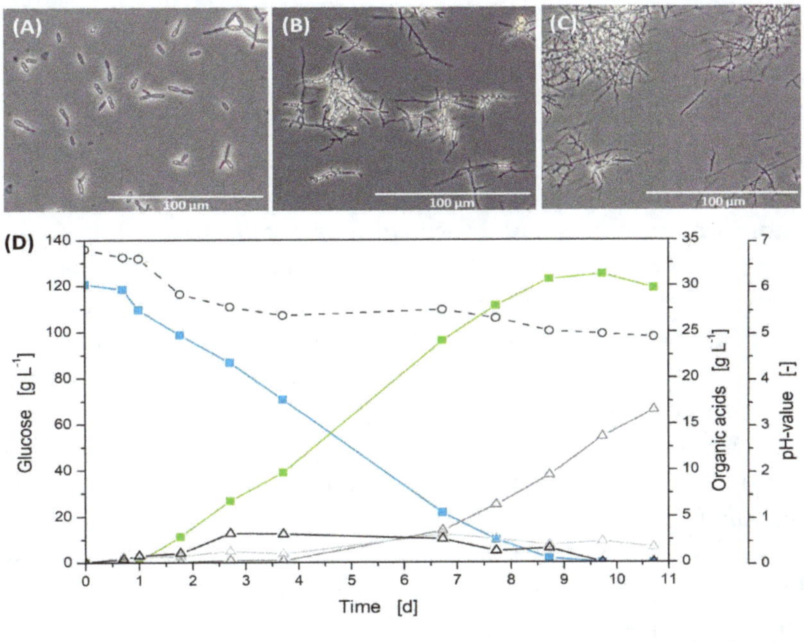

Figure 1. Cultivation of *U. rabenhorstiana* in 250 mL shake flasks in standard Tabuchi-medium (**D**) and its corresponding morphology (**A**): 0–2 days; (**B**): 2–7 days; (**C**): 7–11 days. Glucose (blue square), itaconic acid (green square), α-ketoglutaric acid (grey triangle), succinic acid (black triangle), malic acid (light grey triangle), pH (black circle), 120 rpm, 30 °C, 1% (*v/v*) inoculum.

3.2. Influence of Media Components

To determine the influence of media components, titer, yield, and productivity of a cultivation with 120 g·L^{-1} initial glucose, performed in shake flasks after 7.8 days, are shown in Figure 2. In the case of the variating initial glucose concentrations (Figure 2A), the point in time of glucose limitation was analyzed. 50 g·L^{-1} glucose were consumed in 3.7 days, 100 g·L^{-1} in 6.7 days, 120 g·L^{-1} in 8.7 days, and 150 g·L^{-1} in 10.7 days. Further, 200 g·L^{-1} glucose was not completely consumed by *U. rabenhorstiana* and 35 g·L^{-1}, and remained while the concentration of itaconic acid was constant after 15 days. With increasing initial glucose concentration, the titer of itaconic acid increased, but the productivity and yield decreased slightly from 0.16 g (L·h)$^{-1}$ to 0.09 g (L·h)$^{-1}$ and from 0.27 (*w/w*) to 0.20 (*w/w*) for glucose concentrations larger than 100 g·L^{-1}. The amount of α-ketoglutaric, succinic, and malic acid of the total organic acid concentration was raised from 15.5% (50 g·L^{-1} glucose) over 33.7% (120 $g \cdot L^{-1}$ glucose) to 50.1% (200 $g \cdot L^{-1}$ glucose). In case of different ammonia chloride concentrations, the highest titer of 31.8 $g \cdot L^{-1}$ itaconic acid, a productivity of 0.17 g $(L \cdot h)^{-1}$, and a yield of 0.26 (*w/w*) was achieved using 1.6 $g \cdot L^{-1}$ NH_4Cl, which corresponded to the used concentration in standard Tabuchi-medium (Figure 2B). With a lower concentration of 1 $g \cdot L^{-1}$ NH_4Cl and higher concentrations between 3 and 7 $g \cdot L^{-1}$ NH_4Cl the titer, productivity and yield decreased up to 35%. Also, the chosen concentration of 0.2 $g \cdot L^{-1}$ $MgSO_4 \cdot 7\ H_2O$ in Tabuchi-medium was optimal for itaconic acid production with *U. rabenhorstiana*, and a titer of 28.9 $g \cdot L^{-1}$ with a productivity of 0.16 g $(L \cdot h)^{-1}$ was reached (Figure 2C). Lower or higher levels of magnesium resulted in a decrease of all target values. In the concentration range of 0.1–1 $g \cdot L^{-1}$ KH_2PO_4, there were no significant differences between the titer, yield, and productivity. All cultivations yielded in titers of 29.3 $g \cdot L^{-1}$ ± 1.2 $g \cdot L^{-1}$ with a productivity between 0.15–0.16 g $(L \cdot h)^{-1}$ and a yield of 0.24–0.25 (*w/w*) (Figure 2D). In the range of 0.5–25 $mg \cdot L^{-1}$ $FeSO_4 \cdot 7\ H_2O$, the itaconic acid decreased from 32.4 $g \cdot L^{-1}$ to 24.2 $g \cdot L^{-1}$ (Figure 2E). The productivity of 0.17 g $(L \cdot h)^{-1}$ was reduced by 17% and the yield of 0.3 (*w/w*) itaconic acid by 25%. With increasing yeast extract concentration (0.25–1.5 $g \cdot L^{-1}$) the titer increased to 27.2 $g \cdot L^{-1}$ with a productivity of 0.15 g $(L \cdot h)^{-1}$ at a concentration of 1.5 $g \cdot L^{-1}$ yeast extract (Figure 2F). A further increase in the yeast extract concentration up to 2 $g \cdot L^{-1}$ resulted in a decreased titer of 24 $g \cdot L^{-1}$ and a lowered productivity of 0.13 g $(L \cdot h)^{-1}$. None of the media components had an influence on the filamentous growth.

3.3. Monosaccharide Utilization

It is intended to produce itaconic acid based on renewable feedstocks, e.g., lignocellulosic biomass, biomass with a high starch content, or molasses. The usability of the monosaccharides based on those feedstocks (arabinose, fructose, galactose, glucose, mannose, rhamnose, and xylose) were investigated with 100 $g \cdot L^{-1}$ of each sugar in test tubes (Table 1). For the precultivation, sucrose was used. Filamentous growth occurred on glucose, fructose, mannose, and xylose. An accumulation of long hyphae, a buildup of pellets with a diameter of 50 μm grew with arabinose as substrate. The yield and productivity were very different depending on the substrate. For the reference cultivation with glucose, the productivity was 0.16 g $(L \cdot h)^{-1}$ with a yield of 0.24 (*w/w*). The productivity of 0.09 g $(L \cdot h)^{-1}$ of itaconic acid with mannose was 44% lower, while the yield was in the same range with 0.22 (*w/w*). Using fructose, the same productivity compared to mannose was achieved, but with a lower yield of 0.17 (*w/w*). *U. rabenhorstiana* was able to use both pentoses for itaconic acid production, whereby the productivity with arabinose with 0.04 g $(L \cdot h)^{-1}$ was twice as high as with xylose. In the cultivation with glactose, only traces of itaconic acid were detected. The yeast was not able to produce itaconic acid or even grow with rhamnose as single substrate.

Table 1. Cultivation of *U. rabenhorstiana* in test tubes with different monosaccharides as substrate (four days, 30 °C, 120 rpm, inclination angle of 30°, and 1% (*v/v*) inoculum).

Monosaccharide	Productivity [g $(L \cdot h)^{-1}$]	$Y_{P/S}$ [*w/w*]
Glucose	0.16	0.24
Mannose	0.09	0.22
Fructose	0.09	0.17
Arabinose	0.04	0.06
Xylose	0.02	0.04
Galactose	<0.01	<0.01
Rhamnose	-	-

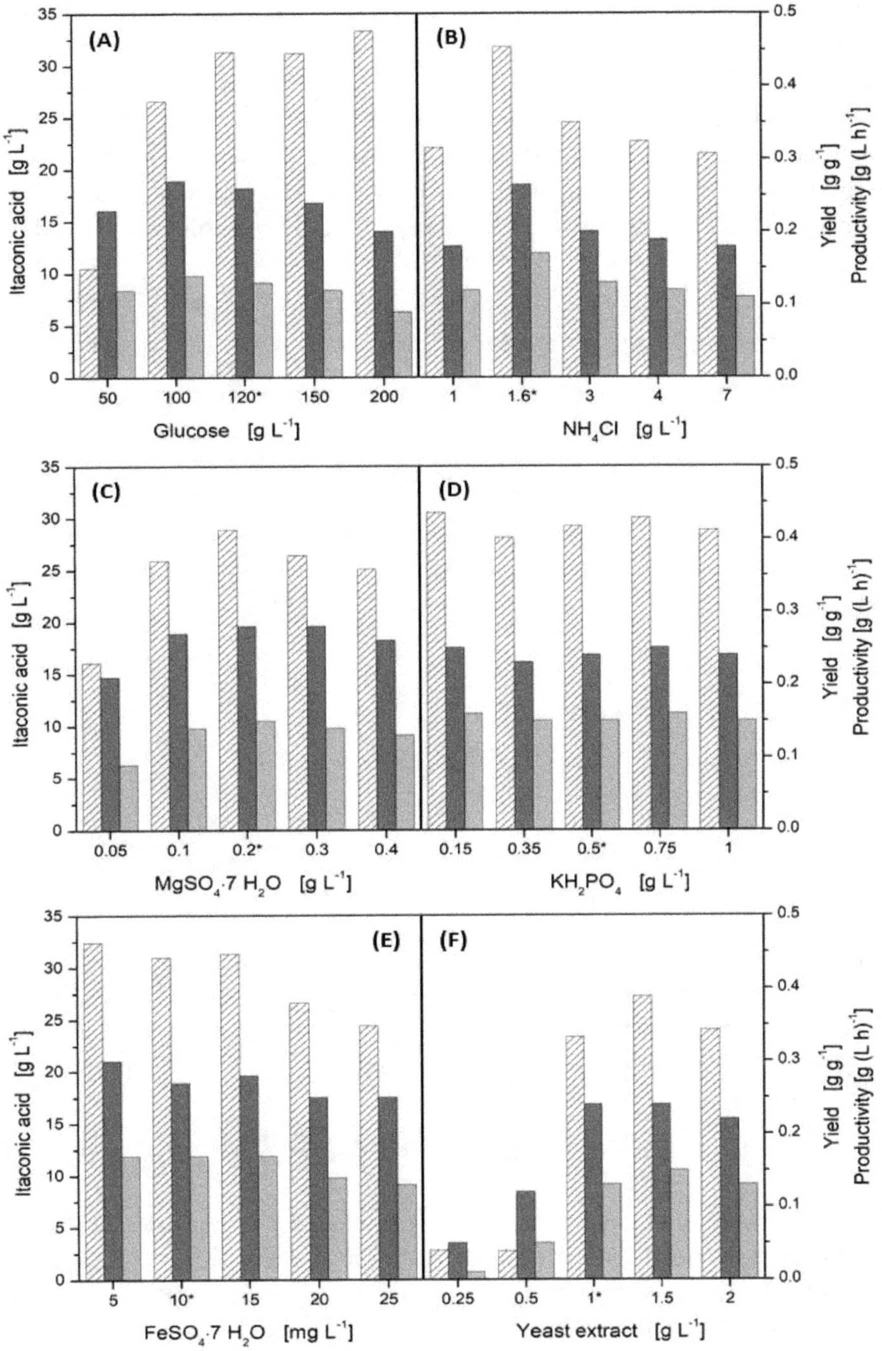

Figure 2. Influence of Tabuchi-medium components in 250 mL shake flasks on the final titer (dashed bar), yield (black bar), and productivity (grey bar) of *U. rabenhorstiana* at 120 rpm and 30 °C after 7.8 days (**B**–**F**). Cultivation time for different initial glucose concentrations depended on the point of glucose limitation (**A**). Asterisks highlight the standard media composition.

3.4. Influence of Sugar Degradation Products

In case of lignocellulosic feedstocks, different sugar degradation products are formed due to the harsh conditions in the pretreatment. To test the inhibitory effect of sugar degradation products like weak acids or furan derivates, the components were added by the lowest expected concentration levels to the media. The effect of 0–2 g·L^{-1} acetic acid, formic acid, furfural, or hydroxymethylfurfural (HMF) was carried out in test tubes (Figure 3). A productivity of 0.15 g (L·h)$^{-1}$ with standard Tabuchi-medium without the addition of inhibitory components was reached. Up to a concentration of 0.5 g·L^{-1} formic acid, the productivity did not differ. The addition of 1 g·L^{-1} resulted in an increased productivity of approximately 1.4 times and decreased to 0.13 g (L·h)$^{-1}$ with 2 g·L^{-1} formic acid. The result was very similar with the addition of acetic acid. The standard productivity increased up to 0.19 g (L·h)$^{-1}$ by adding 0.5 g·L^{-1} acetic acid and was reduced to 0.13 g (L·h)$^{-1}$ by increasing the acetic acid concentration. Both furan derivates influenced the microorganism very strongly, amounts of 0.1 g·L^{-1} of HMF or 0.5 g·L^{-1} furfural already resulted in a growth inhibition. If the growth was not inhibited, the yeast grew filamentous and stored intracellular lipid droplets comparable with the cultivation without addition of inhibitors.

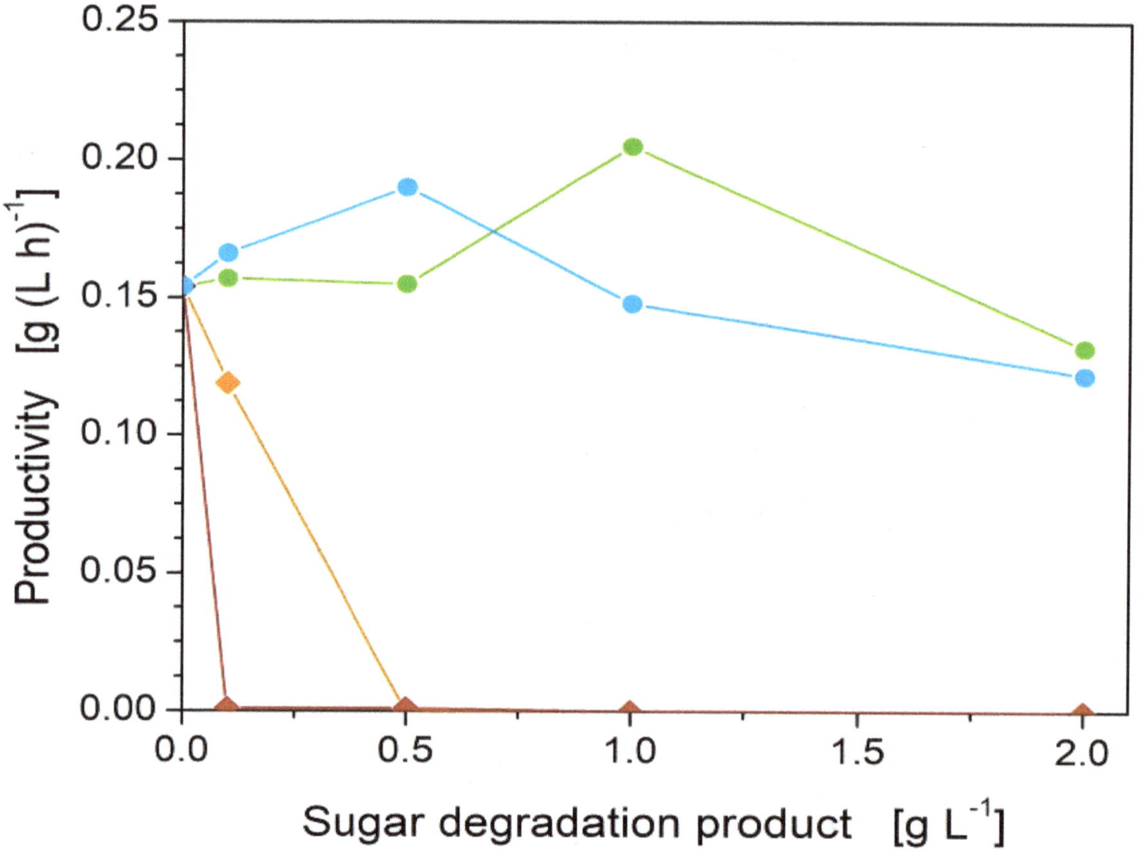

Figure 3. Inhibition effects of sugar degradation products on the itaconic acid productivity with *U. rabenhorstiana* in test tubes with Tabuchi-medium after four days, 30 °C, 120 rpm, inclination angle of 30°, pH > 5.5, and 1% (*v/v*) inoculum. Acetic acid (blue circle), formic acid (green circle), HMF (red diamond), furfural (orange diamond).

3.5. Influence of the pH-Value in 1 L-Bioreactor

The pH-value dropped in shake flask cultivations from pH 6.7 to pH 4.9 with $CaCO_3$ as buffer. To estimate the influence of the pH-value, the cultivation was transferred in 1 L-bioreactors with pH-control using 4 M NaOH (Table 2). Moreover, the modified Tabuchi-medium with 100 g·L^{-1} glucose and 1.5 g·L^{-1} yeast extract was used, based on the findings regarding the tested media components. The highest titer of 31.7 g·L^{-1} itaconic acid with a productivity of 0.23 g (L·h)$^{-1}$ and a yield of 0.34 (*w/w*) was reached with a controlled pH of 6.0. Beside itaconic acid, 0.4 g·L^{-1} α-ketoglutaric acid, 2 g·L^{-1} malic acid, 2.9 g·L^{-1} succinic acid, and the unknown product (<1 g·L^{-1}) were produced. The rate of the byproducts did not differ among the tested pH-values; also, the pH-value did not have any influence on the filamentous growth of the yeast and formation of intracellular lipids.

Table 2. Cultivation results of *U. rabenhorstiana* with pH-control in modified Tabuchi-medium with 100 g·L^{-1} glucose and 1.5 g·L^{-1} yeast extract in 1 L-bioreactor (30 °C, 500 rpm, 0.1 vvm, 4 M NaOH).

pH-Value [-]	Itaconic Acid [g·L^{-1}]	Productivity [g (L·h)$^{-1}$]	$Y_{P/S}$ [*w*/*w*]
5.5	23.7	0.15	0.25
6.0	31.7	0.23	0.34
6.5	18.6	0.12	0.19
7.0	15.9	0.10	0.16

3.6. *Influence of Aeration in 1 L-Bioreactor*

The effect of aeration was tested in 1 L-bioreactors by different aeration rates between 0.1–1 vvm, the stirring rate was kept constant, and the pH-value was regulated to pH 6.0 (Table 3). An increasing aeration rate from 0.1 to 1 vvm resulted in a decreased yield and titer of 20%, as well as in a 45% lower productivity. In contrast, the formed biomass increased from 15.7 g·L^{-1} at 0.1 vvm to 21.3 g·L^{-1} at 1 vvm. There were no significant differences between the by-product concentrations depending on the aeration rate, which corresponds to the concentrations described in Section 3.5.

Table 3. Influence of aeration on the cultivation of *U. rabenhorstiana* in modified Tabuchi-medium with 100 g·L^{-1} glucose and 1.5 g·L^{-1} yeast extract in 1 L-bioreactor (30 °C, 500 rpm, pH 6.0).

Aeration [vvm]	Itaconic Acid [g·L^{-1}]	Productivity [g·(L·h)$^{-1}$]	$Y_{P/S}$ [*w*/*w*]	CDW [g·L^{-1}]
0.1	29.8	0.22	0.30	15.7
0.5	26.1	0.16	0.26	17.5
1.0	23.6	0.12	0.24	21.3

3.7. *Fed-Batch Mode in 1 L-Bioreactor with Glucose*

Glucose concentrations larger than 150 g·L^{-1} resulted in a decreased yield and productivity in shake flasks (Figure 2A). For this reason, a fed batch with glucose was realized at a constant pH of pH 6.0 in a 1 L-bioreactor (Figure 4). An initial glucose concentration of 100 g·L^{-1} was chosen. After five days, 73 g·L^{-1} and after 10 days, 25 g·L^{-1} glucose were added into the cultivation broth, in which the average glucose consumption rate was 0.73 g (L·h)$^{-1}$ for the first batch (0–5 days), 0.53 g (L·h)$^{-1}$ for the second batch (5–10 days), and 0.28 for the third batch (10–15 days). The yield amounted to 0.31 (*w/w*) in the first batch and was constant with 0.26 (*w/w*) in the second and third batches. The DO decreased to 2% within the first day and varied between 2%–20% during the further cultivation. After one day, the itaconic acid production started and rose to a final titer of 50.3 g·L^{-1} within 15 days. Beside itaconic acid, 3.6 g·L^{-1} malic acid, 13.6 g·L^{-1} succinic acid, 2.5 g·L^{-1} α-ketoglutaric acid, and the unknown product (<10 g·L^{-1}) were formed by 17.2 g·L^{-1} filamentous biomass (Appendix A, Figure A2). This cultivation resulted in a productivity of 0.14 g (L·h)$^{-1}$ with an overall yield of 0.27 (*w/w*) after 15 days. Further, 175 mL of a 4 M NaOH was used to keep the pH-value constant at pH 6.0. After 15 days, the cell dry weight was analyzed regarding the fatty acids (Appendix A, Table A1); C16:0, C18:0, and C18:2 were the main elements.

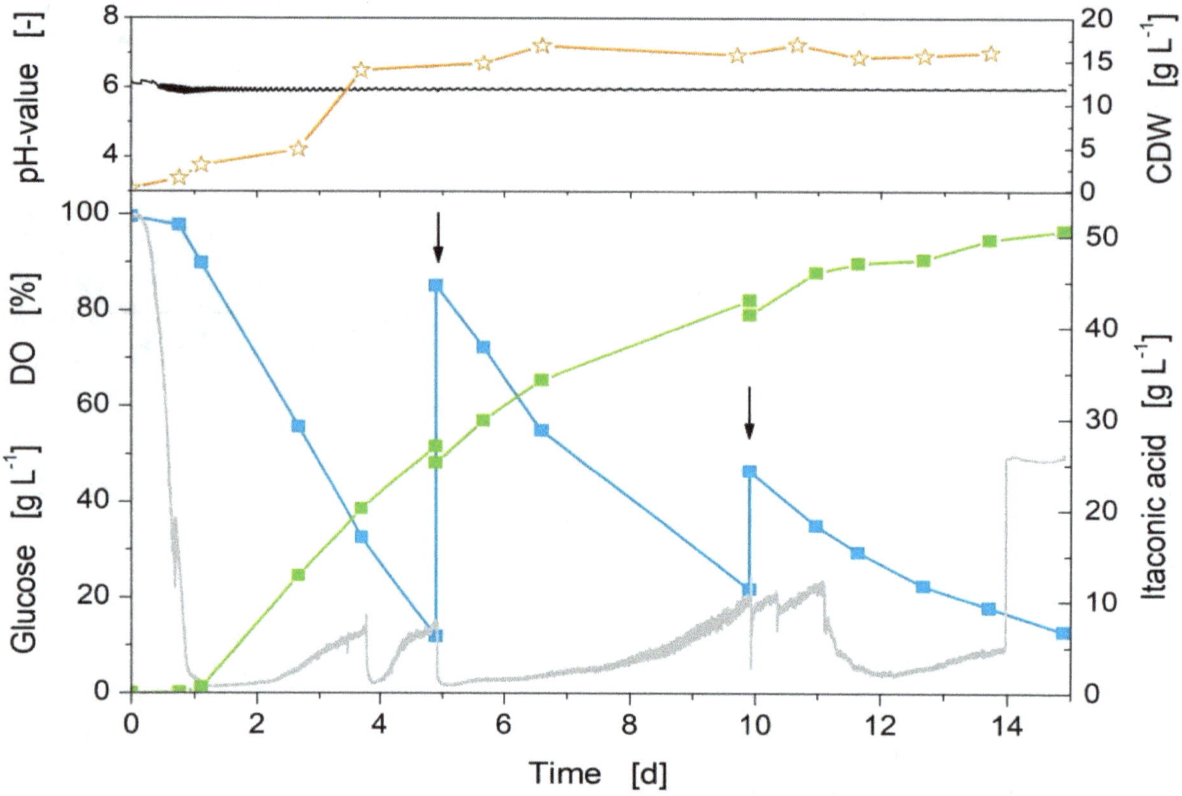

Figure 4. Fed batch with glucose in 1 L-bioreactor with modified Tabuchi-medium (initial glucose concentration 100 g·L^{-1} and 1.5 g·L^{-1} yeast extract) at 30 °C, 500 rpm, 0.1 vvm, pH 6.0 with 4 M NaOH as base and 1% (*v/v*) inoculum. Glucose (blue square), itaconic acid (green square), pO_2 (grey line), pH (black line), cell dry weight (CDW) (orange asterisk), arrows symbolize the addition of glucose.

4. Discussion

In a standard cultivation in shake flasks, a final titer of 31.3 g·L^{-1} itaconic acid was achieved after 9.7 days without media and process optimization of *U. rabenhorstiana*. Additionally, succinic acid, malic acid, α-ketoglutaric acid, an unknown product, and intracellular lipids were formed. Guevarra and Tabuchi reached a titer of about 16 g·L^{-1} itaconic acid in the same media after seven days and verified 2-hydroxyparaconic acid, itatartaric acid, and erythritol as byproducts with a total concentration of 30 g·L^{-1} [18]. Furthermore, they reduced the byproduct concentration to 19 g·L^{-1} with a constant itaconic acid titer by using an unbuffered media, whereby the final pH-value was pH 2.8 [30]. Erythritol as the unknown byproduct could be excluded in this work. In comparison with the literature, the unknown product (<1 g·L^{-1}) could either be 2-hydroxyparaconic acid or itatartaric acid (Appendix A, Figure A1) [18,23,30] and must be characterized precisely in a further study. Moreover, the formation of different organic acids and intracellular lipid bodies by *Ustilaginaceae* besides itaconic acid production are very well known [5,19], as well as the production of cellobiose lipids and mannosylerythritol lipids [20,29], which were not proved in this study.

A point-by-point analysis of each media component indicated that the chosen concentrations of each component in the Tabuchi-medium were nearly optimal for itaconic acid production with *U. rabenhorstiana*. Only the glucose and yeast extract concentrations were adjusted from 120 g·L^{-1} to 100 g·L^{-1} for glucose and from 1 g·L^{-1} to 1.5 g·L^{-1} for yeast extract in the modified Tabuchi-medium. Increased glucose concentration resulted in an increased final titer, but the yield and productivity decreased with glucose concentrations >100 g·L^{-1}. The decreased yield and productivity can be explained by a higher osmotic stress and a higher number of byproducts. A very similar result was obtained by the production of itaconic acid with *U. maydis* [19]. This microorganism also showed lower yields at higher initial glucose concentrations, explained by the formation of other organic acids, polyols, and glycolipids. The increase of yeast extract by 0.5 g·L^{-1} raised the titer, productivity, and yield. A further increase of yeast extract did not result in an improved titer, suggesting that a limitation of vitamins, amino acids, salts, trace elements or nucleic acids [33] is prevented with a concentration >1.5 g·L^{-1} yeast extract. The itaconic acid overproduction by *U. maydis* or *P. antarctica* are mainly triggered by an ammonium limitation, higher concentrations of NH_4Cl > 4 g·L^{-1} resulted in lower itaconic acid yields and an increase in biomass for *U. maydis* [5,14,19]. What is more, in this study, the ammonium concentration had the strongest influence on the fermentation performance of *U. rabenhorstiana*. A concentration of 1.6 g·L^{-1} NH_4Cl was optimal for itaconic acid production, and higher concentrations resulted in reduced yields and productivities as described already for *U. maydis*. For this reason, it can be assumed that an ammonium limitation caused the itaconic acid overproduction of *U. rabenhorstiana*. Further, the nitrogen limitation caused the accumulation of intracellular lipid droplets, which are mostly triacylglycerols [21,34]. A secretion of cellobiose lipids and mannosylerythritol lipids in the form of needle-like crystals or oily droplets of *U. maydis* [20], could not be verified for *U. rabenhorstiana*.

To use alternative, low-cost, or lignocellulosic feedstocks for itaconic acid production, it is important to use a microorganism, which is able to consume different monosaccharides and is robust towards varying impurities. In this study, it was worked out that *U. rabenhorstiana* is able to grow and produce itaconic acid from different monosaccharides like glucose, fructose, mannose, xylose, arabinose, and galactose. The highest productivity was reached with glucose, followed by fructose and mannose. Because the precultivation was based on sucrose, composed of glucose and fructose, it can be assumed that the cultivation with fructose was therefore such successful. An adaption to the used monosaccharide of the microorganisms in the preculture or a mixture of several monosaccharides with glucose would probably lead to higher productivity and yield using that single monosaccharide in the main culture. Furthermore, the plant pathogen *U. rabenhorstiana* is supposed to degrade a range of biomass-based polymers [35–37]. For industrial itaconic acid production, the filamentous fungus *A. terreus* is used, which is very sensitive to weak acids, furan derivates, metal ions, and other impurities, which are contained in such substrates [11,12,38]. *U. maydis* is described as a very robust microorganism [5,22,39]. *U. rabenhorstiana* was also not influenced by the addition of weak acid concentrations up to 2 g·L^{-1} in the main culture. A major advantage of the cultivation of *Ustilaginaceae* is the pH range of 5.0–6.5, whereby the dissociated weak acids cannot cross the plasma membrane into the cytosol and affect the intracellular pH-value [40,41]. A positive effect on the itaconic acid productivity was even achieved with the addition of 0.5 g·L^{-1} acetic acid or 1 g·L^{-1} formic acid. This positive influence of low weak acid concentrations in cultivation media is already known from itaconic acid production with *A. terreus*, ethanol production with *S. cerevisiae*, or enzyme production with *T. reesei* [12,42–44]. In contrast, low concentrations of 0.1 g·L^{-1} HMF or 0.5 g·L^{-1} furfural are growth limiting factors; both furan derivates reduce the activity of a number of important intracellular enzymes of the maintenance metabolism, e.g., pyruvate dehydrogenase [45]. In particular, the activity of the pyruvate dehydrogenase is essential for cells, because this enzyme links glycolysis and citric acid cycle, which supplies the cell with energy intermediates. When using lignocellulosic biomass as feedstock for *U. rabenhorstiana*, it should be taken into account, that some robustness in relation to weak acids exists, but furan derivates influence the microorganisms mostly up to growth inhibition.

For further characterization of *U. rabenhorstiana*, the fermentation was transferred in 1 L-bioreactors to investigate the influence of pH-value and aeration. A constant pH of 6.0 and modified Tabuchi-medium yielded in the highest itaconic acid titer of 31.7 g·L^{-1}, which is comparable with the titer in standard shake flask cultivation with Tabuchi-medium. However, the overall productivity was 1.7 times higher at a constant pH-value than in shake flask experiments with $CaCO_3$, whereby the pH-value continuously decreased to pH 4.9. Moreover, the total concentration of by-products decreased by 66%, which resulted in an increased itaconic acid yield. These suggested that the buffer capacity in shake flasks is insufficient. Buffer systems like $CaCO_3$ or MES and its buffer capacity have a significant impact on the organic acid production of *Ustilaginaceae* not only on titer, but also on the ratio of products [16,24,30]. The higher the buffer capacity, the better is the itaconic acid titer in small-scale experiments, but *U. maydis* achieved the highest itaconic acid titer of 45.5–63.2 $g \cdot L^{-1}$ in cultivations with a constant pH of 6.0–6.5 in bioreactors [19,23]. Consequently, *U. maydis* and *U. rabenhorstiana* have nearly the same requirements in pH and productivity. Titer and yield are positively influenced by a constant pH-value.

In the literature, no detailed studies regarding influence of oxygen levels on the organic acid production with *Ustilaginaceae* exist. Only cultivation parameters like high shaking frequencies or stirrer speeds led to the conclusion that a high input of oxygen is necessary [10]. Contrary results were achieved for itaconic acid production with *U. rabenhorstiana*; the lowest aeration rate of 0.1 vvm and a constant stirring rate of 500 rpm yielded the best result regarding titer, productivity, and yield. Presumably, the increase in these values is related to the formation of 36% more biomass at higher aeration rates, because of a better supply of oxygen. In batch experiments, a maximum itaconic acid concentration of 33.3 $g \cdot L^{-1}$ with an initial glucose concentration of 200 $g \cdot L^{-1}$ was achieved, proving that initial glucose concentration ≥100 $g \cdot L^{-1}$ and a constant pH-value of pH 6.0 have a significant impact on the itaconic acid production with *U. rabenhorstiana*. Therefore, a cultivation in fed-batch mode with glucose was realized at pH 6.0 and resulted in 50.3 $g \cdot L^{-1}$ itaconic acid. Comparing the batch cultivation with 200 $g \cdot L^{-1}$ glucose with the fed-batch cultivation in a bioreactor, the productivity and yield were 1.4 times and the final titer 1.5 times higher. Also, the formation of organic acid as byproducts was reduced. The number of byproducts of the overall organic acid concentration decreased from 50% in batch cultivation to 28% in fed-batch mode. Thereby, the amount of organic acids shifted from α-ketoglutaric acid as the main byproduct in batch mode to succinic acid in fed-batch mode, due to the pH-value of each cultivation [16]. The final titer and yield of this study are slightly increased compared to a wildtype strain of *U. maydis*, which reached 44.5 $g \cdot L^{-1}$ [19]. A higher final titer up to 63.2 $g \cdot L^{-1}$ or yield of 0.48 (*w/w*) was only obtained by a genetical modification of *U. maydis* [23]. All in all, the fermentation broth was diluted by addition of 175 mL NaOH as base. Considering the dilution, the wildtype of *U. rabenhorstiana* demonstrates the potential to produce up to 68 $g \cdot L^{-1}$ itaconic acid. However, in all main cultures, the unicellular growth of *U. rabenhorstiana* shifted to filamentous cells with depots of intracellular lipids. Neither the variation of media components nor the investigation of process parameters influenced the filamentous growth. The typical unicellular yeast-like growth of *U. maydis* for itaconic acid could not be achieved for *U. rabenhorstiana* in this study. This morphology would be a great advantage compared to the filamentous *A. terreus* regarding oxygen supply or viscosity, especially in large-scale fermentations [5,22], but was not focused in this study. In general, it is possible to generate a stable unicellular growth by deleting several genes [24]. The filamentous growth involved the accumulation of intracellular lipid droplets, which is initiated by nitrogen limitation, which in turn is needed for itaconic acid overproduction [5,19,34]. Furthermore, 89% of all fatty acids in *U. rabenhorstiana* were long-chain fatty acids C16:0, C18:0, and C18:2 and suggested the accumulation of triacylglycerols in the cells. The lipid bodies in *U. maydis* mainly contain triacylglycerols consisting of palmitic, linoleic, and oleic acids [21].

5. Conclusions

This study describes a known, but so far unspecified, itaconic acid producer—*U. rabenhorstiana*. The cultivation in shake flasks with a maximal final titer of 33.3 $g \cdot L^{-1}$ itaconic acid was transferred in a bioreactor. With a controlled pH-value, a low initial glucose concentration, and fed-batch mode, a final titer of 50.3 $g \cdot L^{-1}$ was achieved, which is comparable with titer of other wildtype strains of *Ustilago* described in literature. However, the productivity and yield are rather low compared to *U. maydis*, which was studied very precisely in the last years regarding cultivation and process parameters as well as metabolic engineering strategies for further improvements in itaconic acid production. Transferring this knowledge from *U. maydis* to *U. rabenhorstiana* could result in a further increased final titer and improved yield and productivity. Particular attention should be paid to the morphology of the yeast and minimization of byproducts, mainly the formation of intracellular lipid droplets. Moreover, the wildtype strain *U. rabenhorstiana* turned out to be a robust and promising alternative itaconic acid producer based on renewable resources. All in all, this study serves a basis for further promising research regarding lignocellulosic hydrolysates.

Author Contributions: S.K. and M.L. designed and performed the experiments. U.P. and A.K. supervised the experiments. S.K. wrote the manuscript. All authors have read and agreed to the published version of the manuscript.

Appendix A

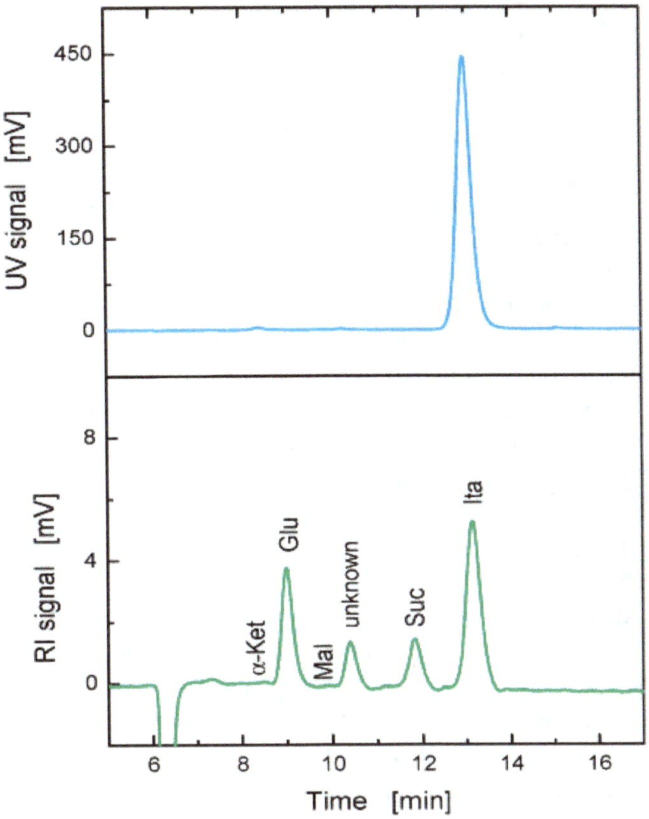

Figure A1. Representative HPLC analysis of a final sample (15 days) of a fed-batch cultivation in 1 L-bioreactor (Section 3.7). α-ketoglutaric acid (α-Ket), glucose (Glu), malic acid (Mal), unknown product, succinic acid (Suc), itaconic acid (Ita).

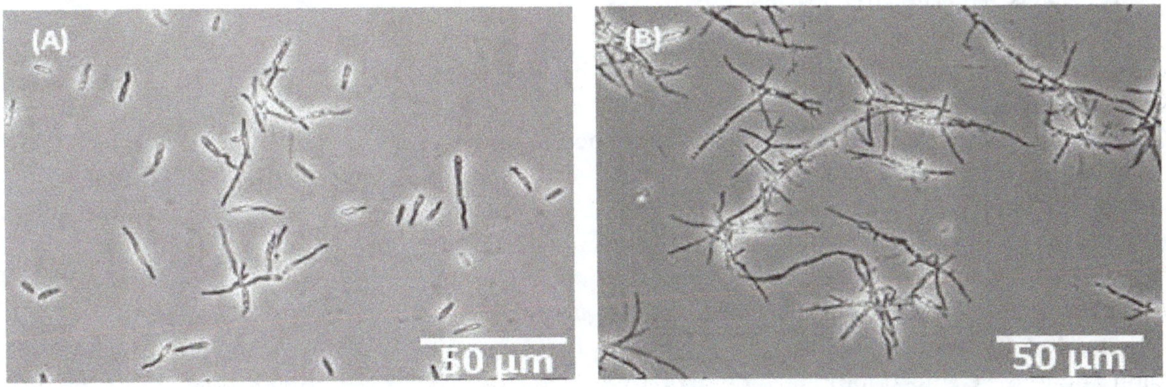

Figure A2. Corresponding morphology of *U. rabenhorstiana* in a fed-batch cultivation in 1 L-bioreactor (Section 3.7) after 1.1 days (**A**) and 6.6 days (**B**).

Table A1. Analysis of fatty acids in intracellular lipid droplets in *U. rabenhorstiana* of a fed-batch cultivation in 1 L-bioreactor (Section 3.7).

Fatty Acid	Concentration [$mg \cdot g^{-1}{}_{CDW}$]
C14:0	1.46
C16:0	30.04
C16:1	1.55
C18:0	16.94
C18:1	1.83
C18:2	48.38
C22:0	4.42
C24:0	2.45

References

1. Werpy, T.; Petersen, G. *Top Value Added Chemicals from Biomass: Volume I—Results of Screening for Potential Candidates from Sugars and Synthesis Gas*; National Renewable Energy Lab.: Golden, CO, USA, 2004.
2. Magalhaes, A.I.; de Carvalho, J.C.; Medina, J.D.C.; Soccol, C.R. Downstream process development in biotechnological itaconic acid manufacturing. *Appl. Microbiol. Biotechnol.* **2017**, *101*, 1–12. [CrossRef]
3. Robert, T.; Friebel, S. Itaconic acid—A versatile building block for renewable polyesters with enhanced functionality. *Green Chem.* **2016**, *18*, 2922–2934. [CrossRef]
4. Okabe, M.; Lies, D.; Kanamasa, S.; Park, E.Y. Biotechnological production of itaconic acid and its biosynthesis in *Aspergillus terreus*. *Appl. Microbiol. Biotechnol.* **2009**, *84*, 597–606. [CrossRef]
5. Klement, T.; Milker, S.; Jäger, G.; Grande, P.M.; de Maria, P.D.; Büchs, J. Biomass pretreatment affects *Ustilago maydis* in producing itaconic acid. *Microb. Cell Factories* **2012**, *11*, 43. [CrossRef]
6. Willke, T.; Vorlop, K.D. Biotechnological production of itaconic acid. *Appl. Microbiol. Biotechnol.* **2001**, *56*, 289–295. [CrossRef]
7. Hevekerl, A.; Kuenz, A.; Vorlop, K.D. Influence of the pH on the itaconic acid production with *Aspergillus terreus*. *Appl. Microbiol. Biotechnol.* **2014**, *98*, 10005–10012. [CrossRef]
8. Karaffa, L.; Diaz, R.; Papp, B.; Fekete, E.; Sandor, E.; Kubicek, C.P. A deficiency of manganese ions in the presence of high sugar concentrations is the critical parameter for achieving high yields of itaconic acid by *Aspergillus terreus*. *Appl. Microbiol. Biotechnol.* **2015**, *99*, 7937–7944. [CrossRef]
9. Krull, S.; Hevekerl, A.; Kuenz, A.; Prüße, U. Process development of itaconic acid production by a natural wild type strain of *Aspergillus terreus* to reach industrially relevant final titers. *Appl. Microbiol. Biotechnol.* **2017**, *101*, 4063–4072. [CrossRef]
10. Kuenz, A.; Krull, S. Biotechnological production of itaconic acid-things you have to know. *Appl. Microbiol. Biotechnol.* **2018**, *102*, 3901–3914. [CrossRef]
11. Kobayashi, T. Production of itaconic acid from wood waste. *Process Biochem.* **1978**, *13*, 15–22.
12. Krull, S.; Eidt, L.; Hevekerl, A.; Kuenz, A.; Prüße, U. Itaconic acid production from wheat chaff by *Aspergillus terreus*. *Process Biochem.* **2017**, *63*, 169–176. [CrossRef]
13. Wu, X.F.; Liu, Q.; Deng, Y.D.; Li, J.H.; Chen, X.J.; Gu, Y.Z.; Lv, X.J.; Zheng, Z.; Jiang, S.T.; Li, X.J. Production of itaconic acid by biotransformation of wheat bran hydrolysate with *Aspergillus terreus* CICC40205 mutant. *Bioresour. Technol.* **2017**, *241*, 25–34. [CrossRef]
14. Levinson, W.E.; Kurtzman, C.P.; Kuo, T.M. Production of itaconic acid by *Pseudozyma antarctica* NRRL Y-7808 under nitrogen-limited growth conditions. *Enzym. Microb. Technol.* **2006**, *39*, 824–827. [CrossRef]
15. Specht, R.; Andreas, A.; Kreyß, E.; Barth, G.; Bodinus, C. Verfahren zur Biotechnologischen Herstellung von Itaconsäure. DE Patent 102008011854 B4, 20 February 2014.
16. Geiser, E.; Wiebach, V.; Wierckx, N.; Blank, L.M. Prospecting the biodiversity of the fungal family *Ustilaginaceae* for the production of value-added chemicals. *Fungal Biol. Biotechnol.* **2014**, *1*, 2. [CrossRef]
17. Tabuchi, T.; Sugisawa, T.; Ishidori, T.; Nakahara, T.; Sugiyama, J. Itaconic Acid Fermentation by a Yeast Belonging to the Genus Candida. *Agric. Biol. Chem.* **1981**, *45*, 475–479. [CrossRef]
18. Guevarra, E.D.; Tabuchi, T. Accumulation of itaconic, 2-hydroxyparaconic, itatartaric, and malic-acids by strains of the genus *Ustilago*. *Agric. Biol. Chem.* **1990**, *54*, 2353–2358. [CrossRef]

19. Maassen, N.; Panakova, M.; Wierckx, N.; Geiser, E.; Zimmermann, M.; Bolker, M.; Klinner, U.; Blank, L.M. Influence of carbon and nitrogen concentration on itaconic acid production by the smut fungus *Ustilago maydis*. *Eng. Life Sci.* **2014**, *14*, 129–134. [CrossRef]
20. Spoeckner, S.; Wray, V.; Nimtz, M.; Lang, S. Glycolipids of the smut fungus *Ustilago maydis* from cultivation on renewable resources. *Appl. Microbiol. Biotechnol.* **1999**, *51*, 33–39. [CrossRef]
21. Aguilar, L.R.; Pardo, J.P.; Lomeli, M.M.; Bocardo, O.I.L.; Oropeza, M.A.J.; Sanchez, G.G. Lipid droplets accumulation and other biochemical changes induced in the fungal pathogen *Ustilago maydis* under nitrogen-starvation. *Arch. Microbiol.* **2017**, *199*, 1195–1209. [CrossRef]
22. Regestein, L.; Klement, T.; Grande, P.; Kreyenschulte, D.; Heyman, B.; Maßmann, T.; Eggert, A.; Sengpiel, R.; Wang, Y.; Wierckx, N. From beech wood to itaconic acid: Case study on biorefinery process integration. *Biotechnol. Biofuels* **2018**, *11*, 1–11. [CrossRef]
23. Geiser, E.; Przybilla, S.K.; Engel, M.; Kleineberg, W.; Büttner, L.; Sarikaya, E.; Den Hartog, T.; Klankermayer, J.; Leitner, W.; Bölker, M. Genetic and biochemical insights into the itaconate pathway of *Ustilago maydis* enable enhanced production. *Metab. Eng.* **2016**, *38*, 427–435. [CrossRef]
24. Tehrani, H.H.; Tharmasothirajan, A.; Track, E.; Blank, L.M.; Wierckx, N. Engineering the morphology and metabolism of pH tolerant *Ustilago cynodontis* for efficient itaconic acid production. *Metab. Eng.* **2019**, *54*, 293–300. [CrossRef]
25. Geiser, E.; Przybilla, S.K.; Friedrich, A.; Buckel, W.; Wierckx, N.; Blank, L.M.; Bolker, M. *Ustilago maydis* produces itaconic acid via the unusual intermediate trans-aconitate. *Microb. Biotechnol.* **2016**, *9*, 116–126. [CrossRef]
26. Zambanini, T.; Hartmann, S.K.; Schmitz, L.M.; Buttner, L.; Hosseinpour Tehrani, H.; Geiser, E.; Beudels, M.; Venc, D.; Wandrey, G.; Buchs, J.; et al. Promoters from the itaconate cluster of *Ustilago maydis* are induced by nitrogen depletion. *Fungal Biol. Biotechnol.* **2017**, *4*, 11. [CrossRef]
27. Wierckx, N.; Agrimi, G.; Lubeck, P.S.; Steiger, M.G.; Mira, N.P.; Punt, P.J. Metabolic specialization in itaconic acid production: A tale of two fungi. *Curr. Opin. Biotechnol.* **2019**, *62*, 153–159. [CrossRef]
28. Zambanini, T.; Hosseinpour Tehrani, H.; Geiser, E.; Merker, D.; Schleese, S.; Krabbe, J.; Buescher, J.M.; Meurer, G.; Wierckx, N.; Blank, L.M. Efficient itaconic acid production from glycerol with *Ustilago vetiveriae* TZ1. *Biotechnol. Biofuels* **2017**, *10*, 131. [CrossRef]
29. Haskins, R.; Thorn, J.; Boothroyd, B. Biochemistry of the Ustilaginales: XI. Metabolic products of *Ustilago zeae* in submerged culture. *Can. J. Microbiol.* **1955**, *1*, 749–756. [CrossRef]
30. Guevarra, E.D.; Tabuchi, T. Production of 2-hydroxyparaconic and itatartaric acids by *Ustilago cynodontis* and simple recovery process of the acids. *Agric. Biol. Chem.* **1990**, *54*, 2359–2365. [CrossRef]
31. Lewis, T.; Nichols, P.D.; McMeekin, T.A. Evaluation of extraction methods for recovery of fatty acids from lipid-producing microheterotrophs. *J. Microbiol. Methods* **2000**, *43*, 107–116. [CrossRef]
32. Klose, J.; Kronstad, J.W. The multifunctional beta-oxidation enzyme is required for full symptom development by the biotrophic maize pathogen *Ustilago maydis*. *Eukaryot. Cell* **2006**, *5*, 2047–2061. [CrossRef]
33. Sommer, R. Yeast extracts: Production, properties and components. *Food Aust.* **1998**, *50*, 181–183.
34. Zavala-Moreno, A.; Arreguin-Espinosa, R.; Pardo, J.P.; Romero-Aguilar, L.; Guerra-Sánchez, G. Nitrogen source affects glycolipid production and lipid accumulation in the phytopathogen fungus *Ustilago maydis*. *Adv. Microbiol.* **2014**, *4*, 934. [CrossRef]
35. Gibson, D.M.; King, B.C.; Hayes, M.L.; Bergstrom, G.C. Plant pathogens as a source of diverse enzymes for lignocellulose digestion. *Curr. Opin. Microbiol.* **2011**, *14*, 264–270. [CrossRef]
36. Geiser, E.; Reindl, M.; Blank, L.M.; Feldbrügge, M.; Wierckx, N.; Schipper, K. Activating intrinsic carbohydrate-active enzymes of the smut fungus *Ustilago maydis* for the degradation of plant cell wall components. *Appl. Environ. Microbiol.* **2016**, *82*, 5174–5185. [CrossRef]
37. Couturier, M.; Navarro, D.; Olivé, C.; Chevret, D.; Haon, M.; Favel, A.; Lesage-Meessen, L.; Henrissat, B.; Coutinho, P.M.; Berrin, J.-G. Post-genomic analyses of fungal lignocellulosic biomass degradation reveal the unexpected potential of the plant pathogen *Ustilago maydis*. *BMC Genom.* **2012**, *13*, 57. [CrossRef]
38. Gyamerah, M. Factors affecting the growth form of *Aspergillus terreus* NRRL 1960 in relation to itaconic acid fermentation. *Appl. Microbiol. Biotechnol.* **1995**, *44*, 356–361. [CrossRef]
39. Benito, B.; Garciadeblás, B.; Pérez-Martín, J.; Rodríguez-Navarro, A. Growth at high pH and sodium and potassium tolerance in media above the cytoplasmic pH depend on ENA ATPases in *Ustilago maydis*. *Eukaryot. Cell* **2009**, *8*, 821–829. [CrossRef]

40. Stratford, M.; Nebe-von-Caron, G.; Steels, H.; Novodvorska, M.; Ueckert, J.; Archer, D.B. Weak-acid preservatives: pH and proton movements in the yeast *Saccharomyces cerevisiae*. *Int. J. Food Microbiol.* **2013**, *161*, 164–171. [CrossRef]
41. Lambert, R.; Stratford, M. Weak-acid preservatives: Modelling microbial inhibition and response. *J. Appl. Microbiol.* **1999**, *86*, 157–164. [CrossRef]
42. Larsson, S.; Palmqvist, E.; Hahn-Hägerdal, B.; Tengborg, C.; Stenberg, K.; Zacchi, G.; Nilvebrant, N.-O. The generation of fermentation inhibitors during dilute acid hydrolysis of softwood. *Enzym. Microb. Technol.* **1999**, *24*, 151–159. [CrossRef]
43. Maiorella, B.; Blanch, H.W.; Wilke, C.R. By-product inhibition effects on ethanolic fermentation by *Saccharomyces cerevisiae*. *Biotechnol. Bioeng.* **1983**, *25*, 103–121. [CrossRef]
44. Szengyel, Z.; Zacchi, G. Effect of acetic acid and furfural on cellulase production of *Trichoderma reesei* RUT C30. *Appl. Biochem. Biotechnol.* **2000**, *89*, 31–42. [CrossRef]
45. Modig, T.; Liden, G.; Taherzadeh, M.J. Inhibition effects of furfural on alcohol dehydrogenase, aldehyde dehydrogenase and pyruvate dehydrogenase. *Biochem. J.* **2002**, *363*, 769–776. [CrossRef]

11

Cytosolic Redox Status of Wine Yeast (*Saccharomyces Cerevisiae*) under Hyperosmotic Stress during Icewine Fermentation

Fei Yang [1], Caitlin Heit [1,2] and Debra L. Inglis [1,2,3,*]

1 Cool Climate Oenology and Viticulture Institute, Brock University, St. Catharines, ON L2S 3A1, Canada; fyang2@brocku.ca (F.Y.); heit.caitlin@gmail.com (C.H.)
2 Centre for Biotechnology, Brock University, St. Catharines, ON L2S 3A1, Canada
3 Department of Biological Sciences, Brock University, St. Catharines, ON L2S 3A1, Canada
* Correspondence: dinglis@brocku.ca

Abstract: Acetic acid is undesired in Icewine. It is unclear whether its production by fermenting yeast is linked to the nicotinamide adenine dinucleotide (NAD^+/NADH) system or the nicotinamide adenine dinucleotide phosphate ($NADP^+$/NADPH) system. To answer this question, the redox status of yeast cytosolic NAD(H) and NADP(H) were analyzed along with yeast metabolites to determine how redox status differs under Icewine versus table wine fermentation. Icewine juice and dilute Icewine juice were inoculated with commercial wine yeast *Saccharomyces cerevisiae* K1-V1116. Acetic acid was 14.3-fold higher in Icewine fermentation than the dilute juice condition. The ratio of NAD^+ to total NAD(H) was 24-fold higher in cells in Icewine fermentation than the ratio from the dilute juice condition. Conversely, the ratio of $NADP^+$ to total NADP(H) from the dilute fermentation was 2.9-fold higher than that in the Icewine condition. These results support the hypothesis that in Icewine, increased NAD^+ triggered the catalysis of NAD^+-dependent aldehyde dehydrogenase(s) (Aldp(s)), which led to the elevated level of acetic acid in Icewine, whereas, in the dilute condition, $NADP^+$ triggered $NADP^+$-dependent Aldp(s), resulting in a lower level of acetic acid. This work, for the first time, analyzed the yeast cytosolic redox status and its correlation to acetic acid production, providing a more comprehensive understanding of the mechanism of acetic acid production in Icewine.

Keywords: Icewine; *Saccharomyces cerevisiae*; hyperosmotic stress; redox status; NAD(H); NADP(H); acetic acid; aldehyde dehydrogenase

---

1. Introduction

Icewine is a sweet dessert wine produced from grapes naturally frozen on the vine. The low temperature during harvesting and pressing (below −8 °C in Canada), traps the water crystals in grape berries, resulting in Icewine juice that is extremely concentrated in soluble solids including sugar, acids, and nitrogen compounds [1]. In Canada, the minimum concentration of soluble solids in Icewine juice for fermentation must reach 35 °Brix [2], but juice between 38 and 42 °Brix is commonly used for Icewine production [3]. Such a high concentration of sugar in Icewine juice places yeast cells under extreme hyperosmotic stress, resulting in prolonged fermentation times and a high concentration of volatile acidity (mainly in the form of acetic acid) [4].

Acetic acid present at high levels is a spoilage compound in wine. In Canada, the maximum allowed acetic acid concentration in Icewine is 2.1 g L^{-1}, whereas in table wines it is 1.3 g L^{-1} [2]. The average acetic acid concentration in commercial Canadian Icewines was found to be 1.30 g L^{-1}, but ranged from 0.49 to 2.29 g L^{-1} [5]. Acetic acid also contributes to the formation of another undesired aromatic compound, ethyl acetate, which is described as having a solvent-like aroma such as that found in nail polish remover [6,7]. Nurgel et al. reported that the ethyl acetate concentration in commercial Canadian Icewines ranged from 0.086 to 0.369 g L^{-1}, with an average value of 0.240 g L^{-1} [5], and some wines testing over the sensory threshold in Icewine of 0.198 g L^{-1} [6].

Acetic acid production appears linked to the osmotic stress yeast are placed under. During Icewine fermentation, the highly concentrated juice places yeast cells under extreme hyperosmotic stress. In response, *Saccharomyces cerevisiae* produces glycerol as an intracellular osmolyte to counteract the high osmolarity [8]. During glycerol formation, the oxidized cofactor NAD^+ is produced, while dihydroxyacetone phosphate is reduced to glycerol-3-phosphate, leading to a potential redox imbalance for the NAD^+/NADH cofactor system. Because of the lack of transhydrogenase in yeast to convert reducing equivalents between the NAD(H) system and the NADP(H) system, yeasts must rely on metabolite formation to maintain the redox balance [9,10]. Acetic acid has been suggested as a potential metabolite that yeast cells can produce to balance the excess NAD^+ produced during glycerol formation under hyperosmotic stress [8,11,12]. The high level of acetic acid present in Icewine may also be due to the downregulation of *ACS* genes encoding for acetyl-CoA synthetases [13]. In *S. cerevisiae*, acetic acid is an intermediate to form the central metabolite, acetyl-CoA, catalyzed by acetyl-CoA synthetases in the pyruvate dehydrogenase (PDH) pathway [14]. Heit et al. found that *ACS1* and *ACS2* were downregulated 19.0-fold and 11.2-fold, respectively, in *S. cerevisicea* K1-V1116 yeasts fermenting Icewine juice compared to diluted juice, suggesting that the lower consumption of acetic acid to form the downstream metabolites during Icewine fermentation contributes to elevated acetic acid production in Icewine [13].

The results from current literature are not in agreement concerning which Aldp isoform is responsible for acetic acid production during sugar fermentations, as different Aldps have distinct cofactor specificities and play different roles for acetic acid production in yeast metabolism [11–13,15]. *S. cerevisiae* contain five aldehyde dehydrogenases [16]. Two cytosolic isoforms are NAD^+-dependent (Ald2p and Ald3p) [16], and one is $NADP^+$-dependent (Ald6p) [17]. Ald2p oxidizes 3-aminopropanal to β-alanine, required for pantothenic acid production [18]. Ald6p has been identified as the main cytosolic aldehyde dehydrogenase responsible for acetic acid production in *S. cerevisiae* strains during the fermentation of glucose [15,19–21], although both mitochondrial $NADP^+$-dependent Ald5p [22] and NAD^+/$NADP^+$-dependent Ald4p [23] have been reported to contribute. Ald6p was linked to acetic acid generation under osmotic stress to NADPH production to compensate for the downregulation of genes in the pentose phosphate pathway [15], thus questioning the linkage of acetic acid production under osmotic stress to glycerol formation and NADH requirements. However, Noti et al. found *ALD6* to be only slightly upregulated in some strains during the first two hours of exposure to 235 g/L sugar [21]. Contrary to these reports, additional studies have linked acetic acid to *ALD3* expression and the NAD^+/NADH cofactor system during fermentation [11,12,24] with a temporal relationship between acetaldehyde increase, *ALD3* expression, and acetic acid production when fermenting juice at 401 g/L sugar [11,12]. *ALD3* was the only isoform upregulated by acetaldehyde stress during fermentation [12]. To reconcile these differences, we hypothesize that the main Aldp responsible for acetic acid production throughout fermentation relates back to the starting sugar concentration in the juice, and the ongoing redox requirements of the cell throughout fermentation.

In *S. cerevisiae*, the NAD(H) and NADP(H) cofactor systems play important roles as electron donors/acceptors in more than 300 biochemical oxidation and reduction reactions [25]. NAD(H) is mainly involved in catabolic reactions such as glycolysis, glycerol production, and nitrogen uptake and consumption [26,27]. NADP(H) is predominantly involved in anabolic processes including the pentose phosphate pathway, which provides substrates for the synthesis of nucleic acids (RNA and DNA) and some amino acids [28], and has been indicated to be growth rate-dependent [29,30].

This study monitored daily changes in the two cofactor systems, NAD(H) and NADP(H), during the first week of fermentation when acetic acid is rapidly produced by *S. cerevisiae* K1-V1116. The fermentations were conducted in both Icewine juice and dilute Icewine juice, which mimics a table wine fermentation. The different ratios of oxidized cofactor to total cofactor for these two systems provides further evidence of which aldehyde dehydrogenases are involved in acetic acid production during Icewine versus dilute juice fermentation.

2. Materials and Methods

2.1. Yeast Strain and Juices

The commercial wine yeast *S. cerevisiae* K1-V1116 (Lalvin, Montreal, QB, Canada) was used for fermentations. Vidal Icewine juice was purchased from Huebel Grapes Estates Ltd. (Niagara-on-the-Lake, ON, Canada) and stored at −35 °C until required for use. The juice was thawed at 7 °C for 24 h, and racked off, followed by filtration through coarse, medium, and fine pore size pad filters using the Bueno Vino Mini Jet filter (Vineco, St. Catharines, ON, Canada). The juice was then filtered through a 0.45-μm membrane cartridge filter (Millipore, Etobicoke, ON, Canada). Sterile filtered Icewine juice (39.4 °Brix, 433 ± 5 g L^{-1} reducing sugars) was used for Icewine fermentation, and diluted to 20.0 °Brix (201 ± 1 g L^{-1} reducing sugars) for dilute juice fermentations, which mimics table wine fermentation.

2.2. Fermentation Setup and Sampling

Dried wine yeast K1-V1116 was prepared using the stepwise acclimatization procedure described in Pigeau and Inglis [11]. A volume of 30 mL of the acclimatized starter culture was used to inoculate 2 L of Icewine juice (39.4 °Brix) and dilute Icewine juice (20.0 °Brix), reaching a final yeast inoculation rate of 0.5 g dry weight L^{-1}. Fermentations were incubated at 17 °C in triplicate and continued until sugar consumption stopped for three consecutive days for Icewine fermentations, or until the sugar level was below 3 g L^{-1} for dilute juice fermentations. After stirring the fermentations for 5 min to ensure homogeneity, 100 mL of sample was removed daily and analyzed for cofactors in the first week of fermentation. Samples (5 mL) were also taken for metabolite analysis every day during the first week, every other day in the second week, and every third day in the following weeks until the fermentations stopped. Samples were centrifuged and the supernatants were stored at −30 °C until metabolite analysis was performed.

2.3. Analysis for Metabolites

Soluble solids were determined by ABBE bench top refractometer (model 10450; American Optical, Buffalo, NY, USA). Acidity was determined by pH measurement using a sympHony pH meter (model B10P; VWR, Mississauga, ON, Canada), and titratable acidity by titration against 0.1 mol L^{-1} NaOH to an endpoint of pH 8.2 [31]. Reducing sugars, ammonia nitrogen, amino nitrogen, glycerol, acetaldehyde, and acetic acid were determined with Megazyme assay kits (K-FRUGL, K-AMIAR, K-PANOPA, K-GCROL, K-ACHD, K-ACET; Megazyme International Ireland, Ltd., Bray, Co. Wicklow, Ireland). Ethanol was determined by gas chromatography (model 6890; Agilent Technologies Inc., Palo Alto, CA, USA) with a flame ionization detector and a capillary column (Agilent 122-7032 DB-Wax, 30 m × 250 μm diameter, 0.25 μm film thickness), using 0.1% 1-butanol as an internal standard. All metabolite measurements were performed in duplicate on each fermentation replicate. Metabolite production during the course of fermentation was calculated by the difference in the respective metabolite concentration measured between the time zero point (immediately after inoculation) and at each sampling time point during the course of fermentations. Normalized metabolite production was determined by dividing the final metabolite production by the final sugar consumed. Metabolite daily production rate was calculated as: daily production rate = (metabolite concentration on the sampling day − metabolite concentration on the previous day) / (sugar concentration on the previous day − sugar concentration on the sampling day).

2.4. Analysis for Enzyme Cofactors

To collect yeast cells, 100 mL of sample was centrifuged at 5108× *g* for 10 min at 4 °C; the pellet was washed once with 2 mL sterile deionized water and re-centrifuged. The pellet was re-suspended in 2 mL g^{-1} (yeast wet weight) of Tris-dithiothreitol (DTT) buffer (100 mM Tris-H_2SO_4, pH 9.4; 10 mM DTT; sterile filtered) and incubated for 30 min at 30 °C with gentle shaking at 0.059× *g*. The treated cells were then collected by centrifugation at 1464× *g* for 5 min at 4 °C. The pellet was re-suspended in 2 mL spheroplast buffer (1.2 M sorbitol; 20 mM potassium phosphate, pH 7.4; sterile filtered) and centrifuged as in the previous step. Cells were then added to a mixture containing 10 mg of Zymolyase-100T g^{-1} of yeast wet weight and 4 mL spheroplast buffer. The mixture was incubated for 60 min at 35 °C with shaking at 0.059× *g*. Following Zymolyase digestion, the spheroplasts were collected by centrifugation at 1464× *g* for 5 min at 4 °C, washed twice in spheroplast buffer, and pelleted as before. The spheroplasts were then re-suspended in ice-cold cell lysis buffer (0.6 M mannitol; 20 mM 4-(2-hydroxyethyl)piperazine-1-ethanesulfonic acid (HEPES)-KOH, pH 7.4; 1 mM phenylmethylsulfonyl fluoride (PMSF); 0.1% *w/v* fatty acid free-bovine serum albumin (BSA); sterile filtered) and disrupted using a pre-chilled Dounce homogenizer on ice. The volume of cell lysis buffer applied was 3.5 mL for yeast samples from diluted fermentation, and 2.0 mL for samples from Icewine fermentation. The lysate was then quickly transferred to pre-chilled centrifuge tubes and fractionated at 9682× *g* for 10 min at 4 °C to separate the cytosolic fraction. A small portion (200 µL) of the supernatant was immediately mixed with 200 µL of Component E from the Amplite Fluorimetric NAD/NADH or NADP/NADPH Ratio Assay Kits (15263, 15264; AAT Bioquest Inc., Sunnyvale, CA, USA) and incubated for 10 min to extract NAD^+ or $NADP^+$. This sample was also diluted serially by a mixture of cell lysis buffer and Component E (at a 1:1 ratio) to ensure that the sample concentration would be within the linear range of the calibration curves of both kits. The rest of the supernatant was kept on ice for the measurement of the sum of NAD^+ and NADH or the sum of $NADP^+$ and NADPH, and serial dilution was also performed for this sample using cell lysis buffer. All samples were then tested by the abovementioned fluorimetric kits in duplicate on each fermentation replicate according to the manuals. The percentage of oxidized cofactor of either the NAD(H) or NADP(H) system was calculated as: percentage of oxidized cofactor = concentration of oxidized cofactor / (concentration of oxidized cofactor + concentration of reduced cofactor) × 100%.

2.5. Statistical Analysis

XLSTAT-Pro by Addinsoft (New York, NY, USA) was used for statistical analysis. Analysis of variance (ANOVA) with Fisher's Least Significant Difference (LSD) test ($p < 0.05$) was used to evaluate differences between variables and mean separation, respectively.

3. Results

3.1. Fermentation Kinetics

The composition of Icewine juice in comparison to the dilute juice indicates that the starting sugar concentration is over 400 g L^{-1} in the Icewine juice and concentrated in all other juice components including acidity, yeast assimilable nitrogen, glycerol, acetaldehyde, and acetic acid (Table 1). Yeast cells exposed to a greater level of hyperosmotic stress from fermentable sugars in the Icewine fermentation consumed sugar very slowly in the first week and stopped before half of the original sugar concentration was consumed on Day 39 (Figure 1). Conversely, in the diluted condition, yeast cells rapidly consumed sugar right after inoculation with only 1 g L^{-1} of sugar left in the final wine on Day 21 (Figure 1 and Table 2).

Table 1. Vidal Icewine juice and diluted juice initial parameters (mean ± SD).

Parameter	Icewine Juice	Diluted Juice
Glucose + fructose (g L^{-1})	433 ± 5	201 ± 1
Titratable acidity (g L^{-1})	6.4 ± 0.0	3.1 ± 0.0
pH	3.76 ± 0.01	3.88 ± 0.01
Free amino nitrogen (mg N L^{-1})	238.4 ± 2.4	114.8 ± 3.0
Ammonia nitrogen (mg N L^{-1})	14.8 ± 0.1	6.6 ± 0.8
Glycerol (g L^{-1})	8.76 ± 0.06	4.04 ± 0
Acetaldehyde (mg L^{-1})	5.3 ± 0.8	2.7 ± 0.2
Acetic acid (g L^{-1})	0.04 ± 0	0.02 ± 0
Ethanol (% *v/v*)	1.3 ± 0	0.6 ± 0

Table 2. Final wine parameters in Icewine and diluted juice fermentation (mean ± SD).

Parameter	Icewine Juice Fermentation	Diluted Juice Fermentation
Glucose + fructose (g L^{-1})	252 ± 3	1 ± 1
Titratable acidity (g L^{-1})	8.3 ± 0.2	5.5 ± 0.1
pH	3.92 ± 0.01	3.63 ± 0.02
Free amino nitrogen (mg N L^{-1})	188.5 ± 1.4	7.4 ± 0.4
Ammonia nitrogen (mg N L^{-1})	3.5 ± 0.1	† ND
Glycerol (g L^{-1})	17.44 ± 0.19	9.43 ± 0.06
Acetaldehyde (mg L^{-1})	48.5 ± 0.8	23.0 ± 0.8
Acetic acid (g L^{-1})	1.48 ± 0.10	0.12 ± 0.01
Ethanol (% *v/v*)	12.0 ± 0.2	13.9 ± 0.1

† ND indicates that no concentration was detected.

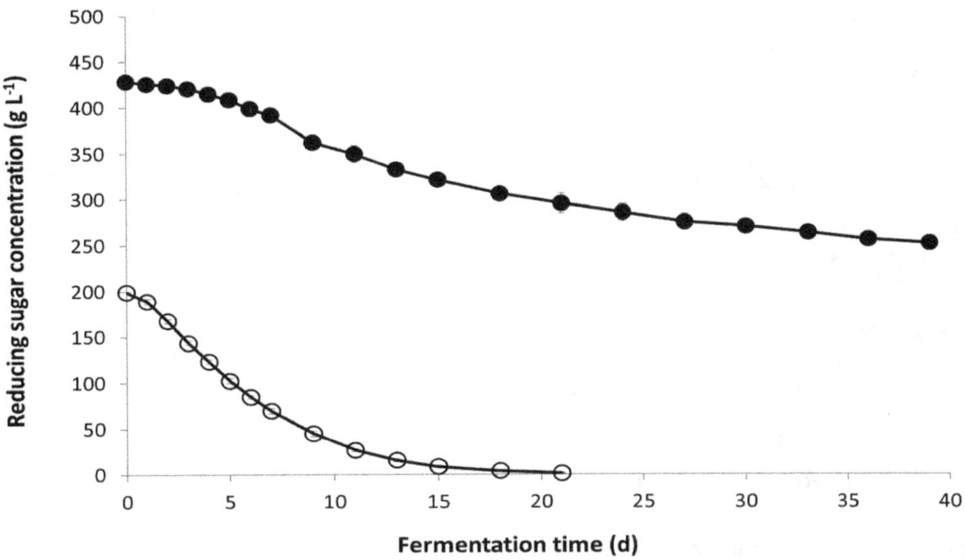

Figure 1. Sugar concentration during fermentation. Fermentations of Icewine juice (●) and diluted juice (○) were inoculated with K1-V1116 and monitored for sugar concentration throughout fermentation. Fermentations were performed in triplicate and samples from each trial were tested in duplicate. Sugar values represent the average ± standard deviation.

3.2. Yeast Metabolite Production

Metabolite production was monitored as a function of time (Figure 2a,c,e), but because yeast consumes sugar at different rates, these metabolites were also plotted as a function of sugar consumed (Figure 2b,d,f) for direct comparison between the two conditions. Yeast produced higher levels of glycerol and acetic acid during Icewine fermentation as a function of sugar consumed, showing a 1.9-fold increase in glycerol and 16.5-fold increase in acetic acid when normalized to sugar consumed (Table 3). The peak concentration of acetaldehyde production was 4.7-fold higher in the Icewine fermentation compared to the dilute fermentation, and then declined when acetaldehyde was consumed for the formation of either ethanol or acetic acid (Figure 2c,d).

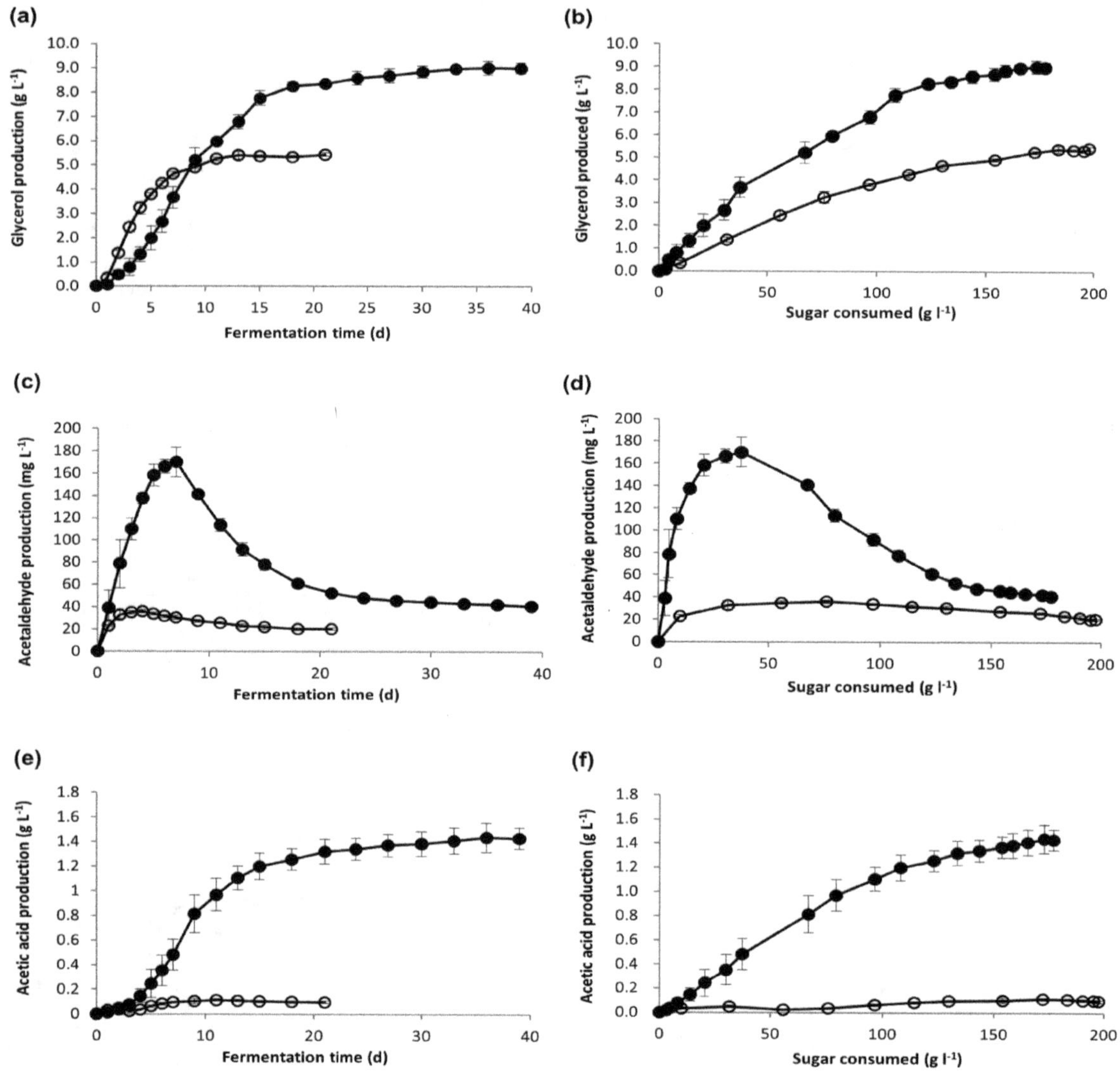

Figure 2. Metabolite production during fermentation. Fermentations of Icewine juice (●) and diluted juice (○) were analyzed for the production of (**a**) glycerol, (**c**) acetaldehyde, and (**e**) acetic acid, and the generation of metabolites were also plotted out as a function of sugar consumed for (**b**) glycerol, (**d**) acetaldehyde, and (**f**) acetic acid. Fermentations were performed in triplicate and samples from each trial were tested in duplicate. Metabolite values represent the average ± standard deviation.

Table 3. Normalized metabolite production by yeast measured at the end of fermentation (mean ± SD).

Fermentation	Glycerol (mg g^{-1} of Sugar Consumed) *	Acetaldehyde (mg g^{-1} of Sugar Consumed) *	Ethanol (g g^{-1} of Sugar Consumed) *	Acetic Acid (mg g^{-1} of Sugar Consumed) *
Diluted juice	27.44 ± 0.30	0.23 ± 0.01	0.53 ± 0	0.49 ± 0.06
Icewine Juice	50.92 ± 0.74	0.10 ± 0	0.47 ± 0	8.10 ± 0.50

* Significantly different as determined by Fisher's Least Significant Difference test ($p < 0.05$).

3.3. *Ratio of Coenzyme Concentration*

The cytosolic coenzymes from both the NAD(H) and NADP(H) systems were analyzed in the first week of fermentation, when yeast cells adapted to the hyperosmotic stress under the two different sugar concentrations. The percentage of the cofactors in their oxidized form were plotted against time (Figure 3). The percentages were then linked to the daily production rate of glycerol and acetic acid (Figure 4) to understand the correlation between redox status and metabolite production under hyperosmotic stress.

During the dilute fermentation, the percent of cytosolic $NADP^+$, which ranged from 30.2–57.2% over the course of fermentation, was consistently higher than the percent of cytosolic NAD^+, which was only 3.6% from Day 1 onward (Figure 3a). The percent of the cytosolic $NADP^+$ was consistently lower in the Icewine condition (between 19.9–31.6%, Figure 3b) compared to the dilute juice condition (between 30.2–57.2%, Figure 3a), showing a 2.9-fold higher level in the dilute juice condition on Day 1. It appears that the cytosolic $NADP^+$-dependent Ald6p has a larger contribution to the low acetic acid production in the dilute juice condition over NAD^+-dependent forms of Aldp (Figure 4b).

However, a very different redox profile resulted during the Icewine fermentation (Figure 3b). During Icewine fermentation, the percent of cytosolic NAD^+ increased over the first two days of fermentation, peaking on Day 2 at 60.1% and remaining much higher over the remainder of the fermentation (Figure 3b) in comparison to the dilute juice fermentation (Figure 3a), showing a 24-fold increase in the Icewine condition. This spike in NAD+ corresponded to the increased production rate of glycerol in the Icewine condition (Figure 4a) and the increased production rate of acetic acid (Figure 4b). The redox profile supports the hypothesis of cytosolic NAD^+-dependent Aldp(s) contributing to the elevated acetic acid in Icewine fermentation along with the cytosolic $NADP^+$-dependent Ald6p.

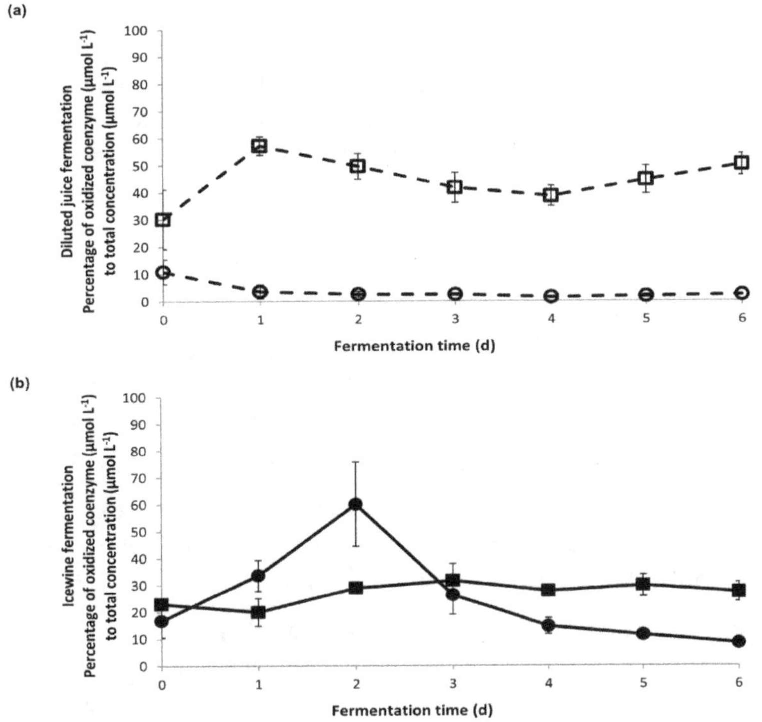

Figure 3. Percentage of NAD^+ and $NADP^+$ in the first week of fermentation. (**a**) The percentage of NAD^+ to total NAD(H) (○) was compared to the percentage of $NADP^+$ to total NADP(H) (□) in diluted fermentation; (**b**) the percentage of NAD^+ to total NAD(H) (●) was compared to the percentage of $NADP^+$ to total NADP(H) (■) in Icewine fermentation. Fermentations were performed in triplicate and samples from each trial were tested in duplicate. Data points represent the average ± standard deviation.

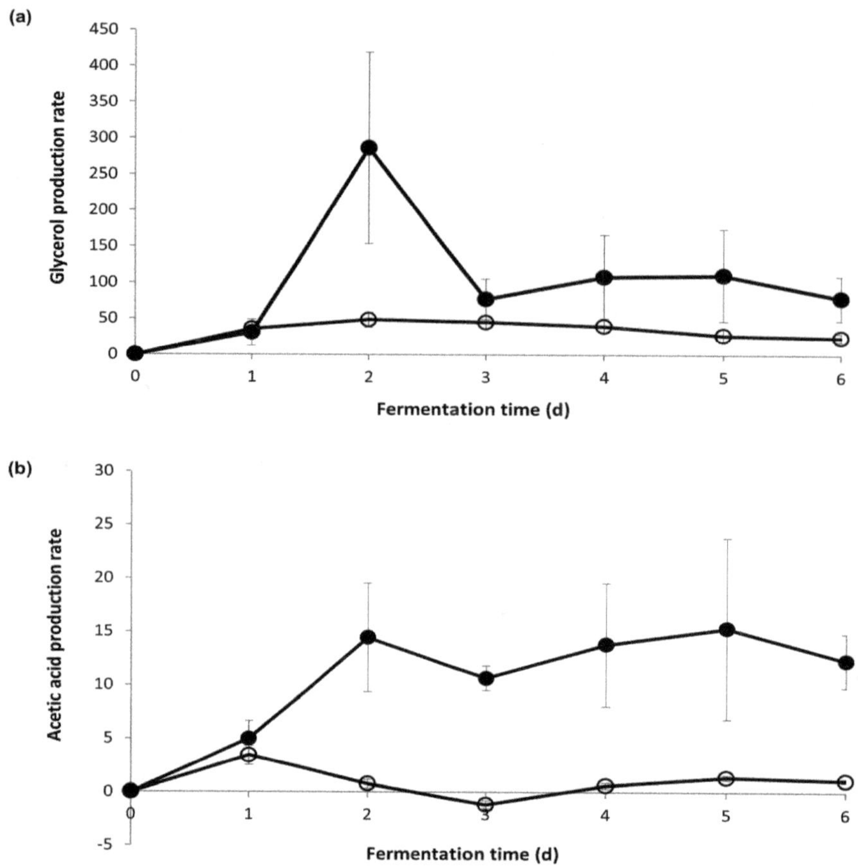

Figure 4. Metabolite production rate during fermentation. The daily metabolite production rate of (**a**) glycerol and (**b**) acetic acid was calculated during the first week of Icewine juice fermentation (●) and diluted juice fermentation (○). Fermentations were performed in triplicate and samples from each trial were tested in duplicate. Metabolite values represent the average ± standard deviation.

4. Discussion

Increased acetic acid production during high sugar fermentation in *S. cerevisiae* may be due to metabolic regulation driven by the redox status of the cell, substrate availability, or product usage, or due to the transcriptional regulation of genes altering enzyme concentrations involved in the synthesis or usage of acetic acid, or a combination of all these factors.

Past studies have linked acetic acid production during fermentation to either the $NADP^+$ cofactor system [15] or the NAD^+ cofactor system [11–13] based on *ALD* gene expression patterns encoding aldehyde dehydrogenases that have different cofactor requirements. We report here, for the first time, the cytosolic redox status of both cofactor systems during fermentation at two different sugar levels.

We have previously reported increased substrate availability for the aldehyde dehydrogenases during high sugar fermentations (acetaldehyde levels quickly rise during Icewine fermentation) [11–13], which was also found in this present study, along with increased expression of *ALD3* encoding a cytosolic NAD^+-dependent isoform over that found during dilute juice fermentation [11–13]. *ALD3* was also the only *ALD* expression stimulated by acetaldehyde stress during fermentation [12]. We now report that the percent of oxidized cofactor NAD^+ on Day 2 of fermentation was 24-fold higher in the Icewine condition versus the dilute juice condition and 2.1-fold higher than the percent of oxidized cofactor $NADP^+$ on Day 2 of Icewine fermentation. Taken together, our results further confirm a role for NAD^+-dependent Aldp isoforms along with $NADP^+$-dependent Aldp isoforms in producing elevated levels of acetic acid during the high sugar fermentations of Icewine production.

In evaluating the cytosolic redox status of fermenting yeast at two different sugar levels (201 versus 433 g L^{-1} of sugar in our present study), the discrepancy in the literature in reporting the Aldp responsible for acetic acid production during fermentation may have been due to the starting sugar concentration in the media used for fermentation. Our data supports other published work where the $NADP^+$-dependent isoform Ald6p was identified as the main aldehyde dehydrogenase responsible for acetic acid production when sugar concentrations were at or below 200 g L^{-1} sugar [15,19–21]. However, under high sugar conditions of 400 g L^{-1}, Erasmus et al. linked acetic acid generation to NADPH production via Ald6p to compensate for the downregulation of genes in the pentose phosphate pathway [15], questioning the linkage of acetic acid production under osmotic stress to glycerol formation and NADH requirements. Based on the cytosolic redox status of yeast fermenting Icewine juice at 433 g L^{-1} sugar, as well as our previous *ALD* expression analysis during Icewine fermentation [11–13], it appears that both Ald6p and Ald3p contribute to acetic acid during Icewine fermentation. Our data supports a linkage of acetic acid production under the high osmotic stress of Icewine fermentation to NADH requirements and glycerol production.

We have also previously reported that elevated acetic acid levels appear due to lack of conversion to acetyl-CoA from the down-regulation in gene expression of cytosolic and mitochondrial acetyl-CoA synthetase [13]. This was further substantiated by 2-fold lower ethyl acetate levels in the Icewines (60 mg/L) compared to that found in the dilute juice fermentations (120 mg/L), whereby acetyl-CoA is required for the esterification to ethanol by alcohol acetyltransferases [13].

We have provided further evidence to support the concept that cytosolic NAD^+-dependent aldehyde dehydrogenase(s) are responsible, in part, for increased levels of acetic acid during Icewine fermentation stimulated by substrate availability through an increase in NAD^+ and acetaldehyde, whereas the $NADP^+$-dependent isoform plays the dominate role in acetic acid production in table wine fermentation stimulated by an increase in $NADP^+$. Although substrate availability [11–13], redox status, and gene expression [11–13] have been evaluated, enzyme activities for the aldehyde dehydrogenases and downstream acetyl CoA synthetases represent missing information that could further confirm the enzymes and cofactor system involved in the high acetic acid accumulation during high sugar fermentations such as that of Icewine.

Acknowledgments: This research was funded by grant #238872-2012 from the Natural Sciences and Engineering Research Council of Canada to Debra L. Inglis.

Author Contributions: Debra L. Inglis, Caitlin Heit, and Fei Yang conceived and designed the experiments; Fei Yang and Caitlin Heit performed the experiments, and Fei Yang along with Debra L. Inglis analyzed the data and wrote the paper.

References

1. Bowen, A. Managing the quality of icewines. In *Managing Wine Quality: Oenology and Wine Quality*, 1st ed.; Reynolds, A.G., Ed.; Woodhead Publishing Ltd.: Cambridge, UK, 2010; pp. 523–552.
2. O. Reg. 406/00: Rules of Vintners Quality Alliance Ontario Relating to Terms for VQA Wine. Available online: https://www.ontario.ca/laws/regulation/000406 (accessed on 14 November 2017).
3. Ziraldo, D.; Kaiser, K. Science. In *Icewine: Extreme Winemaking*, 1st ed.; Key Porter Books: Toronto, ON, Canada, 2007; pp. 73–107.
4. Kontkanen, D.; Inglis, D.; Pickering, G.; Reynolds, A. Effect of yeast inoculation rate, acclimatization, and nutrient addition on Icewine fermentation. *Am. J. Enol. Vitic.* **2004**, *55*, 363–370.
5. Nurgel, C.; Pickering, G.J.; Inglis, D.L. Sensory and chemical characteristics of Canadian ice wines. *J. Sci. Food Agric.* **2004**, *84*, 1675–1684. [CrossRef]
6. Cliff, M.A.; Pickering, G.J. Determination of odour detection thresholds for acetic acid and ethyl acetate in ice wine. *J. Wine Res.* **2006**, *17*, 45–52. [CrossRef]
7. Lilly, M.; Lambrechts, M.G.; Pretorius, I.S. Effect of increased acetyltransferase activity on flavour profiles of wine and distillates. *Appl. Environ. Microbiol.* **2000**, *66*, 744–753. [CrossRef] [PubMed]

8. Blomberg, A.; Adler, L. Roles of glycerol and glycerol-3-phosphate dehydrogenase (NAD+) in acquired osmotolerance of *Saccharomyces cerevisiae*. *J. Bacteriol.* **1989**, *171*, 1087–1092. [CrossRef] [PubMed]
9. Lagunas, R.; Ganedo, J.M. Reduced pyridine nucleotide balance growing on *Saccharomyces cerevisiae*. *Eur. J. Biochem.* **1973**, *37*, 90–94. [CrossRef] [PubMed]
10. Van Dijken, J.; Scheffers, W. Redox balances in the metabolism of sugars by yeasts. *FEMS Microbiol. Rev.* **1986**, *32*, 199–224. [CrossRef]
11. Pigeau, G.M.; Inglis, D.L. Upregulation of ALD3 and GPD1 in *Saccharomyces cerevisiae* during Icewine fermentation. *J. Appl. Microbiol.* **2005**, *99*, 112–125. [CrossRef] [PubMed]
12. Pigeau, G.M.; Inglis, D.L. Response of wine yeast (*Saccharomyces cerevisiae*) aldehyde dehydrogenases to acetaldehyde stress during Icewine fermentation. *J. Appl. Microbiol.* **2007**, *103*, 1576–1586. [CrossRef] [PubMed]
13. Heit, C.; Martin, S.J.; Yang, F.; Inglis, D.L. Osmoadaptation of wine yeast (*Saccharomyces cerevisiae*) to hyperosmotic stress during Icewine fermentation. *J. Appl. Microbiol.* **2017**. under review.
14. Pronk, J.T.; Steensma, H.Y.; van Dijken, J.P. Pyruvate metabolism in *Saccharomyces cerevisiae*. *Yeast* **1996**, *12*, 1607–1633. [CrossRef]
15. Erasmus, D.J.; van Vuuren, H.J.J. Genetic basis for osmosensitivity and genetic instability of the wine yeast *Saccharomyces cerevisiae* V1N7. *Am. J. Enol. Vitic.* **2009**, *60*, 145–154.
16. Navarro-Avino, J.P.; Prasad, R.; Miralles, V.J.; Benito, R.M.; Serreno, R. A proposal of nomenclature of aldehyde dehydrogenases in *Saccharomyces cerevisiae* and characterization of the stress-inducible ALD2 and ALD3 genes. *Yeast* **1999**, *15*, 829–842. [CrossRef]
17. Meaden, P.G.; Dickinson, F.M.; Mifsud, A.; Tessier, W.; Westwater, J.; Bussey, H.; Midgley, M. The *ALD6* gene of *Saccharomyces cerevisiae* encodes a cytosolic, Mg^{2+}-activated acetaldehyde dehydrogenase. *Yeast* **1997**, *13*, 1319–1327. [CrossRef]
18. White, H.W.; Skatrud, P.L.; Xue, Z.; Toyn, J.H. Specialization of function among aldehyde dehydrogenases: The *ALD2* and *ALD3* genes are required for B-alanine biosynthesis in *S. cerevisiae*. *Genetics* **2003**, *163*, 69–77. [PubMed]
19. Eglinton, J.M.; Heinrich, A.J.; Pollnitz, A.P.; Langridge, P.; Henschke, P.A.; de Barros Lopes, M. Decreasing acetic acid accumulation by a glycerol overproducing strain of *Saccharomyces cerevisiae* by deleting the *ALD6* aldehyde dehydrogenase gene. *Yeast* **2002**, *19*, 295–301. [CrossRef] [PubMed]
20. Luo, Z.; Walkey, C.; Madilao, L.; Measday, V.; Van Vuuren, H. Functional improvement of *Saccharomyces cerevisiae* to reduce volatile acidity in wine. *FEMS Yeast Res.* **2013**, *13*, 485–494. [CrossRef] [PubMed]
21. Noti, O.; Vaudano, E.; Pessione, E.; Garcia-Moruno, E. Short-term response of different *Saccharomyces cerevisiae* strains to hyperosmotic stress caused by inoculation in grape must: RT-qPCR study and metabolite analysis. *Food Microbiol.* **2015**, *52*, 49–58. [CrossRef] [PubMed]
22. Saint-Prix, F.; Bönquist, L.; Dequin, S. Functional analysis of the *ALD* gene family of *Saccharomyces cerevisiae* during anaerobic growth on glucose: The $NADP^+$-dependent Ald6p and Ald5p isoforms play a major role in acetate formation. *Microbiology* **2004**, *150*, 2209–2220. [CrossRef] [PubMed]
23. Remize, F.; Andrieu, E.; Dequin, S. Engineering of the pyruvate dehydrogenase bypass in *Saccharomyces cerevisiae*: Role of the cytosolic Mg^{2+} and mitochondrial K^+ acetaldehyde dehydrogenases Ald6p and Ald4p in acetate formation during alcoholic fermentation. *Appl. Environ. Microbiol.* **2000**, *66*, 3151–3159. [CrossRef] [PubMed]
24. Akamatsu, S.; Kamiya, H.; Yamashita, N.; Motoyoshi, T.; Goto-Yamamoto, N.; Ishikawa, T.; Okazaki, N.; Nishimura, A. Effects of aldehyde dehydrogenase and acetyl-CoA synthetase on acetate formation in sake mash. *J. Biosci. Bioeng.* **2000**, *90*, 555–560. [CrossRef]
25. Forster, J.; Famili, I.; Fu, P.; Palsson, B.O.; Nielsen, J. Genome-scale reconstruction of *Saccharomyces cerevisiae* metabolic network. *Genome Res.* **2003**, *13*, 244–253. [CrossRef] [PubMed]
26. Bruinenberg, P.M. The NADP(H) redox couple in yeast metabolism. *Antonie Van Leeuwenhoek* **1986**, *52*, 411–429. [CrossRef] [PubMed]
27. Quirós, M.; Martínez-Moreno, R.; Albiol, J.; Morales, P.; Vázquez-Lima, F.; Barreiro-Vázquez, A.; Ferrer, P.; Gonzalez, R. Metabolic flux analysis during the exponential growth phase of *Saccharomyces cerevisiae* in wine fermentations. *PLoS ONE* **2013**, *8*, e71909. [CrossRef]
28. Villadsen, J.; Nielsen, J.; Lidén, G. Chemicals from metabolic pathways. In *Bioreaction Engineering Principles*, 3rd ed.; Springer: New York, NY, USA, 2011; pp. 7–62.

29. Nissen, T.L.; Schulze, U.; Nielsen, J.; Villadsen, J. Flux distributions in anaerobic, glucose-limited continuous cultures of *Saccharomyces cerevisiae*. *Microbiology* **1997**, *143*, 203–218. [CrossRef] [PubMed]
30. Varela, C.; Pizarro, F.; Agosin, E. Biomass content governs fermentation rate in nitrogen-deficient wine musts. *Appl. Environ. Microbiol.* **2004**, *70*, 3392–3400. [CrossRef] [PubMed]
31. Zoecklein, B.W.; Fugelsang, K.C.; Gump, B.H. Laboratory Procedures. In *Wine Analysis and Production*, 1st ed.; Chapman and Hall, International Thomson Publishing: New York, NY, USA, 1995; pp. 374–378.

12

Recombinant Diploid *Saccharomyces cerevisiae* Strain Development for Rapid Glucose and Xylose Co-Fermentation

Tingting Liu, Shuangcheng Huang and Anli Geng *

School of Life Sciences and Chemical Technology, Ngee Ann Polytechnic, Singapore 599489, Singapore; lttf7@sina.com (T.L.); higle945@yahoo.com (S.H.)

* Correspondence: gan2@np.edu.sg

Abstract: Cost-effective production of cellulosic ethanol requires robust microorganisms for rapid co-fermentation of glucose and xylose. This study aims to develop a recombinant diploid xylose-fermenting *Saccharomyces cerevisiae* strain for efficient conversion of lignocellulosic biomass sugars to ethanol. Episomal plasmids harboring codon-optimized *Piromyces* sp. E2 xylose isomerase (*PirXylA*) and *Orpinomyces* sp. ukk1 xylose (*OrpXylA*) genes were constructed and transformed into *S. cerevisiae*. The strain harboring plasmids with tandem *PirXylA* was favorable for xylose utilization when xylose was used as the sole carbon source, while the strain harboring plasmids with tandem *OrpXylA* was beneficial for glucose and xylose cofermentation. *PirXylA* and *OrpXylA* genes were also individually integrated into the genome of yeast strains in multiple copies. Such integration was beneficial for xylose alcoholic fermentation. The respiration-deficient strain carrying episomal or integrated *OrpXylA* genes exhibited the best performance for glucose and xylose co-fermentation. This was partly attributed to the high expression levels and activities of xylose isomerase. Mating a respiration-efficient strain carrying the integrated *PirXylA* gene with a respiration-deficient strain harboring integrated *OrpXylA* generated a diploid recombinant xylose-fermenting yeast strain STXQ with enhanced cell growth and xylose fermentation. Co-fermentation of 162 g L^{-1} glucose and 95 g L^{-1} xylose generated 120.6 g L^{-1} ethanol in 23 h, with sugar conversion higher than 99%, ethanol yield of 0.47 g g^{-1}, and ethanol productivity of 5.26 g $L^{-1}\cdot h^{-1}$.

Keywords: *Saccharomyces cerevisiae*; diploid; xylose isomerase; xylose fermentation; glucose and xylose co-fermentation; biomass hydrolysate; cellulosic ethanol

---

1. Introduction

Ethanol can be produced from renewable resources such as crops or agricultural waste. It is therefore a sustainable and clean fuel. Further growth in bioethanol production largely depends on the effective conversion of lignocellulosic feedstock such as agricultural and forestry wastes to bioethanol because they are the most abundant polymers of fermentable sugars [1–3].

Saccharomyces cerevisiae is the most effective microorganism for fermenting sugars to ethanol due to its rapid sugar consumption rate, high sugar and ethanol tolerance, and resistance to biomass-derived inhibitors [4,5]. Much research has been done to genetically engineer *S. cerevisiae* strains for xylose fermentation [6–10].

Two xylose-assimilating pathways were heterologously engineered in *S. cerevisiae* for xylose-fermenting yeast construction. One focused on the D-xylose isomerase (XI) pathway [11,12], the other focused on the D-xylose reductase (XR) and xylitol dehydrogenase (XDH) pathway [10,13–15]. In the XI pathway, xylose is first isomerized into xylulose by XI and xylulose was then phosphorylated into xylulose 5-phosphate by xylulokinase. Xylulose is subsequently metabolized to ethanol through glycolysis in the pentose phosphate pathway [11]. As no xylitol is produced in this pathway, much research focused on the XI pathway construction in xylose-fermenting yeast [12].

However, for all the engineered strains developed, rapid glucose and xylose co-fermentation is still challenging, in particular when sugar concentration is high [4,8–10]. Because fermentation time on mixed-substrate hydrolysates is still not cost-effective, strategies in evolutionary engineering were used to improve fermentation kinetics [16–18] and much research focused on the search for new or engineered sugar transporters [19–21]. More recently, robust diploid *S. cerevisiae* strains were developed for rapid xylose-fermentation [22–25].

This study aims to develop a recombinant *S. cerevisiae* strain for rapid glucose and xylose co-fermentation through metabolic engineering, evolutionary engineering and strain mating. Firstly, four episomal plasmids containing the two-copy codon-optimized *Piromyces* sp. E2 XI gene (*PirXylA*, GenBank accession number AJ249909.1), one-copy codon-optimized *Orpinomyces* sp. ukk1 XI gene (*OrpXylA*, GenBank accession number EU411046), one-copy *PirXylA* and *OrpXylA* in tandem, and two-copy *OrpXylA*, were individually constructed. They were subsequently transformed to an evolved respiration-deficient yeast strain. Four engineered strains with episomal XI genes were generated and optimal XI functional expression was identified. Afterwards, *PirXylA* and *OrpXylA* were separately integrated into the genome of two evolved *S. cerevisiae* strains (one respiration-efficient and the other respiration-deficient) in multiple copies according the methods described previously [26,27]. Four engineered yeasts with integrated XI genes were generated and screened for xylose fermentation or glucose/xylose co-fermentation. In the end, a diploid recombinant xylose-fermenting *S. cerevisiae* was constructed by mating a respiration-efficient haploid strain with a respiration-deficient haploid strain. To the best of our knowledge, this is the first report on diploid xylose-fermenting yeast strain construction by such strain mating. The resulted diploid yeast strain displayed superior glucose and xylose co-fermentation performance, which far exceeded that by engineered *S. cerevisiae* reported to-date [28,29].

2. Materials and Methods

2.1. Plasmid Construction

All plasmids used in this work are listed in Table 1. All primers used in this study are listed in Table S1.

E. coli strain DH5α (Life Technologies, Rockville, MD, USA) were used as the transformation host for plasmid construction. *E. coli* were grown in LB medium containing 100 μg/mL ampicillin at 30 °C or 37 °C for plasmid maintenance [25]. The *PGK1* promoter was amplified from genomic DNA of *S. cerevisiae* strain ATCC 24860 and was ligated to pYES2 (Thermo-Fischer Scientific, Singapore) to replace the *GAL1* promoter, resulting in pPY1. *PirXylA* and *OrpXylA* were synthesized by Integrated DNA Technologies Singapore). Cassettes *PGK1*p-*OrpXylA-CYC1*t, *PGK1*p-*PirXylA-CYC1*t-*PGK1*p-*OrpXylA-CYC1*t, *PGK1*p-*OrpXylA-CYC1*t-*PGK1*p-*OrpXylA-CYC1*t and *PGK1*p-*PirXylA-CYC1*t-*PGK1*p-*PirXylA-CYC1*t were cloned into pPY1 individually, resulting in plasmids pPYXo, pPYXpXo, pPYXoXo and pPYXpXp (Figure 1A–D).

NTS2-2 partial fragment (pNTS) was obtained by overlap extension polymerase chain reaction (OE-PCR) of *S. cerevisiae* ATCC 24860 genomic DNA over 274 bp upstream and 245 bp downstream homologous regions of NTS2-2. The XXUN plasmid (Figure 1E) is an integrating yeast plasmid based on pPYXpXp whereby the 2 μ origin was replaced with pNTS. The *loxP-KanMX4-loxP* cassette was amplified from the plasmid pUG6 [30]. Cassettes *PGK1*p-*OrpXylA-CYC1*t and *loxP-KanMX4-loxP*-pNTS were obtained by OE-PCR and then subsequently constructed into plasmid pUC19 [12], resulting in plasmid XoNK (Figure 1F). Both plasmids were digested with *Swa*I for XI genome integration using NTS2-2 homologous recombinant arms.

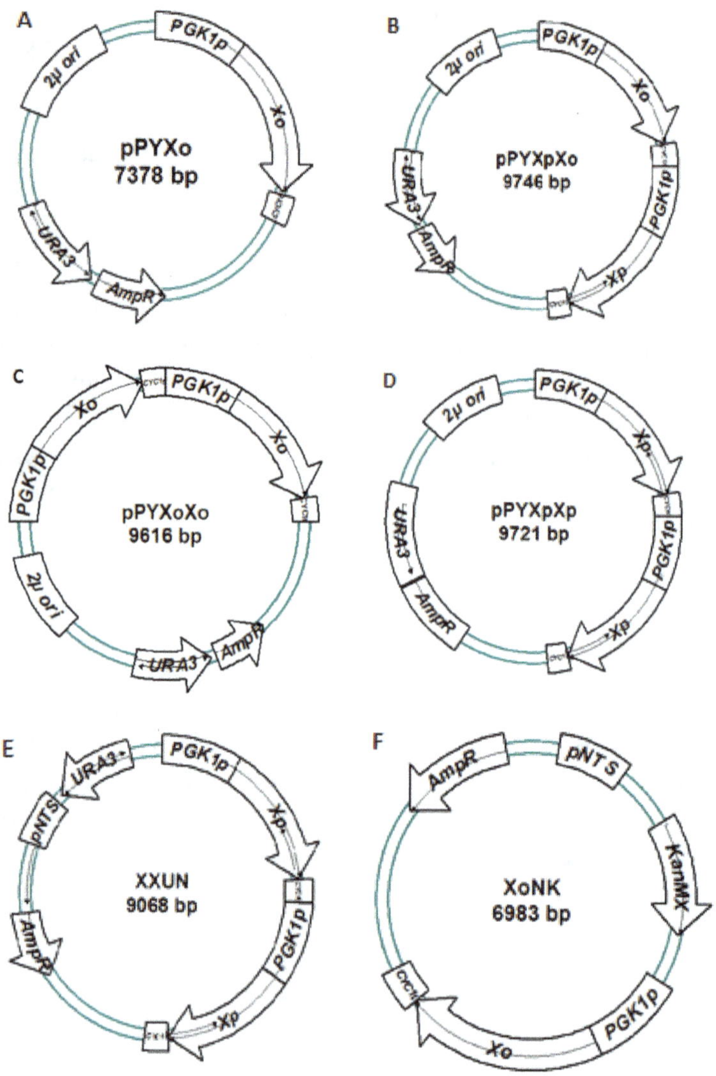

Figure 1. Map of plasmids. (**A**) pPYXo; (**B**) pPYXpXo; (**C**) pPYXoXo; (**D**) pPYXpXp; (**E**) XXUN; and (**F**) XoNK.

2.2. Strain Construction and Adaptive Evolution

All strains used in this work are listed in Table 1. *S. cerevisiae* haploid strains JUK36α and JUK39a were isolated from the diploid strain *S. cerevisiae* ATCC 24860. They were both overexpressed with the non-oxidative pentose phosphate pathway (PPP) genes and xylulokinase gene, *XKS1*. *URA3* and *GRE3* genes were disrupted in both haploid strains. In addition, the *CYC3* gene, encoding cytochrome c heme lyase, was knocked out in strain JUK39a to eliminate respiration [12].

Plasmid pPYXpXp was transformed into strain JUK39a using the LiAc/SS carrier DNA/PEG method [31], resulting in recombinant *S. cerevisiae* 39aXpXp. The respiration-deficient strain 39aXpXp was evolved by continuous transfer and cultivation in a synthetic medium (SM) containing 6.7 g L^{-1} yeast nitrogen base without amino acids (YNB) (Difco Laboratories Inc., Detroit, MI, USA) and 20 g L^{-1} xylose (SMX) under oxygen-limited conditions according to the protocols described in our earlier report [12]. After 75-day continuous transfer, cell doubling time was reduced from 15.9 h to 6.4 h. Samples were taken on day 75 and streaked on SMX plates containing 20 g L^{-1} xylose and 20 g L^{-1} agar. Fifteen large single colonies were selected. They were then incubated in 50 mL SMX medium in 100 mL Erlenmeyer flasks capped with rubber stoppers, shaken at 200 rpm and 30 °C. Weight loss of the cultures from the 15 colonies was individually measured on Day 4. The best ethanol-producing strain was indicated by the highest weight loss [18]. This strain was denoted 39aXpXp2415, and plasmid pPYXpXp in this strain was removed by streaking the culture on 5-FOA plates [12]. This generated the background strain $39a_2$ (Table 1). On the other hand, strain JUK51a_2 (Table 1) was evolved anaerobically in a chemostat with an increase in the dilution rate from 0.01 to 0.14 h^{-1} on xylose using the method described in our earlier report [12]. The fastest growing strain was selected, and the plasmid pJFX11 was removed according to the above-described method on 5-FOA plates. The background strain $36\alpha_2$ was later obtained (Table 1).

Table 1. Strains and plasmids used in this study.

Strains/Plasmid	Genotype/Phenotype	References
Saccharomyces cerevisiae strains		
Saccharomyces cerevisiae ATCC 24860	Obtained from American Type Culture Collection (ATCC)	
JUK36α	*S. cerevisiae* ATCC 24860 segregant; *MAT; ura3::loxP; TKL1::RKIt-RKI1-ADH1p-RPE1t-RPE1-TPI1p-loxP-XKS1t-XKS1-PGK1p-PDC1p-TAL1-TAL1t-FBA1p; gre3::loxP*	[12]
JUK39a	*S. cerevisiae* ATCC 24860 segregant; *MATa; ura3::loxP; TKL1::RKIt-RKI1-ADH1p-RPE1t-RPE1-TPI1p-loxP-XKS1t-XKS1-PGK1p-PDC1p-TAL1-TAL1t-FBA1p; gre3::loxP; cyc3::loxP*	[12]
JUK51a_2	JUK36α derivative; {pJFX11}/(*BvuXylA, XK, PPP, gre3Δ*)	[12]
39aXpXp	JUK39a derivative; {pPYXpXp}/(two-copy *PirXylA, XK, PPP, gre3Δ, cyc3Δ*)	This work
39aXpXp2415	39aXpXp derivative; {pPYXpXp}/(two-copy *PirXylA, XK, PPP, gre3Δ, cyc3Δ*, AE)	This work
$36\alpha_2$	Isolate from chemostat anaerobic and adaptive evolution at a dilution rate of 0.15 h^{-1} on xylose of JUK51a_2 and loss of plasmid pJFX11	This work
$39a_2$	Isolate from 39aXpXp2415 and loss of plasmid pPYXpXp	This work
$39a_2$XpXp	$39a_2$ derivative; {pPYXpXp}/(two-copy *PirXylA, XK, PPP, gre3Δ, cyc3Δ*)	This work
$39a_2$XpXo	$39a_2$ derivative; {pPYXpXo}/(*OrpXylA, PirXylA, XK, PPP, gre3Δ, cyc3Δ*)	This work
$39a_2$Xo	$39a_2$ derivative; {pPYXo}/(*OrpXylA, XK, PPP, gre3Δ, cyc3Δ*)	This work
$39a_2$XoXo	$39a_2$ derivative; {pPYXoXo}/}/(two-copy *OrpXylA, XK, PPP, gre3Δ, cyc3Δ*)	This work
$36\alpha_2$XpXpUN	$36\alpha_2$ derivative;*NTS2-2*::two-copy *PirXylA, ura3, XK, PPP, gre3Δ*	This work
$36\alpha_2$XoNK	$36\alpha_2$ derivative;*NTS2-2::OrpXylA-KanMX4, XK, PPP, gre3Δ, ura3Δ*	This work
$39a_2$XpXpUN	$39a_2$ derivative; *NTS2-2*::two-copy *PirXylA, ura3, XK, PPP, gre3Δ, cyc3Δ*	This work
$39a_2$XoNK	$39a_2$ derivative; *NTS2-2::OrpXylA-KanMX4, XK, PPP, gre3Δ, cyc3Δ, ura3Δ*	This work
STXQ	Isolate from mating of $36\alpha_2$XpXpUN with $39a_2$XoNK	This work
Plasmids		
pUG6	*E. coli* plasmid with segment *loxP–KanMX4–loxP*	[30]
pJFX11	YEp, *TEF1p-BvuXylA-CYC1t*	[12]
pPY1	pPYES2; *GAL1p* replaced by *PGK1p*	This work
pPYXo	pPY1; *PGK1p-OrpXylA-CYC1t*	This work
pPYXpXp	pPY1; 2 copies of *PGK1p-PirXylA-CYC1t* in tandem	This work
pPYXpXo	pPY1; *PGK1p-PirXylA-CYC1t-PGK1p-OrpXylA-CYC1t*	This work
pPYXoXo	pPY1; 2 copies of *PGK1p-OrpXylA-CYC1t* in tandem	This work
XXUN	pPYXpXp-based yeast integration plasmid; 2 μ and *ura3* were replaced with *ura3* and *NTS2-2* partial fragment	This work
XoNK	pUC19-based yeast integration plasmid; *loxP-KanMX4-loxP-pNTS-PGK1p-OrpXylA-CYC1t*	This work

Plasmids pPYXo, pPYXpXo, pPYXoXo and pPYXpXp were individually transformed into $39a_2$, resulting in strains $39a_2$Xo, $39a_2$XpXo, $39a_2$XoXo and $39a_2$XpXp, respectively. Plasmids XoNK and XXUN were digested with *Swa*I and linearized. The linearized fragments were integrated into the genome of $39a_2$ and $36\alpha_2$ at the NTS2-2 site, individually, resulting in recombinant strains $39a_2$XoNK, $39a_2$XpXpUN, $36\alpha_2$XoNK and $36\alpha_2$XpXpUN. For each plasmid transformation, a pool of transformants were generated. The best transformant was isolated based on its cell growth and ethanol production in xylose medium as described in the isolation of 39aXpXp2415. Recombinant strains, $36\alpha_2$XpXpUN and $39a_2$XoNK, were later mated to obtain strain STXQ through screening on SMX agar plates containing 20 g L^{-1} xylose at pH 6 followed by fermentation in SMX medium containing up to 250 g L^{-1} xylose.

2.3. Enzyme Activity Assay

Cells were grown to the exponential phase in SMX medium containing 20 g L^{-1} xylose. After centrifugation, cells were washed twice with chilled distilled water, and then lysed in chilled extraction buffer (100 mM Tris-HCl, 10 mM $MgSO_4$, pH 7.5) by vortex mixing using a Vortex Mixer (Mixer UZUSIO, Tokyo, Japan) with 0.5 mg of 0.5 mm glass beads (Sigma-Aldirich, Singapore). Protease inhibitor cocktail set V (Merck Millipore, Singapore) was added to inhibit serine and cysteine generated in the lysis process. Cell debris was removed by centrifugation (Microcentrifuge D3024, DR. LAB Technology Hong Kong, Hong Kong, China) at 4000× *g* for 10 min at 4 °C, and the crude extract was stored for enzyme activity assay. The protein concentration of the cell extract was determined by the Bradford Assay using a Coomassie Protein Assay Kit (Thermo Scientific, Singapore), and bovine serum albumin (BSA) was used as the standard. Extraction of raw proteins from the yeast strains was performed in duplicate.

The in-vitro XI activity was determined by measuring NADH absorbance using sorbitol dehydrogenase (SDH) (Sigma Aldrich, Singapore). The assay mixture (1 mL) contained extraction buffer, 0.15 mM NADH, 1 U SDH, and 50 μL crude extract. It was equilibrated at 30 °C for 2 min. The reaction was started by the addition of D-xylose to a final concentration of 500 mM. The change of NADH concentration within 3 min was detected using a UV-visible spectrophotometer (Shimadzu, Tokyo, Japan) at wavelength 340 nm, and the specific activity of XI in the recombinant strains was determined [32]. A molar extinction coefficient of 6.25 $(mM\ cm)^{-1}$ at 340 nm for NADH was used to calculate specific activity. Specific activity was expressed as units per mg protein. One unit of enzyme activity was defined as the amount of enzyme required to oxidize 1 μmol of coenzyme/min, under the specified reaction conditions [12].

2.4. Glucose and Xylose Fermentation by the Recombinant Strains

The preculture of the evolved engineered strains was prepared by growing them in 40 mL SMX medium containing 20 g L^{-1} xylose in 100 mL Erlenmeyer flaks at 200 rpm and 30 °C for 24 h. The oxygen-limited conditions in the flasks was maintained by capping the flasks with rubber stoppers pierced with a needle to allow the release of CO_2. Cells in the exponential phase were harvested by centrifugation (Microcentrifuge D3024, DR. LAB Technology Hong Kong, Hong Kong, China) at 14,000× *g* for 1 min. Cell pellets were washed twice and were then inoculated into SM medium supplemented with 20 g L^{-1} xylose with (SMGX) or without 20 g L^{-1} glucose (SMX). The initial optical cell density at 600 nm (OD600) of the culture was about 2 unless otherwise stated. Fermentation was conducted in 100 mL Erlenmeyer shaking flasks under oxygen-limited conditions with a working volume of 40 mL at 200 rpm and 30 °C. The pH value was adjusted at 5.0–6.0 using 3 M NaOH during fermentation. All fermentation experiments were performed in duplicate. Samples were taken periodically to measure OD600, sugar and metabolite concentration.

2.5. Analytical Methods

Cell densities (OD600) were determined using a UV-visible spectrophotometer (Shimadzu, Tokyo, Japan). Fifty-mL cell cultures with varying OD600 (1–5) were filtered with 0.22 μm glass fiber filter membrane (Merck Millipore, Singapore) using Aspirator A-3S (Fisher Scientific, Tokyo, Japan). Cells were washed twice with distilled water, dried at 105 °C in an oven for 24 h, and then weighed. One OD600 unit corresponded to 0.241 g L^{-1} dry cell weight (DCW). Concentrations of glucose, xylose, xylitol, acetate, glycerol and ethanol produced in fermentation were determined by Agilent 1200 series HPLC system (Agilent Technologies, Santa Clara, CA, USA) equipped with a refractive index detector RID-10A using an Aminex HPX-87H ion exchange column (Bio-Rad Laboratories, Woodinville, WA, USA). The column was eluted at 60 °C with 5 mM of sulfuric acid as the mobile phase at a flow rate of 0.6 mL min^{-1}.

2.6. Quantitative Reserve Transcription Polymerase Chain Reaction (RT-PCR)

Recombinant yeast strains $39a_2$Xo, $39a_2$XpXo, $39a_2$XoXo and $39a_2$XpXp were individually cultivated in 40 mL SMX medium containing 20 g L^{-1} xylose under oxygen-limited conditions at 200 rpm and 30 °C. The expression of XI gene transcripts was determined by quantitative reverse transcription polymerase chain reaction (qRT-PCR). Primers for RT-PCR are listed in Table S1. Samples were taken at 24 h, and cells were harvested by centrifuging 2-mL culture at 14,000× *g* and 4 °C for 1 min. Cell pellets were washed twice using double distilled water and total RNA was isolated by using the E.Z.N.A™ Yeast RNA Kit (Omega Bio-tek, Norcross, GA, USA). First-strand cDNA was obtained by using the SuperScript® First-Strand Synthesis System for RT-PCR (Invitrogen, Carlsbad, CA, USA). Such cDNA was then used as the template for qRT-PCR using iCycler iQ™ Real-time PCR Detection System (Bio-Rad Laboratories, Woodinville, WA, USA) and FastStart Universal SYBR Green Master (Roche Applied Science, Penzberg, Germany). The cycle threshold values were calculated with the Optical System Software Version 3.1 (Bio-Rad Laboratories, Woodinville, WA, USA), and the detection threshold over the cycle range was set at 2 to 10. Each PCR was carried out in duplicate. All kits were used under conditions recommended by the manufacturers. The $2^{-\Delta\Delta Ct}$ method [33] was used to analyze the relative changes in gene expression using the housekeeping *ACT1* gene as the reference.

2.7. Biomass Hydrolysate Fermentation Using Strain S. cerevisiae STXQ

The diploid recombinant *S. cerevisiae* strain STXQ (Table 1) was inoculated into YP medium (10 g L^{-1} yeast extract, 20 g L^{-1} peptone, pH 5.0) containing 132 g L^{-1} xylose or mixture of 162 g L^{-1} glucose and 95 g L^{-1} xylose at an initial OD600 of about 13. Fermentation was conducted in 40 mL YP medium in 100 mL shaking flasks under oxygen limited conditions at 200 rpm and 30 °C.

Oil palm empty fruit bunch (OPEFB) hydrolysate was obtained using crude cellulase from *Trichoderma reesei* Rut-C30 according to the protocols described in our earlier report [34]. OPEFB hydrolysate was sterilized using 0. 22 μm filter membrane (Merck Millipore, Singapore) and was supplemented with 7 g L^{-1} yeast extract 2 g L^{-1} peptone, 2 g L^{-1} $(NH_4)_2SO_4$, 2.05 g L^{-1} KH_2PO_4, and 0.25 g L^{-1} Na_2HPO_4. The diploid recombinant *S. cerevisiae* strain STXQ was inoculated into the above OPEFB hydrolysate medium with an initial OD600 about 5. Fermentation was carried out in 40 mL fermentation medium in 100 mL shaking flasks under oxygen-limited conditions at 200 rpm and 30 °C with an initial pH of 4.48. Samples were taken periodically for OD600, sugar and metabolite analysis. Experiments were conducted in duplicate.

3. Results

3.1. Expression of XIs with Various Combinations

XI activities were assayed for $39a_2$ recombinant strains with episomal XI genes (Table 2). XI activity of $39a_2$XoXo was three times higher than that of $39a_2$XpXp and 2.5 times higher than that of $39a_2$Xo. Strain $39a_2$XoNK presented the highest XI activity among all the yeast strains with integrated XI genes, followed by $39a_2$XpXpUN. Quantitative RT-PCR showed that the transcription level of XI gene in the engineered strains significantly increased compared to that in the parent strain $39a_2$ (Table 3). In $36\alpha_2$XpXpUN and $39a_2$XoNK, respective 1.04-fold and 1.41-fold increases in XI gene transcription levels were observed compared to those in $39a_2$XpXp and $39a_2$XoXo. Inconsistent XI activity and transcription level were observed.

Table 2. Specific activity of xylose isomerase in the recombinant strains.

Strains	Specific Activity (U mg^{-1} Protein)
$39a_2XpXp$	0.10 ± 0.003
$39a_2XpXo$	0.11 ± 0.019
$39a_2Xo$	0.12 ± 0.041
$39a_2XoXo$	0.30 ± 0.079
$36\alpha_2XpXpUN$	0.11 ± 0.007
$36\alpha_2$ XoNK	0.04 ± 0.005
$39a_2XpXpUN$	0.26 ± 0.004
$39a_2XoNK$	0.72 ± 0.006

The results represent the mean ± standard deviation of duplicate independent experiments.

Table 3. Fold-change in xylose isomerase (XI) mRNA expression.

Strain	Fold-Change [a]	
	PirXylA	*OrpXylA*
$39a_2XpXp$	50.21 (47.81–52.74)	nil
$39a_2XpXo$	29.86 (28.43–31.36)	11.71 (10.62–12.92)
$39a_2Xo$	nil	59.71 (51.98–68.59)
$39a_2XoXo$	nil	59.71 (55.72–64)
$36\alpha_2XpXpUN$	51.98 (49.50–54.60)	nil
$39a_2XoNK$	nil	84.45 (59.71–119.43)

[a] Fold-change of XI mRNA level in the recombinant yeast strains compared to the parent strain $39a_2$. Results were based on duplicate independent real-time RT-PCR reactions.

3.2. Glucose and Xylose Fermentation by the Engineered $39a_2$ Strains Harboring Episomal XI Genes

Fermentation performance of $39a_2$ recombinant strains with episomal XI genes was tested in SMX medium containing 20 g L^{-1} xylose under oxygen-limited conditions. Apparently, strain $39a_2XpXp$ displayed the fastest xylose utilization rate, followed by $39a_2XoXo$ (Figure 2). Strain $39a_2XpXo$ utilized xylose more slowly than $39a_2XpXp$, though it demonstrated almost the same XI activity (Table 2). On the other hand, strain $39a_2XoXo$ showed a faster xylose utilization rate than strain $39a_2Xo$; however, the former displayed much higher activity than the latter. On the contrary, xylose utilization results accorded quite well with the results of qRT-PCR analysis showing that strain $39a_2XpXp$ had higher XI gene transcription levels than strain $39a_2XpXo$ (Table 3) and a faster xylose utilization rate (Figure 2). However, strains $39a_2XoXo$ and $39a_2Xo$ displayed identical XI gene transcription levels (Table 3), almost the same xylose consumption rate (Figure 2) and xylose conversion (Table 4). Interestingly, strain $39a_2XpXp$ also exhibited the maximal ethanol yield of 0.472 g g^{-1} (Table 4). Such results suggest that expression of two-copy *PirXylA* is favorable for xylose alcoholic fermentation when xylose is used as the sole carbon source.

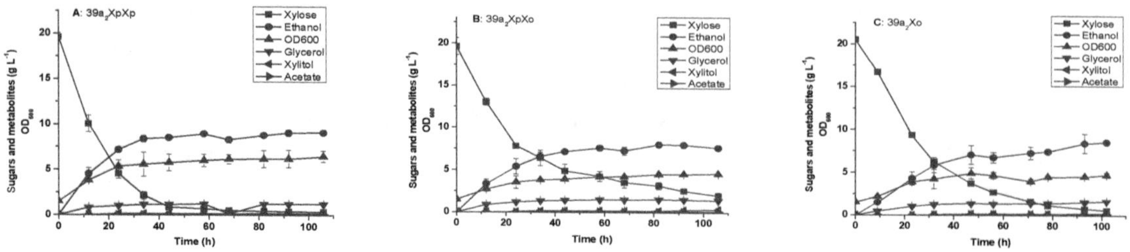

Figure 2. *Cont.*

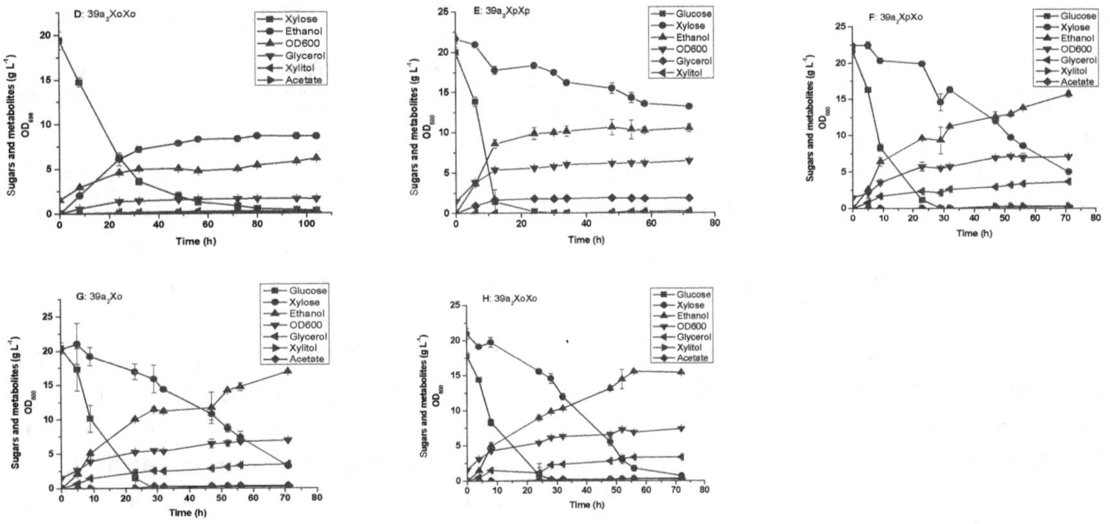

Figure 2. Sugar fermentation under oxygen-limited conditions by the engineered *S. cerevisiae* strains in SM medium containing 20 g L^{-1} xylose (SMX) or 20 g L^{-1} glucose and 20 g L^{-1} xylose (SMGX). (**A**) 39a$_2$XpXp SMX; (**B**) 39a$_2$XpXo SMX; (**C**) 39a$_2$Xo SMX; (**D**) 39a$_2$XoXo SMX; (**E**) 39a$_2$XpXp SMGX; (**F**) 39a$_2$XpXo SMGX; (**G**) 39a$_2$Xo SMGX; (**H**) 39a$_2$XoXo SMGX. The results represent the mean ± standard deviation of duplicate independent experiments ($p < 0.01$).

Table 4. Glucose and xylose fermentation by recombinant xylose-fermenting yeast strains.

Strains	Initial glucose (g L^{-1})	Initial Xylose (g L^{-1})	Xylose Conversion (%)	Ethanol Yield (g g^{-1})	Ethanol Productivity (g h^{-1} L^{-1})	Specific Growth Rate (h^{-1})
39a$_2$xpXp [a]	-	20	98% ± 0.004	0.472 ± 0.013	0.098 ± 0.004	0.014 ± 0.001
39a$_2$XpXo [a]	-	20	89.8% ± 0.011	0.434 ± 0.017	0.098 ± 0.004	0.011 ± 5.89 × 10^{-6}
39a$_2$Xo [a]	-	20	97.3% ± 0.0138	0.428 ± 0.026	0.084 ± 0.001	0.011 ± 0.0003
39a$_2$XoXo [a]	-	20	98.5% ± 0.0057	0.449 ± 0.038	0.108 ± 0.001	0.013 ± 2.29 × 10^{-5}
36α$_2$XpXpUN [a]	-	20	86.4% ± 0.052	0.318 ± 0.01	0.069 ± 0.007	0.014 ± 0.0003
36α$_2$XoNK [a]	-	20	6.42% ± 0.057	ND	ND	0.003 ± 0.0006
39a$_2$XpXpUN[a]	-	20	67.0% ± 0.089	0.398 ± 0.048	0.067 ± 0.001	0.013 ± 0.0002
39a$_2$XoNK [a]	-	20	99.0% ± 0.002	0.368 ± 0.029	0.107 ± 0.007	0.0098 ± 0.0003
39a$_2$XpXp [a]	20	20	22.0% ± 0.003	0.444 ± 0.029	0.221 ± 0.020	0.020 ± 0.0007
39a$_2$XpXo [a]	20	20	77.5% ± 0.017	0.392 ± 0.001	0.081 ± 0.006	0.022 ± 0.0001
39a$_2$Xo [a]	20	20	84.3% ± 0.0007	0.428 ± 0.01	0.238 ± 0.003	0.021 ± 0.0004
39a$_2$XoXo [a]	20	20	96.7% ± 0.011	0.449 ± 0.022	0.213 ± 0.004	0.021 ± 0.001
36α$_2$XpXpUN [a]	20	20	73.8% ± 0.011	0.348 ± 0.049	0.225 ± 0.007	0.029 ± 0.0006
36α$_2$XoNK [a]	20	20	2.62% ± 0.000	0.390 ± 0.043	0.222 ± 0.003	0.016 ± 0.0013
39a$_2$XpXpUN [a]	20	20	25.6% ± 0.097	0.405 ± 0.049	0.213 ± 0.002	0.013 ± 0.0002
39a$_2$XoNK [a]	20	20	97.3% ± 0.006	0.387 ± 0.002	0.243 ± 0.003	0.020 ± 0.0005
36α$_2$XpXpUN [b]	-	40	75.9% ± 0.041	0.384 ± 0.033	0.108 ± 0.011	0.013 ± 0.0004
39a$_2$XoNK [b]	-	40	81.6% ± 0.004	0.421 ± 0.004	0.131 ± 0.000	0.010 ± 0.0003
STXQ [b]	-	40	93.3% ± 0.019	0.393 ± 0.024	0.146 ± 0.009	0.017 ± 0.0002
STXQ [b]	-	132	100%	0.498 ± 0.006	1.13 ± 0.01	0.014 ± 0.0004
STXQ [c]	162	95	99.27% ± 0.002	0.475 ± 0.01	5.24 ± 0.02	0.024 ± 0.0001

[a] Fermentation in SM medium in 72 h; [b] Fermentation in YP medium in 102 h; [c] Fermentation in YP medium. The results represent the mean ± standard deviation of duplicate independent experiments.

Co-fermentation of 20 g L^{-1} glucose and 20 g L^{-1} xylose was carried out in SM medium under oxygen-limited conditions. It can be seen that for all strains, glucose was almost completely consumed at 24 h (Figure 2). However, xylose utilization varied greatly for the four engineered 39a$_2$ strains. Noticeably, in the presence of 20 g L^{-1} glucose, strain 39a$_2$XpXp utilized xylose quite slowly and about 22% xylose was consumed at 72 h (Table 4). On the other hand, xylose utilization was improved to 77.5% by strain 39a$_2$XpXo. Moreover, xylose consumption was improved to 84.3% by strain 39a$_2$Xo and it was further improved to 97% by 39a$_2$XoXo (Table 4). It is worth noting that strain 39a$_2$XoXo presented the highest ethanol yield (0.449 g g^{-1}). The above results suggest that the expression of two-copy of *OrpXylA* is beneficial for glucose and xylose co-fermentation.

3.3. *Glucose and xylose Fermentation by the Engineered Yeast Strains Harboring Integrated XI Genes*

Fermentation performance of $39a_2$ recombinant strains with integrated XI genes was tested in SM medium containing 20 g L^{-1} xylose (SMX) or 20 g L^{-1} glucose and 20 g L^{-1} xylose (SMGX) under oxygen-limited conditions (Figure 3). In SMX fermentation, the specific growth rate of $36\alpha_2$XpXpUN on xylose was 0.014 h^{-1}, much higher than that of $36\alpha_2$XoNK (0.003 h^{-1}), whereas xylose conversion by strain $39a_2$XoNK was 99.0%, much higher than that by $39a_2$XpXpUN (67.0%) (Table 4). In SMGX fermentation, the specific growth rate of $36\alpha_2$XpXpUN on xylose was 0.029 h^{-1}, much higher than that of $36\alpha_2$XoNK (0.016 h^{-1}), whereas xylose conversion by strain $39a_2$XoNK was 97.30%, much higher than that by $39a_2$XpXpUN (73.8%) (Table 4). In both SMX and SMGX fermentation, strain $36\alpha_2$XpXpUN presented the best cell growth, whereas strain $39a_2$XoNK displayed the best xylose conversion. Furthermore, strain $39a_2$XoNK exhibited the highest ethanol productivity in both SMX and SMGX fermentation. Despite the possible XI gene copy number variation in these strains, the above results demonstrated that the elimination of respiration was favorable for xylose fermentation, which corroborated quite well with previous reports [12,14]. Strains $36\alpha_2$XpXpUN and $39a_2$XoNK were therefore selected for mating to generate the diploid recombinant strain STXQ to attain good cell growth and xylose fermentation.

3.4. *Glucose and xylose Fermentation by Diploid Recombinant Strain STXQ*

Fermentation performance of STXQ was tested in YP medium containing 40 g L^{-1} xylose under oxygen-limited conditions. As expected, strain STXQ inherited the advantages of both parent strains, $36\alpha_2$XpXpUN and $39a_2$XoNK. It presented 93.3% xylose conversion and a specific growth rate of 0.017 h^{-1}; both were higher than those of its parent strains (Table 4). In addition, ethanol productivity was greatly improved (Figure 3, Table 4). A very minimum amount of glycerol was produced by STXQ, although glycerol production was notable for the respiration-deficient parent strain $39a_2$XoNK (Figure 3). Moreover, production of xylitol and acetate was almost undetectable.

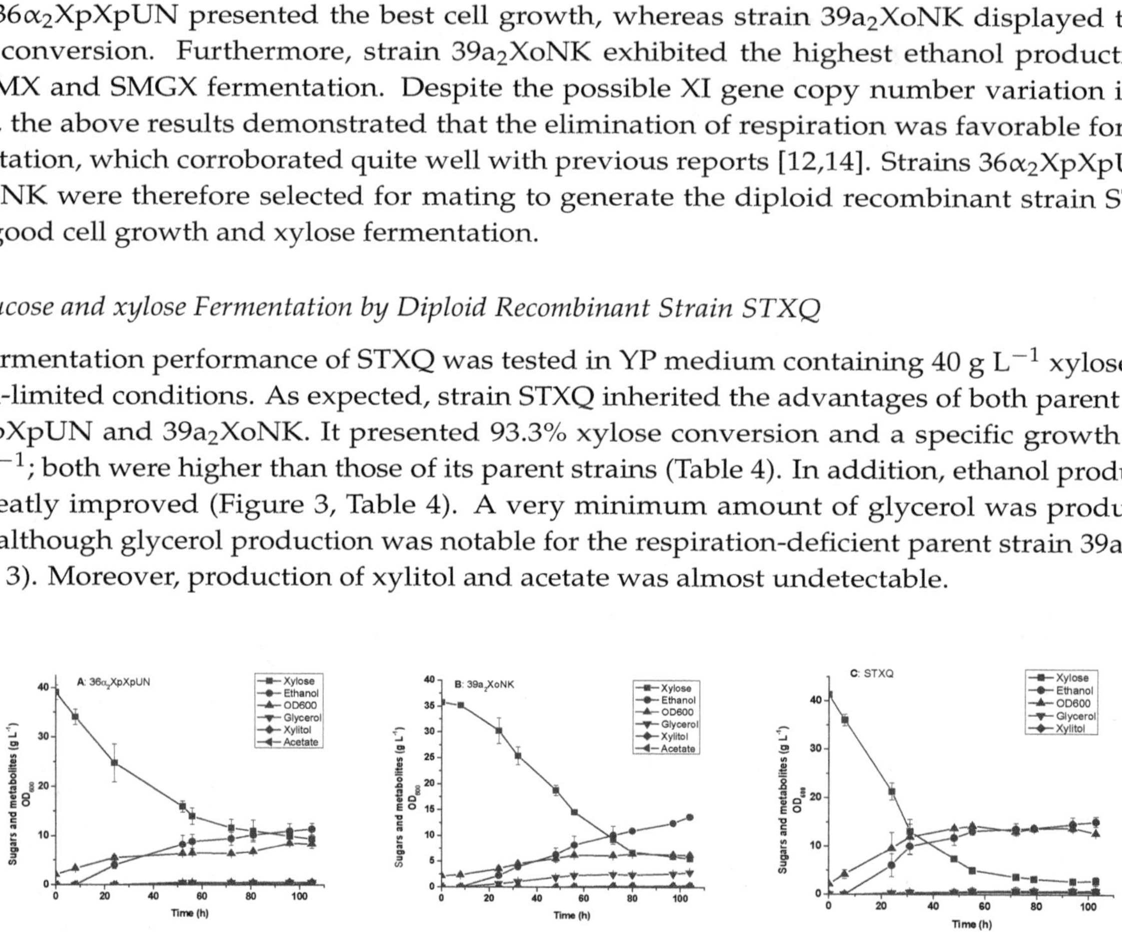

Figure 3. Xylose fermentation in YPX medium containing 40 g L^{-1} xylose under oxygen-limited conditions. (**A**) $36\alpha_2$XpXpUN; (**B**) $39a_2$XoNK; (**C**) STXQ. The results represent the mean ± standard deviation of duplicate independent experiments ($p < 0.01$).

Fermentation performance of strain STXQ was further invested in YP medium containing high-concentration sugar. With an initial OD600 of about 13, strain STXQ consumed 100% xylose in fermenting 132 g L^{-1} xylose and produced 65.8 g L^{-1} ethanol at 46 h. The corresponding ethanol yield was 0.498 g g^{-1} (Figure 4A, Table 4). In fermentation of 162 g L^{-1} glucose and 95 g L^{-1} xylose with about the same initial OD600, more than 99% glucose and xylose were co-utilized within 23 h (Figure 4B, Table 4). Ethanol titer reached 120.6 g L^{-1} corresponding to an ethanol volumetric productivity of 5.26 g L^{-1} h^{-1}. These are so far the highest values compared to those reported in the literature. The above results indicate that the strain development strategy elaborated in this study is efficient in generating a robust *S. cerevisiae* strain with improved xylose fermentation capabilities.

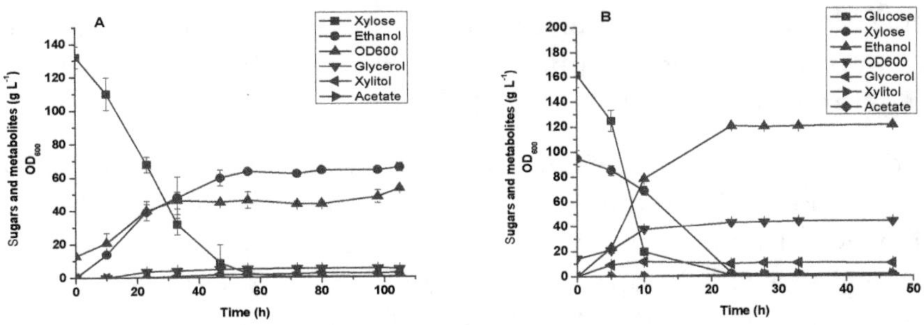

Figure 4. High-titer sugar fermentation under oxygen-limited conditions by strain STXQ in YP media. (**A**) Xylose fermentation in YPX medium containing 132 g L^{-1} xylose; (**B**) Glucose and xylose co-fermentation in YPGX medium containing 162 g L^{-1} glucose and 95 g L^{-1} xylose. The results represent the mean $\pm$ standard deviation of duplicate independent experiments ($p < 0.01$).

3.5. Oil Palm Empty Fruit Bunch Hydrolysate Fermentation by Diploid Recombinant Strain STXQ

Fermentation performance of strain STXQ was further tested in OPEFB hydrolysate containing 41.81 g L^{-1} glucose, 30.00 g L^{-1} xylose, 7 g L^{-1} yeast extract, 2 g L^{-1} peptone, 2 g L^{-1} $(NH4)_2SO_4$, 2.05 g L^{-1} KH_2PO_4, and 0.25 g L^{-1} Na_2HPO_4 under oxygen-limited conditions with an initial OD600 of about 10. Strain STXQ consumed 95.3% glucose and 88.9% xylose (Figure 5). The pH value decreased from 4.48 to 4.00 within 72 h. Strain STXQ consumed 94.0% total sugar from the OPEFB hydrolysate without detoxification. The specific cell growth rate (μ_{max}) reached 0.013 h^{-1} and the ethanol yield was 0.420 g g^{-1}. Glucose was quickly consumed, and glucose-xylose co-fermentation was clearly observed within 24 h. Subsequently, ethanol concentration kept increasing with almost the sole consumption of xylose until it reached the final titer of 28.4 g L^{-1} at 72 h.

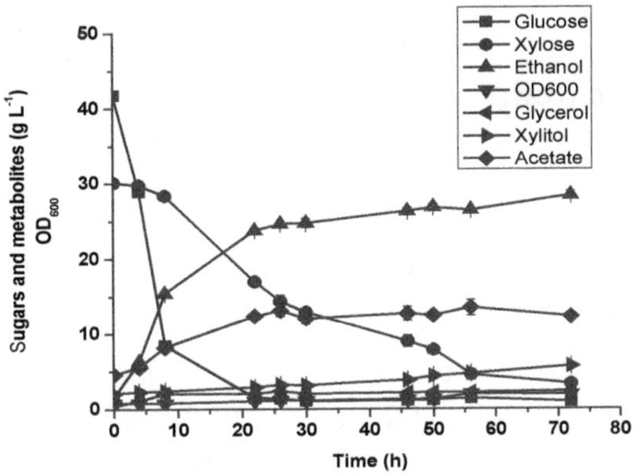

Figure 5. Sugar fermentation under oxygen-limited conditions by strain STXQ in oil palm empty fruit bunch hydrolysate supplemented with 7 g L^{-1} yeast extract, 2 g L^{-1} peptone, 2 g L^{-1} $(NH4)_2SO_4$, 2.05 g L^{-1} KH_2PO_4, and 0.25 g L^{-1} Na_2HPO_4. The results represent the mean $\pm$ standard deviation of duplicate independent experiments ($p < 0.01$).

4. Discussion

Episomal plasmids containing *OrpXylA*, *PirXylA* and *OrpXylA* in tandem, two-copy *PirXylA* and two-copy *OrpXylA* expression cassettes (Figure 1) were transformed into the background strain $39a_2$ individually. Among all the $39a_2$ strains harboring episomal XI genes, strain $39a_2$XpXp presented the fastest rate of xylose utilization when xylose was used as the sole carbon source. However, it did not exhibit the highest XI activity and transcription level (Tables 2 and 3). Inconsistent XI activity and transcription level in recombinant xylose-fermenting *S. cerevisiae* strain was reported and it could be due to rearrangement of pentose phosphate pathway (PPP) genes, decreased glycolysis activity, repressed respiration activity, and enhanced gluconeogenesis [35]. Besides XI activity, the enhanced xylose utilization and fermentation could be associated with the elevated expression of sugar transporter genes, non-oxidative pentose phosphate pathway (PPP) genes such as *TAL1*, *TKL1*, *RKI1*, and *RPE1* and xylulokinase gene, *XKS1* [19,20,25,36].

For glucose and xylose co-fermentation, strain $39a_2$XoXo presented the best glucose and xylose co-utilization (Figure 2) and exhibited the highest xylose conversion (96.7%) (Table 4). *S. cerevisiae* does not contain specific xylose transporters. Xylose was therefore transported by glucose transporters. Xylose affinity for the glucose transporters was very low at a high glucose concentration; however, it could increase to a similar level of glucose affinity at low glucose concentration [19,20]. As a result, diauxic lag is still a practical problem associated with mixed sugar utilization by xylose-fermenting yeasts. However, the diauxic growth was not significant for strains expressing *OrpXylA*, $39a_2$Xo and $39a_2$XoXo (Figure 2). The above results and analysis suggest that expression of *OrpXylA* is beneficial for glucose and xylose co-fermentation. This is consistent with a previous report [37].

Chromosomal integration of the XI gene into the *S. cerevisiae* genome has received a significant amount of attention in recent years [25,26]. It allows the recombinant strain to retain its physiological characteristics in nonselective medium. In the present study, multiple copies of XI genes were integrated at the 18S rDNA sites based on random homologous recombination. Such genome integration led to stable recombinant yeast strains (Figure S1) and boosted XI activity to 0.72 U mg^{-1} protein in $39a_2$XoNK (Table 2). Such improvement might be associated with the multiple copies of XI integrated in the yeast genome [25]. This further led to high XI transcription levels (Table 3), high xylose conversion and ethanol production (Table 4). Such results accorded quite well with earlier reports [25,26,36]. Notably, among these strains, $39a_2$ strains with integrated XI genes presented higher XI activity and transcription level than $36\alpha_2$ strains (Tables 2 and 3). This further confirmed that the elimination of respiration was favorable for xylose fermentation [12,14].

S. cerevisiae strains are regarded as industrial working horses for ethanol production owing to their high ethanol titer and sugar tolerance. Mating is one of the traditional yeast breeding methods to develop improved *S. cerevisiae* strains without genetic modifications [38]. Through strain mating, robust diploid *S. cerevisiae* strains were developed for enhanced xylose-fermentation and inhibitor resistance [23,24]. In the present work, a recombinant diploid *S. cerevisiae* strain STXQ was obtained by mating respiration-dependent strain $36\alpha_2$XpXpUN with the respiration-deficient strain $39a_2$XoNK. Such strain mating enabled strain STXQ to present improved cell growth, xylose utilization and ethanol production (Figure 4, Table 4).

For glucose-xylose co-fermentation by engineered laboratory *S. cerevisiae* strains, higher ethanol concentration of ~60 g L^{-1} was reported by Ho et al. [39] (Table 5). An ethanol titer of 47.5 g L^{-1} was obtained by recombinant *S. cerevisiae* RWB218 expressing *PirXylA* [40]. About 53 g L^{-1} ethanol titer was reported by Diao and his colleagues using the diploid recombinant *S. cerevisiae* strain CIBTS0735 [22]. Demeke et al. obtained an inhibitor-resistant recombinant *S. cerevisiae* through metabolic engineering and adaptive evolution [18]. It utilized glucose and xylose rapidly with ethanol titer up to 46 g L^{-1} and ethanol productivity of 2.58 g $L^{-1}\cdot h^{-1}$. This is the highest volumetric ethanol productivity reported to-date in the literature. More recently, about 58 g L^{-1} ethanol titer was obtained by a diploid recombinant yeast strain LF1 developed from a wild-type *S. cerevisiae* strain in YP medium [25]. Strain STXQ yielded 65.8 g L^{-1} ethanol with an ethanol yield of 0.50 g g^{-1} at 56 h in YPX medium containing 132 g L^{-1} xylose (Figure 4A, Table 4). Further glucose-xylose co-fermentation produced 120.6 g L^{-1} ethanol with an ethanol yield of 0.48 g g^{-1} at 23 h in YP medium containing 162 g L^{-1} glucose and 95 g L^{-1} xylose. The corresponding volumetric ethanol productivity reached 5.24 g $L^{-1}\cdot h^{-1}$. (Figure 4B, Table 4). Both ethanol titer and volumetric productivity far exceeded the results in previous reports.

Table 5. Comparison of fermentation performance of engineered xylose-fermenting *S. cerevisiae*.

Strain	Description	Inoculum Biomass (g DCW L^{-1})	Initial Glucose (g L^{-1})	Initial Xylose (g L^{-1})	Final Ethanol (g L^{-1})	Ethanol Yield (g g^{-1})	Volumetric Ethanol Productivity (g L^{-1}·h^{-1})	Reference
CIBTS0735	*PirXylA*; *XKS1*; PPP; ci*GXF1*; adaptive evolution	Rich medium, initial inoculum size at 0.63 g DCW L^{-1}		40	17.47	0.44	1.09	[18]
		Rich medium, initial inoculum size at 0.63 g DCW L^{-1}	80	40	53	0.45	2.22	
RWB218	*PirXylA*; *XKS1*; PPP; *gre3Δ*; adaptive evolution)	defined synthetic medium, initial inoculum size at 1.1 g DCW L^{-1}	100	25	47.5	0.38	1.98	[40]
1400 (pLNH32)	*XR*; *XDH*; *XK*; adaptive evolution	Rich medium; OD 40–45	50		24	0.48	0.52	[39]
		Rich medium; OD 40–45	80	40	60	0.45	1.3	
H31-A3-AL^{CS}	*PirXylA*; *XKS1*; PPP; *gre3Δ*; adaptive evolution	Defined medium;initial inoculum size at 0.05 g DCW L^{-1}		40	16.4	0.41	0.55	[28]
GS1.11-26	*cpXylA*; *XKS1*; PPP; *HXT7*; *AraT*; *AraA*; *AraB*; *AraD*; *TAL2*; *TKL2* mutagenesis; genome shuffling; adaptive evolution	Rich medium; initial inoculum size at 1.3 g DCW L^{-1}	36	37	33.6	0.46	2.58	[18]
LF1	*Ru-XylA*, *XKS1*, PPP, *gre3::MGT05196*^{N360F}, adaptive evolution	Rich medium, initial inoculum size at 1.00 g DCW L^{-1}	80	42	58.0	0.47	3.60	[25]
STXQ	*OrpXylA*, *PirXylA*, *XKS1*, PPP, *gre3Δ*; adaptive evolution	Rich medium, initial inoculum size at 3.13 g DCW L^{-1}		132	65.8	0.50	1.13	This study
		Rich medium, initial inoculum size at 3.43 g DCW L^{-1}	162	95	120.6	0.48	5.24	This study

The OPEFB hydrolysate fermentation result demonstrated that strain STXQ could co-ferment glucose and xylose without detoxification, with 95.3% glucose conversion and 88.9% xylose conversion at 72 h (Figure 5). Ethanol titer reached 28.4 g L^{-1} at 72 h with an ethanol yield of 0.420 g g^{-1}. Strain STXQ has potential in the conversion of lignocellulosic biomass hydrolysate to ethanol. These results suggest that the strain development protocols outlined in this study are effective in obtaining robust xylose-fermenting yeast strains for industrial applications.

5. Conclusions

A recombinant haploid strain containing plasmids harboring two-copy *PirXylA* presented the best xylose utilization among the engineered yeast strains harboring episomal XI genes when xylose was used as the sole carbon source. On the other hand, the strain containing plasmids with two-copy *OrpXylA* exhibited the best glucose and xylose co-fermentation. Respiration-deficient $39a_2$ strains harboring the *OrpXylA* gene were favorable for xylose fermentation and glucose-xylose co-fermentation in the engineered yeast strains. Chromosome integration of XI genes in *S. cerevisiae* resulted in high XI activity, high XI transcription levels, and improved xylose fermentation. Mating the respiration-efficient strain $36\alpha_2$XpXpUN with the respiration-deficient strain $39a_2$XoNK resulted in a diploid recombinant *S. cerevisiae* strain STXQ with enhanced cell growth and xylose fermentation. Strain STXQ demonstrated superior glucose and xylose co-fermentation performance. It produced 120.6 g L^{-1} ethanol with a volumetric productivity of 5.24 g L^{-1} h^{-1}, the highest among those reported to-date. Such superior performance by strain STXQ is largely associated with its development process, in particular with strain adaptive evolution, XI gene chromosome integration and strain mating.

Author Contributions: Conceptualization, A.G. and T.L; Methodology, T.L.; Investigation, T.L.; Data Curation, T.L. and S.H.; Writing—Original Draft Preparation, T.L.; Writing—Review & Editing, A.G.; Supervision, A.G.; Project Administration, A.G.; Funding Acquisition, A.G.

Acknowledgments: The authors would like to thank Ngee Ann Polytechnic for providing the internship opportunities to Tingting Liu and Shuangcheng Huang during the period of this project.

References

1. Somerville, C.; Youngs, H.; Taylor, C.; Davis, S.C.; Long, S.P. Feedstocks for lignocellulosic biofuels. *Science* **2010**, *329*, 790–792. [CrossRef] [PubMed]
2. Miao, Z.; Grift, T.E.; Hansen, A.C.; Ting, K.C. An overview of lignocellulosic biomass feedstock harvest, processing and supply for biofuel production. *Biofuels* **2013**, *4*, 5–8. [CrossRef]
3. Fatma, S.; Hameed, A.; Noman, M.; Ahmed, T.; Shahid, M.; Tariq, M.; Sohail, I.; Tabassum, R. Lignocellulosic biomass: A sustainable bioenergy source for the future. *Protein Pept. Lett.* **2018**, *25*, 148–163. [CrossRef] [PubMed]
4. Moyses, D.N.; Reis, V.C.; de Almeida, J.R.; de Moraes, L.M.; Torres, F.A. Xylose fermentation by *Saccharomyces cerevisiae*: Challenges and prospects. *Int. J. Mol. Sci.* **2016**, *17*, 207. [CrossRef] [PubMed]
5. Li, Y.C.; Gou, Z.X.; Zhang, Y.; Xia, Z.Y.; Tang, Y.Q.; Kida, K. Inhibitor tolerance of a recombinant flocculating industrial *Saccharomyces cerevisiae* strain during glucose and xylose co-fermentation. *Braz. J. Microbiol.* **2017**, *48*, 798–800. [CrossRef] [PubMed]
6. Hahn-Hägerdal, B.; Karhumaa, K.; Jeppsson, M.; Gorwa-Grauslund, M.F. Metabolic engineering for pentose utilization in *Saccharomyces cerevisiae*. *Adv. Biochem. Eng. Biotechnol.* **2007**, *108*, 147–177. [CrossRef] [PubMed]
7. Madhavan, A.; Srivastava, A.; Kondo, A.; Bisaria, V.S. Bioconversion of lignocellulose-derived sugars to ethanol by engineered *Saccharomyces cerevisiae*. *Crit. Rev. Biotechnol.* **2012**, *32*, 22–48. [CrossRef] [PubMed]
8. Zhang, G.C.; Liu, J.J.; Kong, I.I.; Kwak, S.; Jin, Y.S. Combining C6 and C5 sugar metabolism for enhancing microbial bioconversion. *Curr. Opin. Chem. Biol.* **2015**, *29*, 49–57. [CrossRef] [PubMed]
9. Kwak, S.; Jin, Y.S. Production of fuels and chemicals from xylose by engineered *Saccharomyces cerevisiae*: A review and perspective. *Microb. Cell Factories* **2017**, *16*, 82. [CrossRef] [PubMed]
10. Hou, J.; Qiu, C.; Shen, Y.; Li, H.; Bao, X. Engineering of *Saccharomyces cerevisiae* for the efficient co-utilization of glucose and xylose. *FEMS Yeast Res.* **2017**, *17*. [CrossRef] [PubMed]

11. Kuyper, M.; Hartog, M.M.; Toirkens, M.J.; Almering, M.J.; Winkler, A.A.; van Dijken, J.P.; Pronk, J.T. Metabolic engineering of a xylose-isomerase-expressing *Saccharomyces cerevisiae* strain for rapid anaerobic xylose fermentation. *FEMS Yeast Res.* **2005**, *5*, 399–409. [CrossRef] [PubMed]
12. Peng, B.; Huang, S.; Liu, T.; Geng, A. Bacterial xylose isomerases from the mammal gut Bacteroidetes cluster function in *Saccharomyces cerevisiae* for effective xylose fermentation. *Microb. Cell Factories* **2015**, *14*, 70. [CrossRef] [PubMed]
13. Johansson, B.; Hahn-Hägerdal, B. The non-oxidative pentose phosphate pathway controls the fermentation rate of xylulose but not of xylose in *Saccharomyces cerevisiae* TMB3001. *FEMS Yeast Res.* **2002**, *2*, 277–282. [PubMed]
14. Peng, B.; Shen, Y.; Li, X.; Chen, X.; Hou, J.; Bao, X. Improvement of xylose fermentation in respiratory-deficient xylose-fermenting *Saccharomyces cerevisiae*. *Metab. Eng.* **2012**, *14*, 9–18. [CrossRef] [PubMed]
15. Lee, S.H.; Kodaki, T.; Park, Y.C.; Seo, J.H. Effects of NADH-preferring xylose reductase expression on ethanol production from xylose in xylose-metabolizing recombinant *Saccharomyces cerevisiae*. *J. Biotechnol.* **2012**, *158*, 184–191. [CrossRef] [PubMed]
16. Sonderegger, M.; Sauer, U. Evolutionary engineering of Saccharomyces cerevisiae for anaerobic growth on xylose. *Appl. Environ. Microbiol.* **2003**, *69*, 1990–1998. [CrossRef] [PubMed]
17. Scalcinati, G.; Otero, J.M.; Vleet, J.R.; Jeffries, T.W.; Olsson, L.; Nielsen, J. Evolutionary engineering of *Saccharomyces cerevisiae* for efficient aerobic xylose consumption. *FEMS Yeast Res.* **2012**, *12*, 582–597. [CrossRef] [PubMed]
18. Demeke, M.M.; Dietz, H.; Li, Y.; Foulquie-Moreno, M.R.; Mutturi, S.; Deprez, S.; Den Abt, T.; Bonini, B.M.; Liden, G.; Dumortier, F.; et al. Development of a D-xylose fermenting and inhibitor tolerant industrial *Saccharomyces cerevisiae* strain with high performance in lignocellulose hydrolysates using metabolic and evolutionary engineering. *Biotechnol. Biofuels* **2013**, *6*, 89. [CrossRef] [PubMed]
19. Sedlak, M.; Ho, N.W. Characterization of the effectiveness of hexose transporters for transporting xylose during glucose and xylose co-fermentation by a recombinant *Saccharomyce*s yeast. *Yeast* **2004**, *21*, 671–684. [CrossRef] [PubMed]
20. Runquist, D.; Hahn-Hagerdal, B.; Radstrom, P. Comparison of heterologous xylose transporters in recombinant *Saccharomyces cerevisiae*. *Biotechnol. Biofuels* **2010**, *17*, 3–5. [CrossRef] [PubMed]
21. Goncalves, D.L.; Matsushika, A.; de Sales, B.B.; Goshima, T.; Bon, E.P.; Stambuk, B.U. Xylose and xylose/glucose co-fermentation by recombinant *Saccharomyces cerevisiae* strains expressing individual hexose transporters. *Enzyme Microb. Technol.* **2014**, *63*, 13–20. [CrossRef] [PubMed]
22. Diao, L.; Liu, Y.; Qian, F.; Yang, J.; Jiang, Y.; Yang, S. Construction of fast xylose-fermenting yeast based on industrial ethanol-producing diploid *Saccharomyces cerevisiae* by rational design and adaptive evolution. *BMC Biotechnol.* **2013**, *13*, 110. [CrossRef] [PubMed]
23. Demeke, M.M.; Dumortier, F.; Li, Y.; Broeckx, T.; Foulquié-Moreno, M.R.; Thevelein, J.M. Combining inhibitor tolerance and D-xylose fermentation in industrial *Saccharomyces cerevisiae* for efficient lignocellulose-based bioethanol production. *Biotechnol. Biofuels* **2013**, *6*, 120. [CrossRef] [PubMed]
24. Kim, S.R.; Lee, K.S.; Kong, I.I.; Lesmana, A.; Lee, W.H.; Seo, J.H.; Kweon, D.H.; Jin, Y.S. Construction of an efficient xylose-fermenting diploid *Saccharomyces cerevisiae* strain through mating of two engineered haploid strains capable of xylose assimilation. *J. Biotechnol.* **2013**, *164*, 105–111. [CrossRef] [PubMed]
25. Li, H.; Shen, Y.; Wu, M.; Hou, J.; Jiao, C.; Li, Z.; Liu, X.; Bao, X. Engineering a wild-type diploid *Saccharomyces cerevisiae* strain for second-generation bioethanol production. *Bioresour. Bioprocess.* **2016**, *3*, 51. [CrossRef] [PubMed]
26. Tanino, T.; Hotta, A.; Ito, T.; Ishii, J.; Yamada, R.; Hasunuma, T.; Ogino, C.; Ohmura, N.; Ohshima, T.; Kondo, A. Construction of a xylose-metabolizing yeast by genome integration of xylose isomerase gene and investigation of the effect of xylitol on fermentation. *Appl. Microbiol. Biotechnol.* **2010**, *88*, 1215–1221. [CrossRef] [PubMed]
27. Bamba, T.; Hasunuma, T.; Kondo, A. Disruption of PHO13 improves ethanol production via the xylose isomerase pathway. *AMB Express* **2016**, *6*, 4. [CrossRef] [PubMed]
28. Zhou, H.; Cheng, J.S.; Wang, B.L.; Fink, G.R.; Stephanopoulos, G. Xylose isomerase overexpression along with engineering of the pentose phosphate pathway and evolutionary engineering enable rapid xylose utilization and ethanol production by *Saccharomyces cerevisiae*. *Metab. Eng.* **2012**, *14*, 611–622. [CrossRef] [PubMed]

29. Ko, J.K.; Um, Y.; Woo, H.M.; Kim, K.H.; Lee, S.M. Ethanol production from lignocellulosic hydrolysates using engineered *Saccharomyces cerevisiae* harboring xylose isomerase-based pathway. *Bioresour. Technol.* **2016**, *209*, 290–296. [CrossRef] [PubMed]
30. Güldener, U.; Heck, S.; Fiedler, T.; Beinhauer, J.; Hegemann, J.H. A new efficient gene disruption cassette for repeated use in budding yeast. *Nucleic Acids Res.* **1996**, *24*, 2519–2524. [CrossRef] [PubMed]
31. Gietz, R.D.; Schiestl, R.H. Large-scale high-efficiency yeast transformation using the LiAc/SS carrier DNA/PEG method. *Nat. Protoc.* **2007**, *2*, 38–41. [CrossRef] [PubMed]
32. Kersters-Hilderson, H.; Callens, M.; Opstal, O.V.; Vangrysperre, W.; Bruyne, C.K.D. Kinetic characterization of d-xylose isomerases by enzymatic assays using d-sorbitol dehydrogenase. *Enzyme Microb. Technol.* **1987**, *9*, 145–148. [CrossRef]
33. Livak, K.J.; Schmittgen, T.D. Analysis of relative gene expression data using real-time quantitative PCR and the 2–ΔΔCT method. *Methods* **2001**, *25*, 402–408. [CrossRef] [PubMed]
34. Wang, Z.; Ong, H.X.; Geng, A. Cellulase production and oil palm empty fruit bunch saccharification by a new isolate of *Trichoderma koningii* D-64. *Proc. Biochem.* **2012**, *47*, 1564–1571. [CrossRef]
35. Qi, X.; Zha, J.; Liu, G.G.; Zhang, W.; Li, B.Z.; Yuan, Y.J. Heterologous xylose isomerase pathway and evolutionary engineering improve xylose utilization in *Saccharomyces cerevisiae. Front. Microbiol.* **2015**, *6*, 1165. [CrossRef] [PubMed]
36. Vilela Lde, F.; de Araujo, V.P.; Paredes Rde, S.; Bon, E.P.; Torres, F.A.; Neves, B.C.; Eleutherio, E.C. Enhanced xylose fermentation and ethanol production by engineered *Saccharomyces cerevisiae* strain. *AMB Express* **2015**, *5*, 16. [CrossRef] [PubMed]
37. Madhavan, A.; Tamalampudi, S.; Srivastava, A.; Fukuda, H.; Bisaria, V.S.; Kondo, A. Alcoholic fermentation of xylose and mixed sugars using recombinant *Saccharomyces cerevisiae* engineered for xylose utilization. *Appl. Microbiol. Biotechnol.* **2009**, *82*, 1037–1047. [CrossRef] [PubMed]
38. Perez-Traves, L.; Lopes, C.A.; Barrio, E.; Querol, A. Evaluation of different genetic procedures for the generation of artificial hybrids in *Saccharomyces* genus for winemaking. *Int. J. Food Microbiol.* **2012**, *156*, 102–111. [CrossRef] [PubMed]
39. Ho, N.W.; Chen, Z.; Brainard, A.P. Genetically engineered *Saccharomyces* yeast capable of effective cofermentation of glucose and xylose. *Appl. Environ. Microbiol.* **1998**, *64*, 1852–1859. [PubMed]
40. Kuyper, M.; Harhangi, H.R.; Stave, A.K.; Winkler, A.A.; Jetten, M.S.; de Laat, W.T.; den Ridder, J.J.; Op den Camp, H.J.; van Dijken, J.P.; et al. High-level functional expression of a fungal xylose isomerase: The key to efficient ethanolic fermentation of xylose by *Saccharomyces cerevisiae*? *FEMS Yeast Res.* **2003**, *4*, 69–78. [CrossRef]

13

Screening and Application of *Cyberlindnera* Yeasts to Produce a Fruity, Non-Alcoholic Beer

Konstantin Bellut [1], Maximilian Michel [2], Martin Zarnkow [2], Mathias Hutzler [2], Fritz Jacob [2], Jonas J. Atzler [1], Andrea Hoehnel [1], Kieran M. Lynch [1] and Elke K. Arendt [1,3,*]

1 School of Food and Nutritional Sciences, University College Cork, College Road, Cork, Ireland
2 Research Center Weihenstephan for Brewing and Food Quality, Technische Universität München, Alte Akademie 3, 85354 Freising-Weihenstephan, Germany
3 APC Microbiome Ireland, University College Cork, College Road, Cork, Ireland
* Correspondence: e.arendt@ucc.ie

Abstract: Non-alcoholic beer (NAB) is enjoying growing demand and popularity due to consumer lifestyle trends and improved production methods. In recent years in particular, research into the application of non-*Saccharomyces* yeasts to produce NAB via limited fermentation has gained momentum. Non-*Saccharomyces* yeasts are known to produce fruity aromas, owing to a high ester production. This trait could be harnessed to mask the often-criticized wort-like off-flavor of NAB produced via limited fermentation. Six *Cyberlindnera* strains were characterized and screened in wort extract. Four of the six strains produced a pleasant, fruity aroma while exhibiting low ethanol production. The strain *Cyberlindnera subsufficiens* C6.1 was chosen for fermentation optimization via response surface methodology (RSM) and a pilot-scale (60 L) brewing trial with subsequent sensory evaluation. A low fermentation temperature and low pitching rate enhanced the fruitiness and overall acceptance of the NAB. The NAB (0.36% ABV) produced on pilot-scale was significantly more fruity and exhibited a significantly reduced wort-like off-flavor compared to two commercial NABs. This study demonstrated the suitability of *Cyberlindnera subsufficiens* to produce a fruity NAB, which can compete with commercial NABs. The outcome strengthens the position of non-*Saccharomyces* yeasts as a serious and applicable alternative to established methods in NAB brewing.

Keywords: brewing; *Cyberlindnera*; NABLAB; non-alcoholic beer; non-conventional yeast; non-*Saccharomyces* yeast; response surface methodology

---

1. Introduction

While the overall market growth of beer is slowing down, non-alcoholic and low alcohol beer (NABLAB) is growing in volume and popularity, owed to stricter legislation, lifestyle trends and improved production methods [1]. The increasing interest has fueled research in NABLAB production methods, especially in recent years, aimed at overcoming taste deficits compared to regular beer and consequently improving consumer acceptance. The two major production methods, physical dealcoholization and limited fermentation, both compromise the taste of the beer. Dealcoholized beer is often criticized for its lack of body and aromatic profile, a consequence of the removal of volatile esters and higher alcohols in conjunction with ethanol. Apart from a sweet taste due to residual sugars, one of the main points of criticism of NAB produced by limited fermentation is its wort-like off-flavor caused by aldehydes present in the wort [2]. In regular beer, ethanol significantly increases aldehyde retention, reducing the perceptibility of the wort-like flavor. However, in NAB produced by limited fermentation, the low ethanol content and higher levels of mono- and disaccharides intensify this undesired off-flavor [3].

It is known that esters, which yeast produce as a by-product of alcoholic fermentation, are extremely important for the flavor profile of beer [4,5]. The lack thereof, as well as their overproduction, can significantly compromise the flavor. Aside from strain-specific differences, the process parameters such as the fermentation temperature, pitching rate and wort gravity have been shown to have a significant influence on ester formation [4,6]. In non-alcoholic beers, ester concentrations are lower compared to regular beer, independent of the production method [7,8]. While physical dealcoholization removes esters that were previously produced, limited fermentation adversely affects the production of substantial amounts in the first place.

Non-*Saccharomyces* yeasts are known for their important contribution to the flavor profile of fermented foods and beverages and have therefore been investigated for their targeted application in bioflavoring and, not least, NABLAB brewing [1,9,10]. Species that have been mentioned in the context of NABLAB production belong to the genera *Cyberlindnera*, *Hanseniaspora*, *Lachancea*, *Mrakia*, *Pichia*, *Torulaspora*, *Saccharomycodes*, *Scheffersomyces* and *Zygosaccharomyces* [1,11–16]. In particular, the *Cyberlindnera* species are known for their high ester production, which was shown in studies with *Cyberlindnera saturnus* (formerly *Williopsis saturnus*), *C. mrakii* (formerly *Williopsis saturnus* var. *mrakii*) and *C. subsufficiens* (formerly *Williopsis saturnus* var. *subsufficiens*) [17–20]. Furthermore, it has been proposed to use yeasts with high production of flavor compounds (i.e., esters, higher alcohols) to mask the wort-like flavor of NAB produced by limited fermentation. However, research in that direction is sparse [21,22]. In addition, such yeasts are capable of reducing aldehydes to their correspondent alcohol, which can also enhance the reduction of the often-criticized wort-like off-flavor [23,24].

In this study, six strains of the genus *Cyberlindnera* were investigated to create a fruity NAB. After identification, the strains were characterized for their substrate utilization, flocculation behavior and stress responses. A screening in diluted wort extract was performed to investigate the strains' potential to produce a pronounced fruity flavor without the production of high concentrations of ethanol. Interspecific differences in sugar consumption and the production of volatile fermentation by-products was investigated by means of high-performance liquid chromatography (HPLC) and gas chromatography (GC). The most promising strain was studied further to determine the optimal fermentation conditions to enhance the fruity flavor, which was performed by means of response surface methodology (RSM). Finally, a non-alcoholic beer was produced on pilot-scale (60 L), and its analytical attributes, aroma, and taste compared to two commercial NABs were examined.

2. Materials and Methods

2.1. Materials

All reagents used in this study were at least analytical grade from Sigma-Aldrich (St Louis MO, USA) unless stated otherwise. The wort extract applied in this study was spray-dried wort from 100% barley malt (Spraymalt Light, Muntons plc, Suffolk, UK). For the pilot-scale brewing, pilsner malt and acidulated malt were sourced from Weyermann (Malzfabrik Weyermann, Bamberg, Germany).

2.2. Yeast Strains

Strain Origin and Identification

Strain 837A was isolated from a brewery cellar, NT Cyb originates from a dried fermentation starter for rice wine, strain C6.1 originates from a coconut, and L1 from "Lulo", the fruit of *Solanum quitoense*. The type strains CBS 1707 and CBS 5763 originate from soil samples. For identification, the D1/D2 domain of the 26S rRNA gene was amplified, sequenced and compared to publicly available sequences in the National Center for Biotechnology Information (NCBI) database using the Basic Local Alignment Search Tool (BLAST; https://blast.ncbi.nlm.nih.gov/Blast.cgi).

The DNA of the yeast isolates was extracted using an extraction kit (Yeast DNA Extraction Kit, Thermo Fisher Scientific, Waltham MA, USA). To amplify the D1/D2 domain of the 26S rRNA gene, the primers NL1 (5′-GCATATCAATAAGCGGAGGAAAAG-3′) and NL4 (5′-GGTCCGTGTTTCAAGACGG-3′) were used. Polymerase chain reaction (PCR) was performed using the temperature protocol: 95 °C/2 min; 30 cycles of 95 °C/30 s, 56 °C/15 s; 72 °C/60 s; 72 °C/5 min. Stock cultures were kept in 50% (*v/v*) glycerol at −80 °C.

2.3. *Yeast Characterization*

2.3.1. Flocculation Assay and Phenolic Off-Flavor (POF) Test

The flocculation test was performed using a slightly modified Helm's assay [25,26]. Essentially, all cells were washed in ethylenediaminetetraacetic acid (EDTA) and the sedimentation period was extended to 10 min. Wort was composed of 75 g/L spray-dried malt extract (Spraymalt Light, Muntons plc, Suffolk, UK) adjusted to 15 International Bitterness Units (IBU) (15 mg/mL iso-α-acids; from 30% stock solution; Barth-Haas Group, Nürnberg, Germany).

The phenolic off-flavor test was performed according to Meier-Dörnberg et al. [27]. In short, yeast strains were spread on yeast and mold agar plates (YM-agar) containing only one of the following precursors: either ferulic acid, cinnamic acid or coumaric acid. After three days of incubation at 25 °C, plates were evaluated by a trained panel by sniffing to detect any of the following aromas: clove-like (4-vinylguajacol), Styrofoam-like (4-vinylstyrene) and medicinal-like (4-vinylphenol). *Saccharomyces cerevisiae* LeoBavaricus—TUM 68®(Research Center Weihenstephan for Brewing and Food Quality, Freising-Weihenstephan, Germany) was used as a positive control.

2.3.2. Substrate Utilization

To analyze substrate utilization by the *Cyberlindnera* strains, the test kit API ID 32C (BioMérieux, Marcy-l'Étoile, France) was used. Preparation of the inoculum and inoculation of the strips were performed according to the manufacturers' instructions. Colonies for the inoculum were grown on yeast extract peptone dextrose (YPD) agar plates for 48 h at 27 °C. After inoculation, API ID 32C strips were incubated for 2 days at 28 °C. The samples were evaluated visually for turbidity in the wells, differentiating positive (+), negative (−), and weak (w) growth.

2.3.3. Stress Tests

Stress tests were performed via the measurement of yeast growth in a microplate, through the repeated measurement of absorbance over a time period of 96 h (Multiskan FC, Thermo Scientific, Waltham, MA, USA). The substrate for the hop sensitivity test was sterile-filtered wort extract (75 g/L Muntons Spraymalt Light) adjusted to 0, 50 and 100 mg/L iso-α-acids (1 mg/L = 1 International Bitterness Unit, IBU), respectively, by using an aliquot of a stock solution of 3% iso-α-acids in 96% (*v/v*) ethanol (Barth-Haas Group, Nürnberg, Germany). For testing ethanol sensitivity, the sterile-filtered wort extract was adjusted to 0%, 2.5%, 5% and 7.5% ABV with an aliquot of 100% (*v/v*) ethanol. For testing pH sensitivity, the sterile-filtered wort extract was adjusted to the following pHs with 2 M HCl: 5.5 (control without addition of HCl): 5.0, 4.0 and 3.0. For inoculation, strains were grown in sterilized wort extract for 24 h at 25 °C under aerobic conditions. The microtiter plate wells were inoculated with a concentration of 10^5 cells/mL. The wells contained 200 μL of the respective wort substrates. Plates were incubated at 25 °C, and absorbance was measured every 30 min at 600 nm without shaking over a time period of 96 h (Multiskan FC, Thermo Scientific, Waltham, MA, USA). Stress tests were performed in triplicate.

2.4. *Yeast Screening*

2.4.1. Propagation

Single colonies of the respective strains were taken from yeast extract peptone dextrose (YPD) agar plates after 72 h growth at 25 °C and transferred into a 250 mL sterile Duran glass bottle (Lennox

Laboratory Supplies Ltd, Dublin, Ireland) containing 150 mL propagation wort consisting of 75 g/L spray-dried malt (Spraymalt light, Muntons plc, Suffolk, UK) and 30 g/L glucose (Gem Pack Foods Ltd., Dublin, Ireland), sterilized at 121 °C for 15 min. The bottles were covered with sterile cotton and placed in an incubator with orbital shaker (ES-80 shaker-incubator, Grant Instruments (Cambridge) Ltd, Shepreth, UK) and incubated for 24 h at an orbital agitation of 170 rpm at 25 °C (Strain 837A was incubated for 48 h). Cell count was performed using a Thoma Hemocytometer with a depth of 0.1 mm (Blaubrand, Sigma-Aldrich, St. Louis, MO, USA).

2.4.2. Fermentation

Fermentation wort was prepared by dissolving 75 g/L spray-dried malt extract (Munton Spraymalt light) in 1 L of brewing water and sterilizing at 121 °C for 15 min, followed by filtration through a sterile grade 1V Whatman filter (Whatman plc, Maidstone, UK) to remove hot trub formed during sterilization. The analytical attributes of the fermentation wort for the yeast screening trial and RSM trial is shown in Table 1.

Table 1. Attributes of screening wort from wort extract.

Attribute	Unit	Value
Real Extract	°P	6.97 ± 0.00
pH	–	5.20 ± 0.01
Free amino nitrogen (FAN)	mg/L	115 ± 1
Maltotriose	g/L	8.12 ± 0.15
Maltose	g/L	32.37 ± 0.57
Sucrose	g/L	0.83 ± 0.04
Glucose	g/L	5.68 ± 0.91
Fructose	g/L	1.45 ± 0.10

Fermentation trials were carried out in 1 L sterile Duran glass bottles, equipped with an air lock. Per yeast strain, triplicate bottles were filled with 400 mL of wort and left untouched throughout the fermentation. Yeast cells for pitching were washed by centrifugation at 900 *g* for 5 min and resuspended in sterile water to ensure no carryover of sugars from the propagation wort into the fermentation wort. Pitching rate was 3×10^7 cells/mL. Fermentation temperature was 25 °C. Fermentation was performed until no change in extract could be measured for two consecutive days.

2.5. Scanning Electron Microscopy (SEM)

Yeast cultures for scanning electron microscopy (SEM) were prepared following the protocol for cultured microorganisms by Das Murtey and Ramasamy [28]. Single colonies were taken from a YPD agar plate and grown in YPD broth for 24 h at 25 °C. One milliliter of sample was centrifuged at 900 g for 2 min for pellet formation and resuspended in 5% glutaraldehyde solution prepared in 0.1 M phosphate buffer (pH 7.2) for fixation. After 30 min, the sample was centrifuged, the supernatant was discarded, and the pellet was washed twice in 0.1 M phosphate buffer. Consequently, the pellet was resuspended in 1% osmium tetroxide prepared in 0.1 M phosphate buffer. After 1 h, cells were again washed twice in 0.1 M phosphate buffer. The sample was then dehydrated through an ethanol series of 35%, 50%, 75%, 95%, absolute ethanol, and hexamethyldisilazane (HDMS), with 30 min per step (last two ethanol steps twice), centrifuging and discarding the supernatant at each change. Lastly, the second HDMS was discarded and the sample left drying overnight in a desiccator.

The dehydrated yeast sample was mounted onto plain aluminum stubs using carbon double surface adhesive and coated with a 5 nm gold-palladium (80:20) layer using a Gold Sputter Coater (BIO-RAD Polaron Division, SEM coating system, England), then observed under a constant accelerating voltage of 5 kV under a JEOL scanning electron microscope type 5510 (JEOL, Tokyo, Japan).

2.6. *Response Surface Modeling (RSM)*

To investigate optimal fermentation conditions for C6.1 to produce a fruity, non-alcoholic beer, response surface methodology (RSM) was performed using DesignExpert 9 software (StatEase, Minneapolis, MN, USA). A two-factorial, face-centered, central composite design with single factorial points and 5 replications of the center point was chosen. The predictor factors were temperature (17, 22, 27 °C), and pitching rate (10, 35, 60 × 10^6 cells/mL).

Spray-dried malt extract (Spraymalt light, Muntons plc, Suffolk, UK) served as the substrate. Wort preparation, propagation and inoculation were carried out as outlined in 2.4.1. The wort used was the same as in the screening (Table 1). Fermentation volume was 150 mL in 250 mL Duran glass bottles equipped with an air lock. Fermentation was performed until no change in extract could be measured for two consecutive days. Table 2 shows the experimental design.

Table 2. Response surface methodology (RSM) experimental design: Two-factorial, face-centered, central composite design with five repetitions of the center point. Factor 1, A: temperature, range 17, 22, 27 °C. Factor 2, B: pitching rate, range 10, 35, 60 × 10^6 cells/mL.

Run	Factor 1	Factor 2
	A: Temperature (°C)	**B: Pitching Rate (×10^6 cells/mL)**
1	22	60
2	22	10
3	17	35
4	27	35
5 *	22	35
6 *	22	35
7	17	60
8 *	22	35
9 *	22	35
10 *	22	35
11	17	10
12	27	10
13	27	60

* Center point.

Models were produced applying backward elimination regression of insignificant model terms with α to exit of 0.1 (detailed report in supplementary Data Sheet S1). For significant models with insignificant lack of fit (LOF), 3D response surface plots were produced. Fermentations for model validation were performed in the same wort with propagation as outlined in 2.4.1 and fermentation as outlined above.

2.7. *Pilot-Scale Brewing*

2.7.1. Wort Production

Wort for the pilot brew was produced in a 60 L pilot-scale brewing plant consisting of a combined mash-boiling vessel, a lauter tun and whirlpool (FOODING Nahrungsmitteltechnik GmbH, Stuttgart, Germany). The grain bill comprised 6.65 kg Weyermann Pilsner Malt and 0.35 kg Weyermann Acidulated Malt (Malzfabrik Weyermann, Bamberg, Germany). Grains were milled with a two-roller mill ("Derby", Engl Maschinen, Schwebheim, Germany) at a 0.8 mm gap size. The crushed malt was mashed-in with 30 L of brewing water at 50 °C. The following mashing regime was employed: 20 min at 50 °C, 20 min at 62 °C, 10 min at 72 °C and mashing out at 78 °C. The mash was pumped into the lauter tun, and lautering was performed after a 15 min lauter rest, employing four sparging steps of 5 L hot brewing water each. Boil volume was 50 L at a gravity of 1.030 (7.0 °P), and total boiling time was 60 min. Thirty minutes into the boil, 15 g of Magnum hop pellets (14% iso-α-acids) were added for a calculated IBU content of 9. After boiling, gravity was readjusted to 1.030 (7.0 °P) with hot brewing water, and hot trub precipitates and hop residue were removed in the whirlpool with a rest of 20 min. Clear wort was pumped through a heat exchanger and filled into 60 L cylindroconical fermentation vessels at a temperature of 17 °C.

2.7.2. Propagation, Fermentation and Aftercare

A first propagation step was employed as described in 2.4.1. A second propagation step was performed by transferring the small-scale propagated wort into a 5 L carboy filled with 2 L of sterile wort extract at 7 °P and closed with sterile cotton. The second propagation step was conducted for 24 h under constant agitation at ambient temperature (20 ± 2 °C).

Yeast was pitched into the fermenter at a pitching rate of 10^7 cells/mL. Fermentation was carried out in cylindroconical fermentation vessels with a capacity of 60 L, at ambient pressure and at a glycol-controlled fermentation temperature of 17 °C. Samples were withdrawn every day. Fermentation was carried out until no change in extract could be measured for two consecutive days. The beer was then filled into a 50 L keg and carbonated by repeated pressurization with CO_2 to 1 bar at 2 °C. After 5 days, the carbonated beer was filled into 330 mL brown glass bottles with a counter-pressure hand-filler (TOPINCN, Shenzen, China) and capped. Bottles were pasteurized in a pilot retort (APR-95; Surdry, Abadiano, Vizcaya, Spain) with spray water at 65 °C for 10 min resulting in approximately 23 pasteurization units (PU). The successful pasteurization was confirmed by plating the pasteurized NAB on agar plates. Beer bottles were stored at 2 °C in a dark place for further analysis and sensory evaluation.

2.8. Sensory Evaluation

The sensory evaluation of the samples produced during yeast screening and RSM trial were judged by a panel of 12–15 experienced tasters. Samples were given at ambient temperature (20 °C) with a three-digit code. Each panelist evaluated the samples in an individual booth at ambient temperature (20 °C). The tasters were asked to desribe the sample in their own words, followed by evaluation of the intensity of a fruity smell and the overall acceptance of the smell of the sample on a hedonic scale from 0 ("not fruity"/"dislike extremely") to 5 ("extremely fruity"/"like extremely") according to MEBAK Sensory Analysis 3.2.1 "Simple Descriptive Test" and 3.2.2 "Profile Test", respectively.

The non-alcoholic beer samples (C6.1 pilot scale and commercial samples) were tasted and judged by a sensory panel of ten experienced and certified (DLG International Certificate for Sensory Analysis—beer and beer-based mixed drinks; Deutsche Landwirtschafts-Gesellschaft e.V.) panelists. A "Simple Descriptive Test" and "Profile Test" were performed according to MEBAK Sensory Analysis 3.2.1 and 3.2.2, respectively. Attributes for the aroma were "wort-like", "floral", "fruity", "citrus-like" and "tropical". A taste attribute "sweet taste" was also included. Panelists were asked to evaluate the attributes in their intensity on a line-marking scale from 0, "not perceptible", to 5, "strongly perceptible". Before the evaluation of the intensity, a descriptive sensory was performed, where the panelists were asked to describe the aroma of the samples in their own words. Samples were provided in dark glasses with a three-digit code and evaluated at a temperature of 20 °C in order to evaluate the full flavor profile (following DLG guidelines). The commercial samples NAB A and NAB B were non-alcoholic beers produced by limited fermentation [29] and "dialysis technology" [30], respectively. Each panelist tasted the samples in an individual booth at ambient temperature (20 °C). The amount of sample tasted was 50 mL per sample.

2.9. Wort and Beer Analyses

2.9.1. HPLC Analyses

Sugars and ethanol were determined by HPLC Agilent 1260 Infinity (Agilent Technologies, Santa Clara CA, USA) equipped with a refractive index detector (RID) and a Sugar-Pak I 10 µm,

6.5 mm × 300 mm column (Waters, Milford MA, USA), with 50 mg/L Ca-EDTA as mobile phase and a flow rate of 0.5 mL/min at 80 °C. Differentiation of maltose and sucrose was achieved with a Nova-Pak 4 µm, 4.6 mm × 250 mm column (Waters, Milford MA, USA), with acetonitrile/water 78:22 (*v/v*) as mobile phase and a flow rate of 1.0 mL/min. Quantification was achieved by external standards in a calibration range of 0.5 to 30 mM.

2.9.2. GC Analyses

Free vicinal diketones were quantified by a Clarus 500 gas chromatograph (Perkin-Elmer, Waltham MA, USA) with a headspace unit and Elite-5 60 m × 0.25 mm, 0.5 µm column using a 2,3-hexandione internal standard. Fermentation by products (esters, higher alcohols) was quantified using a Clarus 580 (Perkin-Elmer, Waltham MA, USA) gas chromatograph with a headspace unit and INNOWAX cross-linked polyethylene-glycol 60 m × 0.32 mm, 0.5 µm column (Perkin-Elmer, Waltham MA, USA). Vials containing beer samples were equilibrated for 25 min at 60 °C. The samples were injected at 50 °C, rising to 85 °C after one minute by heating at 7 °C/min. A temperature of 85 °C was maintained for one minute and then elevated to 190 °C at a heating rate of 25 °C/min.

2.9.3. Other

Glycerol was determined via enzymatic assay kit (glucokinase method), following the recommended procedure (K-GCROLGK, Megazyme, Bray Co. Wicklow, Ireland). The method is based on the use of ADP-glucokinase and an increase in absorbance on conversion of NAD^+ to NADH, and is performed at ambient temperature at a sample volume of 2 mL.

Free amino nitrogen (FAN) was measured using a ninhydrin-based dying method, where absorbance is measured at 570 nm against a glycine standard (ASBC Method Wort-12 A). The method is performed at a total volume of 10 mL. Following the color reaction at 95 °C, the samples are measured at ambient temperature.

Extract (apparent and real) and ethanol (for fermentation monitoring) were analyzed via density meter DMA 4500M with Alcolyzer Beer ME (Anton-Paar GmbH, Graz, Austria) at 20 °C and a sample volume of 30 mL.

The pH was determined using a digital pH meter (Mettler Toledo LLC, Columbus, OH, USA).

2.10. Statistical Analyses

Screening fermentations and analyses were carried out in triplicate. Statistical analysis was performed using RStudio, Version 1.1.463 with R version 3.5.2 (RStudio Inc, Boston, MA, USA; R Core Team, r-project). One-way analysis of variance (ANOVA) was used to compare means, and Tukey's post hoc test with 95% confidence intervals was applied for the pairwise comparison of means. When available, values are given as the mean ± standard deviation. Statistical analyses during the RSM trials were performed using the DesignExpert 9 software (StatEase, Minneapolis, MN, USA).

3. Results and Discussion

3.1. Yeast Strain Characterization

To identify the species of the yeast strains, amplification of the D1/D2 domain via PCR was performed and sequenced. The obtained sequences were compared to publicly available sequences in the NCBI nucleotide database via BLAST. The results of the strain identification are shown in Table 3.

Table 3. Yeast strain designation, species and origin of yeast strains used in this study.

Strain Designation	Species	Origin	Yeast Bank
837A	*Cyberlindnera misumaiensis*	Brewery cellar	FZW BLQ [1], Weihenstephan, Germany
NT Cyb	*Cyberlindnera fabianii*	Dried yeast starter for rice wine	FZW BLQ [1], Weihenstephan, Germany
L1	*Cyberlindnera jadinii*	Fruit of *Solanum quitoense*, "Lulo"	UCC Culture Collection, Cork, Ireland
C6.1	*Cyberlindnera subsufficiens*	Coconut	UCC Culture Collection, Cork, Ireland
CBS 1707 [T]	*Cyberlindnera mrakii*	Soil	Westerdijk Fungal Biodiversity Institute, Utrecht, Netherlands
CBS 5763[T]	*Cyberlindnera subsufficiens*	Soil	Westerdijk Fungal Biodiversity Institute, Utrecht, Netherlands

[1] Research Centre Weihenstephan for Brewing and Food Quality, Technische Universität München; [T] Type strain.

The yeast strains were found to belong to the species *Cyberlindnera misumaiensis* (837A), *C. fabianii* (NT Cyb), *C. jadinii* (L1), and *C. subsufficiens* (C6.1). The *Cyberlindnera mrakii* type strain CBS 1707 (former *Williopsis saturnus* var. *mrakii*; synonym NCYC 500) was included in this study as a strain that has previously been investigated for the production of a low alcohol beer with high levels of esters [20]. The *Cyberlindnera subsufficiens* type strain CBS 5763 was included as an example to investigate potential intraspecific differences from C6.1.

3.2. API Substrate Utilization

Before considering non-conventional yeasts for NABLAB brewing, their behavior regarding utilization of important wort sugars like maltose and sucrose should be investigated. An API ID 32C test was performed to investigate the utilization of those sugars and to show general, interspecific differences between the strains. The results of the API test are shown in Table 4.

Table 4. Results of the API ID 32C substrate utilization test of the individual strains. Substrates without brewing relevance, which were negative for all strains, are not shown. "+" positive, "−" negative, "w" weak.

Substrate	837A	NT Cyb	L1	C6.1	CBS 1707	CBS 5763
Cycloheximide (Actidione)	+	−	−	−	−	−
D-Cellobiose	+	+	+	+	+	+
D-Galactose	−	−	w	−	−	−
D-Glucose	+	+	+	+	+	+
D-Maltose	−	+	+	−	+[1]	−[1]
D-Mannitol	+	+	w	w	w	w
D-Melibiose	−	−	−	−	−	−
D-Melezitose	−	+	+	−	+	−
D-Raffinose	−	+	+	+	+	+
D-Sorbitol	+	+	w	+	−	−
D-Sucrose	−	+	+	+	−	+
D-Trehalose	−	+	−	−	+	−
D-Xylose	−	+	+	+	+	+
Esculin Ferric Acid	+	+	+	+	+	+
Glucosamine	−	−	−	w	−	−
Glycerol	+	+	+	+	+	+
Lactic Acid	−	+	+	+	+	+
Levulinic Acid	−	w	w	w	w	+
L-Sorbose	−	−	−	−	−	+
Methyl-αD-Glucopyranoside	−	+	−	−	−	−
N-Acetyl-Glucosamine	−	−	w	−	w	−
Palatinose	−	+	+	−	+	−
Potassium Gluconate	w	w	−	+	w	+

[1] Growth "variable" according to Kurtzman et al. [31].

Maltose utilization was positive for NT Cyb, L1 and CBS 1707, in accordance with the reported literature, although assimilation of maltose by CBS 1707 is classified as "variable" [31]. Sucrose utilization was positive for four of the six strains and negative for 837A and CBS1707. The results suggest that in brewers' wort, where maltose is the most abundant fermentable sugar, only NT Cyb, L1 and CBS 1707 have the capability to achieve high attenuations. However, the API test investigates substrate utilization under aerobic conditions. Sugar consumption during fermentation, under anaerobic conditions, can differ significantly [31], which is also known as the Kluyver effect [32]. Due to the inability of 837A and CBS 1707 to utilize sucrose, lower attenuations in fermentations in wort could be expected.

3.3. Stress Tests

When considering non-*Saccharomyces* yeast strains for brewing purposes, several brewing-relevant parameters such as flocculation behavior, POF production and stress responses should be investigated [33]. The flocculation behavior can give initial indications regarding yeast handling in terms of potential bottom cropping. POF behavior is important because in most beer styles, POF is not desired. Substances like hop-derived iso-α-acids, ethanol content, or the pH value of the wort can have significant influences on yeast activity, manifesting mainly in a prolonged lag time, and even complete growth inhibition [33–35]. With the investigated yeast strains, iso-α-acid concentrations of up to 100 IBU had no significant effect on the yeast growth (data not shown), which is in accordance with previous reports on seven different non-*Saccharomyces* species [34,35]. However, Michel et al. [33] reported a minor prolongation in the lag time of *Torulaspora delbrueckii* strains in concentrations of up to 90 IBU. The results of the investigated characterization attributes are shown in Table 5.

Table 5. Characterization of yeast strains for flocculation behavior, phenolic off-flavor (POF) production and lag time in wort with and without a stressor at different concentrations. "—" no growth.

Characterization Attributes		Unit	837A	NT Cyb	L1	C6.1	CBS 1707	CBS 5763
Flocculation		%	78 ± 3	22 ± 2	35 ± 4	32 ± 1	85 ± 2	51 ± 4
POF		-	negative	negative	negative	negative	negative	negative
Ethanol	0% ABV	h	18	6	9	6	9	9
	2.5% ABV	h	120	12	18	18	12	18
	5% ABV	h	—	24	36	24	48	—
	7.5% ABV	h	—	42	—	—	126	—
pH	5.5	h	18	6	9	6	9	9
	5	h	18	6	9	6	9	9
	4	h	66	6	9	6	9	9
	3	h	—	12	24	18	78	42

CBS 1707 exhibited the strongest flocculation behavior, at 85%, followed by 837A and CBS 5763, at 78% and 51%, respectively. NT Cyb, L1 and C6.1 exhibited very low flocculation of below 35%. All strains were negative for POF behavior. NT Cyb and C6.1 exhibited the fastest growth in wort (without a stress factor), overcoming the lag time after only 6 hours, followed by L1 and the CBS strains after 9 hours. Strain 837A exhibited a long lag phase of 18 hours (Figure 1). Concentrations of 2.5% ABV ethanol in the wort affected the lag time of all investigated strains. 837A was especially susceptible, with a prolonged lag phase of 120 hours. The remainder of the strains showed an extension of the lag phase of 3 to 12 hours. At 5% ABV, growth was fully inhibited for 837A and CBS 5763, while the other strains again exhibited an extension of the lag phase, of up to a maximum of 48 hours in CBS 1707. Complete growth inhibition was observed for L1 and C6.1 at 7.5% ABV, while the lag phase of NT Cyb and CBS 1707 was prolonged to 42 and 126 hours, respectively. All strains except 837A, which showed a significant extension of the lag phase to 66 hours, remained unaffected by a lower pH of 4. Only at pH 3 were lag times affected, while 837A was fully inhibited. Growth at low pH is important when considering the yeast for sour beer production, where the yeast must withstand pH values of below 4 [36]. However, it has been shown that organic acids like lactic acid can have a stronger inhibitory effect on yeasts and other microorganisms than HCl, which is caused by its chemical properties as a weak acid [35,37]. Inhibition by lactic acid could therefore be more pronounced than the HCl inhibition observed in this study. Figure 1 shows the growth of the investigated yeast strains in wort without the addition of a stressor.

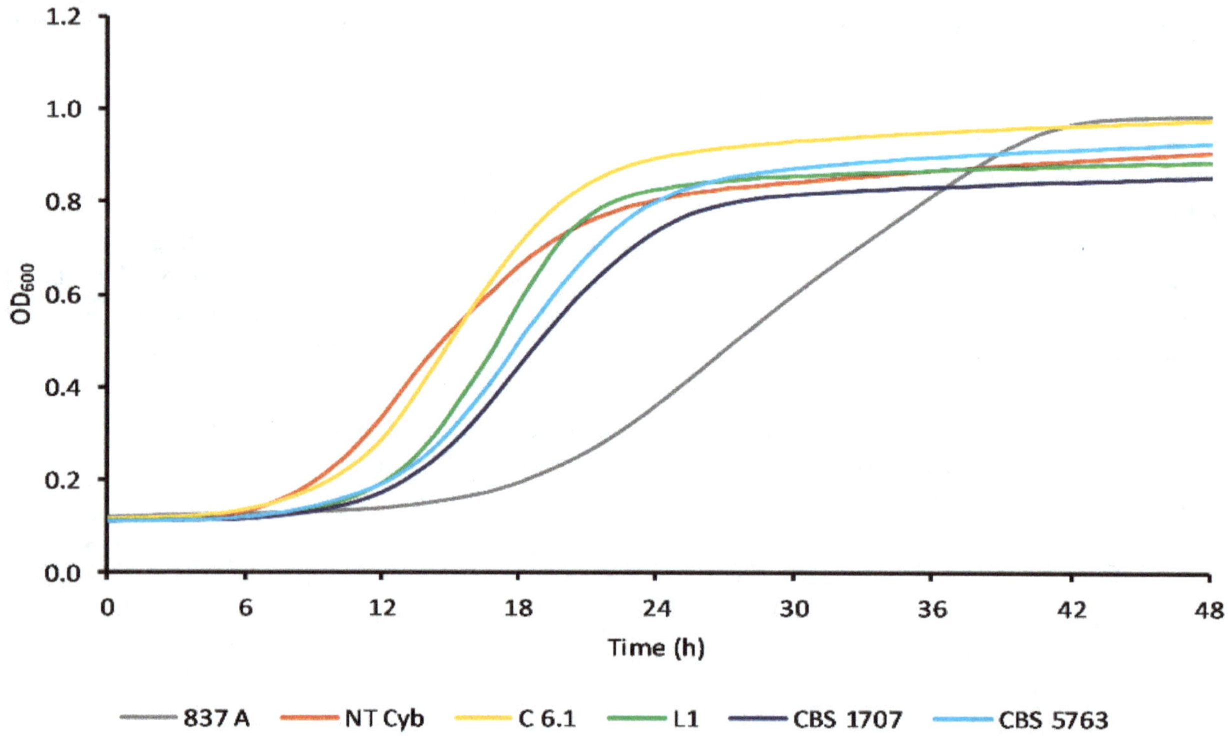

Figure 1. Growth of yeast strains in 7 °P wort extract at 25 °C without a stressor. Growth curves shown are the mean of a triplicate.

3.4. Screening

To investigate interspecific differences in the fermentation of wort, fermentation trials were performed in a diluted wort extract of 7 °P. Previous studies have shown that extract contents of around 7 °P will yield ethanol concentrations of around 0.5% ABV, a popular legal limit for NAB [7], in fermentations with maltose-negative yeast strains [1,14,34,38]. After aerobic propagation for 24 hours, NT Cyb exhibited the highest number of cells, at 2×10^9 cells/mL, more than four-fold the amount of cells compared to L1, C6.1, and the CBS strains with counts between 3.4 and 4.9×10^8 cells/mL (Table 6). Due to a delayed growth (compare Figure 1), 837A had to be propagated for 48 hours, reaching a cell count of 6.1×10^8 cells/mL. For the screening in wort, yeast cells were added at a concentration of 3×10^7 cells/mL, after a gentle washing step in water to prevent carry-over of propagation wort sugars. The results from the yeast screening are shown in Table 6. The fermentations were carried out until no change in extract could be measured for two consecutive days.

Table 6. Results of the screening of the investigated *Cyberlindnera* strains in wort extract.

Non-Alcoholic Beer (NAB) Attributes		Unit	837A	NT Cyb	L1	C6.1	CBS 1707	CBS 5763
Propagation	Cell count (24 h)[1]	× 10^6 cells/mL	611 ± 34 [1]	2055 ± 21	486 ± 27	445 ± 4	338 ± 25	386 ± 48
Fermented wort	Real Extract	°P	6.53 ± 0.03 ^{b}	6.40 ± 0.04 ^{ab}	6.45 ± 0.05 ^{ab}	6.36 ± 0.03 ^{a}	6.57 ± 0.10 ^{b}	6.35 ± 0.10 ^{a}
	Attenuation	%	18 ± 1 ^{ab}	23 ± 1 ^{bc}	21 ± 2 ^{abc}	24 ± 1 ^{c}	17 ± 3 ^{a}	24 ± 3 ^{c}
	Ethanol	% ABV	0.55 ± 0.01 ^{a}	0.63 ± 0.01 ^{b}	0.66 ± 0.00 ^{cd}	0.63 ± 0.00 ^{bc}	0.54 ± 0.01 ^{a}	0.67 ± 0.02 ^{d}
	pH	-	4.41 ± 0.02 ^{ab}	4.51 ± 0.02 ^{c}	4.44 ± 0.01 ^{bc}	4.38 ± 0.03 ^{b}	4.37 ± 0.06 ^{ab}	4.33 ± 0.01 ^{a}
	FAN	mg/L	88 ± 1 ^{b}	83 ± 2 ^{ab}	80 ± 1 ^{a}	81 ± 4 ^{a}	78 ± 1 ^{a}	84 ± 5 ^{ab}
Sugar consumption	Maltotriose	%	3 ± 2	6 ± 1	4 ± 1	5 ± 1	3 ± 1	5 ± 3
	Maltose	%	4 ± 2	4 ± 0	3 ± 1	4 ± 0	4 ± 1	4 ± 4
	Sucrose	%	2 ± 10	100	100	100	2 ± 2	100
	Glucose	%	100	100	100	100	100	100
	Fructose	%	81 ± 1	75 ± 1	100	80 ± 1	73 ± 8	83 ± 2
Fermentation by-products	Glycerol	g/L	0.25 ± 0.05 ^{ab}	0.23 ± 0.04 ^{a}	0.36 ± 0.03 ^{b}	0.30 ± 0.04 ^{ab}	0.18 ± 0.05 ^{a}	0.21 ± 0.05 ^{a}
	Acetaldehyde	mg/L	9.70 ± 2.83 ^{b}	8.05 ± 1.48 ^{b}	2.60 ± 0.14 ^{a}	3.37 ± 0.71 ^{a}	3.83 ± 0.45 ^{a}	2.57 ± 0.21 ^{a}
	Ethyl acetate	mg/L	65.70 ± 14.57 ^{b}	22.55 ± 2.90 ^{a}	9.27 ± 3.23 ^{a}	4.90 ± 0.85 ^{a}	8.10 ± 0.28 ^{a}	5.17 ± 0.29 ^{a}
	Isoamyl acetate	mg/L	0.90 ± 0.14 ^{ab}	<LOD	0.15 ± 0.07 ^{a}	1.60 ± 0.62 ^{b}	1.67 ± 0.12 ^{b}	1.03 ± 0.23 ^{ab}
	Ethyl formate	mg/L	0.53 ± 0.04 ^{a}	0.31 ± 0.06 ^{a}	0.57 ± 0.09 ^{a}	0.25 ± 0.03 ^{a}	2.70 ± 0.57 ^{c}	1.45 ± 0.07 ^{b}
	Ethyl propionate	mg/L	0.13 ± 0.04 ^{a}	0.13 ± 0.01 ^{a}	<LOD	<LOD	0.16 ± 0.01 ^{a}	0.17 ± 0.03 ^{a}
	Isoamyl alcohols	mg/L	11.20 ± 0.14 ^{a}	16.40 ± 0.57 ^{b}	23.15 ± 0.92 ^{c}	11.67 ± 1.74 ^{a}	11.93 ± 0.93 ^{a}	10.50 ± 0.14 ^{a}
	n-Propanol	mg/L	4.03 ± 0.84 ^{ab}	3.73 ± 0.21 ^{ab}	4.40 ± 0.62 ^{b}	3.27 ± 0.15 ^{ab}	2.93 ± 0.29 ^{a}	3.33 ± 0.15 ^{ab}
	Isobutanol	mg/L	7.57 ± 1.24 ^{ab}	7.70 ± 0.36 ^{b}	8.27 ± 1.38 ^{b}	8.03 ± 0.40 ^{b}	5.33 ± 0.55 ^{a}	7.20 ± 0.20 ^{ab}
	Σ Esters	mg/L	67.26 ± 14.79 ^{b}	22.99 ± 2.87 ^{a}	9.99 ± 3.31 ^{a}	6.75 ± 0.61 ^{a}	12.62 ± 1.48 ^{a}	7.82 ± 0.30 ^{a}
	Σ Alcohols	mg/L	22.80 ± 0.14 ^{a}	27.83 ± 0.64 ^{b}	35.82 ± 1.48 ^{c}	22.97 ± 1.97 ^{a}	20.20 ± 0.17 ^{a}	21.03 ± 0.21 ^{a}
Sensory	Aroma	-	Solvent-like, unpleasant	Cabbage-like, unpleasant	Fruity, pleasant	Fruity, pleasant	Fruity, pleasant	Fruity, pleasant

[1] Cell count after 48 h due to delayed growth compared to other strains (compare Figure 1). LOD 'limit of detection'. Different superscripts of values within a row indicate a significant difference ($p \leq 0.05$).

Strains 837A and CBS 1707 exhibited the lowest attenuation of only 18% and 17%, respectively, owing to their inability to utilize sucrose (Table 4), which was confirmed by the lack of sucrose consumption. Liu and Quek [20] also reported the absence of sucrose utilization by CBS 1707. The other strains, which depleted sucrose completely, reached attenuations of 21% to 24%. Consequently, 837A and CBS 1707 also produced, at 0.55% and 0.56% ABV, the lowest amounts of ethanol ($p \leq 0.05$) compared to the remaining strains, where ethanol concentrations ranged from 0.63% to 0.67% ABV. The final pH of the fermented samples ranged from 4.33 (CBS 5763) to 4.51 (NT Cyb). Residual FAN ranged from 78 (CBS 1707) to 88 mg/L (837A). As expected, none of the strains consumed maltotriose. Maltose consumption was also neglectable in all strains, although the species *Cyberlindnera fabianii* (like NT Cyb) has been reported to be able to ferment maltose [31,39]. The observations also underlined that results from the API substrate utilization test (where NT Cyb, L1 and CBS 1707 were positive for maltose) are not necessarily reflected in practice, especially since sugar utilization during respiration and fermentation can differ [31,32,40]. While glucose was depleted by all strains, fructose was only fully depleted by L1. The remaining strains exhibited glucophilic behavior and consumed only 73% to 83% of fructose during fermentation. Regarding fermentation by-products, glycerol concentrations were low, ranging from 0.18 to 0.36 g/L. The strains 837A and NT Cyb accumulated significantly higher amounts of acetaldehyde, at 9.7 and 8.1 mg/L, respectively, compared to 2.6 to 3.8 mg/L in the remaining samples. The sample fermented with *Cyberlindnera misumaiensis* 837A exhibited extremely high values of ethyl acetate, at 65.7 mg/L, twice the flavor threshold concentration in beer [2,41]. Ethyl acetate is described to have a fruity, estery character but also solvent-like, especially in high concentrations. The remaining strains exhibited ethyl acetate production between 4.9 (C6.1) and 22.6 mg/L (NT Cyb). Isoamyl acetate, which is predominantly described as having a fruity, banana-like aroma, has a much lower flavor threshold of only 1.4–1.6 mg/L [2,41]. The strains C6.1 and CBS 1707 produced the highest amounts of isoamyl acetate, at 1.67 and 1.60 mg/L, followed by CBS 5763, 837A and L1, at 1.03, 0.90 and 0.15 mg/L, respectively. NT Cyb did not produce detectable amounts of isoamyl acetate. Concentrations of ethyl formate and ethyl propionate in the fermented samples were low, ranging from undetectable to 2.7 mg/L. Ethyl butyrate and ethyl caproate were not detected in either of the samples (data not shown). The strain L1 produced a significantly higher amount of higher alcohols, at 35.8 mg/L, followed by NT Cyb, at 27.8 mg/L, and the remaining strains at 20–23 mg/L. During sensory evaluation, the high ethyl acetate concentration in the sample fermented with 837A was indeed perceptible and described as an unpleasant, solvent-like aroma. The sample fermented with NT Cyb was described as having an unpleasant, cabbage-like aroma. The remaining samples were characterized by a pleasant, fruity aroma.

The unpleasant, solvent-like aroma in the sample fermented with 837A was attributed to the very high ethyl acetate concentration, well above the flavor threshold. However, the cabbage-like aroma, which is generally associated with sulfides or thiol compounds [41], that was detected in the sample fermented with NT Cyb could not be linked to the volatile by-products that were measured. Interestingly, ethyl acetate concentrations in the remaining samples, characterized by a pleasant, fruity aroma, were low, at only 2.6–3.8 mg/L. However, C6.1, CBS 1707 and CBS 5763 exhibited higher amounts of isoamyl acetate, a desired ester in beer (particularly ales) [42], when compared to the samples with unpleasant aroma. The concentrations of 1.0–1.6 mg/L are within the reported flavor threshold in beer of 0.5–2.0 mg/L [43]. Additionally, it is well known that synergistic effects between esters occur that can push the concentration of perception below their individual flavor thresholds [42,44,45]. Isoamyl acetate could therefore have been a cause of the fruity aroma in the samples fermented with C6.1, CBS 1707 and CBS 5763. However, the sample fermented with L1, which was also characterized by a fruity aroma, only contained a very low isoamyl acetate concentration of 0.15 mg/L. It is noteworthy, however, that the L1 sample contained a significantly higher amount of isoamyl alcohol, at 23.2 mg/L, which is described as having an alcoholic, fruity and banana-like flavor [2]. The results have confirmed that not a high amount of esters, but rather a balanced profile will lead to a pleasant, fruity aroma [5].

Based on the results from the screening, *Cyberlindnera subsufficiens* C6.1 was chosen for optimization of fermentation conditions by means of response surface methodology, followed by an up-scaled brewing trial at 60 L to create a fruity, non-alcoholic beer (≤0.5% ABV). Strains 837A and NT Cyb were eliminated because of their poor flavor characteristics. CBS 1707 was eliminated due to its inability to ferment sucrose, which apart from the lower attenuation, would remain in the wort after fermentation, acting as an additional sweetening agent and potential contamination risk. *Cyberlindnera jadinii* strain L1 was eliminated due to its very low isoamyl acetate production (Table 6) and due to its maltose utilization when oxygen was present (Table 4). The decision between the two similarly performing *Cyberlindnera subsufficiens* strains C6.1 and CBS 5763 was made in favor of C6.1 due to a more pleasant fruitiness. In addition, C6.1 showed increased tolerance towards stress caused by ethanol or low pH (Table 5).

3.5. Response Surface Methodology (RSM)

To find the optimal fermentation conditions for C6.1 for an up-scaled application to produce a fruity, non-alcoholic beer, RSM was performed. Michel et al. [46] applied RSM to optimize the fermentation conditions of a *Torulaspora delbrueckii* strain for brewing purposes. They found that the pitching rate and fermentation temperature were crucial parameters, which influenced the flavor character of the final beer. The optimal fermentation conditions were shown to be at 21 °C with a high pitching rate of 60×10^6 cells/mL. Especially for non-*Saccharomyces* yeasts, the pitching rate can be crucial since most non-*Saccharomyces* species have comparably smaller cell sizes [46]. Figure 2 shows an example of the differing cell size between *Cyberlindnera subsufficiens* strain C6.1 (A) and the brewers' yeast strain *Saccharomyces cerevisiae* WLP001 (B) at identical magnification.

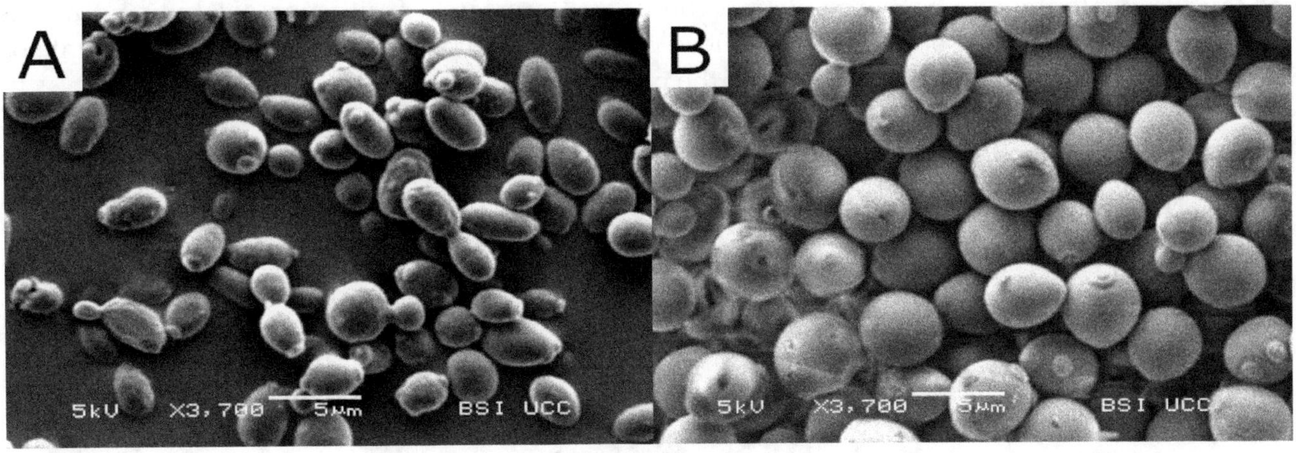

Figure 2. Scanning electron microscopy (SEM) picture of *Cyberlindnera subsufficiens* strain C6.1 (**A**) and the brewers' yeast strain *Saccharomyces cerevisiae* WLP001 (**B**) at a magnification of × 3700. Size of bar: 5 µm.

It is also known that temperature and pitching rate have an influence on ester production, though strain-specific differences also play a role [4,6]. Previously reported fermentation temperatures of *Cyberlindnera subsufficiens* and other *Cyberlindnera* spp. range from 20 to 25 °C [12,17,19,20,47]. Consequently, a two-factorial, face-centered central composite design was chosen with Factor A: fermentation temperature (17, 22, 27 °C), and Factor B: pitching rate (10, 35, 60×10^6 cells/mL). The individual experiment runs are listed in Table 2. The wort extract applied in the RSM trial was the same as that used for the screening, at an extract content of 7 °P (Table 1). Fermentation was conducted until no change in extract could be measured for two consecutive days. With the measured response values, significant models could be produced. The significant response models, with their respective minima and maxima and a summary of the model statistics, are shown in Table 7. Insignificant response models are not shown, and response models with a significant lack of fit will not be discussed in this study but are included in the visualized data for the sake of a complete picture. For a full report on model statistics and response values, refer to the supplementary Data Sheet S1.

Table 7. Analysis of variance (ANOVA) results for response models of the response surface methodology (RSM) trial.

Response	Unit	Minimum	Maximum	Model	*P*-Value	LOF *P*-Value
Ethanol	% ABV	0.41	0.60	RQuadratic	2.80×10^{-3} **	0.648
Ethyl acetate	mg/L	3.4	9.3	2FI	3.12×10^{-2} *	0.007 **
Isoamyl acetate	mg/L	0.8	2.2	RQuadratic	1.42×10^{-2} *	0.046 *
Acetaldehyde	mg/L	1.9	3.4	RLinear	1.35×10^{-3} **	0.337
n-Propanol	mg/L	3.2	4.5	2FI	9.03×10^{-3} **	0.029 *
Isobutanol	mg/L	3.2	6.7	RQuadratic	4.30×10^{-9} ***	0.145
Isoamyl alcohols	mg/L	7.3	13.3	Quadratic	2.67×10^{-5} ***	0.270
Σ Esters	mg/L	4.2	11.1	RQuadratic	1.48×10^{-2} *	0.018 *
Σ Alcohols	mg/L	13.7	22.9	RQuadratic	3.28×10^{-8} ***	0.339
Glycerol	g/L	0.17	0.37	RQuadratic	4.85×10^{-5} ***	0.034 *
Acceptance	-	1.08	3.38	Linear	1.31×10^{-2} *	0.377
Fruitiness	-	1.13	3.38	Linear	7.31×10^{-3} **	0.484

Model terminology: "RQuadratic" Reduced Quadratic; "2FI" Two-Factor Interaction; "RLinear" Reduced Linear. "LOF" Lack of Fit. ANOVA significance codes: *** $p \leq 0.001$, ** $p \leq 0.01$, * $p \leq 0.05$.

It was possible to create significant models for 12 responses (Table 7). However, five also exhibited significant lack of fit (LOF), rendering them unusable for predictions. The aim of the RSM was to investigate the optimal fermentation conditions to create a fruity, non-alcoholic beer. The three-dimensional response surface plots of the interactive effects of temperature and pitching rate on the final ethanol content and the fruitiness of the produced NAB are shown in Figures 3 and 4.

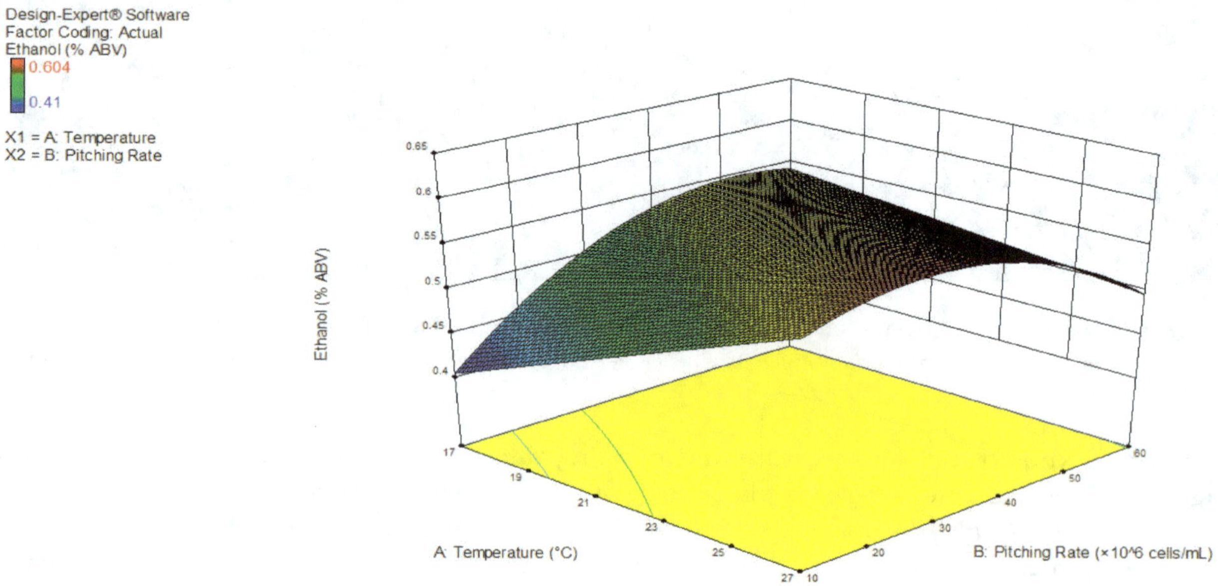

Figure 3. Three-dimensional response surface plot of the interactive effects of temperature and pitching rate on the ethanol content of the produced non-alcoholic beer ($p < 0.01$).

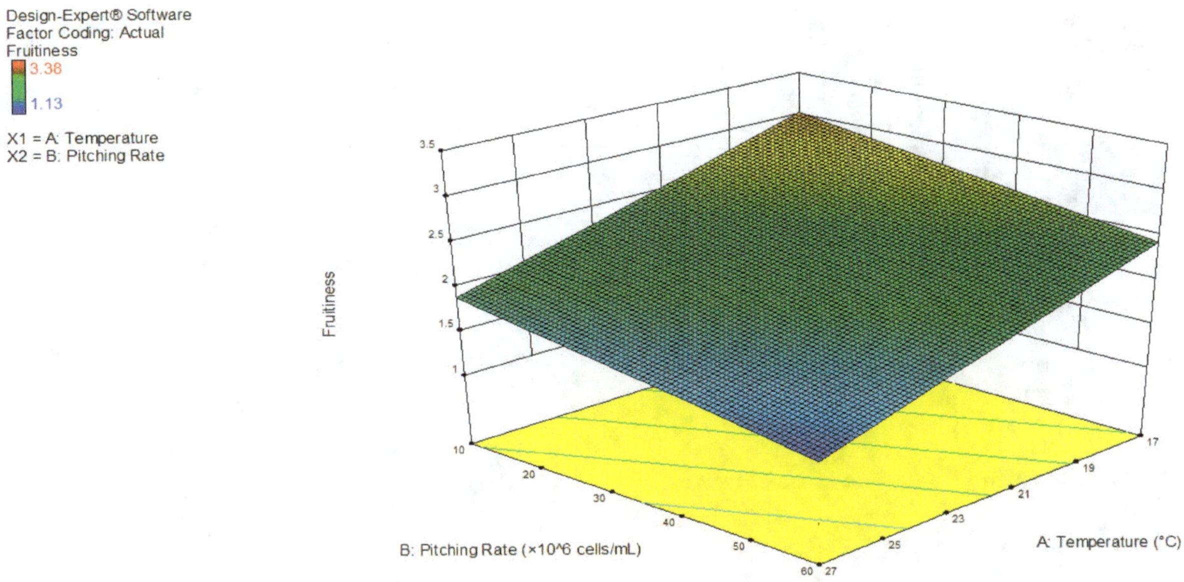

Figure 4. Three-dimensional response surface plot of the effects of temperature and pitching rate on the fruitiness of the produced non-alcoholic beer ($p < 0.01$).

Ethanol content was lowest at a low temperature of 17 °C and low pitching rate (10^7 cells/mL), and it went up with increasing temperature and pitching rate, but lowered again at a high pitching rate combined with a high fermentation temperature (Figure 3). The minium and maximum values were 0.41% and 0.60% ABV. Sugar analysis revealed that at 17 °C and 10^7 cells/mL, about 0.5 g/L of glucose was remaining after fermentation, while it was fully depleted in worts fermented at higher pitching rates and higher temperatures (data not shown). The residual sugar explained the lower final ethanol concentration. Fructose was only fully depleted in the samples that were fermented at 27 °C. At 22 °C, fermented samples exhibited residual fructose concentrations between 0.2 and 0.5 g/L, and at 17 °C, fermented samples showed remaining fructose concentrations between 0.2 and 0.7 g/L. Acetaldehyde concentrations were only dependent on the pitching rate, with increasing amounts of acetaldehyde found at lower pitching rates (Figure A1). This result correlates with other studies that found a decrease in acetaldehyde with increasing pitching rate in wort fermentations with brewers' yeasts [48,49]. However, overdosing yeast ($>5 \times 10^7$ cells/mL) can lead to an increase in acetaldehyde again, as observed by Erten et al. [50]. The temperature did not have a significant effect on the acetaldehyde concentration and was therefore excluded from the model ($p = 0.39$; supplementary Data Sheet S1). However, regarding higher alcohols, the fermentation temperature had a stronger effect, with increasing amounts of higher alcohols found at higher temperatures (Figures 5 and A2), which is consistent with the literature [4,5]. Isoamyl acetate concentrations were generally high and ranged from 0.8 to 2.2 mg/L. Although the model was significant ($p < 0.05$), it was unsuitable for value prediction due to a significant lack of fit ($p = 0.046$).

Interestingly, the production of the esters ethyl acetate and isoamyl acetate did not show a clear correlation to temperature, which underlines that the general rule of thumb, that higher fermentation temperatures lead to increased ester production, is not valid for all yeast strains (Figure 5) [4]. Furthermore, the amount of esters that were quantified in this study did not correlate with the perceived fruitiness of the NAB, which tentatively suggests that the fruity flavor profile was caused by yet unidentified compounds (Figure 5).

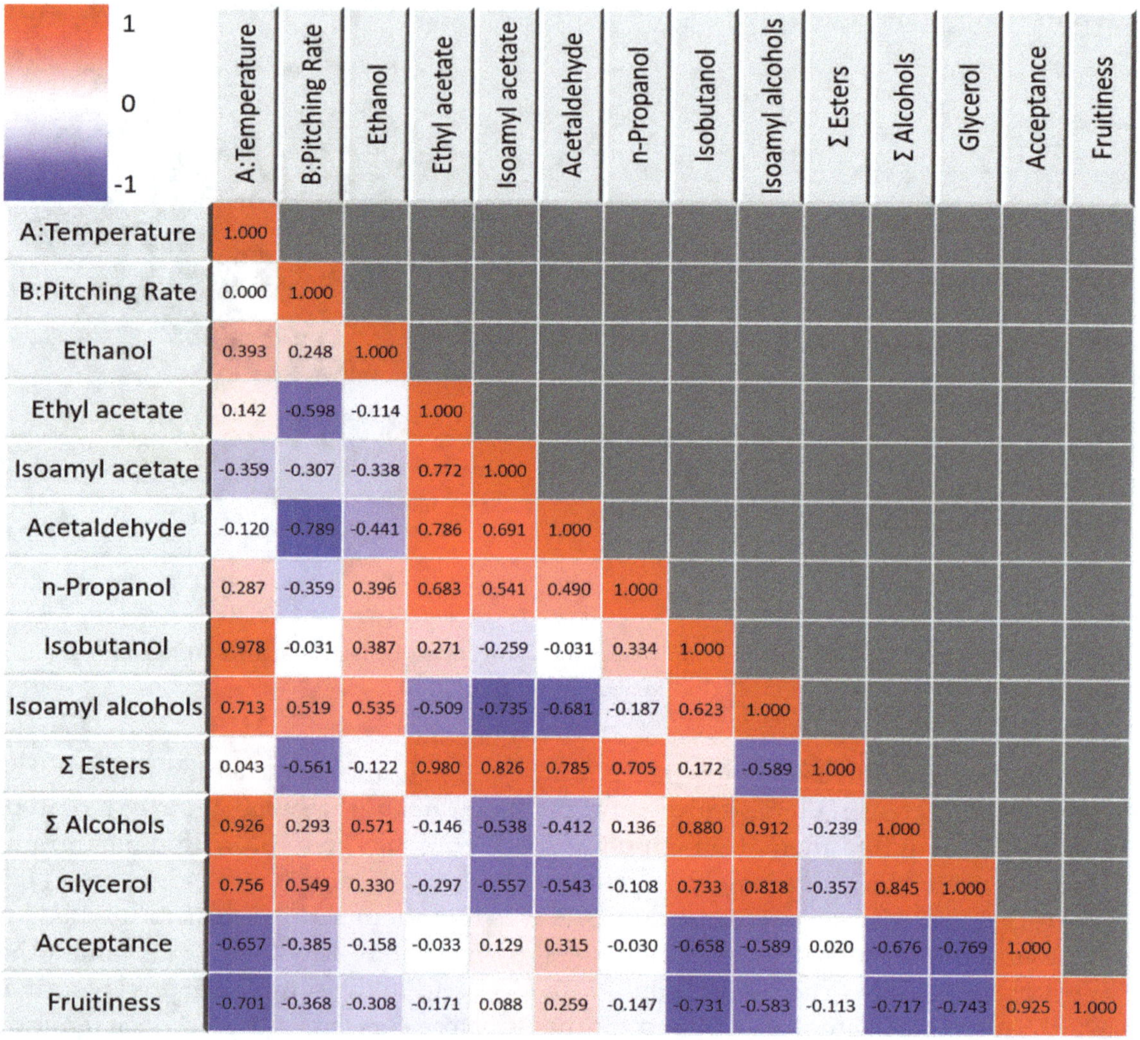

	A:Temperature	B:Pitching Rate	Ethanol	Ethyl acetate	Isoamyl acetate	Acetaldehyde	n-Propanol	Isobutanol	Isoamyl alcohols	Σ Esters	Σ Alcohols	Glycerol	Acceptance	Fruitiness
A:Temperature	1.000													
B:Pitching Rate	0.000	1.000												
Ethanol	0.393	0.248	1.000											
Ethyl acetate	0.142	-0.598	-0.114	1.000										
Isoamyl acetate	-0.359	-0.307	-0.338	0.772	1.000									
Acetaldehyde	-0.120	-0.789	-0.441	0.786	0.691	1.000								
n-Propanol	0.287	-0.359	0.396	0.683	0.541	0.490	1.000							
Isobutanol	0.978	-0.031	0.387	0.271	-0.259	-0.031	0.334	1.000						
Isoamyl alcohols	0.713	0.519	0.535	-0.509	-0.735	-0.681	-0.187	0.623	1.000					
Σ Esters	0.043	-0.561	-0.122	0.980	0.826	0.785	0.705	0.172	-0.589	1.000				
Σ Alcohols	0.926	0.293	0.571	-0.146	-0.538	-0.412	0.136	0.880	0.912	-0.239	1.000			
Glycerol	0.756	0.549	0.330	-0.297	-0.557	-0.543	-0.108	0.733	0.818	-0.357	0.845	1.000		
Acceptance	-0.657	-0.385	-0.158	-0.033	0.129	0.315	-0.030	-0.658	-0.589	0.020	-0.676	-0.769	1.000	
Fruitiness	-0.701	-0.368	-0.308	-0.171	0.088	0.259	-0.147	-0.731	-0.583	-0.113	-0.717	-0.743	0.925	1.000

Figure 5. Map visualizing correlations of response surface methodology (RSM) factors and responses based on the Pearson Correlation Coefficient. 1 signifies strong positive correlation, 0 signifies no correlation, and −1 signifies a strong negative correlation.

In terms of fruitiness, a low fermentation temperature paired with a low pitching rate led to the highest perceived fruitiness. Indeed, the highest fruitiness was recorded at 17 °C and 1×10^7 cells/mL and the lowest at 27 °C and 6×10^7 cells/mL, following a linear model. General acceptance showed a strong positive correlation with the fruitiness, indicating that the panel preferred fruity samples (Figures 5 and A3).

Due to the ideal combination of lowest ethanol content and highest fruitiness and acceptance, the fermentation temperature of 17 °C and pitching rate of 1×10^7 cells/mL were chosen as the optimal fermentation conditions for application to produce a fruity, non-alcoholic beer.

A small-scale fermentation at the optimal conditions (17 °C, 10^7 cells/mL) was conducted to validate the RSM model. Table 8 shows the predicted mean including 95% prediction intervals (PI) and the measured ("observed") mean with standard deviation.

Although predicted by a significant model, the observed means for ethanol, acetaldehyde and isobutanol values were not within the 95% prediction interval. Sugar analysis revealed the complete depletion of glucose in the experimental fermentation trial at optimal conditions compared to the RSM model prediction, which explained the increased ethanol production (data not shown). The moderate success in model validation demonstrates the limitations in the application of RSM to optimize fermentations, where small differences in substrate and process conditions can have significant influences on the outcome. Because wort is a very complex substrate, comprising a complex mixture of different sugars, nitrogen sources, minerals and vitamins, among others, any interpretation or the transfer of the RSM results to other substrates (even different wort substrates) should be made with caution. In particular, a different sugar composition will have a significant effect on the responses when applying maltose-negative yeasts. However, the improved fruitiness and therefore higher acceptance of the NAB produced at low temperature and low pitching rate, the main goal from the optimization, was significant and reproducable (Table 8).

Table 8. Response surface methodology (RSM) model validation via predicted value vs. observed value.

Response	95% PI Low	Predicted Mean	95% PI High	Observed Mean	Std. Dev.
Ethanol *	0.33	0.40	0.48	0.53	0.01
Ethyl acetate	0.89	4.74	8.60	6.83	0.59
Isoamyl acetate	0.78	1.63	2.47	2.50	0.10
Acetaldehyde *	2.19	2.97	3.74	1.27	0.29
n-Propanol	2.68	3.28	3.88	3.57	0.06
Isobutanol *	2.91	3.23	3.54	2.80	0.10
Isoamyl alcohols	5.78	7.03	8.29	4.10	0.10
SUM Esters	3.01	7.10	11.19	9.33	0.68
SUM Alcohols *	12.84	13.74	14.64	10.47	0.31
Glycerol	0.13	0.18	0.22	0.27	0.01
Acceptance *	2.12	3.23	4.34	3.75	0.62
Fruitiness *	2.02	3.03	4.05	3.58	0.87

* Significant model with insignificant lack of fit. 'PI' Prediction interval.

3.6. Pilot-Scale Brewing

Despite the limited model validation, the fermentation parameters were successfully optimized to enhance the fruity character of the NAB. Therefore, the pilot-scale brewing trial was conducted with the optimized conditions of 17 °C fermentation temperature and a pitching rate of 10^7 cells/mL.

The grain bill of the wort for the pilot-scale brewing trial consisted of 95% pilsner malt and 5% acidulated malt to lower the starting pH of the wort, to account for the reduced pH drop during fermentations with non-*Saccharomyces* yeasts compared to brewers' yeast. A low beer pH is desired to prevent microbial spoilage and to ensure good liveliness of the beer [51,52]. The analytical attributes of the wort produced at pilot-scale are shown in Table 9.

Table 9. Attributes of the wort produced on pilot-scale.

Wort attributes	Unit	Value
Extract	°P	7.00 ± 0.01
pH		4.86 ± 0.01
FAN	mg/L	107 ± 3
Glucose	g/L	6.01 ± 0.08
Fructose	g/L	0.80 ± 0.01
Sucrose	g/L	2.13 ± 0.03
Maltose	g/L	31.59 ± 0.44
Maltotriose	g/L	9.32 ± 0.13

To assess the suitability of *Cyberlindnera subsufficiens* C6.1 to produce a fruity NAB, it was compared to two commercial NABs. NAB A was a commercial non-alcoholic beer produced by limited fermentation [29], and NAB B was a non-alcoholic beer produced by "dialysis technology" [30]. The NABs were analyzed for their extract, ethanol, FAN and glycerol content as well as their sugar composition and concentration of volatile fermentation by-products. The results are shown in Table 10.

Table 10. Attributes of the non-alcoholic beer (NAB) produced with C6.1 compared to two commercial NABs, NAB A and NAB B.

NAB Attributes	Unit	C6.1 NAB	NAB A	NAB B
Extract (real)	°P	6.60 ± 0.01	6.76 ± 0.07	7.05 ± 0.03
Extract (apparent)	°P	6.46 ± 0.02	6.57 ± 0.06	6.86 ± 0.01
Ethanol	% ABV	0.36 ± 0.00	0.50 ± 0.03	0.49 ± 0.04
pH		4.45 ± 0.01	4.29 ± 0.02	4.29 ± 0.04
FAN	mg/L	96 ± 2	86 ± 6	24 ± 0
Glycerol	g/L	0.30 ± 0.02	0.33 ± 0.01	1.40 ± 0.03
Glucose	g/L	2.77 ± 0.05	2.74 ± 0.04	5.61 ± 0.04
Fructose	g/L	1.65 ± 0.03	1.96 ± 0.03	0.19 ± 0.00
Sucrose	g/L	<LOD	<LOD	<LOD
Maltose	g/L	30.27 ± 0.62	30.11 ± 0.50	17.69 ± 0.24
Maltotriose	g/L	8.67 ± 0.24	8.31 ± 0.21	1.84 ± 0.03
Acetaldehyde	mg/L	10.55	2.40	0.70
Ethyl acetate	mg/L	12.00	<0.10	2.70
Isoamyl acetate	mg/L	0.80	<0.1	0.70
Isoamyl alcohols	mg/L	4.00	4.80	17.40
n-Propanol	mg/L	2.20	<0.5	2.50
Isobutanol	mg/L	3.60	1.00	4.90
Diacetyl	mg/L	<0.01	0.02	0.04
2,3-Pentandione	mg/L	<0.01	<0.01	<0.01
Σ Esters	mg/L	12.8	<0.1	3.4
Σ Alcohols	mg/L	9.8	5.8	24.8

The C6.1 NAB reached final attenuation after 13 days of fermentation at 17 °C, at an ethanol content of 0.36% ABV. At the end of fermentation, 2.77 g/L glucose was remaining in the wort and sucrose was fully depleted. Compared to the initial sugar concentration of the wort (Table 9), fructose concentrations in the final beer were significantly higher, at 1.65 g/L, twice as high as the starting concentration in the wort. Since sucrose was fully depleted, it can be assumed that it was converted to glucose and fructose by the yeast's invertase. The high residual fructose could therefore be attributed to the previously observed glucophilic character of the C6.1 strain in the screening and RSM trial. As a result, fructose was not consumed by the yeast due to the permanent presence of glucose until fermentation came to a halt. As expected, maltose and maltotriose consumption was negligible. Despite the limited fermentation, C6.1 produced a relatively high amount of esters, at 12.8 mg/L, the majority of which was ethyl acetate (12 mg/L). NAB A had an ethanol content of 0.50% ABV. Interestingly, the sugar composition was very similar to that of the C6.1 NAB. Regarding fermentation by-products, however, NAB A exhibited very low concentrations, at about half the amount of higher alcohols and a total lack of the esters ethyl acetate and isoamyl acetate. NAB B had an ethanol content of 0.49% ABV. Owing to its fundamentally different production method, the analyzed attributes were very different from those of the two NABs produced solely by limited fermentation. The low FAN content together with a high glycerol content compared to the other NABs were indicators of a more extensive fermentation, with subsequent removal of ethanol. However, NAB B still exhibited high amounts of monosaccharides, which suggested that the production of the NAB either also entailed a limited fermentation, or the dealcoholized beer was blended with wort (or other means of sugar addition). The increased amounts of higher alcohols in NAB B, at 24.8 mg/L, are uncommon for beers dealcoholized via dialysis, since the process commonly reduces their content in the final NAB by 90%–95% [7]. Despite the addition of acid malt during the wort production for the C6.1 NAB, the final pH after fermentation was, at 4.45, higher compared to 4.29 in the commercial NABs.

Due to the high amounts of residual sugars, proper pasteurization is essential for non-alcoholic beers produced by limited fermentation to avoid microbial spoilage [1,38,53]. After bottling, C6.1 NAB was therefore pasteurized with approximately 23 PU, and the successful pasteurization was confirmed by plating the pasteurized NAB on agar to check for microorganism growth, which was found to be negative.

3.7. Sensory Evaluation

For a holistic evaluation of the C6.1 NAB compared to the two commercial NABs, a sensory trial was conducted with 10 trained and experienced panelists. The panel was asked to describe the flavor of the beer in their own words, followed by an assessment of several intensity attributes. The mean score values of the parameters wort-like, floral, fruity, citrus-like and tropical aroma, as well as sweet taste, of the NABs are shown in Figure 6.

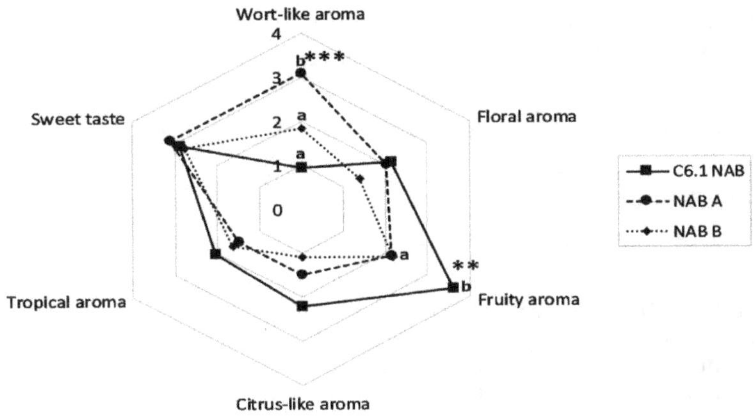

Figure 6. Spider web with the means of the descriptors from the sensory trial of the NAB produced with *Cyberlindnera subsufficiens* C6.1 and the two commercial NABs. Different letters next to data points indicate a significant difference as per Tukey's *post hoc* test. Significance codes: *** $p \leq 0.001$, ** $p \leq 0.01$.

The NAB produced with C6.1 was described as very fruity with aromas of pear, banana, mango and maracuja together with a slightly wort-like character. NAB A was described as malty, wort-like and hoppy, while NAB B was described as wort-like and caramel-like. The C6.1 NAB was indeed evaluated as being significantly more fruity than the commercial NABs ($p \leq 0.01$), at an average of 3.6 out of 5 compared to 2.1 and 2.2 out of 5, scoring also higher in citrus-like and tropical aromas. Consequently, the wort-like aroma, one of the most criticized flaws of NABs produced by limited fermentation [1,2,52], was least pronounced in the NAB produced with C6.1 with an average of 1 out of 5, followed by NAB B with 1.8 out of 5. NAB A exhibited, at an average of 3.2, a significantly more pronounced wort-like aroma ($p \leq 0.001$). A sweet taste, caused by a high amount of residual sugars, is another major point of criticism for NABs produced by limited fermentation [1,2,52]. All NABs scored similarly in sweet taste without significant differences. NAB B scored lower for "floral" compared to the other NABs. However, the difference was not statistically significant. When the panelists were asked for their favorite sample, 40% chose C6.1 NAB, 40% chose NAB A, and 20% chose NAB B. Similarly, Strejc et al. [3] investigated the production of a non-alcoholic beer (0.5% ABV) by a cold contact process (characterized by a low temperature and high pitching rate) with a mutated lager yeast strain (*Saccharomyces pastorianus*). The strain's targeted mutation resulted in an overproduction of isoamyl acetate and isoamyl alcohols. The authors reported that the fruity flavour of the NAB produced with the mutated strain was "partially able to disguise" the typical wort-like off-flavor [21]. However, the isoamyl acetate concentration of the resulting NAB was, at 0.5 mg/L, lower than the concentration in the C6.1 NAB in this study (Table 10). Furthermore, the complex mutation and isolation procedure paired with a potentially limited stability of the mutation limits its applicability in practice. Saerens and Swiegers [22] reported the successful production of a NAB at 1000 L scale with a *Pichia kluyveri* strain, owing to its high production of isoamyl acetate (2–5 mg/L), which reportedly gave the NAB a fruity flavor that was more like that of a regular beer than commercial NABs. In accordance, the results of the sensory indicated that a strong fruity aroma can mask the wort-like off flavor, and that the non-*Saccharomyces* yeasts, which produce a pronounced fruity character, can therefore be a means to produce NAB with improved flavor characteristics.

4. Conclusions

The *Cyberlindnera* genus was found to be a promising non-*Saccharomyces* genus for application in the production of a fruity, non-alcoholic beer. Four of the six investigated species produced a fruity character, despite the limited fermentative capacity, which resulted in a low ethanol concentration. It was shown that through optimization of the fermentation parameters of temperature and pitching rate, the fruity character could be enhanced. Process up-scaling with *Cyberlindnera subsufficiens* strain C6.1 produced a NAB that was significantly more fruity compared to two commercial NABs. Owing to the strong fruity aroma, the often-criticized wort-like aroma could successfully be masked. Yeast handling throughout the process (i.e., propagation, yeast pitching, fermentation) proved to be suitable for pilot-scale brewing, with potential for application at industrial scale. Further studies should investigate if the masking effect was enhanced by a reduction of wort aldehydes via yeast metabolism.

This study demonstrated the suitability of the non-*Saccharomyces* species *Cyberlindnera subsufficiens* for the production of non-alcoholic beer (<0.5% ABV) with novel flavor characteristics that can compete with commercial NABs. The successful pilot-scale (60 L) brewing trial gives prospect to future studies with diverse non-*Saccharomyces* yeasts and strengthens their position as a serious and applicable alternative to established methods in non-alcoholic and low alcohol beer brewing.

Author Contributions: Conceptualization, K.B., M.M., M.H., M.Z. and E.K.A.; methodology, K.B. and M.M.; investigation, K.B., J.J.A., and A.H.; resources, M.H., F.J. and E.K.A.; formal analysis, K.B.; writing—original draft preparation, K.B.; writing—review and editing, K.B., M.M., M.Z., M.H., K.M.L. and E.K.A.; visualization, K.B.; supervision, E.K.A.; project administration, E.K.A.; funding acquisition, E.K.A.

Acknowledgments: The authors would like to thank Philippe Lösel for his valuable contribution to this study, Muntons plc for supplying the sprayed wort extract, the Barth-Haas Group for supplying the iso-α-acids extract, and David De Schutter and Luk Daenen for their review.

Appendix A

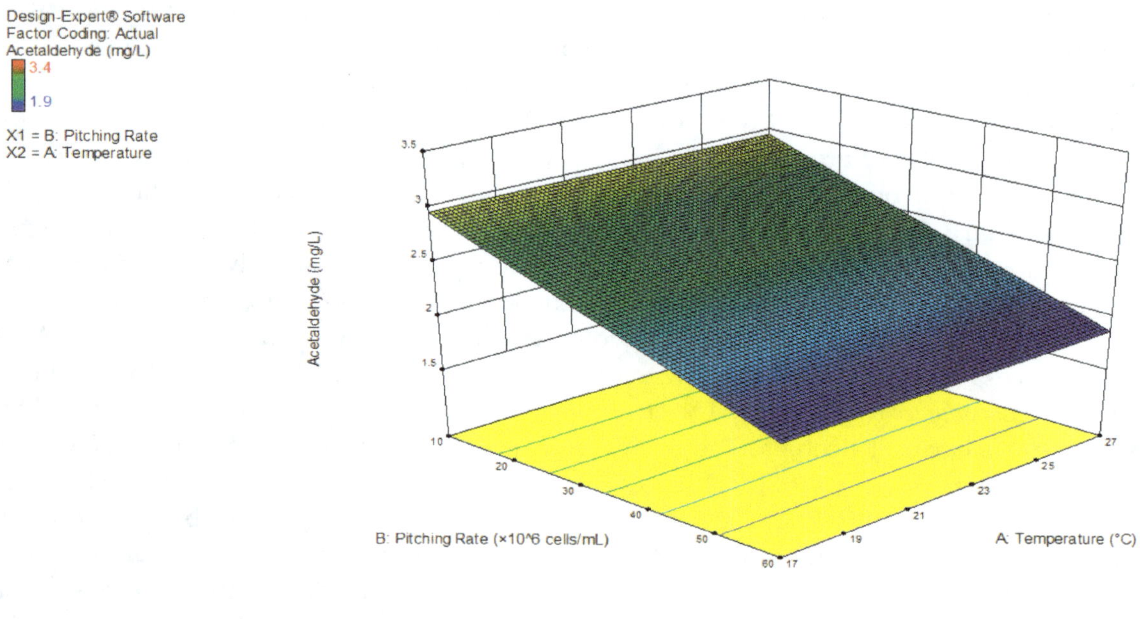

Figure A1. Three-dimensional response surface plot of the effect of pitching rate on the acetaldehyde content of the produced NAB ($p < 0.01$). The factor temperature was excluded from the model due to insignificance ($p = 0.39$; supplementary Data Sheet 1).

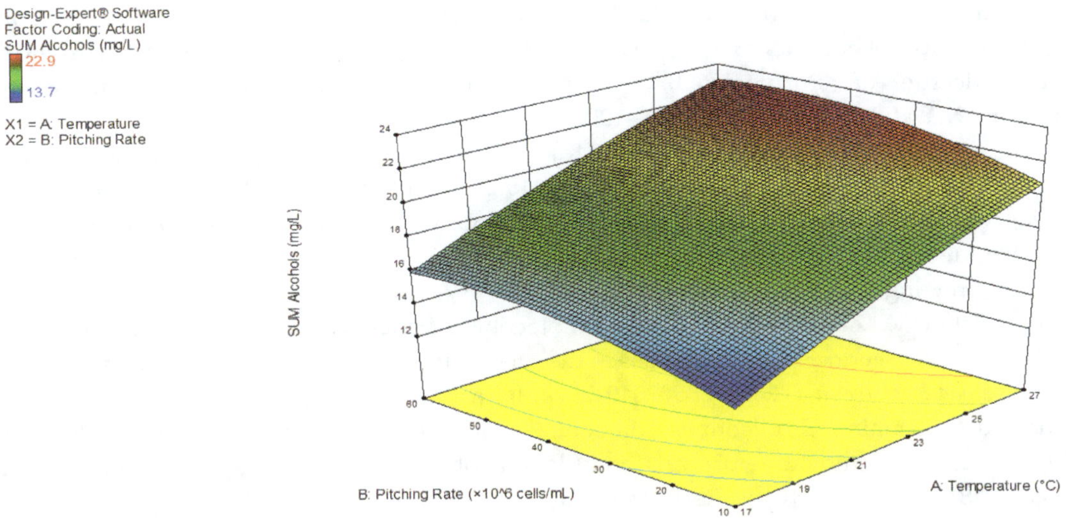

Figure A2. Three-dimensional response surface plot of the interactive effects of temperature and pitching rate on the sum of higher alcohols of the produced NAB ($p < 0.001$).

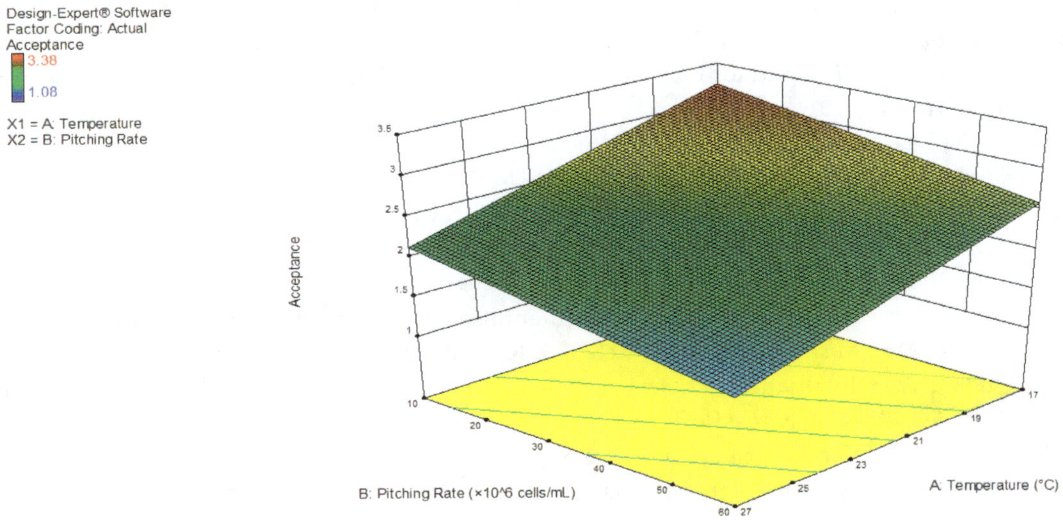

Figure A3. Three-dimensional response surface plot of the effects of temperature and pitching rate on the overall acceptance of the produced NAB ($p < 0.05$).

References

1. Bellut, K.; Arendt, E.K. Chance and Challenge: Non-*Saccharomyces* yeasts in nonalcoholic and low alcohol beer brewing: A Review. *J. Am. Soc. Brew. Chem.* **2019**, *77*, 77–91. [CrossRef]
2. Blanco, C.A.; Andrés-Iglesias, C.; Montero, O. Low-alcohol Beers: Flavor Compounds, Defects, and Improvement Strategies. *Crit. Rev. Food Sci. Nutr.* **2016**, *56*, 1379–1388. [CrossRef] [PubMed]
3. Perpète, P.; Collin, S. Influence of beer ethanol content on the wort flavour perception. *Food Chem.* **2000**, *71*, 379–385. [CrossRef]
4. Verstrepen, K.J.; Derdelinckx, G.; Dufour, J.P.; Winderickx, J.; Thevelein, J.M.; Pretorius, I.S.; Delvaux, F.R. Flavor-active esters: Adding fruitiness to beer. *J. Biosci. Bioeng.* **2003**, *96*, 110–118. [CrossRef]
5. Pires, E.J.; Teixeira, J.A.; Brányik, T.; Vicente, A.A. Yeast: The soul of beer's aroma—A review of flavour-active esters and higher alcohols produced by the brewing yeast. *Appl. Microbiol. Biotechnol.* **2014**, *98*, 1937–1949. [CrossRef]
6. Olaniran, A.O.; Hiralal, L.; Mokoena, M.P.; Pillay, B. Flavour-active volatile compounds in beer: Production, regulation and control. *J. Inst. Brew.* **2017**, *123*, 13–23. [CrossRef]
7. Müller, M.; Bellut, K.; Tippmann, J.; Becker, T. Physical Methods for Dealcoholization of Beverage Matrices and their Impact on Quality Attributes. *Chem. Ing. Tech.* **2016**, *88*. [CrossRef]

8. Riu-Aumatell, M.; Miró, P.; Serra-Cayuela, A.; Buxaderas, S.; López-Tamames, E. Assessment of the aroma profiles of low-alcohol beers using HS-SPME-GC-MS. *Food Res. Int.* **2014**, *57*, 196–202. [CrossRef]
9. Basso, R.F.; Alcarde, A.R.; Portugal, C.B. Could non-*Saccharomyces* yeasts contribute on innovative brewing fermentations? *Food Res. Int.* **2016**, *86*, 112–120. [CrossRef]
10. Holt, S.; Mukherjee, V.; Lievens, B.; Verstrepen, K.J.; Thevelein, J.M. Bioflavoring by non-conventional yeasts in sequential beer fermentations. *Food Microbiol.* **2018**, *72*, 55–66. [CrossRef]
11. Bellut, K.; Michel, M.; Hutzler, M.; Zarnkow, M.; Jacob, F.; De Schutter, D.P.; Daenen, L.; Lynch, K.M.; Zannini, E.; Arendt, E.K. Investigation into the application of *Lachancea fermentati* strain KBI 12.1 in low alcohol beer brewing. *J. Am. Soc. Brew. Chem.* **2019**, *77*, 157–169.
12. Bellut, K.; Michel, M.; Zarnkow, M.; Hutzler, M.; Jacob, F.; Lynch, K.M.; Arendt, E.K. On the suitability of alternative cereals, pseudocereals and pulses in the production of alcohol-reduced beers by non-conventional yeasts. *Eur. Food Res. Technol.* **2019**, *245*, 2549–2564. [CrossRef]
13. De Francesco, G.; Sannino, C.; Sileoni, V.; Marconi, O.; Filippucci, S.; Tasselli, G.; Turchetti, B. *Mrakia gelida* in brewing process: An innovative production of low alcohol beer using a psychrophilic yeast strain. *Food Microbiol.* **2018**, *76*, 354–362. [CrossRef] [PubMed]
14. Narziss, L.; Miedaner, H.; Kern, E.; Leibhard, M. Technology and composition of non-alcoholic beers—Processes using arrested fermentation. *Brauwelt Int.* **1992**, *4*, 396–410.
15. Michel, M.; Meier-Dörnberg, T.; Jacob, F.; Methner, F.J.; Wagner, R.S.; Hutzler, M. Review: Pure non-*Saccharomyces* starter cultures for beer fermentation with a focus on secondary metabolites and practical applications. *J. Inst. Brew.* **2016**, *122*, 569–587. [CrossRef]
16. Li, H.; Liu, Y.; Zhang, W. Method for manufacturing alcohol-free beer through *Candida shehatae*. China Patent CN102220198B, 13 May 2011.
17. Yilmaztekin, M.; Erten, H.; Cabaroglu, T. Production of Isoamyl Acetate from Sugar Beet Molasses by *Williopsis saturnus* var. *saturnus*. *J. Inst. Brew.* **2008**, *114*, 34–38. [CrossRef]
18. Inoue, Y.; Fukuda, K.; Wakai, Y.; Sudsai, T.; Kimura, A. Ester Formation by Yeast *Hansenula mrakii* IFO 0895: Contribuition of Esterase fir Iso-Amyl Acetate Production in Sake Brewing. *LWT Food Sci. Technol.* **1994**, *27*, 189–193. [CrossRef]
19. Aung, M.T.; Lee, P.R.; Yu, B.; Liu, S.Q. Cider fermentation with three *Williopsis saturnus* yeast strains and volatile changes. *Ann. Microbiol.* **2015**, *65*, 921–928. [CrossRef]
20. Liu, S.Q.; Quek, A.Y.H. Evaluation of Beer Fermentation with a Novel Yeast *Williopsis saturnus*. *Food Technol. Biotechnol* **2016**, *54*, 403–412. [CrossRef]
21. Strejc, J.; Siříšťová, L.; Karabín, M.; Almeida e Silva, J.B.; Brányik, T. Production of alcohol-free beer with elevated amounts of flavouring compounds using lager yeast mutants. *J. Inst. Brew.* **2013**, *119*, 149–155. [CrossRef]
22. Saerens, S.; Swiegers, J.H. Production of low-alcohol or alcohol-free beer with *Pichia kluyveri* yeast strains. International Patent WO2014135673A2, 7 March 2014.
23. Gibson, B.; Geertman, J.-M.A.; Hittinger, C.T.; Krogerus, K.; Libkind, D.; Louis, E.J.; Magalhães, F.; Sampaio, J.P. New yeasts—New brews: Modern approaches to brewing yeast design and development. *Fems. Yeast Res.* **2017**, *17*, 1–32. [CrossRef] [PubMed]
24. Saison, D.; De Schutter, D.P.; Vanbeneden, N.; Daenen, L.; Delvaux, F.; Delvaux, F.R. Decrease of aged beer aroma by the reducing activity of brewing yeast. *J. Agric. Food Chem.* **2010**, *58*, 3107–3115. [CrossRef] [PubMed]
25. Bendiak, D.; Van Der Aar, P.; Barbero, F.; Benzing, P.; Berndt, R.; Carrick, K.; Dull, C.; Dunn-Default, S.; Eto, M.; Gonzalez, M.; et al. Yeast Flocculation by Absorbance Method. *J. Am. Soc. Brew. Chem.* **1996**, *54*, 245–248.
26. D'Hautcourt, O.; Smart, K.A. Measurement of Brewing Yeast Flocculation. *J. Am. Soc. Brew. Chem.* **1999**, *57*, 123–128. [CrossRef]
27. Meier-Dörnberg, T.; Hutzler, M.; Michel, M.; Methner, F.-J.; Jacob, F. The Importance of a Comparative Characterization of *Saccharomyces Cerevisiae* and *Saccharomyces Pastorianus* Strains for Brewing. *Fermentation* **2017**, *3*, 41. [CrossRef]
28. Das Murtey, M.; Ramasamy, P. Sample Preparations for Scanning Electron Microscopy—Life Sciences. *Intech Open* **2018**, *2*, 64.

29. Stiegl-Freibier Alkoholfrei. Available online: https://www.stiegl-shop.at/braushop/en/shop/beer/non-alcoholic/stiegl-freibier-alcohol-free/?card=4193 (accessed on 12 October 2019).
30. Baltika 0 Alcohol Free Beer. Available online: https://eng.baltika.ru/products/baltika/baltika-0-alcohol-free-beer/ (accessed on 11 October 2019).
31. Kurtzman, C.P.; Fell, J.W.; Boekhout, T. *The Yeasts, A Taxonomic Study*, 5th ed.; Kurtzman, C.P., Fell, J.W., Boekhout, T., Eds.; Elsevier: London, UK, 2011; ISBN 9780444521491.
32. Fukuhara, H. The Kluyver effect revisited. *Fems. Yeast Res.* **2003**, *3*, 327–331. [CrossRef]
33. Michel, M.; Kopecká, J.; Meier-Dörnberg, T.; Zarnkow, M.; Jacob, F.; Hutzler, M. Screening for new brewing yeasts in the non-*Saccharomyces* sector with *Torulaspora delbrueckii* as model. *Yeast* **2016**, *33*, 129–144. [CrossRef]
34. Bellut, K.; Michel, M.; Zarnkow, M.; Hutzler, M.; Jacob, F.; De Schutter, D.P.; Daenen, L.; Lynch, K.M.; Zannini, E.; Arendt, E.K. Application of Non-*Saccharomyces* Yeasts Isolated from Kombucha in the Production of Alcohol-Free Beer. *Fermentation* **2018**, *4*, 66. [CrossRef]
35. Bellut, K.; Krogerus, K.; Arendt, E.K. *Lachancea fermentati* strains isolated from kombucha: fundamental insights, and practical application in low alcohol beer brewing. manuscript in preparation.
36. Peyer, L.C.; Zarnkow, M.; Jacob, F.; De Schutter, D.P. Sour Brewing: Impact of *Lactobacillus amylovorus* FST2.11 on Technological and Quality Attributes of Acid Beers. *J. Am. Soc. Brew. Chem.* **2017**, *75*, 207–216. [CrossRef]
37. Peyer, L.C.; Bellut, K.; Lynch, K.M.; Zarnkow, M.; Jacob, F.; De Schutter, D.P.; Arendt, E.K. Impact of buffering capacity on the acidification of wort by brewing-relevant lactic acid bacteria. *J. Inst. Brew.* **2017**, *123*, 497–505. [CrossRef]
38. Meier-Dörnberg, T.; Hutzler, M. Alcohol-Free Wheat Beer with Maltose Negative Yeast Strain *Saccharomycodes ludwigii*. In 3rd Young Scientists Symposium. Poster No. P.3.5. Available online: https://www.researchgate.net/publication/307992436_AlcoholFree_Wheat_Beer_with_Maltose_Negative_Yeast_Strain_Saccharomycodes_ludwigii (accessed on 14 November 2019).
39. Nyanga, L.K.; Nout, M.J.R.; Gadaga, T.H.; Theelen, B.; Boekhout, T.; Zwietering, M.H. Yeasts and lactic acid bacteria microbiota from masau (*Ziziphus mauritiana*) fruits and their fermented fruit pulp in Zimbabwe. *Int. J. Food Microbiol.* **2007**, *120*, 159–166. [CrossRef] [PubMed]
40. Goffrini, P.; Ferrero, I.; Donnini, C. Respiration-dependent utilization of sugars in yeasts: A determinant role for sugar transporters. *J. Bacteriol.* **2002**, *184*, 427–432. [CrossRef] [PubMed]
41. Meilgaard, M.C. Flavor chemistry in beer: Part II: Flavor and flavor threshold of 239 aroma volatiles. *Master Brew. Assoc. Am. Tech. Q.* **1975**, *12*, 151–168.
42. Holt, S.; Miks, M.H.; de Carvalho, B.T.; Foulquié-Moreno, M.R.; Thevelein, J.M. The molecular biology of fruity and floral aromas in beer and other alcoholic beverages. *Fems Microbiol. Rev.* **2019**, *43*, 193–222. [CrossRef] [PubMed]
43. American Society of Brewing Chemists ASBC Beer Flavors Database. Available online: http://methods.asbcnet.org/extras/flavors_database.pdf (accessed on 10 October 2019).
44. Meilgaard, M.C. Flavor chemistry in beer: Part I: Flavor interaction between principal volatiles. *Master Brew. Assoc. Am. Tech. Q.* **1975**, *12*, 107–117.
45. Lytra, G.; Tempere, S.; Le Floch, A.; De Revel, G.; Barbe, J.C. Study of sensory interactions among red wine fruity esters in a model solution. *J. Agric. Food Chem.* **2013**, *61*, 8504–8513. [CrossRef]
46. Michel, M.; Meier-Dornberg, T.; Jacob, F.; Schneiderbanger, H.; Haselbeck, K.; Zarnkow, M.; Hutzler, M. Optimization of Beer Fermentation with a Novel Brewing Strain *Torulaspora delbrueckii* Using Response Surface Methodology. *Master Brew. Assoc. Am. Tech. Q.* **2017**, *54*, 23–33.
47. Van Rijswijck, I.M.H.; Wolkers-Rooijackers, J.C.M.; Abee, T.; Smid, E.J. Performance of non-conventional yeasts in co-culture with brewers' yeast for steering ethanol and aroma production. *Microb. Biotechnol.* **2017**, *10*, 1591–1602. [CrossRef]
48. Kucharczyk, K.; Tuszyński, T. The effect of pitching rate on fermentation, maturation and flavour compounds of beer produced on an industrial scale. *J. Inst. Brew.* **2015**, *121*, 349–355. [CrossRef]
49. Jonkova, G.N.; Georgieva, N.V. Effect of some technological factors on the content of esters in beer. *Sci. Study Res.* **2009**, *10*, 271–276.
50. Erten, H.; Tanguler, H.; Cakiroz, H. The effect of pitching rate on fermentation and flavour compounds in high gravity brewing. *J. Inst. Brew.* **2007**, *113*, 75–79. [CrossRef]
51. Vriesekoop, F.; Krahl, M.; Hucker, B.; Menz, G. 125th Anniversary review: Bacteria in brewing: The good, the bad and the ugly. *J. Inst. Brew.* **2012**, *118*, 335–345. [CrossRef]

52. Brányik, T.; Silva, D.P.; Baszczyňski, M.; Lehnert, R.; Almeida, E.; Silva, J.B. A review of methods of low alcohol and alcohol-free beer production. *J. Food Eng.* **2012**, *108*, 493–506. [CrossRef]
53. Rachon, G.; Rice, C.J.; Pawlowsky, K.; Raleigh, C.P. Challenging the assumptions around the pasteurisation requirements of beer spoilage bacteria. *J. Inst. Brew.* **2018**, *124*, 443–449. [CrossRef]

14

Starter Cultures for Sparkling Wine

Carmela Garofalo [1], Mattia Pia Arena [1], Barbara Laddomada [2], Maria Stella Cappello [2], Gianluca Bleve [2], Francesco Grieco [2], Luciano Beneduce [1], Carmen Berbegal [1], Giuseppe Spano [1,*] and Vittorio Capozzi [1]

1 Dipartimento di Scienze Agrarie, degli Alimenti e dell'Ambiente, Università di Foggia, via Napoli 25, Foggia 71100, Italy; carmela.garofalo@unifg.it (C.G.); mattiapia.arena@unifg.it (M.P.A.); luciano.beneduce@unifg.it (L.B.); carmen.berbegal@unifg.it (C.B.); vittorio.capozzi@unifg.it (V.C.)

2 Istituto di Scienze delle Produzioni Alimentari, Consiglio Nazionale delle Ricerche, Unità Operativa di Supporto di Lecce, Lecce 73100, Italy; barbara.laddomada@ispa.cnr.it (B.L.); maristella.cappello@ispa.cnr.it (M.S.C.);gianluca.bleve@ispa.cnr.it (G.B.); francesco.grieco@ispa.cnr.it (F.G.)

* Correspondence: giuseppe.spano@unifg.it

Academic Editor: Ronnie G. Willaert

Abstract: The sparkling wine market has expanded in recent years, boosted by the increasing demand of the global market. As for other fermented beverages, technological yeasts and bacteria selected to design commercial starter cultures represent key levers to maximize product quality and safety. The increasing economic interest in the sector of sparkling wine has also implied a renewed interest in microbial resource management. In this review, after a brief introduction, we report an overview of the main characterization criteria in order to select *Saccharomyces cerevisiae* strains suitable for use as starter cultures for the production of base wines and to drive re-fermentation of base wines to obtain sparkling wines. Particular attention has been reserved to the technological characterization aspects of re-fermenting phenotypes. We also analysed the possible uses of selected non-*Saccharomyces* and malolactic strains in order to differentiate specific productions. Finally, we highlighted the main safety aspects related to microbes of enological interest and underlined some microbial-based biotechnological applications helpful to pursue product and process innovations. Overall, the sparkling wine industry may find a relevant benefit from the exploitation of the wide resources associated with vineyard/wine microbial diversity.

Keywords: sparkling wine; starter cultures; *Saccharomyces cerevisiae*; non-*Saccharomyces* autolysis; flocculation; alcoholic fermentation; re-fermentation

---

1. Introduction

Humans have produced alcoholic beverages for millennia and these products have been traditionally used for medicinal, nutritional, and social purposes [1,2]. During the centuries, the technical procedures for their production have continuously evolved, since the discovery of spontaneous fermentations to the industrial application of starter cultures. The microbial strains mainly used for this last purpose belong to the *Saccharomyces* species. However, non-*Saccharomyces* species, previously considered spoilage yeasts, have also been recently used as fermentation starters, in the perspective of wines designed to respond to consumer demands [3–6].

During alcoholic fermentation, yeasts produce several compounds, mainly ethanol and carbon dioxide, with the latter released directly into the atmosphere if the process is conducted in an open vessel. This is what happens during the production of most wines, normally defined "still" wines for the negligible amounts of carbon dioxide that they still contain. Contrariwise, wines containing a relevant concentration of carbon dioxide are referred as "effervescent" wines, distinguished into semi-sparkling (1–2.5 atmospheres of pressure in the bottle) and sparkling varieties (3–6 atmospheres of pressure in the bottle) [7].

Almost all sparkling wines are the result of two fermentation steps. During the first fermentation, the must is converted into wine (usually denoted as "base" wine), whereas in the second fermentation step, the base wine is firstly added with several ingredients (e.g., sucrose, yeasts cells, nitrogen source) and then re-fermented in a cellar for, at least, 9–12 months [8].

The sparkling wine market has expanded in recent years, boosted by a high global consumer demand [9,10]. The production of sparkling wine has significantly increased, showing a rise of 40% in the last ten years, while that of non-sparkling wines only increased by 7% over the same period [11].

Sparkling wines have an important economic impact due to their high added value. Moreover, considering the consumers' attention toward quality and safety of fermented beverages, research in the sparkling wine sector is nowadays also focuses on biotechnological innovations to improve product qualities, to simplify the production process and, at the same time, to enhance the preservation of typical and unique product characteristics [12–14].

Among the steps of production of sparkling wine, an important phase for ensuring the quality of the final product is the aging. In fact, at the end of second fermentation, sparkling wines undergo an aging period during which yeast autolysis occurs, with the consequent release of several cellular compounds, such as amino acids, proteins, carbohydrates, and lipids, that improve the quality of sparkling wines [15].

Modern biotechnologies can be used not only to improve the quality of sparkling wines, but also to reduce their production time and costs [8,14,16,17]; for example, some authors suggest new methods to accelerate the above-described autolysis process. Starter cultures for sparkling wine production need to be selected in order to produce either quality base wine or to vigorously promote the second fermentation, which occurs in a harsh environment, mainly due to the elevated ethanol content, low pH (2.8–3.3), and carbon dioxide-induced pressure [18].

2. Sparkling Wine: Production Process, Legislation, and Classification

Sparkling wines can be produced according to two main procedures, the traditional, also called as "champenoise", and the Charmat methods. The traditional method, is performed by an in-bottle secondary fermentation, while in the Charmat method, secondary fermentation is carried out in hermetically-sealed tanks [11].

2.1. Production of Sparkling Wine Using the Charmat Method

The Charmat method is characterized by sealed tanks provided with agitating mechanisms, with the aim to mix the yeast uniformly into the base wine during secondary fermentation [8]. Base wines are usually corrected with 20–24 g/L of sugar and then pasteurized with the aim to accelerate sucrose hydrolysis. Briefly, the base wine integrated with sugar is added with yeast into a pressure tank made of stainless steel, built to resist the pressure. When all of the sugar is transformed into alcohol and carbon dioxide, the yeasts are removed and the wine is bottled in an isobaric, refrigerated environment. The duration of fermentation usually influences the quality of the final product and a prolonged fermentation protects the wine aroma and allows maintaining bubbles that are more durable. At the end of secondary fermentation and after clarification, sparkling wine is bottled. After that, it should be aged at least 20 days before sale, during which aging wine remains in contact with yeast lees [19]. The Charmat method is simpler and cheaper than the traditional one. However, the process cannot be used for sparkling wines with specific regional designations.

2.2. Production of Sparkling Wine Using the Traditional Method

The traditional method is referred to as "méthode champenoise", but this expression can be officially used only for sparkling wines produced in the Champagne region (EU regulation number 3309/85). Consequently, all other sparkling wines can be identified by the expression "traditional method", "classic method", or similar terms.

Production of sparkling wine using traditional method includes two steps (Figure 1): primary fermentation and secondary fermentation. During primary fermentation, the grape juice is converted into base wine, while during secondary fermentation (also known as *"prise de mousse"*) the alcoholic fermentation of the sucrose-base-wine mixture produces the sparkling wine, with its typical characteristics, flavour, and foam.

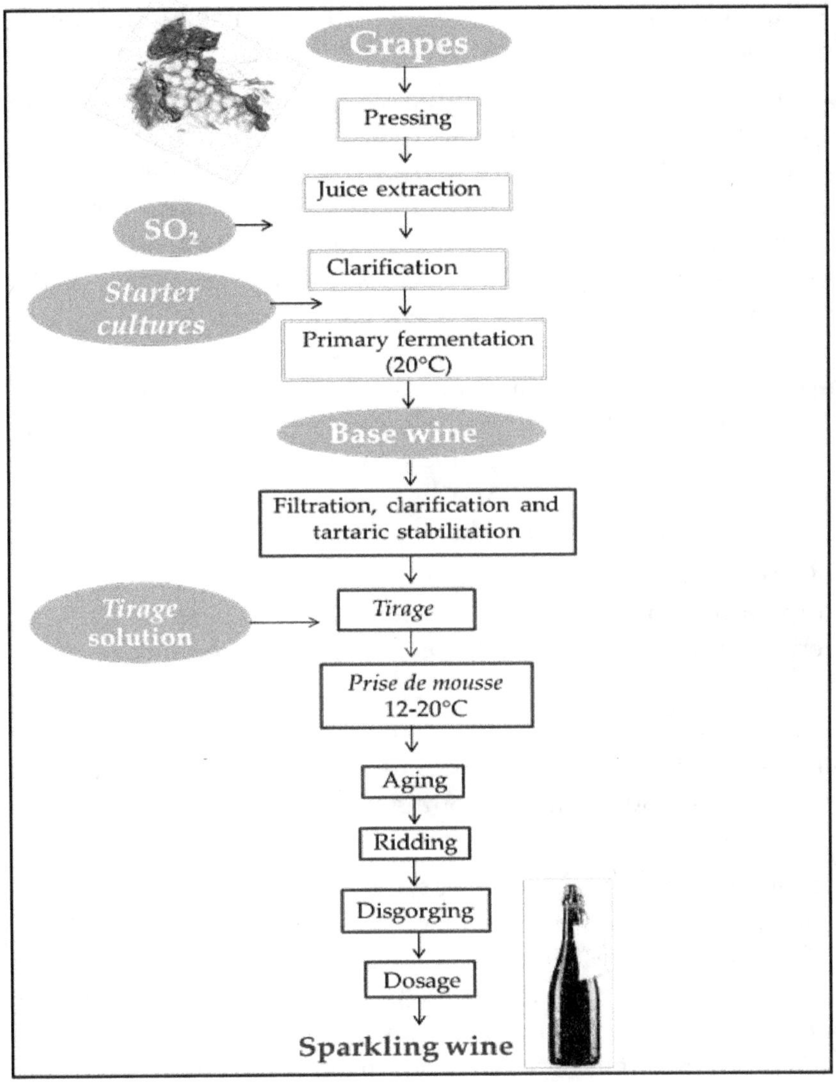

Figure 1. Flowchart for sparkling wine production using the traditional, or *champenoise*, method.

A typical base wine usually presents a moderate alcohol concentration (10%–11% v/v), low levels of sugar and acetic acid (volatile acidity), and a high concentration of other organic acids (total acidity) [8]. In particular, a significant value of the total acidity of base wine (about 12–18 g/L measured as tartaric acid), represents a critical point for sparkling wine production, because during the production process, the total acidity may decrease due to several factors, such as malic acid degradation by yeasts and lactic acid bacteria and potassium bitartrate precipitations [20].

A common practice to avoid insufficient total acidity is to slightly anticipate harvesting with respect to traditional white wine grapes. Harvesting, in fact, is a very delicate step during base wine production. Grapes are usually hand-picked and collected into small tanks to avoid berry breakage. Only sound grapes are collected and quickly transported to the cellar, in order to prevent spontaneous fermentation. Usually, grape berries are immediately pressed, without crushing to avoid oxidations, macerations, and the development of flat aromas and browning, to obtain a good quality must. Grape juice extraction must be extremely meticulous to avoid vegetal and bitter defects, due to an excessive maceration of skins. Base wines should also have a fruity flavor and a pale color. Indeed, color extraction should be avoided, especially for white sparkling wines obtained from red grapes, also named "*blanc de noirs*" [21]. After must extraction, the next steps are (i) the addition of sulphur dioxide (apart from that added before crushing) to prevent oxidation and undesired fermentation; (ii) the clarification with pectolytic enzymes to remove solids and minimize phenol oxidations; (iii) the inoculation of starter cultures of *Saccharomyces cerevisiae*, which drive the alcoholic fermentation process in stainless steel tanks under temperature control.

At the end of primary fermentation base wines are subjected to several manipulations, such as further clarification, decantation, and filtration. Common practices also include the addition of bentonite and cold precipitation, to promote protein precipitation and potassium bitartrate precipitation, respectively. After such treatments, the base wine is ready for the secondary fermentation and selected starter cultures (*S. cerevisiae*) are added within a so-called "*tirage*" solution (containing ingredients such as saccharose 20–25 g/L, yeasts, grape must or wine, and bentonite) [22]. After the addition of the *tirage* solution, the wine is bottled and the bottles are sealed with a crown cap, that underneath it has a "*bidule*", i.e., a plastic cylinder where the lees will accumulate. The bottles are then horizontally stacked in special aging rooms at low temperature (12–15 °C). Sparkling aging usually takes place horizontally, as this position promotes an efficient contact between wine and yeast sediment, with a slow release of several compounds originating both from yeasts and wine. After the production of the desired CO_2 concentration and aging, a disgorging procedure, traditionally denoted as "dégorgement" [22], carries out the removal of lees.

The kinetics of the secondary fermentation depend on various factors, such as the yeast species/strains, temperature, and chemical composition of base wine. Usually, at 12–15 °C, the secondary fermentation takes almost 15–45 days and can be monitored by checking the internal pressure using an afrometer.

Aging duration is regulated by national legislation; as a consequence, it may vary according to country. Nevertheless, sparkling wine maturation is a slow process and it takes from a minimum of nine months for the "Cava" (Spain) to 12 months for "Talento" (Italy) or "Champagne" (France) wines [23].

A prolonged aging is essential to improve and develop the organoleptic properties of sparkling wine since it is correlated with roundness, flavor, complexity, and foaming [14,15,24]. During aging, the characteristics of sparkling wines change due to the release of yeast cytoplasmic and cell wall compounds into the wine, by the autolysis process promoted by the activity of hydrolytic enzymes. Several authors, who suggested its positive effect on sparkling wine quality, have investigated this biological event [14,24]. Proteins released by yeast into sparkling wine show a positive correlation on "body sensation" and foam stability, while polysaccharides should improve wine stability against protein haze [14]. Contrariwise, yeasts wall should adsorb volatile compounds affecting the aroma of sparkling wine [25].

When the aging is complete, the next step is the riddling or "*remuage*", i.e., a kind shaking of the bottles to convey yeast lees into the *bidule*. This step is improved by adding a small amount of bentonite to the *tirage* solution and by yeast's aptitude to flocculate. During riddling, each bottle should be hand-rotated one eighth (of the total rotation) each day for 15 days until bottles are practically perpendicular to the floor. Riddling should promote the subsequent disgorging process, during which, lees collected at the neck of the bottle are removed, thanks to freezing and internal pressure in the bottle. Then, it is a common practice to add a dosage solution, traditionally called "*liqueur d'expédition*", to compensate liquid lost during disgorging. It consists of a mixture of variable composition with pure sparkling wine, sparkling wine containing sucrose, grape must, brandy, SO_2, or other components typical of a determined production area. The dosage solution and its composition influence sparkling wine characteristics, and will give to each sparkling wine a distinctive structure and aroma.

3. Yeast Characterization for Wine Base Production

Wine organoleptic properties are strictly correlated with the physiological and metabolic characteristics of *S. cerevisiae* and non-*Saccharomyces* strains used as starters. Indeed, such microbial component influences the production of several compounds and transform grape compounds with positive or negative effects on fermentative or secondary aromas [26,27].

A starter culture is a microbial strain that is characterized and selected for its fermentation properties. Specific criteria have been indicated to select *S. cerevisiae* starter cultures with exquisite oenological properties. However, the first fundamental step for the selection of oenological starters is the availability of genetics and molecular diagnostic tools that allow a quick and accurate yeast identification, at either species or strain level, and their monitoring during wine fermentation [28–32].

3.1. *Yeast Genotypic Characterization: Methods to Differentiate Saccharomyces Cerevisiae Strains*

Several molecular techniques have been developed and successfully applied to the identification and characterization of yeasts that allow to differentiate *S. cerevisiae* at the strain level [28,33–38] (Table 1).

Table 1. *Saccharomyces cerevisiae* genotypic characterization.

Molecular Method	Reference
Random amplified polymorphic DNA (RAPD) PCR	[39]
Interdelta sequences analysis	[40]
Pulse field electrophoresis (PFGE) electrophoretic karyotypes	[38]
Mitochondrial DNA (mtDNA) restriction analysis	[33,34,36,41]
Polymorphic microsatellite loci (SSRs, simple sequence repeats)	[40,42]
Multilocus sequence typing (MLST)	[43,44]

The first technique used to reveal *Saccharomyces* strain diversity is pulsed field electrophoresis (PFGE) i.e., separation of intact chromosomes by pulsed field agarose gel electrophoresis, also called electrophoretic karyotyping [38].

Random amplified polymorphic DNA (RAPD-PCR) also has been powerful to differentiate *S. cerevisiae* strains; nevertheless, other methods are more discriminating [41].

Some authors have suggested that mitochondrial DNA (mtDNA) restriction analysis (mtDNA RFLP) could be an efficient technique to differentiate at the strain level [33,34,36,41]. In particular, this molecular technique has been used to check the dominance of *S. cerevisiae* starter cultures, thanks to the marked mtDNA polymorphism of wine *Saccharomyces* strains [45–47].

Another commonly used molecular approach relies on sequencing the interdelta element, whose amplification by PCR allows differentiating at the strain level *S. cerevisiae* strains [33,48]. Other powerful molecular tools for *S. cerevisiae* strain differentiation are the amplification of polymorphic microsatellite loci, also called simple sequence repeats (SSRs) [40,42], the multilocus sequence typing (MLST) and yeast killer virus (virus dsRNA) [43,44,49,50].

3.2. *Yeast Technological and Qualitative Characterization for Starter Culture Production*

Yeasts, mainly *S. cerevisiae* strains, have a fundamental role during winemaking and alcoholic fermentation. Grape sugars, in particular hexoses, must be rapidly and completely converted into ethanol and CO_2, with the associated production, by the yeasts, of several metabolites important to confer wine typical organoleptic properties (but also the possible release of off-flavors) [51–54]. Nevertheless, it is important to underline that a complete transformation of sugar occurs on dry wine but not on sweet wine.

Must/wine system represents a hostile environment due to several factors, such as high sugar concentration (average 200 g/L), growing ethanol and glycerol amount, low pH (3–3.5), the presence of sulphites, and progressive consumption of nutrients (such as nitrogen sources, vitamins, and lipids) [55].

Usually, starter cultures or autochthonous strains should be selected on the basis of typical oenological traits due to the peculiar characteristics of grape juice, base wine, and to desirable qualities of wines. Indeed, an efficient procedure to characterize *S. cerevisiae* starter strains selected from natural fermenting needs biotechnological tools/criteria to optimize global wine quality [51–55]. Several authors proposed technological and qualitative criteria to select yeast strains with desirable features. Among these, tolerance to alcohol, resistance to sulphur dioxide, several enzymatic activities, osmotic properties, killer factor, and low production of H_2S are determinants [26]. Table 2 reports the most important technological and qualitative criteria to select yeast starter cultures.

Table 2. Yeast technological and qualitative characteristics for starter cultures production.

Technological and Qualitative Characteristics	Reference
Resistance to low pH, sugars, ethanol, and sulphur dioxide contents	[26,46,51,55–57]
Low volatile acidity production	[52,58–60]
Low production of sulphur compounds (H_2S, SO_2)	[26,46,47,51–53,61]
Fermentation vigour	[51]
Desired enzymatic activities (e.g., β-glucosidase, β-xylosidase, protease, polygalacturonase, pectinase, glucanase, xylanase, and decarboxylase activities)	[46,47,57,62,63]
Desired fermentation-associated metabolites (glycerol, succinic acid, acetic acid, acetaldehyde, n-propanol, iso-butanol, isoamyl alcohol, and β-phenylethanol)	[46,47,57,62–64]
Implantation aptitude	[65–67]

Other relevant features include the strain-specific formation of fermentation-associated metabolites (such as glycerol, succinic acid, acetaldehyde, n-propanol, isobutanol, isoamyl alcohol, and β-phenylethanol) and the presence of specific extracellular enzymatic activities (β-glucosidase, β-xylosidase, protease, polygalacturonase, pectinase, glucanase, xylanase, and decarboxylase) [46,47,57,62–64]. Obviously, implantation aptitude of starter cultures also is a criterion to be checked during starter technological selection programs. Indeed, several studies suggested that starter cultures dominance is not always guaranteed, as a function of the diversity associated with the naturally present microbial consortia, and that during winemaking indigenous yeasts can survive and grow, affecting starter dominance [65–67].

4. Yeast Technological Characterization for Secondary Fermentation of Sparkling Wine Production

The starter cultures used for the secondary fermentation in the traditional method need to possess several additional technological properties to those of the yeast used in the primary fermentation (Figure 2).

Chemical composition of the base wine and the sparkling wine production process represent a hostile environment for yeast growth and fermentation efficiency [23]. Base wine usually is characterized by consistent ethanol concentration (about 10%–12% v/v), low pH (2.8–3.5), high total acidity (5–7 g/L H_2SO_4), and total SO_2 contents (50–80 mg/L). In addition to these critical factors, we have to consider low temperatures occurring during the secondary fermentation (10–15 °C) and the high amount of CO_2/high pressure (usually 6 atm) associated with this process [68]. Hence, yeast starter cultures for secondary fermentation have to be selected in order to survive the above-described stresses, and, in particular, to high ethanol concentration and low pH value [69]. Ethanol in base wines affects yeast growth, viability, and ability to carry on the secondary fermentation [70]. Analysis for ethanol tolerance involves the yeast exposure to several increasing concentrations of ethanol and the monitoring of its growth [71]. It is generally recognized that yeasts should be adapted prior to inoculation for secondary fermentation, by exposing them to increasing trends of ethanol concentrations [72]. This step (known also as "*prise de mousse*") is essential for an efficient and successful secondary fermentation because it allows the yeast acclimatization to low pH and high amounts of ethanol [73]. Cell acclimatization can improve cell viability, biomass accumulation, and time required for complete secondary fermentation. In addition, improving tolerance to stress by starter yeast may reduce the production of off flavors [21,74,75]. The low pH value is one of the main negative factors that can affect secondary fermentation of a typical base wine. In fact, base wines contain generally high amounts of organic acids (such as tartaric, malic, succinic, and acetic acids) in the undissociated form (at the common pH), that are susceptible to acidify the yeast cytosol, thus leading to sluggish or stuck fermentation [68]. In particular, the parameter to be considered during the selection procedure is the yeast resistance to high concentrations of acetic acid. In fact, acetic acid combined with ethanol can affect yeast fermentative behavior by decreasing cell pH, fermentation rate, and enolase activity [76].

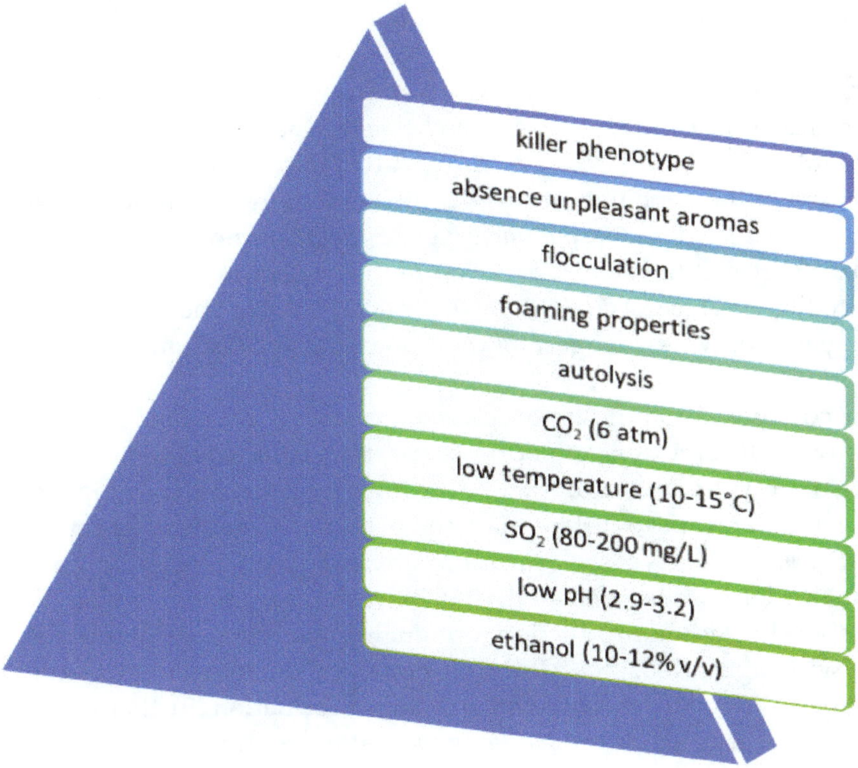

Figure 2. Yeast technological characterization for sparkling wine production.

Another biological parameter that should be considered during the selection of starter cultures for the secondary fermentation is their ability to undergo autolysis. A strain-dependent phenotype that, as a consequence of cell disruption, implies the release in the fermenting wine of several yeast-associated compounds able to influence organoleptic and foaming properties of sparkling wine [77,78]. Several reports suggested the positive effect of yeast strains with high autolytic capacity on sparkling wine quality and foaming properties [24,79,80]. This characteristic is desired at the end of the re-fermentation in sparkling wines, generally two or four months after the end of secondary fermentation. During autolysis, several compounds, such as peptides, amino acids, polysaccharides, higher alcohols, and aldehydes, are released, [69,81,82] and may improve sparkling wine flavor, because some of them (e.g., amino acids) are often precursors of aroma compounds [78]. A further technological characteristic of interest in sparkling wine production is the flocculation capacity of the candidate strain. High flocculation ability is an important criterion among sparkling starter culture selection, because yeast high flocculation aptitude facilitates the removal of sediment at disgorging [23]. At the same time, flocculation prevent yeast sediments from remaining attached to the bottle, thus avoiding turbidity of the final product [7].

Moreover, yeast autolysate compounds could positively influence foaming properties [1] since several macromolecules released by yeast (e.g., mannoproteins) are involved in both foam formation and stabilization [80,83].

Linked to this lytic phenotype, another yeast property to be assessed during starter selection is the presence of the killer phenotype. A killer yeast is a yeast strain which is able to secrete one or more toxic proteins, which are lethal to sensitive yeast strains. This biological phenomenon should be analyzed, checking both killer and sensible phenotypes, because yeast autolysis can be improved with a mixed inoculum of "killer" and sensitive yeast strains for the secondary fermentation [84].

5. Sparkling Wine Production: Role of Non-*Saccharomyces* and Lactic Acid Bacteria

In light of the existing innovative trends in the field of microbial resources in enology, a specific insight is needed for non-*Saccharomyces* yeasts and malolactic lactic acid bacteria (LAB) that might be involved in sparkling wine production [3,85].

Even though non-*Saccharomyces* yeasts are usually involved in winemaking to increase wine's organoleptic properties, little is known on non-*Saccharomyces* application for sparkling wine production. Gonzalez-Royo et al. [9] investigated the chemical and organoleptic properties of base wines accomplished by sequential inoculation of two different non-*Saccharomyces* strains (belonging to *Metschnikowia pulcherrima* and *Torulaspora delbrueckii* species) and *S. cerevisiae* during the alcoholic fermentation of base wines of the AOC Cava. Sequential inoculation of *T. delbrueckii* and *S. cerevisiae* on base wines led to higher glycerol content, lower volatile acidity, and higher foaming properties than their corresponding control wines, suggesting their potential application to innovate specific sparkling wine production [9].

Sequential or co-inoculation of non-*Saccharomyces* and *S. cerevisiae* could be a powerful tool to make base wines with different organoleptic properties, in particular following recent trends on winemaking that suggest the important properties of non-*Saccharomyces* in order to solve specific technological challenges and/or to differentiate the production in terms of sensorial quality [3–6,53].

Among non-*Saccharomyces* yeasts, *Schizosaccharomyces pombe*, usually recognized as a spoilage yeast, presents a malic dehydrogenase activity [86,87] that might have a role in the induction of yeast autolysis during sparkling wine production, enhancing mannoprotein and polysaccharide release during sparkling wine aging [86,87]. Although non-*Saccharomyces* biodiversity could represent a considerable resource of innovation for sparkling wines production, further investigations are needed to understand the possible role of non-*Saccharomyces* starters in the base wine re-fermentation and sparkling wine aging.

The malolactic fermentation (MLF), i.e., the decarboxylation of L-malic acid into L-lactic acid, if desired, takes place post-first fermentation or simultaneously with alcoholic fermentation (co-fermentation of malolactic LAB with yeasts). MLF attenuates acidity, enhances wine biological stability, and modifies sparkling wine sensorial qualities [88]. Several LAB are involved in MLF, the most important is *Oenococcus oeni*, while other LAB species can produce off-flavors and, for this reason, are considered spoilage LAB [89]. The induction of MLF during sparkling wine aging is a common practice in the Champagne region and it allows the production of wine denoted by a higher pH value that also has a reduction in the time requested for their maturation [90]. However, malic and lactic acids were shown to have controversial effects on sparkling wine foaming properties. In fact, malic acid can improve foaming height, while lactic acid produces an opposite action [91–93].

6. Safety Aspects Correlated to Base and Sparkling Wine

Base and sparkling wine safety can be affected by several compounds derived from grapes (e.g., pesticides, phytosanitary products, or trace metal compounds) or from microbial metabolism (e.g., biogenic amines (BA) and the mycotoxin ochratoxin A) [94]. Concerning the compounds of microbial origin, BA are low molecular weight compounds formed in foods by fermentative processes and during aging and storage as a consequence of microbial amino acid decarboxylation [55,95,96]. Sensitive consumers can be intoxicated by BA, which produce several physiological and toxic effects on human health, such as rash, edema, headaches, hypotension, vomiting, palpitations, diarrhea, and heart problems. BA are usually recovered from all fermented foods. Nevertheless, in alcoholic drinks, especially in wine, ethanol and acetaldehyde can enhance the negative effects of biogenic amines by affecting the efficiency of their detoxification by the human body [97–99]. In wine, putrescine represents the major biogenic amine, followed by histamine, tyramine, and cadaverine [100–102], and several authors suggested that BA are produced by LAB metabolism during MLF [98,103–105]. Nevertheless, recent studies also demonstrated that some yeast strains are able to produce BA in wines [53,106,107]. From this point of view, the selection of suitable yeast might represent a fundamental phase to 'build' wine safety.

Another toxic microbial by-product is ocratoxin A (OTA), one of the most common naturally occurring mycotoxins in wine, with several toxic effects [108]. OTA is present in grape musts as a consequence of fungal growth on grapes, particularly attributed to *Aspergillus* and *Penicillium* metabolism [109]. Several OTA-elimination methods have been proposed, from physical to biotechnological mechanisms. The latter represent the best methods to remove OTA from wine without affecting wine organoleptic properties or using toxic chemical compound [110]. Among biologically-based approaches, several studies suggested that *Saccharomyces* and non-*Saccharomyces* strains can remove OTA from wines [111–117].

Therefore, it is very important to monitor the presence of indigenous yeast or LAB during sparkling wine production to reduce the risks to consumer health due to the presence of this toxic microbial compound [118].

7. Biotechnological Applications

Several authors studied biotechnological applications to improve sparkling wine quality, with particular attention to autolysis, flocculation, and sparkling wine flavor [12,14,72,119]. The yeast autolysis, for example, is a slow process that can occur in a few months, or several years, thanks to different environmental conditions, such as temperature, pH, ethanol concentration, nutrient availability, redox potential, and yeast strain [14]. Usually, a slow sparkling wine aging corresponds to an increase of wine organoleptic properties; nevertheless, a long sparkling wine aging can affect entrepreneurial costs. However, several methods, such as adding yeast autolysates to sparkling wine, increasing the aging temperature to accelerate yeast autolysis or microbial genetic improvement, are reported [8,78,84,120]. In particular, the use of yeast autolysates or the increase of aging temperature can affect sparkling wine organoleptic properties [121], while a mixture of killer and killer-sensitive yeast seems to be more promising [32]. In fact, co-inoculation of killer and sensitive *S. cerevisiae* strains allows for increased autolysis, shortening sparkling wine aging time without affecting wine flavor [14,17,32,84]. Genetic improvement can help to design yeast strains with an increased autolytic and flocculation capacity [23]. This method allows improving autolysis and to shorten aging, without affecting wine flavor and needing to introduce modifications in the production process. Among genetic-based methods, the main techniques used are random mutagenesis (for example UV mutagenesis) and genetic engineering [24]. In addition, studies reported in the literature proposed several methods for the development of new *S. cerevisiae* strains with improved flocculation aptitudes, such as clonal selection, recombinant DNA, and hybridization [57,122–124].

With this concern, it is interesting to underline the possibility to hybridize *S. cerevisiae* and *S. uvarum* (*S. bayanus* var. *uvarum*) that led to hybrids with improved technological characteristics (higher fermentative rate, tolerance to low and high temperature, better flocculation capacity, excellent aromatic properties) compared to those of the parental strains [124–128].

8. Conclusions

Both traditional and Charmat methods for sparkling wine productions can benefit from using appropriate starter cultures that could allow the increase of both production efficiency and product quality. Important advances have already been performed by investigating the benefit associated with microbe biodiversity in the vineyard/wine environments. Furthermore, the sector of sparkling wine production will benefit from the development of novel procedures for a renewed exploitation of the enormous opportunities associated with natural microbial biodiversity.

Acknowledgments: This research was supported by the Apulia Region with a grant from Project "Innovazioni di processo e di prodotto nel comparto dei vini spumanti da vitigni autoctoni pugliesi"—IPROVISP (Bando "Aiuti a Sostegno Cluster Tecnologici Regionali"; Project code VJBKVF4). Vittorio Capozzi was supported by 'Fondo di Sviluppo e Coesione 2007–2013—APQ Ricerca Regione Puglia "Programma regionale a sostegno della specializzazione intelligente e della sostenibilità sociale ed ambientale—FutureInResearch".

References

1. Blasco, L.; Viñas, M.; Villa, T.G. Proteins influencing foam formation in wine and beer: The role of yeast. *Int. Microbiol.* **2011**, *14*, 61–71. [PubMed]
2. Legras, J.-L.; Merdinoglu, D.; Cornuet, J.-M.; Karst, F. Bread, beer and wine: *Saccharomyces cerevisiae* diversity reflects human history. *Mol. Ecol.* **2007**, *16*, 2091–2102. [CrossRef] [PubMed]
3. Jolly, N.P.; Varela, C.; Pretorius, I.S. Not your ordinary yeast: Non-*Saccharomyces* yeasts in wine production uncovered. *FEMS Yeast Res.* **2014**, *14*, 215–237. [CrossRef] [PubMed]
4. Garofalo, C.; El Khoury, M.; Lucas, P.; Bely, M.; Russo, P.; Spano, G.; Capozzi, V. Autochthonous starter cultures and indigenous grape variety for regional wine production. *J. Appl. Microbiol.* **2015**, *118*, 1395–1408. [CrossRef] [PubMed]
5. Garofalo, C.; Tristezza, M.; Grieco, F.; Spano, G.; Capozzi, V. From grape berries to wine: Population dynamics of cultivable yeasts associated to "Nero di Troia" autochthonous grape cultivar. *World J. Microbiol. Biotechnol.* **2016**, *32*, 59. [CrossRef] [PubMed]
6. Capozzi, V.; Garofalo, C.; Chiriatti, M.A.; Grieco, F.; Spano, G. Microbial terroir and food innovation: The case of yeast biodiversity in wine. *Microbiol. Res.* **2015**, *181*, 75–83. [CrossRef] [PubMed]
7. Carrascosa, A.V.; Martinez-Rodiguez, A.; Cebollero, E.; Gonzalez, R. *Saccharomyces* Yeasts II: Secondary fermentation. In *Molecular Wine Microbiology*; Elsevier: London, UK, 2011; Volume 2, pp. 33–48.
8. Cebollero, E.; Gonzalez, R. Induction of autophagy by second-fermentation yeasts during elaboration of sparkling wines. *Appl. Environ. Microbiol.* **2006**, *72*, 4121–4127. [CrossRef] [PubMed]
9. González-Royo, E.; Pascual, O.; Kontoudakis, N.; Esteruelas, M.; Esteve-Zarzoso, B.; Mas, A.; Canals, J.M.; Zamora, F. Oenological consequences of sequential inoculation with non-*Saccharomyces* yeasts (*Torulaspora delbrueckii* or *Metschnikowia pulcherrima*) and *Saccharomyces cerevisiae* in base wine for sparkling wine production. *Eur. Food Res. Technol.* **2014**, *240*, 999–1012. [CrossRef]
10. Ody-Brasier, A.; Vermeulen, F. The price you pay price-setting as a response to norm violations in the market for champagne grapes. *Adm. Sci. Q.* **2014**, *59*, 109–144. [CrossRef]
11. Torresi, S.; Frangipane, M.T.; Anelli, G. Biotechnologies in sparkling wine production. Interesting approaches for quality improvement: A review. *Food Chem.* **2011**, *129*, 1232–1241. [CrossRef] [PubMed]
12. Perpetuini, G.; Di Gianvito, P.; Arfelli, G.; Schirone, M.; Corsetti, A.; Tofalo, R.; Suzzi, G. Biodiversity of autolytic ability in flocculent *Saccharomyces cerevisiae* strains suitable for traditional sparkling wine fermentation. *Yeast* **2016**, *33*, 303–312. [CrossRef] [PubMed]
13. Pérez-Magariño, S.; Martínez-Lapuente, L.; Bueno-Herrera, M.; Ortega-Heras, M.; Guadalupe, Z.; Ayestarán, B. Use of commercial dry yeast products rich in mannoproteins for white and rosé sparkling wine elaboration. *J. Agric. Food Chem.* **2015**, *63*, 5670–5681. [CrossRef] [PubMed]
14. Lombardi, S.J.; de Leonardis, A.; Lustrato, G.; Testa, B.; Iorizzo, M. Yeast autolysis in sparkling wine aging: Use of killer and sensitive *Saccharomyces cerevisiae* strains in co-culture. *Recent Pat. Biotechnol.* **2015**, *9*, 223–230. [CrossRef] [PubMed]

15. Alexandre, H.; Guilloux-Benatier, M. Yeast autolysis in sparkling wine—A review. *Aust. J. Grape Wine Res.* **2006**, *12*, 119–127. [CrossRef]
16. Tita, O.; Jascanu, V.; Tita, M.; Sand, C. Economical comparative analysis of different bottle fermentation methods. In Proceedings of the AVA 2003, International Conference on Agricultural Economics, Rural Development and Informatics in the New Millennium, Debrecen, Hungary, 1–2 April 2003.
17. Pozo-Bayón, M.A.; Andujar-Ortiz, I.; Alcaide-Hidalgo, J.M.; Martín-Alvarez, P.J.; Moreno-Arribas, M.V. Characterization of commercial inactive dry yeast preparations for enological use based on their ability to release soluble compounds and their behavior toward aroma compounds in model wines. *J. Agric. Food Chem.* **2009**, *57*, 10784–10792. [CrossRef] [PubMed]
18. Ganss, S.; Kirsch, F.; Winterhalter, P.; Fischer, U.; Schmarr, H.-G. Aroma changes due to second fermentation and glycosylated precursors in Chardonnay and Riesling sparkling wines. *J. Agric. Food Chem.* **2011**, *59*, 2524–2533. [CrossRef] [PubMed]
19. Pozo-Bayón, M.A.; Hernández, M.T.; Martín-Alvarez, P.J.; Polo, M.C. Study of low molecular weight phenolic compounds during the aging of sparkling wines manufactured with red and white grape varieties. *J. Agric. Food Chem.* **2003**, *51*, 2089–2095. [CrossRef] [PubMed]
20. Ribéreau-Gayon, P.; Dubourdieu, D.; Donèche, B.; Lonvaud, A. *Trattato di Enologia I, Microbiologia del Vino*; Edagricole: Bologna, Italy, 2004.
21. Zoecklein, B. *A Review of Méthode Champenoise Production*, 2nd ed.; Virginia Tech: Blacksburg, VA, USA, 2002.
22. Martínez-Rodríguez, A.J.; Pueyo, E. Sparkling Wines and Yeast Autolysis. In *Wine Chemistry and Biochemistry*; Moreno-Arribas, M.V., Polo, M.C., Eds.; Springer: New York, NY, USA, 2009.
23. Bidan, P.; Feuillat, M.; Moulin, J.P. Les vins mousseux. *Bull. l'OIV* **1986**, *59*, 663–664.
24. Nunez, Y.P.; Carrascosa, A.V.; González, R.; Polo, M.C.; Martínez-Rodríguez, A.J. Effect of accelerated autolysis of yeast on the composition and foaming properties of sparkling wines elaborated by a champenoise method. *J. Agric. Food Chem.* **2005**, *53*, 7232–7237. [CrossRef] [PubMed]
25. Luguera, C.; Moreno-Arribas, M.V.; Pueyo, E.; Bartolomé, B.; Polo, M.C. Fractionation and partial characterization of protein fractions present at different stages of the production of sparkling wines. *Food Chem.* **1998**, *63*, 465–471. [CrossRef]
26. Nikolaou, E.; Soufleros, E.H.; Bouloumpasi, E.; Tzanetakis, N. Selection of indigenous *Saccharomyces cerevisiae* strains according to their oenological characteristics and vinification results. *Food Microbiol.* **2006**, *23*, 205–211. [CrossRef] [PubMed]
27. Šuranská, H.; Vránová, D.; Omelková, J. Isolation, identification and characterization of regional indigenous *Saccharomyces cerevisiae* strains. *Braz. J. Microbiol.* **2016**, *47*, 181–190. [CrossRef] [PubMed]
28. Mercado, L.; Sturm, M.E.; Rojo, M.C.; Ciklic, I.; Martínez, C.; Combina, M. Biodiversity of *Saccharomyces cerevisiae* populations in Malbec vineyards from the "Zona Alta del Río Mendoza" region in Argentina. *Int. J. Food Microbiol.* **2011**, *151*, 319–326. [CrossRef] [PubMed]
29. Pérez, F.; Regodón, J.; Valdés, M.; de Miguel, C.; Ramírez, M. Cycloheximide resistance as marker for monitoring yeasts in wine fermentations. *Food Microbiol.* **2000**, *17*, 119–128. [CrossRef]
30. Ambrona, J.; Maqueda, M.; Zamora, E.; Ramírez, M. Sulfometuron Methyl Resistance as Genetic Marker for Monitoring Yeast Populations in Wine Fermentations. *J. Agric. Food Chem.* **2005**, *53*, 7438–7443. [CrossRef] [PubMed]
31. Ambrona, J.; Vinagre, A.; Maqueda, M.; Ramírez, M. Rhodamine-pink as a genetic marker for yeast populations in wine fermentation. *J. Agric. Food Chem.* **2006**, *54*, 2977–2984. [CrossRef] [PubMed]
32. Velázquez, R.; Zamora, E.; Álvarez, M.L.; Álvarez, M.L.; Ramírez, M. Using mixed inocula of new killer strains of *Saccharomyces cerevisiae* to improve the quality of traditional sparkling-wine. *Food Microbiol.* **2016**, *59*, 150–160. [CrossRef] [PubMed]
33. Martínez, C.; Cosgaya, P.; Vásquez, C.; Gac, S.; Ganga, A. High degree of correlation between 464 molecular polymorphism and geographic origin of wine yeast strains. *J. Appl. Microbiol.* **2007**, *103*, 2185–2195. [CrossRef] [PubMed]
34. Schuller, D.; Valero, E.; Dequin, S.; Casal, M. Survey of molecular methods for the typing of wine yeast strains. *FEMS Microbiol. Lett.* **2004**, *231*, 19–26. [CrossRef]
35. Querol, A.; Barrio, E.; Huerta, T.; Ramón, D. Molecular monitoring of wine fermentations conducted by active dry yeast strains. *Appl. Environ. Microbiol.* **1992**, *58*, 2948–2953. [PubMed]

36. Guillamon, J.M.; Sabaté, J.; Barrio, E.; Cano, J.; Querol, A. Rapid identification of wine yeast species based on RFLP analysis of the ribosomal internal transcribed spacer (ITS) region. *Arch. Microbiol.* **1998**, *169*, 387–392. [CrossRef] [PubMed]
37. Fernández-Espinar, M.T.; López, V.; Ramón, D.; Bartra, E.; Querol, A. Study of the authenticity of commercial wine yeast strains by molecular techniques. *Int. J. Food Microbiol.* **2001**, *70*, 1–10. [CrossRef]
38. Torija, M.J.; Rozès, N.; Poblet, M.; Guillamón, J.M.; Mas, A. Yeast population dynamics in spontaneous fermentations: Comparison between two different wine-producing areas over a period of three years. *Antonie Van Leeuwenhoek* **2001**, *79*, 345–352. [CrossRef] [PubMed]
39. Blondin, B.; Vezinhet, F. Identification de souches de levures oenologiques par leurs caryotypes obtenus en électrophorèse en champ pulsé. *Rev. Franç. Oenol.* **1988**, *28*, 7–11.
40. Ness, F.; Lavallée, F.; Dubourdieu, D.; Aigle, M.; Dulau, L. Identification of yeast strains using the polymerase chain reaction. *J. Sci. Food Agric.* **1993**, *62*, 89–94. [CrossRef]
41. Quesada, M.P.; Cenis, J.L. Use of Random Amplified Polymorphic DNA (RAPD-PCR) in the Characterization of Wine Yeasts. *Am. J. Enol. Vitic.* **1995**, *46*, 204–208.
42. Strand, M.; Prolla, T.A.; Liskay, R.M.; Petes, T.D. Destabilization of tracts of simple repetitive DNA in yeast by mutations affecting DNA mismatch repair. *Nature* **1993**, *365*, 274–276. [CrossRef] [PubMed]
43. Pérez, M.A.; Gallego, F.J.; Martínez, I.; Hidalgo, P. Detection, distribution and selection of microsatellites (SSRs) in the genome of the yeast *Saccharomyces cerevisiae* as molecular markers. *Lett. Appl. Microbiol.* **2001**, *33*, 461–466. [CrossRef] [PubMed]
44. Fay, J.C.; Benavides, J.A. Evidence for domesticated and wild populations of *Saccharomyces cerevisiae*. *PLoS Genet.* **2005**, *1*, 66–71. [CrossRef] [PubMed]
45. Guillamón, J.M.; Barrio, E.; Querol, A. Characterization of Wine Yeast Strains of the *Saccharomyces* genus on the basis of molecular markers: Relationships between genetic distance and geographic or ecological origin. *Syst. Appl. Microbiol.* **1996**, *19*, 122–132. [CrossRef]
46. Capece, A.; Romaniello, R.; Siesto, G.; Pietrafesa, R.; Massari, C.; Poeta, C.; Romano, P. Selection of indigenous *Saccharomyces cerevisiae* strains for Nero d'Avola wine and evaluation of selected starter implantation in pilot fermentation. *Int. J. Food Microbiol.* **2010**, *144*, 187–192. [CrossRef] [PubMed]
47. Capece, A.; Romaniello, R.; Siesto, G.; Romano, P. Diversity of *Saccharomyces cerevisiae* yeasts associated to spontaneously fermenting grapes from an Italian "heroic vine-growing area". *Food Microbiol.* **2012**, *31*, 159–166. [CrossRef] [PubMed]
48. Barrajón, N.; Arévalo-Villena, M.; Rodríguez-Aragón, L.J.; Briones, A. Ecological study of wine yeast in inoculated vats from La Mancha region. *Food Control.* **2009**, *20*, 778–783. [CrossRef]
49. Maqueda, M.; Zamora, E.; Rodríguez-Cousiño, N.; Ramírez, M. Wine yeast molecular typing using a simplified method for simultaneously extracting mtDNA, nuclear DNA and virus dsRNA. *Food Microbiol.* **2010**, *27*, 205–209. [CrossRef] [PubMed]
50. Muñoz, R.; Gómez, A.; Robles, V.; Rodríguez, P.; Cebollero, E.; Tabera, L.; Carrascosa, A.V.; Gonzalez, R. Multilocus sequence typing of oenological *Saccharomyces cerevisiae* strains. *Food Microbiol.* **2009**, *26*, 841–846. [CrossRef] [PubMed]
51. Tristezza, M.; Fantastico, L.; Vetrano, C.; Bleve, G.; Corallo, D.; Grieco, F.; Mita, G.; Grieco, F. Molecular and technological characterization of *Saccharomyces cerevisiae* strains isolated from natural fermentation of Susumaniello grape must in Apulia, Southern Italy. *Int. J. Microbiol.* **2014**, *2014*, 897427. [CrossRef] [PubMed]
52. Tristezza, M.; Vetrano, C.; Bleve, G.; Grieco, F.; Tufariello, M.; Quarta, A.; Mita, G.; Spano, G.; Grieco, F. Autochthonous fermentation starters for the industrial production of Negroamaro wines. *J. Ind. Microbiol. Biotechnol.* **2011**, *39*, 81–92. [CrossRef] [PubMed]
53. Tristezza, M.; Vetrano, C.; Bleve, G.; Spano, G.; Capozzi, V.; Logrieco, A.; Mita, G.; Grieco, F. Biodiversity and safety aspects of yeast strains characterized from vineyards and spontaneous fermentations in the Apulia Region, Italy. *Food Microbiol.* **2013**, *36*, 335–342. [CrossRef] [PubMed]
54. Pretorius, I.S. Tailoring wine yeast for the new millennium: Novel approaches to the ancient art of winemaking. *Yeast* **2000**, *16*, 675–729. [CrossRef]
55. Cappello, M.S.; Poltronieri, P.; Blaiotta, G.; Zacheo, G. Molecular and physiological characteristics of a grape yeast strain containing atypical genetic material. *Int. J. Food Microbiol.* **2010**, *144*, 72–80. [CrossRef] [PubMed]

56. Bely, M.; Stoeckle, P.; Masneuf-Pomarède, I.; Dubourdieu, D. Impact of mixed *Torulaspora delbrueckii-Saccharomyces cerevisiae* culture on high-sugar fermentation. *Int. J. Food Microbiol.* **2008**, *122*, 312–320. [CrossRef] [PubMed]
57. Rainieri, S.; Pretorius, I.S. Selection and improvement of wine yeasts. *Ann. Microbiol.* **2000**, *50*, 15–32.
58. Henschke, P.A. Wine yeast. In *Yeast Sugar Metabolism*; Zimmermann, F.K., Entian, K.D., Eds.; Technomic Publishing Co Inc.: Lancaster, PA, USA, 1997; pp. 527–560.
59. Salmon, J.M.; Vincent, O.; Mauricio, J.C.; Bely, M.; Barre, P. Sugar transport inhibition and apparent loss of activity in *Saccharomyces cerevisiae* as a major limiting factor of enological fermentations. *Am. J. Enol. Vitic.* **1993**, *44*, 56–64.
60. Blomberg, A.; Adler, L. Physiology of osmotolerance in fungi. *Adv. Microb. Physiol.* **1992**, *33*, 145–212. [PubMed]
61. Romano, P. Lievito starter e qualità aromatica del vino. *Inf. Agric.* **2006**, *62*, 27–31.
62. Csoma, H.; Zakany, N.; Capece, A.; Romano, P.; Sipiczki, M. Biological diversity of *Saccharomyces* yeasts of spontaneously fermenting wines in four wine regions: Comparative genotypic and phenotypic analysis. *Int. J. Food Microbiol.* **2010**, *140*, 239–248. [CrossRef] [PubMed]
63. Romano, P.; Fiore, C.; Paraggio, M.; Caruso, M.; Capece, A. Function of yeast species and strains in wine flavour. *Int. J. Food Microbiol.* **2003**, *86*, 169–180. [CrossRef]
64. Manzanares, P.; Ramón, D.; Querol, A. Screening of non-*Saccharomyces* wine yeasts for the production of b-D-xylosidase activity. *Int. J. Food Microbiol.* **1999**, *46*, 105–112. [CrossRef]
65. Beltran, G.; Torija, M.J.; Novo, M.; Ferrer, N.; Poblet, M.; Guillamón, J.M.; Rozès, N.; Mas, A. Analysis of yeast populations during alcoholic fermentation: A six year follow-up study. *Syst. Appl. Microbiol.* **2002**, *25*, 287–293. [CrossRef] [PubMed]
66. Cocolin, L.; Bisson, L.F.; Mills, D.A. Direct profiling of the yeast dynamics in wine fermentations. *FEMS Microbiol. Lett.* **2000**, *189*, 81–87. [CrossRef] [PubMed]
67. Santamaría, P.; Garijo, P.; López, R.; Tenorio, C.; Rosa Gutiérrez, A. Analysis of yeast population during spontaneous alcoholic fermentation: Effect of the age of the cellar and the practice of inoculation. *Int. J. Food Microbiol.* **2005**, *103*, 49–56. [CrossRef] [PubMed]
68. Borrull, A.; Poblet, M.; Rozès, N. New insights into the capacity of commercial wine yeasts to grow on sparkling wine media. Factor screening for improving wine yeast selection. *Food Microbiol.* **2015**, *48*, 41–48. [CrossRef] [PubMed]
69. Kemp, B.; Alexandre, H.; Robillard, B.; Marchal, R. Effect of production phase on bottle-fermented sparkling wine quality. *J. Agric. Food Chem.* **2015**, *63*, 19–38. [CrossRef] [PubMed]
70. Kumar, G.R.; Goyashiki, R.; Ramakrishnan, V.; Karpel, J.E.; Bisson, L.F. Genes required for ethanol tolerance and utilization in *Saccharomyces cerevisiae*. *Am. J. Enol. Vitic.* **2008**, *59*, 401–411.
71. Bisson, L.F.; Block, D.E. Ethanol tolerance in *Saccharomyces*. In *Biodiversity and Biotechnology of Wine Yeasts*; Ciani, M., Ed.; Research Signpost: Kerala, India, 2002.
72. Martí-Raga, M.; Martín, V.; Gil, M.; Sancho, M.; Zamora, F.; Mas, A.; Beltran, G. Contribution of yeast and base wine supplementation to sparkling wine composition. *J. Sci. Food Agric.* **2016**. [CrossRef] [PubMed]
73. Laurent, M.; Valade, M. La propagation des levains de tirage. *Vign. Champen.* **1998**, *3*, 29–52.
74. Monk, P.R.; Storer, R.J. The kinetics of yeast growth and sugar utilization in tirage: The influence of different methods of starter culture preparation and inoculation levels. *Am. J. Enol. Vitic.* **1986**, *37*, 72–76.
75. Benucci, I.; Liburdi, K.; Cerreti, M.; Esti, M. Characterization of active dry wine yeast during starter culture (Pied de cuve). Preparation for sparkling wine production. *J. Food Sci.* **2016**, *81*, 2015–2020. [CrossRef] [PubMed]
76. Pampulha, M.E.; Loureiro-Dias, M.C. Activity of glycolytic enzymes of *Saccharomyces cerevisiae* in the presence of acetic acid. *Appl. Microbiol. Biotechnol.* **1989**, *34*, 375–380. [CrossRef]
77. Martínez-Rodríguez, A.J.; Polo, M.C. Characterization of the Nitrogen Compounds Released during Yeast Autolysis in a Model Wine System. *J. Agric. Food Chem.* **2000**, *48*, 1081–1085. [CrossRef] [PubMed]
78. Charpentier, C.; Feuillat, M. Yeast autolysis. In *Wine Microbiology and Biotechnology*; Fleet, G.H., Ed.; Harwood Academic Publishers: Chur, Switzerland, 1993; pp. 225–242.
79. Martínez-Rodríguez, A.J.; Polo, M.C.; Carrascosa, A.V. Structural and ultrastructural changes in yeast cells during autolysis in a model wine system and in sparkling wines. *Int. J. Food Microbiol.* **2001**, *71*, 45–51. [CrossRef]

80. Martínez-Rodriguez, A.J.; Carrascosa, A.V.; Polo, M.C. Release of nitrogen compounds to the extracellular medium by three strains of *Saccharomyces cerevisiae* during induced autolysis in a model wine system. *Int. J. Food Microbiol.* **2001**, *68*, 155–160. [CrossRef]
81. Molnar, I.; Oura, E.; Suomalainen, H. Study of volatile substrates produced during the autolysis of champagne yeast. *Acta Aliment.* **1981**, *10*, 27–36.
82. Torresi, S.; Frangipane, M.T.; Garzillo, A.M.V.; Massantini, R.; Contini, M. Effects of a β-glucanase enzymatic preparation on yeast lysis during aging of traditional sparkling wines. *Food Res. Int.* **2014**, *55*, 83–92. [CrossRef]
83. Moreno-Arribas, M.V.; Polo, M.C. Amino Acids and Biogenic Amines. In *Wine Chemistry and Biochemistry*; Moreno-Arribas, M.V., Polo, M.C., Eds.; Springer: New York, NY, USA, 2009; pp. 163–189.
84. Todd, B.E.N.; Fleet, G.H.; Henschke, P.A. Promotion of autolysis through the interaction of killer and sensitive yeasts: Potential application in sparkling wine production. *Am. J. Enol. Vitic.* **2000**, *51*, 65–72.
85. Jolly, J.; Augustyn, O.P.H.; Pretorius, I.S. The role and use of non-Saccharomyces yeasts in wine production. *S. Afr. J. Enol. Vitic.* **2006**, *27*, 15–39.
86. Suárez-Lepe, J.A.; Morata, A. New trends in yeast selection for winemaking. *Trends Food Sci. Technol.* **2012**, *23*, 39–50. [CrossRef]
87. Benito, S.; Gálvez, L.; Palomero, F.; Calderón, F.; Morata, A.; Palmero, D.; Suárez-Lepe, J.A. *Schizosaccharomyces* selective differential media. *Afr. J. Microbiol. Res.* **2013**, *7*, 3026–3036.
88. Lasik, M. The application of malolactic fermentation process to create good-quality grape wine produced in cool-climate countries: A review. *Eur. Food Res. Technol.* **2013**, *237*, 843–850. [CrossRef]
89. Bartowsky, E.J.; Borneman, A.R. Genomic variations of *Oenococcus oeni* strains and the potential to impact on malolactic fermentation and aroma compounds in wine. *Appl. Microbiol. Biotechnol.* **2011**, *92*, 441–447. [CrossRef] [PubMed]
90. Auge, D.; Valade, M.; Moncomble, D. Acidity of Champagne wines: Use of malolactic fermentation? *Vign. Champen. Epernay* **2000**, *121*, 44–56.
91. Girbau-Solà, T.; López-Barajas, M.; López-Tamames, E.; Buxaderas, S. Foam aptitude of Trepat and Monastrell red varieties in cava elaboration. 2. Second fermentation and aging. *J. Agric. Food Chem.* **2002**, *50*, 5600–5604. [CrossRef] [PubMed]
92. Girbau-Solà, T.; López-Tamames, E.; Buján, J.; Buxaderas, S. Foam aptitude of Trepat and Monastrell red varieties in cava elaboration. 1. Base wine characteristics. *J. Agric. Food Chem.* **2002**, *50*, 5596–5599. [CrossRef] [PubMed]
93. Lopez-Barajas, M.; Lopez-Tamames, E.; Buxaderas, S.; de la Torre-Boronat, M.C. Effect of vinification and variety on foam capacity of wine. *Am. J. Enol. Vitic.* **1998**, *49*, 397–402.
94. Pozo-Bayón, M.Á.; Monagas, M.; Bartolomé, B.; Moreno-Arribas, M.V. Wine features related to safety and consumer health: An integrated perspective. *Crit. Rev. Food Sci. Nutr.* **2012**, *52*, 31–54. [CrossRef] [PubMed]
95. Ferreira, I.M.; Pinho, O. Biogenic amines in Portuguese traditional foods and wines. *J. Food Prot.* **2006**, *69*, 2293–2303. [CrossRef] [PubMed]
96. Ancín-Azpilicueta, C.; González-Marco, A.; Jiménez-Moreno, N. Current knowledge about the presence of amines in wine. *Crit. Rev. Food Sci. Nutr.* **2008**, *48*, 257–275. [CrossRef] [PubMed]
97. Maynard, L.S.; Schenker, V.J. Monoamine-oxidase inhibition by ethanol in vitro. *Nature* **1996**, *196*, 575. [CrossRef]
98. Coton, M.; Romano, A.; Spano, G.; Ziegler, K.; Vetrana, C.; Desmarais, C.; Lonvaud-Funel, A.; Lucas, P.; Coton, E. Occurrence of biogenic amine-forming lactic acid bacteria in wine and cider. *Food Microbiol.* **2010**, *27*, 1078–1085. [CrossRef] [PubMed]
99. Landete, J.M.; de Las Rivas, B.; Marcobal, A.; Muñoz, R. Molecular methods for the detection of biogenic amine-producing bacteria on foods. *Int. J. Food Microbiol.* **2007**, *117*, 258–269. [CrossRef] [PubMed]
100. Spano, G.; Russo, P.; Lonvaud-Funel, A.; Lucas, P.; Alexandre, H.; Grandvalet, C.; Coton, E.; Coton, M.; Barnavon, L.; Bach, B.; et al. Biogenic amines in fermented foods. *Eur. J. Clin. Nutr.* **2010**, *64*, S95–S100. [CrossRef] [PubMed]
101. Beneduce, L.; Romano, A.; Capozzi, V.; Lucas, P.; Barnavon, L.; Bach, B.; Vuchot, P.; Grieco, F.; Spano, G. Biogenic amine in wines. *Ann. Microbiol.* **2010**, *60*, 573–578. [CrossRef]
102. Landete, J.M.; Ferrer, S.; Polo, L.; Pardo, I. Biogenic Amines in Wines from Three Spanish Regions. *J. Agric. Food Chem.* **2005**, *53*, 1119–1124. [CrossRef] [PubMed]

103. Smit, A.; Moses, S.G.; Pretorius, I.S.; Cordero Otero, R.R. The Thr505 and Ser557 residues of the AGT1-encoded alpha-glucoside transporter are critical for maltotriose transport in *Saccharomyces cerevisiae*. *J. Appl. Microbiol.* **2008**, *104*, 1103–1111. [CrossRef] [PubMed]
104. Marcobal, A.; Martín-Alvarez, P.J.; Polo, M.C.; Muñoz, R.; Moreno-Arribas, M.V. Formation of biogenic amines throughout the industrial manufacture of red wine. *J. Food Prot.* **2006**, *69*, 397–404. [CrossRef] [PubMed]
105. Torrea, D.; Ancín, C. Content of biogenic amines in a Chardonnay wine obtained through spontaneous and inoculated fermentations. *J. Agric. Food Chem.* **2002**, *50*, 4895–4899. [CrossRef] [PubMed]
106. Caruso, M.; Capece, A.; Salzano, G.; Romano, P. Typing of *Saccharomyces cerevisiae* and *Kloeckera apiculata* strains from Aglianico wine. *Lett. Appl. Microbiol.* **2002**, *34*, 323–328. [CrossRef] [PubMed]
107. Granchi, L.; Romano, P.; Mangani, S.; Guerrini, S.; Vincenzini, M. Production of biogenic amines by wine microorganisms. *Bull. l'OIV Off. Int. Vigne Vin* **2005**, *78*, 595–610.
108. Delage, N.; d'Harlingue, A.; Colonna Ceccaldi, B.; Bompeix, G. Occurrence of mycotoxins in fruit juices and wine. *Food Control* **2003**, *14*, 225–227. [CrossRef]
109. Esti, M.; Benucci, I.; Liburdi, K.; Acciaro, G. Monitoring of ochratoxin A fate during alcoholic fermentation of wine-must. *Food Control* **2012**, *27*, 53–56. [CrossRef]
110. Anli, E.; Bayram, M. Ochratoxin A in Wines. *Food Rev. Int.* **2009**, *25*, 214–232. [CrossRef]
111. Angioni, A.; Caboni, P.; Garau, A.; Farris, A.; Orro, D.; Budroni, M.; Cabras, P. In vitro interaction between ochratoxin A and different strains of *Saccharomyces cerevisiae* and *Kloeckera apiculata*. *J. Agric. Food Chem.* **2007**, *55*, 2043–2048. [CrossRef] [PubMed]
112. Caridi, A. Enological functions of parietal yeast mannoproteins. *Antonie Van Leeuwenhoek* **2006**, *89*, 417–422. [CrossRef] [PubMed]
113. Caridi, A.; Cufari, J.A.; Ramondino, D. Isolation and clonal pre-selection of enological *Saccharomyces*. *J. Gen. Appl. Microbiol.* **2002**, *48*, 261–267. [CrossRef] [PubMed]
114. Csutorás, C.; Rácz, L.; Rácz, K.; Fűtő, P.; Forgó, P.; Kiss, A. Monitoring of ochratoxin A during the fermentation of different wines by applying high toxin concentrations. *Microchem. J.* **2013**, *107*, 182–184. [CrossRef]
115. Meca, G.; Blaiotta, G.; Ritieni, A. Reduction of ochratoxin A during the fermentation of Italian red wine Moscato. *Food Control* **2010**, *21*, 579–583. [CrossRef]
116. Piotrowska, M.; Nowak, A.; Czyzowska, A. Removal of ochratoxin A by wine *Saccharomyces cerevisiae* strains. *Eur. Food Res. Technol.* **2013**, *236*, 441–447. [CrossRef]
117. Petruzzi, L.; Bevilacqua, A.; Corbo, M.R.; Garofalo, C.; Baiano, A.; Sinigaglia, M. Selection of autochthonous *Saccharomyces cerevisiae* strains as wine starters using a polyphasic approach and ochratoxin A Removal. *J. Food Prot.* **2014**, *77*, 1168–1177. [CrossRef] [PubMed]
118. Russo, P.; Capozzi, V.; Spano, G.; Corbo, M.R.; Sinigaglia, M.; Bevilacqua, A. Metabolites of microbial origin with an impact on health: Ochratoxin A and biogenic amines. *Front. Microbiol.* **2016**, *7*, 482. [CrossRef] [PubMed]
119. Penacho, V.; Valero, E.; Gonzalez, R. Transcription profiling of sparkling wine second fermentation. *Int. J. Food Microbiol.* **2012**, *153*, 176–182. [CrossRef] [PubMed]
120. Tini, V.; Zambonelli, C.; Benevelli, M.; Castellari, L. The autolysogenic *Saccharomyces cerevisiae* strains for the sparkling wines production. *Ind. Bevande* **1995**, *24*, 113–118.
121. Peppler, H.J. Yeast extracts. In *Fermented Foods. Economic Microbiology*; Rose, A.H., Ed.; Academic Press: London, UK, 1982; Volume 7, pp. 293–311.
122. Romano, P.; Soli, M.G.; Suzzi, G.; Grazia, L.; Zambonelli, C. Improvement of a Wine *Saccharomyces cerevisiae* Strain by a Breeding Program. *Appl. Environ. Microbiol.* **1985**, *50*, 1064–1067. [PubMed]
123. Giudici, P.; Solieri, L.; Pulvirenti, A.M.; Cassanelli, S. Strategies and perspectives for genetic improvement of wine yeasts. *Appl. Microbiol. Biotechnol.* **2005**, *66*, 622–628. [CrossRef] [PubMed]
124. Coloretti, F.; Zambonelli, C.; Tini, V. Characterization of flocculent *Saccharomyces* interspecific hybrids for the production of sparkling wines. *Food Microbiol.* **2006**, *23*, 672–676. [CrossRef] [PubMed]
125. Winge, O.; Lausten, O. Artificial species hybridisation in yeast. *Comp. Rend. Trav. Lab. Carslberg. Sér. Physiol.* **1938**, *22*, 235–244.
126. Castellari, M.; Ferruzzi, A.; Magrini, P.; Giudici, P.; Passarelli, C. Zambonelli Unbalanced wine fermentation by cryotolerant vs. non-cryotolerant. *Saccharomyces* strains. *Vitis* **1994**, *33*, 49–52.

127. Massoutier, C.; Alexandre, H.; Feuillat, M.; Charpentier, C. Isolation and characterization of cryotolerant *Saccharomyces* strains. *Vitis* **1998**, *37*, 55–59.
128. Suzzi, G.; Romano, P.; Zambonelli, C. Flocculation of wine yeasts: Frequency, differences, and stability of the character. *Can. J. Microbiol.* **1984**, *30*, 36–39. [CrossRef]

15

The Smell of Synthetic Biology: Engineering Strategies for Aroma Compound Production in Yeast

Niël van Wyk [1,2,*], **Heinrich Kroukamp** [1] **and Isak S. Pretorius** [3]

1 Department of Molecular Sciences, Faculty of Science and Engineering, Sydney, NSW 2109, Australia; Heinrich.kroukamp@mq.edu.au
2 Institut für Mikrobiologie und Biochemie Zentrum Analytische Chemie und Mikrobiologie, Hochschule Geisenheim University, 65366 Geisenheim, Germany
3 Chancellery, Macquarie University, Sydney, NSW 2109, Australia; sakkie.pretorius@mq.edu.au
* Correspondence: niel.vanwyk@mq.edu.au

Abstract: Yeast—especially *Saccharomyces cerevisiae*—have long been a preferred workhorse for the production of numerous recombinant proteins and other metabolites. *S. cerevisiae* is a noteworthy aroma compound producer and has also been exploited to produce foreign bioflavour compounds. In the past few years, important strides have been made in unlocking the key elements in the biochemical pathways involved in the production of many aroma compounds. The expression of these biochemical pathways in yeast often involves the manipulation of the host strain to direct the flux towards certain precursors needed for the production of the given aroma compound. This review highlights recent advances in the bioengineering of yeast—including *S. cerevisiae*—to produce aroma compounds and bioflavours. To capitalise on recent advances in synthetic yeast genomics, this review presents yeast as a significant producer of bioflavours in a fresh context and proposes new directions for combining engineering and biology principles to improve the yield of targeted aroma compounds.

Keywords: aroma; bioflavour; *Saccharomyces cerevisiae*; synthetic biology; yeast; Yeast 2.0

---

1. Introduction

An overarching definition for the term "aroma compound" is one that provides a sensorial stimulus to the olfactory senses and, in certain cases, also the gustatory senses. In literature, it shares overlapping designations with words like "flavours", "scents", "odorants" and "fragrances" and these terms are often used interchangeably. Aroma compounds have various applications in the food, feed, cosmetic and pharmaceutical industries [1]. Some compounds have applications beyond their sense-activating properties, including potential as a biofuel [2], the improvement of the shelf-life of certain fruit varieties [3] and antimicrobial activities [4]. They can either be desirable or unwanted in a given product and significant efforts can be made to either eliminate or increase levels depending on the application. Aroma compounds are rarely perceived in isolation (especially in fermented foodstuffs) and thus its interaction with other compounds can greatly affect how they are identified.

Although not discussed in this review, a crucial component in the perception of aroma compounds is the olfactory receptors that recognize odorous ligands. Seminal work done by Nobel laureates Richard Axel and Linda Buck show the large and diverse nature of these membrane-bound receptors present in our olfactory neurons which are responsible for the detection of odorants and give rise to the sense of smell [5]. These receptors can be variably expressed among individuals resulting in the different perceptions of the same compound by individuals—a key consideration of consumer preference of foodstuffs [6]. Often neglected and poorly understood are the psychological aspects of odour perception as it can relate to the associative memory of the individual [7].

Aroma compounds are structurally remarkably heterogeneous. They can have cyclic or non-cyclic, saturated or unsaturated, straight-chain or branched-chain structures bearing all kinds of functional groups (e.g., alcohols, aldehydes, ketones, esters and ethers) and, in some cases, have nitrogen and sulphur within the structure. Certain aroma compounds are even inorganic in nature. If made enzymatically, aroma compounds are derived from the pool of precursor molecules from the core metabolism of the cell (i.e., the carbohydrates, fatty acid, nucleotides and amino acids). Odour thresholds (i.e., the concentration ranges at which a given aroma compound is detected or sensed) are key parameters in aroma compound studies.

Most aroma compounds on the market are produced by isolating natural compounds from plant or animal or by chemical synthesis. However, there is a clear swing away from chemically-produced aroma compounds and aroma compounds that require extensive extraction from plants or animals towards the production and use of aroma compounds of (micro) biological origin—also called bioflavours. This is despite the fact that the chemically produced compounds are identical to their natural counterparts. Reasons for such a change in market preferences include the fact that chemical synthesis can often result in environmentally detrimental production processes and in undesired racemic mixtures. Also, extraction of aroma compounds from plants or animal sources can be resource-intensive and cost-inefficient because of low yields. In addition, multiple purification steps often lead to product loss and degradation. Consumer aversion toward chemical compounds relates especially to food and home-care products. Despite changing preferences in consumer markets, the financial implication of aroma compound generation remains a strong consideration as those derived from chemical synthesis are, in general, markedly less expensive than those derived from natural sources.

In this context, researchers are directing their research efforts toward producing aroma compounds from microbial sources. This usually involves *Escherichia coli* or *S. cerevisiae* as cell factories by incorporating genes that code for enzymes that are relevant to the production of the given compound in a recombinant host [8]. Despite the campaigns against genetically-modified organisms (GMOs) in some sections of global consumer markets, there are numerous food ingredients derived from GMOs that are commercially-available the world over. However, in the case of such GM food-ingredients that comply with regulatory safeguards, high yields using cost-effective substrates have not yet been achieved in many instances.

This review primarily focusses on recent advances in research aimed at the production of aroma compounds in yeast. This paper is distinct from other published reviews, including those that extensively covered the use of flavour-active brewing and wine yeasts for the enhancement of the aroma of beer and wine [9–11]. Here, we focus on the exploitation of two types of yeast precursors which are responsible for a variety of aroma compounds, namely the aromatic amino acids L-tyrosine and L-phenylalanine, which are derived from the shikimate pathway and the mevalonate pathway-derived isoprenoid precursors dimethylallyl pyrophosphate and isopentenyl pyrophosphate.

2. Yeast as a Recombinant Host for Bioflavour Production

Various yeasts—with *S. cerevisiae* being the model organism—have long been harnessed for the expression of recombinant genes to enhance endogenous aroma-active metabolites of the host cells or to produce novel recombinant compounds. The initial reasons why researchers opted for *S. cerevisiae* remains true, that is, this yeast species is by far the best-studied unicellular eukaryote with the genomes of several of its strains fully sequenced [12]; it is a non-pathogen that enjoys GRAS (generally recognised as safe) status; and it is amenable to genetic manipulation with a wide range of genetic tools available to alter the genetic make-up of the yeast. *S. cerevisiae* also possesses an efficient homologous recombination machinery, which greatly assists stable integration of genetic elements. This yeast is also the most robust fermenter and laboratory-scale processes can be scaled up to industrial-level set-ups with relative ease. Some of the abovementioned attributes also hold true for *E. coli*. However, as a prokaryote, this bacterium lacks a sophisticated protein-folding mechanism.

This often leads to the recombinant proteins being insoluble and most likely non-functional and that might require additional recovery steps for refolding of the protein of interest.

S. cerevisiae is, however, by no means a perfect host; for example, it is not a prolific biomass producer and the way secreted proteins are glycosylated sometimes lead to pronounced reduction in bioactivity. There are also reports of recombinant genes that cannot be successfully expressed for unknown reasons. Regardless of the whether *S. cerevisiae* turns out to be appropriate as a host to produce a particular recombinant product, it remains the best starting point to move onwards to other organisms. A prudent strategy is to examine the expression levels in multiple yeast hosts and to compare titres of a protein (or metabolite) of interest. Often the methylotrophic yeast *Pichia pastoris* (now reclassified as *Komagataella phaffii*) and *Hansenula polymorpha* (now reclassified as *Ogataea polymorpha*) have shown superior protein and/or metabolite production capabilities owing to their unusually high biomass production [13]. Many other yeast species with their own special attributes can (and have) been utilised as a recombinant host with varying outcomes. Examples of such yeasts include *Kluyveromyces lactis*, *Yarrowia lipolytica* and *Schizosaccharomyces pombe*. The usefulness of non-*Saccharomyces* yeasts in the biotransformations of certain substrates into aroma compounds with whole-cell or resting cell systems are well-documented [14]. This has been a popular way of producing aroma compounds as it can allow for the assembly of regio- and stereoselective compounds under mild and mostly solvent-free conditions.

Identification of natural variation within a yeast strains and species has undeniably created a valuable source of flavour-active strains [15]. The underlying molecular determinants for a particular phenotype has been elucidated through the advances in 'omics' capability. Effective mining of genes and alternative alleles responsible for a desired phenotype have become common practice, with access to comprehensive conventional yeast libraries based on mutagenesis, breeding [16], single gene deletions [17] and overexpression [18]. Yeast libraries have become more sophisticated and, in many cases, combine the genomic variation generation with a selection for the particular characteristic of interest. This includes biosensor-enabled directed evolution (discussed in later section below), rapid genome-wide editing (YOGE) or the complete reconstruction of pathways (VEGAS) and genomes (Yeast 2.0).

Yeast Oligo-mediated Genome Engineering (YOGE) enables rapid genome engineering by introducing allele variation by sequential oligonucleotide recombination [19]. Designer synthetic DNA oligonucleotides allow the combinatorial alteration of pathway genes and, with successive rounds of transformation, gradually remodel the yeast genome toward the production of a metabolite or to embody a specific phenotype. Smaller, directed libraries, only altering the pathway(s) of interest, have been demonstrated with techniques like Versatile Genetic Assembly System (VEGAS). VEGAS uses the yeast's innate preference for homologous recombination to assemble complex pathways, allowing different combinations of the pathway genes to be assembled and subsequently screened for the best production [20].

A new generation of yeasts might allow us to greatly expand yeast strain diversity beyond what has resulted to date with directed breeding and natural selection. The revolutionary synthetic biology initiative known as the Yeast 2.0 project (also known as Sc2.0) was initiated in 2007 [21] to deepen our understanding of the molecular mechanisms that drives this versatile organism. Upon completion, the Sc2.0 strain will be world's first eukaryote with a streamlined chemically-synthesised genome. In addition to the removal of repetitive sequences, the liberation of a codon and the introduction of hundreds of watermark sequences, LoxPsym sequences were introduced at the 5′-ends of all genes considered individually non-essential [22,23]. These sites allow for inducible homologous recombination downstream of all non-essential genes, mediated by the action of the site-specific Cre-recombinase. Upon activation of the site-specific Cre-recombinase, homologous recombination is promoted between these LoxPsym sites, resulting in rapid gene deletion, duplication or inversion. This process—known as SCRaMbLE (Synthetic Chromosome Rearrangement and Modification by LoxPsym-mediated Evolution)—allows for the rapid synthetic rearrangement and evolution of the yeast genome [24] (Figure 1). In addition to this novel way of producing large libraries of genomically-divergent yeasts, SCRaMbLE also allows us to produce and explore minimum eukaryotic genomes for the first time. These libraries will be valuable assets in the screening for interesting phenotypes, like aroma compound production and the elucidation of the underlying principles governing these production pathways.

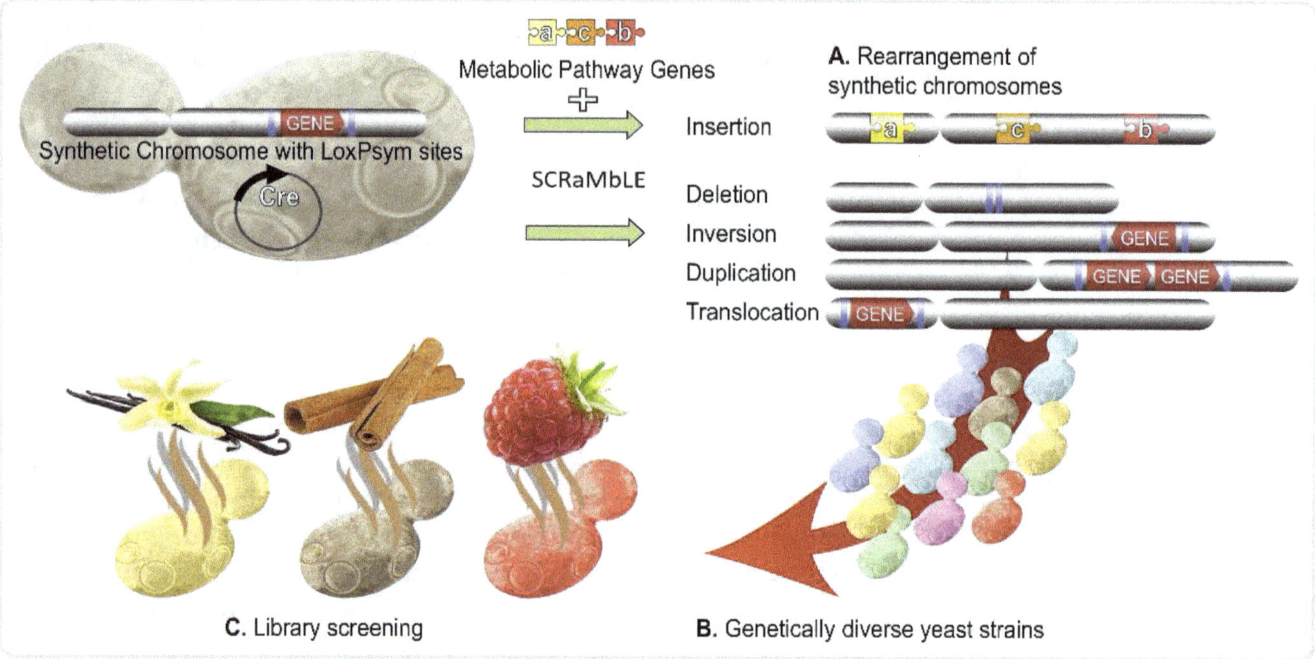

Figure 1. Depiction of aroma compound pathway optimisation through loxPsym-mediated rearrangement of synthetic chromosomes (SCRaMbLE) in yeast. (**A**) A yeast containing synthetic versions of their respective chromosomes with multiple loxPsym sequences would be subjected to the actions of the loxPsym-specific Cre recombinase. (**B**) The subsequent insertions, duplications, deletions, inversions and other genetic alterations will allow for the generation of an instantly-made library of yeast that have tremendous diversity in their respective genetic backgrounds (**C**) allowing for the screening of yeast with preferred phenotypes. By introducing metabolite pathway genes, flanked by loxP sequences, copy number optimised pathways can be assembled into the generated library. At the time of writing this review, 6 of the 16 chromosomes have been fully synthesized, with the rest at various stages of construction and debugging [25]. The strains harbouring these chromosomes (or combinations thereof) can currently be used for SCRaMbLE-based phenotype generation experiments.

Irrespective of the specific yeast strain used, optimisation of the recombinant production of a given protein or metabolite would require the systematic improvement of the properties of the recombinant host using analytical and computational methods to quantify fluxes and their regulation. The following guiding principle questions, regarding global and pathway-specific metabolic engineering, have been proposed previously [26]: (i) can the precursor and/or cofactor supply be increased?; (ii) can the heterologous expression of non-native genes be different or the expression thereof be improved?; (iii) can pathways that compete for the same precursors and co-factors be blocked or down-regulated?; (iv) are transcriptional regulators known and what would be the effect if they are overexpressed?; and (v) can the enzyme specificity be improved? Most of these questions are directly applicable in improving a yeast's ability to produce aroma compounds. Below we will discuss the work researchers have undertaken in addressing these questions in order to increase the levels of phenylpropanoid and terpenoid production in yeast.

3. Yeast Precursors Utilised

3.1. *Phenylpropanoids*

The aromatic amino acids L-phenylalanine, L-tyrosine and L-tryptophan serve as the precursors to many compounds of commercial interest [27]. More specifically, L-phenylalanine, L-tyrosine provide the precursors for a large group of compounds called phenylpropanoids—of which many have aroma-active properties. The biosynthesis of the aromatic amino acids proceeds via the shikimate pathway [28] (Figure 2). It is a seven-step metabolic pathway leading to the production of chorismate, the common aromatic precursor to all three amino acids. The shikimate pathway is initiated with the condensation of phosphoenolpyruvate (PEP)—an intermediate in the glycolysis pathway—and erythrose-4-phosphate (E4P)—an intermediate in the pentose phosphate pathway—to generate 3-deoxy-D-arabino-heptulosonate-7-phosphate (DAHP). Chorismate is the branching node, where L-tryptophan is separated from the other two amino acids as chorismate is converted to prephenic acid (the precursor molecule of L-phenylalanine and L-tyrosine) by a chorismate mutase. Subsequent decarboxylation and transamination events lead to the production of L-tyrosine and L-phenylalanine. In general, intracellular L-tyrosine levels in *S. cerevisiae* are about ten-fold higher than L-phenylalanine with L-tryptophan 10 times less than L-phenylalanine [29].

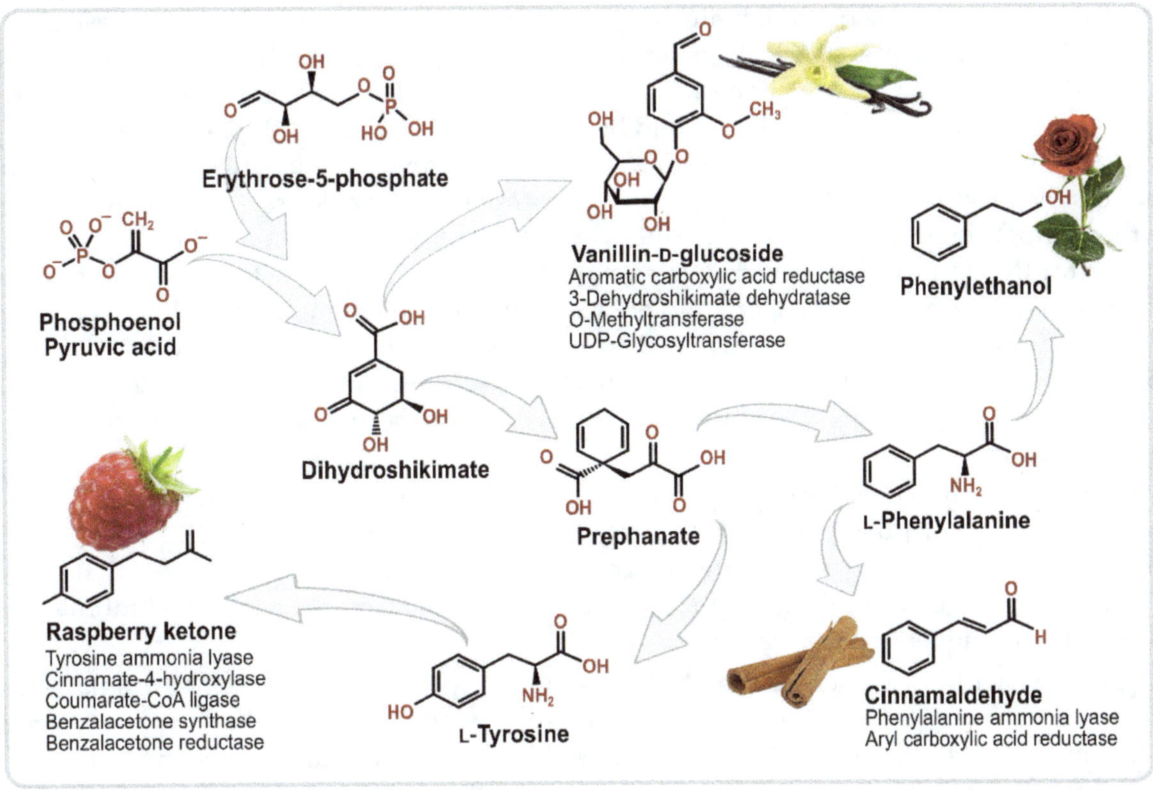

Figure 2. Biosynthetic pathway for phenylpropanoids. Yeast can synthesize all three aromatic amino acids (L-phenylalanine, L-tyrosine and L-tryptophan) via the shikimate pathway but have few processing capabilities beyond utilising them in peptide synthesis or their catabolism via the Ehrlich pathway (which can produce the aroma compound 2-phenylethanol). *S. cerevisiae* and other yeast have been exploited to convert their free aromatic amino acids to compounds with aroma properties. The recombinant enzymes that have been incorporated in yeast to convert precursors to aroma compounds of commercial value are shown below each recombinant metabolite.

S. cerevisiae has a limited capacity to process aromatic amino acids beyond using them for protein synthesis. Pathways involved in using L-tyrosine and L-phenylalanine as precursors have been incorporated into yeast to produce a multitude of compounds and of these, the phenyl ring structure represents a central feature (Figure 2). A key aroma compound derived from the shikimate pathway is that of vanillin (imparting vanilla flavour) and has been the subject of many investigations in the past due to its high value and wide use. Vanillin is not synthesised from any of the aromatic amino acids, but from an intermediate in the shikimate pathway, namely dehydroshikimate. The first report of vanilla production by yeast used three recombinant genes in the fission yeast *S. pombe* to transform dehydroshikimate to vanillin [30]. In the same study, *S. cerevisiae* was also used, but an additional activation enzyme was needed. Vanillin is moderately toxic to yeast cells (it represses translational processes [31]). It was shown that adding a glycosyl moiety, by expressing a 1-UDP-glycosyltransferase, leads to the conversion of vanillin to vanillin-glucoside (VG), which markedly increased production levels. Remarkable improvements in VG titres have been achieved with rational engineering design approaches: in silico metabolic engineering algorithms have been implemented to identify yeast target genes that could enhance productivity [32]. Manipulations of two of the identified targets (*PDC1* and *GDH1*) led to a five-fold improvement of VG yields and was attributed to the recycling of the supply of cofactors. Additional modelling-based methodologies underlined the utility of in silico design for improvement in VG levels [33,34].

The pathway for the production of *p*-hydroxycinnamic acid (also known as *p*-coumaric acid), which imparts a cinnamon aroma, has been incorporated in *S. cerevisiae* [35]. This simply involved the incorporation of various phenylalanine ammonia-lyases (PAL)/tyrosine ammonia-lyases (TAL) which deaminate L-tyrosine. Several metabolic engineering strategies have proven successful in enhancing *p*-hydroxycinnamic acid along with the levels of so-called *trans*-cinnamic derivatives (which include cinnamaldehyde, cinnamyl alcohol and hydrocinnamyl alcohol) [36]. These strategies involved removing known feedback-regulated steps of aromatic amino acid biosynthesis and directing the flux towards the production of these *trans*-cinnamic compounds by side-tracking the decarboxylation step of the competing Ehrlich pathway. A phenylacrylic acid decarboxylase (PAD1) is thought to be responsible for the decarboxylation of *trans*-cinnamic derivatives, as a *pad1* knockout strain showed no endogenous activity on trans-cinnamic acid and *p*-hydroxycinnamic acid [37]. Nevertheless, the *trans*-cinnamic derivatives are converted to less toxic compounds by the yeast via unknown mechanisms [36]. It was found, similar to vanillin, that by adding a glycosyl moiety to trans-cinnamic acid catalysed by an UDP-glucose:cinnamate glucosyltransferase reduces its toxicity and led to increased levels.

A recent addition to the phenylpropanoid aroma compounds that are recombinantly produced in yeast is that of raspberry ketone [4-(4-hydroxyphenyl)butan-2-one] [38]. This involved the incorporation of a four-gene pathway from various organisms into yeast that converted L-phenylalanine and L-tyrosine to raspberry ketone. Testing various enzyme combinations and fusions resulted in higher levels of raspberry ketone.

Improving yields of 2-phenylethanol (2-PE)—a compound with a rose-like aroma—has been investigated extensively. 2-PE is the fusel alcohol of L-phenylalanine and of the four phenylpropanoid aroma compounds discussed, 2-PE does not require the expression of recombinant genes as it arises from the catabolism of L-phenylalanine via the Ehrlich pathway. This includes its deamination, decarboxylation and reduction that are conducted by ARO9, ARO10 and various alcohol dehydrogenases (ALD1-5) in *S. cerevisiae* respectively. Metabolic engineering efforts to increase 2-PE levels included the streamlining the Ehrlich pathway which involved the overexpression of *ARO9* and *ARO10* with the concomitant removal of a competing phenylacetaldehyde oxidase (*ALD3*) [39]. A transcription factor (ARO80) is known as an activator of the *ARO9* and *ARO10* genes and its overexpression, together with *ARO9* and *ARO10*, led to a four-fold increase in 2-PE levels.

Efforts have also been made to increase the intracellular levels of the precursors of the shikimate pathway PEP and E4P. Especially targeting E4P, which has lower intracellular concentrations than PEP [40] would result in an equal balance of the two precursors and could facilitate improved flux toward aromatic amino acid production. Attempts thus far to increase levels of E4P have involved alternations within the pentose phosphate pathway [41]. It was shown that the deletion of the glucose-6-phosphate dehydrogenase (*ZWF1*) gene and overexpression of the transketolase (*TKL1*) gene reversed the flux from the glycolytic intermediates and led to a higher (~eight-fold) increase in E4P levels [42].

Some non-*Saccharomyces* yeasts like *Ashbya gossypii* [43], *Kluyveromyces marxianus* [44] and *Candida glycerinogenes* [45] have been investigated for 2-PE production with yields reported that were greater than for *S. cerevisiae*. Similarly, as with *S. cerevisiae*, overexpressing the genes involved in the Ehrlich pathway have led to increased levels in *K. marxianus* [44], but not *A. gossypii* [43].

Media composition, especially adding L-phenylalanine, have shown in many cases to enhance the production of polypropanoids [46,47]. This implies that the yeast precursor is still a major bottleneck for phenylpropanoid production. High-throughput mass spectrometry experiments conducted on a yeast gene deletion library, which determined the intracellular concentration of each amino acid, revealed that certain gene knock-outs resulted in a two to four times higher intracellular concentration of L-phenylalanine than the wild type [29]. Many of the strains carrying these respective gene deletions also had an increased level of L-tyrosine and L-tryptophan suggesting that these gene products might have a putative role in regulating the shikimate pathway, but in most cases no obvious connection has ever been reported.

3.2. Terpenoids

Terpenoids (also called terpenes or isoprenoids) are the largest and most diverse group of natural compounds. They are derived from the basic five-carbon (C5) precursor unit isopentenyl diphosphate (IPP) and its double-bond isomer dimethylallyl diphosphate (DMAPP) that can be assembled and modified in over 60,000 different types of terpene-like structures. Apart from the exceptional flavour qualities of many terpenoids, certain terpenoids have promising applications in biofuel and antimicrobial research [48].

Terpenoids are either produced via the mevalonate biosynthesis pathway (MVA) or the 2-C-methyl-D-erythritol-4-phosphate pathway (MEP) with the former being the best-studied and found in yeast. In this pathway, acetyl-CoA is condensed to produce the universal isoprene building unit (C5), isopentenyl diphosphate (IPP). Subsequent condensations conducted by prenyltransferases of IPP and DMAPP result in terpenoid precursors called polyisoprenoid diphosphates of different lengths: geranyl diphosphate (GPP) for monoterpenoids (C10), farnesyl diphosphate (FPP) for sesquiterpenoids (C15), geranylgeranyl diphosphate (GGPP) for diterpenoids (C20), 2 units of FPP for triterpenoids (C30) and 2 units of GGPP for tetraterpenoids (C40). The C30 and C40 precursors lead to the biosynthesis of sterols and carotenoids, respectively. Cyclisation of the abovementioned polyisoprenoid diphosphates are catalysed by a large group of enzymes called terpene synthases to generate terpenoids with single or multiple ring structures (with some remaining open). These enzymes often display a high level of promiscuity with regards to their substrate preference leading to the large diversity among terpenoid structures. In addition, tailoring enzymes like oxygenases, methyltransferases, acetyltransferases and glycosyltransferases can add functional groups to different positions of the terpenoid structure.

As mentioned, yeast including *S. cerevisiae*, do possess an MVA pathway but does not have terpene synthases that are able to produce monoterpenoids, sesquiterpenoids and diterpenoids. *S. cerevisiae* implements the MVA pathway to produce sterols (specifically ergosterol) that are structural elements of the cell membrane and impart modulation to the membrane fluidity. It is an essential pathway for yeast as strains with mutated genes in this pathway require exogenous sterol for survival [49].

Significant efforts have been made to create a yeast platform that would be able to produce terpenoids as discussed in multiple papers and reviews [50–54] (Figure 3). A yeast without any modification within its MVA pathway would produce negligible levels of recombinant terpenoids. A key strategy to improve levels would be to direct carbon flux away from producing sterols without the complete elimination of the pathway. On a transcriptional level, the native promoter of the squalene synthase *ERG9* gene (that encodes the enzyme that catalyses the first reaction of converting farnesyl diphosphate to ergosterol) was replaced with repressible promoters which led to subsequent lower concentrations of ERG9, facilitating the increased levels of recombinant terpenoid production [55]. Recently, a degradation tag attached to ERG9 was shown to destabilise the protein, which also led to a dramatic improvement in recombinant terpenoid production without compromising the cell viability to any significant extent [56].

The overexpression of a truncated version of the hydroxymethylglutaryl-CoA reductase (tHMG1), that is devoid of its transmembrane moiety and is thus present in the cytosol, led to an increased amount of squalene. This confirmed that HMG-CoA reductase, the enzyme that produces mevalonate, is a rate-limiting step [57]. Many subsequent attempts using yeast to produce recombinant terpenoids contain this feature [58,59]. Similarly, the overexpression of the *ERG20* gene encoding an enzyme with both dimethylallyltranstransferase and geranyltransferase activities and the sterol regulator *upc2-1* [60] has generally led to improved yields in recombinant terpenoid production.

Improving the intracellular levels of the precursor molecule acetyl-CoA—an intermediate central to many metabolic pathways—has also been investigated. In one study by Meadows et al. [61], a *S. cerevisiae* strain was developed by completely overhauling the native metabolic network involved in acetyl-CoA supply by incorporating several synthetic pathways which directly resulted in increased recombinant terpenoid levels.

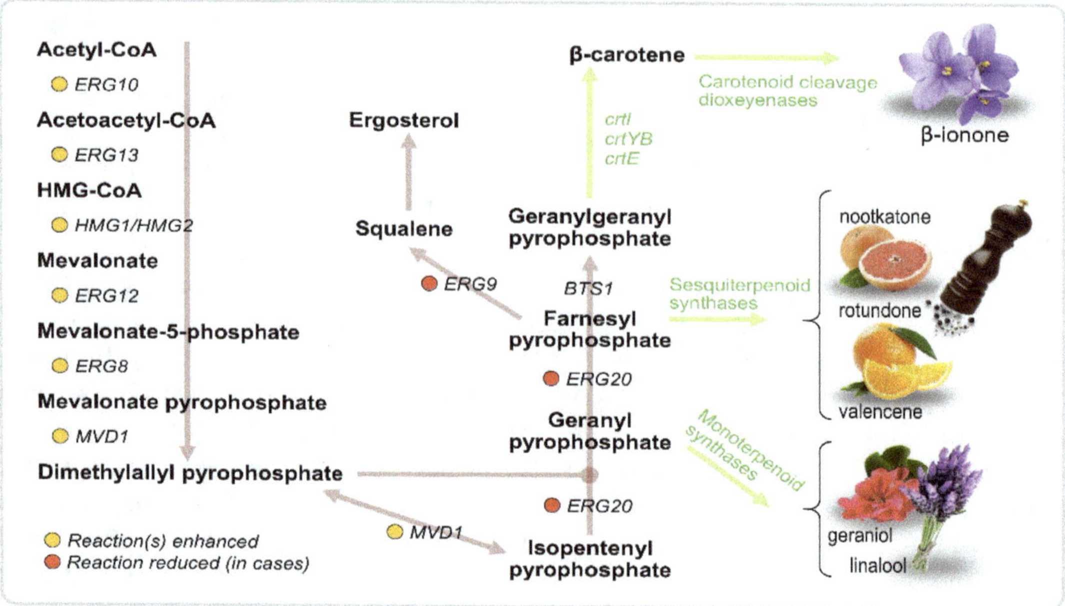

Figure 3. Manipulations in mevalonate pathway for recombinant terpenoid production. Enzymes directly involved in the MVA pathway of *S. cerevisiae* have been upregulated, downregulated or altered to increase the flux towards the production of various terpenoid-based aroma compounds. Additional recombinant enzymes needed for the catalysis of the terpenoid production are shown in orange with the respective aroma compounds being produced. Adapted from [53]. ERG10, acetyl-CoA *C*-acetyltransferase; ERG13, hydroxymethylglutaryl-CoA synthase; HMG1/HMG2, hydroxymethylglutaryl-CoA reductase 1/2; ERG12, mevalonate kinase; ERG8, Phosphomevalonate kinase; ERG19, mevalonate diphosphate decarboxylase; IPP, isopentenyl diphosphate; DMAPP, dimethylallyl diphosphate; IDI1, isopentenyl diphosphate isomerase; ERG20, geranyl/farnesyl diphosphate synthase; BTS1, GGPP synthase; ERG9, squalene synthase; crtYB, phytoene synthase and lycopene cyclase; crtI, crtE phytoene desaturase; geranylgeranyl diphosphate (GGPP) synthase.

It was shown that by fusing mitochondria signals to sesquiterpenoid synthases, along with the introduction of a recombinant farnesyl diphosphate synthase, resulted in a marked increase of the citrus aroma compound valencene [62]. This demonstrates the prowess of compartmentalisation approaches where enzymes, substrates and intermediates are close to each other and competing pathways in the cytosol are avoided.

Other successful attempts to produce terpenoids in yeast include the highly sought-after nootkatone, which imparts a strong grapefruit aroma [63]. This was achieved in *P. pastoris* with a similar engineering strategy which has been proven fruitful for *S. cerevisiae* to overexpress a truncated version of its *HMG1* gene to improved levels.

Carotenoids (specifically β-carotene) have also been produced in *S. cerevisiae* by expressing carotenogenic genes from strains from the ascomycete *Xanthophyllomyces dendrorhous* (previously *Phaffia rhodozyma*) [64]. Increased concentrations of β-carotene were observed when the *BTS1* gene encoding a geranylgeranyl diphosphate synthase was overexpressed. These strains appear bright orange and although itself not an aroma compound, β-carotenes do serve as a substrate for carotenoid cleavage oxygenase, which releases compounds known as apocarotenoids. The expression of a carotenoid cleavage oxygenase from *Petunia hybrida* in a strain already producing β-carotene, led to the release of detectable levels of a compound known as β-ionone, which has a highly desired violet scent [65]. Interestingly, a polycistronic version (genes separated from each other by viral T2A sequences) of the abovementioned carotenogenic genes was successfully expressed in yeast [66].

Assessing the feasibility of replacing elements of the MVA with that of the MEP pathway—theorized to yield a higher stoichiometric maximum plus having a lower requirement for oxygen—has been investigated. The yeast strain was developed where its MVA pathway was replaced with an MEP pathway. The resulting strain showed a slight growth defect made less biomass compared to the wild type, implying slight incompatibility [67].

4. Biosensing Aroma Compounds in Yeast

Rational engineering strategies to redistribute carbon flux towards the production of a specific metabolite or the introduction of novel synthetic biosynthesis pathways, have been successfully employed in the past to produce the desired molecule of interest [68,69]. Aroma compound production has also benefitted from these methodologies when ample precursor molecules are available [70]. However, the inherent volatile nature of some precursor molecules have excluded them from many directed evolution endeavours, as these are mostly limited to growth-selectable phenotypes, with limited high-throughput possibilities for the rapid screening of mutant libraries [71].

A promising, more recent addition to the synthetic biologist's toolbox is biosensors [72]. These genetic circuits translate a metabolic 'input' into a measurable 'output' signal like fluorescence; decoupling metabolite production from cellular growth. Biosensor designs are becoming increasingly more complex, with higher order circuits combining multiple interacting components and logic gate arrays to allow enhanced pathway regulation and output sensitivity. Table 1 shows some recent examples of biosensors developed for the direct or indirect detection of aroma compounds produced in yeast.

In one, early biosensor study [73], it was shown how biosensors can be employed to develop a high-throughput screen for yeast strains producing high concentrations of β-phenylethanol. This indirect method used the flux through the Ehrlich pathway as an indicator of high end-product concentrations. Endogenous biosensors have previously been employed to increase the flux toward the production of precursors of aroma compounds in yeast [74–77]. Synthetic biosensors have been constructed to allow feedback-regulated evolution of high IPP-producing strains [74].

There is much more scope for the development of aroma biosensors for yeast when compared to the diversity of available bacterial sensors, with *E. coli* sensors allowing for the detection of benzaldehyde, cinnamaldehyde, salicylaldehyde, syringaldehyde and vanillin [78,79]. Attempts have been made to translate some of these concepts in an endeavour to sense aroma compounds produced by bioengineered yeast, using encapsulated *p*-coumaric acid-sensing *E. coli* to screen for yeast cells producing *p*-coumaric acid [80].

Table 1. Examples of yeast biosensors to detect phenylpropanoids and terpenoids produced in *Saccharomyces cerevisiae*.

Aroma Compound	Molecule(s) Sensed	Description	Reference
		Shikimate pathway	
β-phenylethanol	Aromatic amino acids	Allosteric transcription factor sensor. Transcriptional regulation *LacZ* reporter gene by the aromatic amino acid responsive *ARO9* promoter. Increased β-galactosidase activity correlated with elevated β-phenylethanol levels.	[73]
Precursor	Betaxanthin	Enzyme-coupled sensor. Highly yeast-active heterologous L-tyrosine hydroxylases were identified, based on increased betaxanthin fluorescence intensities in yeast expressing the plant DOPA dioxygenase.	[70]
***p*-Coumaric acid**	*p*-Coumaric acid	Exogenous bacterial sensor. Droplet sorting of encapsulated *p*-coumaric acid producing yeast cells and *p*-coumaric acid sensing *E. coli* cells, to select producers based on bacterial YFP-fluorescence output.	[80]
Precursor	Muconic acid	Heterologous allosteric transcription factor. Used an *Acinetobacter* sp. transcriptional regulator to drive GFP expressing in the presence of muconic acid.	[77]
		Mevalonate pathway	
Precursor	Malonyl-CoA	Recombinant allosteric transcription factor sensor. Used a bacterial FapR transcription factor and FapO operator pair to identify strains from a genome-wide overexpression library that produce high levels of malonyl-CoA.	[75,76]
Precursor	Isopentenyl diphosphate	Synthetic transcription factor to allow feedback-regulated evolution of phenotype. Higher intracellular IPP concentrations resulted in increased *GAL10* transcription, generating an evolvable growth phenotype on galactose	[74]

5. Future Outlook

In only the past decade has research been focussed on incorporating biochemical pathways in yeast to produce key aroma compounds. Beyond proving the concept—as with all biotechnological approaches—researchers will always try to find ways to improve the overall yield of a given aroma compound. Simple fermentation optimisation has proven, on many occasions, to be impactful but the most profound improvements were achieved with the metabolic engineering of the yeast. Metabolic engineering can be defined and categorised in many ways, but, it is generally regarded as any kind of genetic modification applied onto a cell that would cause a preferential change in its phenotype. It is a cornerstone characteristic of synthetic biology: a confluence of many different streams of science and engineering with the scope of building artificial biological systems. Elements that appear to be central to the field of synthetic biology have been part of science for decades: the earliest recombinant genes that were expressed in *E. coli* (that of human insulin and the human growth hormone) were indeed chemically synthesised in the 1970s (before the advent of PCR-based cloning). Even slightly more complex genetic elements like synthetic promoters have been described to function in yeast since the late 1980s [81]. Yet, it is well-recognised that the dramatic drop in the cost for synthesising DNA-sequences in the past decade was the catalyst to jump-start the current wave of synthetic biology. Large pieces of DNA can now be designed and purchased from a multitude of companies which has drastically alleviated the often-time-consuming effort of constructing genetic elements.

As has been explained in instructive reviews [82,83], central for the rapid advancement of synthetic biology is the application of the concept of design-build-test-learn (or DBTL) (Figure 4) to address biological and/or engineering questions. This idea, adapted from other engineering fields, aims to streamline and accelerate the iterative process of improving a biological system as well as to minimise human input to eliminate bias in the interpretation of the output. As mentioned by Hollywood et al. [84], the testing part is often reliant on a high-throughput screen which makes this component—if available—often the most time-consuming and costly. This is particularly true for the determination of most aroma compounds, as the screening of millions of cells and assessing individual cells for aroma compound production is not yet feasible. This is why the development of aroma biosensors with appropriate sensitivity and specificity is most desired. A desired attribute of the biosensor would be for it to have a 'dynamic range,' that is, besides detecting aroma compounds it would have the ability to provide a differing signal depending on the concentration of the aroma. This might be achieved, although not yet tested in such a capacity, by coupling either a bioluminescence resonance energy transfer (BRET)-based sensor [85] or metal oxide [86] sensor, developed to detect volatile organic compounds, within a high-throughput screening process to test a yeast library.

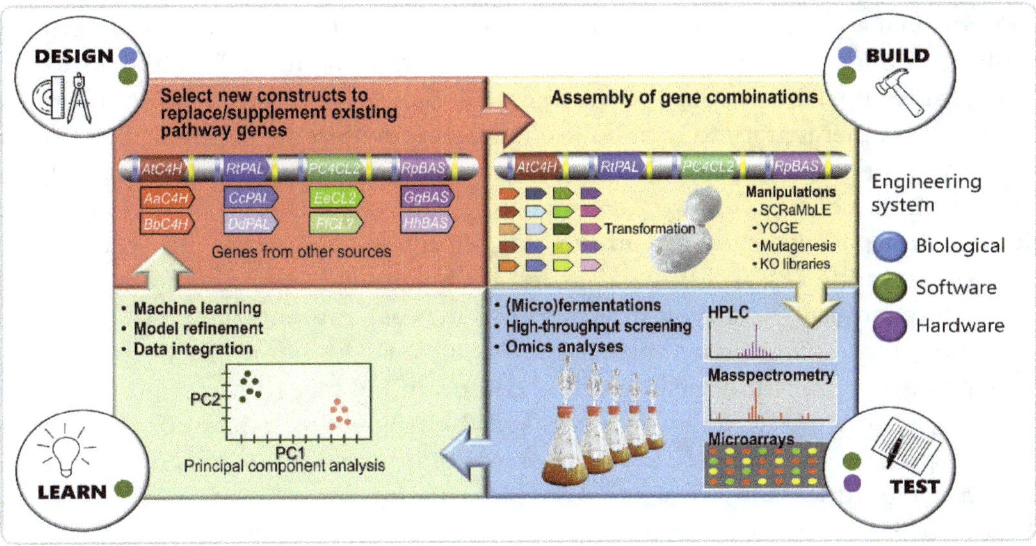

Figure 4. An example of how DBTL could be applied to build a yeast with increased production levels of raspberry ketone. During the "design" part of the project, researchers would need to plan all the experiments with a particular focus on the types of metabolic engineering tactics that will be employed to alter the target pathway within the yeast. The raspberry ketone synthesis pathway has already been introduced in *S. cerevisiae* [29], thus an avenue to follow is to test several similar enzymes from different organisms to assess their compatibility within the pathway. The "build" part encompasses all the aspects of constructing the strains that need to be tested. Multiplex CRISPR-based techniques for yeast have been developed and could easily be adapted to introduce combinations of the recombinant raspberry ketone pathway genes [76,77]. CRISPR-based techniques enjoy the benefit of its precision in editing the genome that would lead to minor modifications being made to the genome. The system can also allow for the strain to be 'markerless'—free of any antibiotic resistance genes normally used in conventional genetic manipulations. Additional mutagenesis, ranging from chemical mutagenesis to using a strain in which one could induce SCRaMbLE, could be included which will add more genetic diversity to the strains that need to be tested. The 'testing' part would, in most cases, involve some level of culturing of the strains coupled with determining the levels of the metabolite of interest. This would require high-throughput screening machinery. Raspberry ketone titres will need to be measured with mass spectrometry-based techniques. Eventual sequencing of high performers would be needed to ascertain what mutagenesis allowed for the superior production levels. Once the testing is concluded, the 'learning' part—where interpretation of the data (in this case identifying strains with superior raspberry ketone titres) and assessment of how one could improve upon the titres based on which strain performed better—is executed before commencing with a new cycle of DBTL. This is a simple example of a single iteration of DBTL with each step reliant on human intervention. The eventual goal—especially with the advent of genome or biofoundries—is to curtail human input and allow for automation and machine learning to dominate proceedings. Adapted from [83,87].

As has been discussed in this review, nearly all of the reported metabolic engineering approaches for improvement in aroma compound production have been by way of rational design principles. This is because the pathways of the precursors have been largely elucidated and known bottlenecks or rate-limiting steps have been identified. Rational design will continue to be a potent pursuit as we learn more about relevant pathways and the regulation thereof. However, for fast-tracking improvements in aroma compound production in yeast, applying a DBTL-type approach holds a lot of potential. Indeed, it is perfectly suited for such an approach to target the production levels of a single metabolite. This is in contrast, however, to recent work in aroma compound development in yeast as part of an alcoholic beverage set-up, where hops flavour (a combination of geraniol and linalool) was introduced in yeast for beer production [87]. An elegant DBTL approach was followed to build strains with ideal hops flavour (similar to commercial beer). Although the researchers were eventually successful in obtaining strains that could ferment malt and impart hops flavour, many of their initial strain building attempts failed as strains selected for ideal hops flavour could not completely consume the malt sugars. This was unintentional and not easily explained but emphasised the limitations of employing DBTL on strains with multiparametric purposes. Nevertheless, to date, no DBTL-like approach has been performed on yeast for the sole purpose of producing enhanced levels of an aroma compound.

A key element complementary to metabolic engineering within the synthetic biology realm, is that of individual enzyme engineering in order to alter the catalytic activity and increase the flux toward the production pathway of a given aroma compound [88]. Few cases exist where mutants of the enzymes within certain pathways were examined for improved production. Site-directed mutagenesis in enzymes involved in the MVA pathway has led to the general increase in terpenoids [89] or just specific classes of terpenoids [90]. Although not studied in yeast, mutants of a sesquiterpenoid synthase (TPS24) from Syrah grape (*Vitis vinifera*) produced significantly higher levels of α-guaiene, which is a non-enzymatic precursor of rotundone, the active compound of black pepper [91].

There are also non-metabolic engineering-based challenges that exist in order to enhance aroma compound development. The overwhelming majority of interesting aroma compounds are derived from plants and it would thus make sense to explore the genes relevant to the production of the given product and assess their suitability within a recombinant yeast host. Yet, genomic data of many plants are still lacking and even though the enzymes involved in the production of a certain aroma compound are known, the genes that encode these enzymes have still not been identified. The complexities in assembling plant genomes explain the current dearth of publicly available fully-annotated whole genome sequences of plants [92,93] but more options will open up as more become available. Many recombinant pathways (like with *p*-coumaric acid and raspberry ketone) employed *Arabidopsis thaliana* genes, mainly due its sequence availability as it is a model organism and hardly because it is the 'best' enzyme candidates. Related to this is the mystery of how certain recombinant genes are expressed in high levels whereas similar ones are expressed poorly [94]. This underlines the utility of still employing a trial-and-error approach and exploring many recombinant genes to assess their ability to be expressed in yeast.

Non-*Saccharomyces* yeast have been shown, on many occasions, to be a more suitable host for the production of aroma compounds. However, the tools available to manipulate these so-called "non-conventional" yeast are not as extensive. As discussed in Wagner and Alper [95] tremendous strides have been made in this field as even CRISPR-based tools have been developed for a number of species. Further development is required, with a possible aim to incorporate large synthetically designed genetic elements in the non-*Saccharomyces* genomes similar to the SCRaMbLE-set-up developed for *S. cerevisiae*.

As already shown with the development of a hoppy yeast [87], many of the strategies employed and discussed here could easily be used to engineer aroma development in strains involved in the fermented beverage industry [96–98]. The engineering strategies should not, however, interfere with all other processes involved in fermentation. Strains tailor-made to produce certain aroma compounds could streamline the fermentation process. An ultimate goal would be to have a winemaker or brewmaster use a specific yeast strain to achieve a more predictable outcome in the aroma profile of the fermented product. An example would be to develop a yeast strain that could synthesise the oak lactones found in oak barrels. Its biosynthesis pathway is complicated and fine-tuning within the yeast strain would be necessary but such a strain would eliminate the use of expensive oak barrels in winemaking and dramatically shorten the production time.

Recombinant aroma compound development in yeast is still in its infancy especially when compared to chemical synthesis or extraction from plants—both methods in their various incarnations are centuries old. The field of synthetic biology is primed to evolve by aiming to achieve levels that can compete with the status quo. An even newer competitor is that of cell-free enzyme pathways [99] and it would be interesting to witness how each approach would evolve to compete in producing aroma compounds in a cost-effective manner. A revolutionary step, as one could imagine, would be a large yeast fermentation set-up producing a given aroma compound eventually replacing the acres and acres of flower fields and all the required inputs, dedicated for eventual aroma compound extraction.

Author Contributions: N.v.W. outlined of the review and wrote the biggest part of the review. N.v.W., H.K. and I.S.P. contributed towards the graphical expansion of the text. All authors contributed to writing specific sections and approved the final version of the manuscript.

Acknowledgments: We thank Macquarie University, Bioplatforms Australia, the New South Wales (NSW) Chief Scientist and Engineer and the NSW Government's Department of Primary Industries for providing the start-up funds for the Synthetic Biology initiative at Macquarie University. We also acknowledge the ongoing support from our Hochschule Geisenheim collaborators, Manfred Grossmann and Christian von Wallbrunn.

References

1. Schrader, J. *Flavours and Fragrances: Chemistry, Bioprocessing and Sustainability*; Berger, R.G., Ed.; Springer-Verlag: Berlin/Heidelberg, Germany, 2007; pp. 507–574. [CrossRef]
2. Kempinski, C.; Jiang, Z.; Bell, S.; Chappell, J. *Biotechnology of Isoprenoids*; Schrader, J., Bohlmann, J., Eds.; Springer-Verlag: Berlin/Heidelberg, Germany, 2015; pp. 161–199. [CrossRef]
3. Lanciotti, R.; Gianotti, A.; Patrignani, F.; Belletti, N.; Guerzoni, M.E.; Gardini, F. Use of natural aroma compounds to improve shelf-life and safety of minimally processed fruits. *Trends Food Sci. Technol.* **2004**, *15*, 201–208. [CrossRef]
4. Belletti, N.; Kamdem, S.S.; Patrignani, F.; Lanciotti, R.; Covelli, A.; Gardini, F. Antimicrobial activity of aroma compounds against *Saccharomyces cerevisiae* and improvement of microbiological stability of soft drinks as assessed by logistic regression. *Appl. Environ. Microbiol.* **2007**, *73*, 5580–5586. [CrossRef] [PubMed]
5. Buck, L.; Axel, R. A novel multigene family may encode odorant receptors: A molecular basis for odor recognition. *Cell* **1991**, *65*, 175–187. [CrossRef]
6. Swiegers, J.H.; Chambers, P.J.; Pretorius, I.S. Olfaction and taste: Human perception, physiology and genetics. *Aust. J. Grape Wine Res.* **2005**, *11*, 109–113. [CrossRef]
7. Stevenson, R.J.; Boakes, R.A. A mnemonic theory of odor perception. *Psychol. Rev.* **2003**, *110*, 340–364. [CrossRef] [PubMed]
8. Carroll, A.L.; Desai, S.H.; Atsumi, S. Microbial production of scent and flavor compounds. *Curr. Opin. Biotechnol.* **2016**, *37*, 8–15. [CrossRef] [PubMed]
9. Steensels, J.; Snoek, T.; Meersman, E.; Nicolino, M.P.; Voordeckers, K.; Verstrepen, K.J. Improving industrial yeast strains: Exploiting natural and artificial diversity. *FEMS Microbiol. Rev.* **2014**, *38*, 947–995. [CrossRef] [PubMed]
10. Dzialo, M.C.; Park, R.; Steensels, J.; Lievens, B.; Verstrepen, K.J. Physiology, ecology and industrial applications of aroma formation in yeast. *FEMS Microbiol. Rev.* **2017**, *41*, S95–S128. [CrossRef] [PubMed]
11. Hirst, M.B.; Richter, C.L. Review of aroma formation through metabolic pathways of *Saccharomyces cerevisiae* in beverage fermentations. *Am. J. Enol. Vitic.* **2016**, *67*, 361–370. [CrossRef]
12. Peter, J.; Chiara, M.D.; Friedrich, A.; Yue, J.; Pflieger, D.; Bergström, A.; Sigwalt, A.; Barre, B.; Freel, K.; Llored, A.; et al. Genome evolution across 1011 *Saccharomyces cerevisiae* isolates. *Nature* **2018**, *556*, 339–344. [CrossRef] [PubMed]

13. Gellissen, G.; Kunze, G.; Gaillardin, C.; Cregg, J.M.; Berardi, E.; Veenhuis, M.; Klei, I.V.D. New yeast expression platforms based on methylotrophic Hansenula polymorpha and Pichia pastoris and on dimorphic Arxula adeninivorans and Yarrowia lipolytica—A comparison. *FEMS Yeast Res.* **2005**, *5*, 1079–1096. [CrossRef] [PubMed]
14. Forti, L.; Mauro, S.D.; Cramarossa, M.R.; Filippucci, S.; Turchetti, B.; Buzzini, P. Non-conventional yeasts whole cells as efficient biocatalysts for the production of flavors and fragrances. *Molecules* **2015**, *20*, 10377–10398. [CrossRef] [PubMed]
15. Verstrepen, K.J.; Derdelinckx, G.; Dufour, J.P.; Winderickx, J.; Thevelein, J.M.; Pretorius, I.S.; Delvaux, F.R. Flavor-active esters: Adding fruitiness to beer. *J. Biosci. Bioeng.* **2003**, *96*, 110–118. [CrossRef]
16. Kroukamp, H.; den Haan, R.; la Grange, D.C.; Sibanda, N.; Foulquié-Moreno, M.R.; Thevelein, J.M.; van Zyl, W.H. Strain breeding enhanced heterologous cellobiohydrolase secretion by *Saccharomyces cerevisiae* in a protein specific manner. *Biotechnol. J.* **2017**, *12*, 1–10. [CrossRef] [PubMed]
17. Brachmann, C.B.; Davies, A.; Cost, G.J.; Caputo, E.; Li, J.; Hieter, P.; Boeke, J.D. Designer deletion strains derived from *Saccharomyces cerevisiae* S288C: A useful set of strains and plasmids for PCR-mediated gene disruption and other applications. *Yeast* **1998**, *14*, 115–132. [CrossRef]
18. Hvorecny, K.; Prelich, G. Characterization of chromosomal integration sites for heterologous gene expression in *Saccharomyces cerevisiae*. *Yeast* **2010**, *27*, 861–865. [CrossRef] [PubMed]
19. DiCarlo, J.E.; Conley, A.J.; Penttilä, M.; Jäntti, J.; Wang, H.H.; Church, G. Yeast gligo-mediated genome Engineering (YOGE). *ACS Synth. Biol.* **2013**, *2*, 742–749. [CrossRef] [PubMed]
20. Mitchell, L.A.; Chuang, J.; Agmon, N.; Khunsriraksakul, C.; Phillips, N.A.; Cai, Y.; Truong, D.M.; Veerakumar, A.; Wang, Y.; Mayorga, M.; et al. Versatile genetic assembly system (VEGAS) to assemble pathways for expression in *S. cerevisiae*. *Nucleic Acids Res.* **2015**, *43*, 6620–6630. [CrossRef] [PubMed]
21. Pennisi, E. Building the ultimate yeast genome. *Science* **2014**, *343*, 1426–1429. [CrossRef] [PubMed]
22. Annaluru, N.; Muller, H.; Mitchell, L.A.; Ramalingam, S.; Stracquadanio, G.; Richardson, S.M.; Dymond, J.S.; Kuang, Z.; Scheifele, L.Z.; Cooper, E.M.; et al. Total synthesis of a functional designer eukaryotic chromosome. *Science* **2014**, *344*, 55–58. [CrossRef] [PubMed]
23. Richardson, S.M.; Mitchell, L.A.; Stracquadanio, G.; Yang, K.; Dymond, J.S.; DiCarlo, J.E.; Lee, D.; Huang, C.L.V.; Chandrasegaran, S.; Cai, Y.; et al. Design of a synthetic yeast genome. *Science* **2017**, *355*, 1040–1044. [CrossRef] [PubMed]
24. Shen, Y.; Stracquadanio, G.; Wang, Y.; Yang, K.; Mitchell, L.A.; Xue, Y.; Cai, Y.; Chen, T.; Dymond, J.S.; Kang, K.; et al. SCRaMbLE generates designed combinatorial stochastic diversity in synthetic chromosomes. *Genome Res.* **2015**, *26*, 36–49. [CrossRef] [PubMed]
25. Pretorius, I.S.; Boeke, J.D. Yeast 2.0—Connecting the dots in the construction of the world's first functional synthetic eukaryotic genome. *FEMS Yeast Res.* **2018**, *18*. [CrossRef] [PubMed]
26. Pickens, L.; Tang, Y.; Chooi, Y.-H. Metabolic engineering for the production of natural products. *Annu. Rev. Chem. Biomol. Eng.* **2014**, *2*, 211–236. [CrossRef] [PubMed]
27. Lee, J.H.; Wendisch, V.F. Biotechnological production of aromatic compounds of the extended shikimate pathway from renewable biomass. *J. Biotechnol.* **2017**, *257*, 211–221. [CrossRef] [PubMed]
28. Braus, G.H. Aromatic amino acid biosynthesis in the yeast *Saccharomyces cerevisiae*: A model system for the regulation of a eukaryotic biosynthetic pathway. *Microbiol. Rev.* **1991**, *55*, 349–370. [PubMed]
29. Mulleder, M.; Calvani, E.; Alam, M.T.; Wang, R.K.; Eckerstorfer, F.; Zelezniak, A.; Ralser, M. Functional metabolomics describes the yeast biosynthetic regulome. *Cell* **2016**, *167*, 553–565. [CrossRef] [PubMed]
30. Hansen, E.H.; Møller, B.L.; Kock, G.R.; Bünner, C.M.; Kristensen, C.; Jensen, O.R.; Okkels, F.T.; Olsen, C.E.; Motawia, M.S.; Hansen, J. De novo biosynthesis of vanillin in fission yeast (*Schizosaccharomyces pombe*) and baker's yeast (*Saccharomyces cerevisiae*). *Appl. Environ. Microbiol.* **2009**, *75*, 2765–2774. [CrossRef] [PubMed]
31. Iwaki, A.; Ohnuki, S.; Suga, Y.; Izawa, S.; Ohya, Y. Vanillin inhibits translation and induces messenger ribonucleoprotein (mRNP) granule formation in *Saccharomyces cerevisiae*: Application and validation of high-content, image-based profiling. *PLoS ONE* **2013**, *8*, 2–11. [CrossRef] [PubMed]
32. Brochado, A.R.; Matos, C.; Møller, B.L.; Hansen, J.; Mortensen, U.H.; Patil, K.R. Improved vanillin production in baker's yeast through in silico design. *Microb. Cell Fact.* **2010**, *9*, 1–15. [CrossRef] [PubMed]

33. Brochado, A.R.; Patil, K.R. Overexpression of *O*-methyltransferase leads to improved vanillin production in baker's yeast only when complemented with model-guided network engineering. *Biotechnol. Bioeng.* **2013**, *110*, 656–659. [CrossRef] [PubMed]
34. Yin, L.H.; Choon, Y.W.; Chai, L.E.; Chong, C.K.; Deris, S.; Illias, R.M.; Mohamad, M.S. *Advances in Biomedical Infrastructure*; Sidhu, A.S., Dhillon, S.K., Eds.; Springer-Verlag: Berlin/Heidelberg, Germany, 2013; Volume 477, pp. 101–116. [CrossRef]
35. Vannelli, T.; Qi, W.W.; Sweigard, J.; Gatenby, A.A.; Sariaslani, F.S. Production of p-hydroxycinnamic acid from glucose in *Saccharomyces cerevisiae* and *Escherichia coli* by expression of heterologous genes from plants and fungi. *Metab. Eng.* **2007**, *9*, 142–151. [CrossRef] [PubMed]
36. Gottardi, M.; Grün, P.; Bode, H.B.; Hoffmann, T.; Schwab, W.; Oreb, M.; Boles, E. Optimisation of trans-cinnamic acid and hydrocinnamyl alcohol production with recombinant *Saccharomyces cerevisiae* and identification of cinnamyl methyl ketone as a by-product. *FEMS Yeast Res.* **2017**, *9*, 142–151. [CrossRef] [PubMed]
37. Jiang, H.; Wood, K.V.; Morgan, J.A. Metabolic engineering of the phenylpropanoid pathway in *Saccharomyces cerevisiae*. *Appl. Environ. Microbiol.* **2005**, *71*, 2962–2969. [CrossRef] [PubMed]
38. Lee, D.; Lloyd, N.D.R.; Pretorius, I.S.; Borneman, A.R. Heterologous production of raspberry ketone in the wine yeast *Saccharomyces cerevisiae* via pathway engineering and synthetic enzyme fusion. *Microb. Cell Fact.* **2016**, *15*. [CrossRef] [PubMed]
39. Kim, B.; Cho, B.R.; Hahn, J.S. Metabolic engineering of *Saccharomyces cerevisiae* for the production of 2-phenylethanol via Ehrlich pathway. *Biotechnol. Bioeng.* **2014**, *111*, 115–124. [CrossRef] [PubMed]
40. Vaseghi, S.; Baumeister, A.; Rizzi, M.; Reuss, M. In vivo dynamics of the pentose phosphate pathway in *Saccharomyces cerevisiae*. *Metab. Eng.* **1999**, *1*, 128–140. [CrossRef] [PubMed]
41. Curran, K.A.; Leavitt, J.M.; Karim, A.S.; Alper, H.S. Metabolic engineering of muconic acid production in *Saccharomyces cerevisiae*. *Metab. Eng.* **2013**, *15*, 55–66. [CrossRef] [PubMed]
42. Deaner, M.; Alper, H.S. Systematic testing of enzyme perturbation sensitivities via graded dCas9 modulation in *Saccharomyces cerevisiae*. *Metab. Eng.* **2017**, *40*, 14–22. [CrossRef] [PubMed]
43. Ravasio, D.; Wendland, J.; Walther, A. Major contribution of the Ehrlich pathway for 2-phenylethanol/rose flavor production in Ashbya gossypii. *FEMS Yeast Res.* **2014**, *14*, 833–844. [CrossRef] [PubMed]
44. Kim, T.Y.; Lee, S.W.; Oh, M.K. Biosynthesis of 2-phenylethanol from glucose with genetically engineered Kluyveromyces marxianus. *Enzym. Microb. Technol.* **2014**, *61–62*, 44–47. [CrossRef] [PubMed]
45. Lu, X.; Wang, Y.; Zong, H.; Ji, H.; Zhuge, B.; Dong, Z. Bioconversion of L-phenylalanine to 2-phenylethanol by the novel stress-tolerant yeast Candida glycerinogenes WL2002-5. *Bioengineered* **2016**, *7*, 418–423. [CrossRef] [PubMed]
46. Mierzejewska, J.; Tymoszewska, A.; Chreptowicz, K.; Krol, K. Mating of 2 laboratory *Saccharomyces cerevisiae* strains resulted in enhanced production of 2-phenylethanol by biotransformation of L-Phenylalanine. *J. Mol. Microbiol. Biotechnol.* **2017**, *27*, 81–90. [CrossRef] [PubMed]
47. Eshkol, N.; Sendovski, M.; Bahalul, M.; Katz-Ezov, T.; Kashi, Y.; Fishman, A. Production of 2-phenylethanol from L-phenylalanine by a stress tolerant *Saccharomyces cerevisiae* strain. *J. Appl. Microbiol.* **2009**, *106*, 534–542. [CrossRef] [PubMed]
48. Zwenger, S.; Basu, C. Plant terpenoids: Applications and future potentials. *Biotechnol. Mol. Biol. Rev.* **2008**, *3*, 1–7. [CrossRef]
49. Rodriguez, R.J.; Low, C.; Bottema, C.D.K.; Parks, L.W. Multiple functions for sterols in *Saccharomyces cerevisiae*. *Biochim. Biophys. Acta* **1985**, *837*, 336–343. [CrossRef]
50. Liao, P.; Hemmerlin, A.; Bach, T.J.; Chye, M.L. The potential of the mevalonate pathway for enhanced isoprenoid production. *Biotechnol. Adv.* **2016**, *34*, 697–713. [CrossRef] [PubMed]
51. Paramasivan, K.; Mutturi, S. Progress in terpene synthesis strategies through engineering of *Saccharomyces cerevisiae*. *Crit. Rev. Biotechnol.* **2017**, *37*, 974–989. [CrossRef] [PubMed]
52. Ignea, C.; Pontini, M.; Maffei, M.E.; Makris, A.M.; Kampranis, S.C. Engineering monoterpene production in yeast using a synthetic dominant negative geranyl diphosphate synthase. *ACS Synth. Biol.* **2014**, *3*, 298–306. [CrossRef] [PubMed]
53. Vickers, C.E.; Williams, T.C.; Peng, B.; Cherry, J. Recent advances in synthetic biology for engineering isoprenoid production in yeast. *Curr. Opin. Chem. Biol.* **2017**, *40*, 47–56. [CrossRef] [PubMed]

54. Wriessnegger, T.; Pichler, H. Yeast metabolic engineering—Targeting sterol metabolism and terpenoid formation. *Prog. Lipid Res.* **2013**, *52*, 277–293. [CrossRef] [PubMed]
55. Asadollahi, M.A.; Maury, J.; Schalk, M.; Clark, A.; Nielsen, J. Enhancement of farnesyl diphosphate pool as direct precursor of sesquiterpenes through metabolic engineering of the mevalonate pathway in *Saccharomyces cerevisiae*. *Biotechnol. Bioeng.* **2010**, *106*, 86–96. [CrossRef] [PubMed]
56. Peng, B.; Plan, M.R.; Chrysanthopoulos, P.; Hodson, M.P.; Nielsen, L.K.; Vickers, C.E. A squalene synthase protein degradation method for improved sesquiterpene production in *Saccharomyces cerevisiae*. *Metab. Eng.* **2017**, *39*, 209–219. [CrossRef] [PubMed]
57. Polakowski, T.; Stahl, U. Overexpression of a cytosolic hydroxymethylglutaryl-CoA reductase leads to squalene accumulation in yeast. *Appl. Microbiol. Biotechnol.* **1998**, *8*, 66–71. [CrossRef]
58. Rico, J.; Pardo, E.; Orejas, M. Enhanced production of a plant monoterpene by overexpression of the 3-hydroxy-3-methylglutaryl coenzyme a reductase catalytic domain in *Saccharomyces cerevisiae*. *Appl. Environ. Microbiol.* **2010**, *76*, 6449–6454. [CrossRef] [PubMed]
59. Pardo, E.; Rico, J.; Gil, J.V.; Orejas, M. De novo production of six key grape aroma monoterpenes by a geraniol synthase-engineered *S. cerevisiae* wine strain. *Microb. Cell Fact.* **2015**, *14*. [CrossRef] [PubMed]
60. Vik, A.; Rine, J. Upc2p and Ecm22p, dual regulators of sterol biosynthesis in *Saccharomyces cerevisiae*. *Mol. Cell. Biol.* **2001**, *21*, 6395–6405. [CrossRef] [PubMed]
61. Meadows, A.L.; Hawkins, K.M.; Tsegaye, Y.; Antipov, E.; Kim, Y.; Raetz, L.; Dahl, R.H.; Tai, A.; Mahatdejkul-meadows, T.; Xu, L.; et al. Rewriting yeast central carbon metabolism for industrial isoprenoid production. *Nature* **2016**, *537*, 694–697. [CrossRef] [PubMed]
62. Farhi, M.; Marhevka, E.; Masci, T.; Marcos, E.; Eyal, Y.; Ovadis, M.; Abeliovich, H.; Vainstein, A. Harnessing yeast subcellular compartments for the production of plant terpenoids. *Metab. Eng.* **2011**, *13*, 474–481. [CrossRef] [PubMed]
63. Wriessnegger, T.; Augustin, P.; Engleder, M.; Leitner, E.; Müller, M.; Kaluzna, I.; Schürmann, M.; Mink, D.; Zellnig, G.; Schwab, H.; et al. Production of the sesquiterpenoid (+)-nootkatone by metabolic engineering of Pichia pastoris. *Metab. Eng.* **2014**, *24*, 18–29. [CrossRef] [PubMed]
64. Verwaal, R.; Wang, J.; Meijnen, J.P.; Visser, H.; Sandmann, G.; Van Den Berg, J.A.; Van Ooyen, A.J.J. High-level production of beta-carotene in *Saccharomyces cerevisiae* by successive transformation with carotenogenic genes from Xanthophyllomyces dendrorhous. *Appl. Environ. Microbiol.* **2007**, *73*, 4342–4350. [CrossRef] [PubMed]
65. López, J.; Essus, K.; Kim, I.-K.; Pereira, R.; Herzog, J.; Siewers, V.; Nielsen, J.; Agosin, E. Production of β-ionone by combined expression of carotenogenic and plant CCD1 genes in *Saccharomyces cerevisiae*. *Microb. Cell Fact.* **2015**, *14*, 84. [CrossRef] [PubMed]
66. Beekwilder, J.; van Rossum, H.M.; Koopman, F.; Sonntag, F.; Buchhaupt, M.; Schrader, J.; Hall, R.D.; Bosch, D.; Pronk, J.T.; van Maris, A.J.A.; et al. Polycistronic expression of a β-carotene biosynthetic pathway in *Saccharomyces cerevisiae* coupled to β-ionone production. *J. Biotechnol.* **2014**, *192*, 383–392. [CrossRef] [PubMed]
67. Kirby, J.; Dietzel, K.L.; Wichmann, G.; Chan, R.; Antipov, E.; Moss, N.; Baidoo, E.E.K.; Jackson, P.; Gaucher, S.P.; Gottlieb, S.; et al. Engineering a functional 1-deoxy-D-xylulose 5-phosphate (DXP) pathway in *Saccharomyces cerevisiae*. *Metab. Eng.* **2016**, *38*, 494–503. [CrossRef] [PubMed]
68. Siddiqui, M.S.; Thodey, K.; Trenchard, I.; Smolke, C.D. Advancing secondary metabolite biosynthesis in yeast with synthetic biology tools. *FEMS Yeast Res.* **2012**, *12*, 144–170. [CrossRef] [PubMed]
69. Jensen, M.K.; Keasling, J.D. Recent applications of synthetic biology tools for yeast metabolic engineering. *FEMS Yeast Res.* **2015**, *15*, 1–10. [CrossRef] [PubMed]
70. Deloache, W.C.; Russ, Z.N.; Narcross, L.; Gonzales, A.M.; Martin, V.J.J.; Dueber, J.E. An enzyme-coupled biosensor enables (*S*)-reticuline production in yeast from glucose. *Nat. Chem. Biol.* **2015**, *11*, 465–471. [CrossRef] [PubMed]
71. Leavitt, J.M.; Wagner, J.M.; Tu, C.C.; Tong, A.; Liu, Y.; Alper, H.S. Biosensor-enabled directed evolution to improve muconic acid production in *Saccharomyces cerevisiae*. *Biotechnol. J.* **2017**, *12*, 1–9. [CrossRef] [PubMed]
72. Adeniran, A.; Sherer, M.; Tyo, K.E.J. Yeast-based biosensors: Design and applications. *FEMS Yeast Res.* **2015**, *15*, 1–15. [CrossRef] [PubMed]
73. Ravasio, D.; Walther, A.; Trost, K.; Vrhovsek, U.; Wendland, J. An indirect assay for volatile compound production in yeast strains. *Sci. Rep.* **2014**, *4*. [CrossRef] [PubMed]

74. Chou, H.H.; Keasling, J.D. Programming adaptive control to evolve increased metabolite production. *Nat. Commun.* **2013**, *4*, 1–8. [CrossRef] [PubMed]
75. Li, S.; Si, T.; Wang, M.; Zhao, H. Development of a synthetic malonyl-CoA sensor in *Saccharomyces cerevisiae* for intracellular metabolite monitoring and genetic screening. *ACS Synth. Biol.* **2015**, *4*, 1308–1315. [CrossRef] [PubMed]
76. David, F.; Nielsen, J.; Siewers, V. Flux Control at the Malonyl-CoA node through hierarchical dynamic pathway regulation in *Saccharomyces cerevisiae*. *ACS Synth. Biol.* **2016**, *5*, 224–233. [CrossRef] [PubMed]
77. Skjoedt, M.L.; Snoek, T.; Kildegaard, K.R.; Arsovska, D.; Eichenberger, M.; Goedecke, T.J.; Rajkumar, A.S.; Zhang, J.; Kristensen, M.; Lehka, B.J.; et al. Engineering prokaryotic transcriptional activators as metabolite biosensors in yeast. *Nat. Chem. Biol.* **2016**, *12*, 951–958. [CrossRef] [PubMed]
78. Ho, J.C.H.; Pawar, S.V.; Hallam, S.J.; Yadav, V.G. An improved whole-cell biosensor for the discovery of lignin-transforming enzymes in functional metagenomic screens. *ACS Synth. Biol.* **2018**, *7*, 392–398. [CrossRef] [PubMed]
79. Fiorentino, G.; Ronca, R.; Bartolucci, S. A novel *E. coli* biosensor for detecting aromatic aldehydes based on a responsive inducible archaeal promoter fused to the green fluorescent protein. *Appl. Microbiol. Biotechnol.* **2009**, *82*, 67–77. [CrossRef] [PubMed]
80. Siedler, S.; Khatri, N.K.; Zsohár, A.; Kjærbølling, I.; Vogt, M.; Hammar, P.; Nielsen, C.F.; Marienhagen, J.; Sommer, M.O.A.; Joensson, H.N. Development of a bacterial biosensor for rapid screening of yeast *p*-coumaric Acid production. *ACS Synth. Biol.* **2017**, *6*, 1860–1869. [CrossRef] [PubMed]
81. West, R.W.; Chen, S.M.; Putz, H.; Butler, G.; Banerjee, M. GAL1-GAL10 divergent promoter region of *Saccharomyces cerevisiae* contains negative control elements in addition to functionally separate and possibly overlapping upstream activating sequences. *Genes Dev.* **1987**, *1*, 1118–1131. [CrossRef] [PubMed]
82. Chao, R.; Mishra, S.; Si, T.; Zhao, H. Engineering biological systems using automated biofoundries. *Metab. Eng.* **2017**, *42*, 98–108. [CrossRef] [PubMed]
83. Nielsen, J.; Keasling, J.D. Engineering cellular metabolism. *Cell* **2016**, *164*, 1185–1197. [CrossRef] [PubMed]
84. Hollywood, K.A.; Schmidt, K.; Takano, E.; Breitling, R. Metabolomics tools for the synthetic biology of natural products. *Curr. Opin. Biotechnol.* **2018**, *54*, 114–120. [CrossRef] [PubMed]
85. Dacres, H.; Wang, J.; Leitch, V.; Horne, I.; Anderson, A.R.; Trowell, S.C. Greatly enhanced detection of a volatile ligand at femtomolar levels using bioluminescence resonance energy transfer (BRET). *Biosens. Bioelectron.* **2011**, *29*, 119–124. [CrossRef] [PubMed]
86. Berna, A. Metal oxide sensors for electronic noses and their application to food analysis. *Sensors* **2010**, *10*, 3882–3910. [CrossRef] [PubMed]
87. Denby, C.M.; Li, R.A.; Vu, V.T.; Costello, Z.; Lin, W.; Chan, L.J.G.; Williams, J.; Donaldson, B.; Bamforth, C.W.; Petzold, C.J.; et al. Industrial brewing yeast engineered for the production of primary flavor determinants in hopped beer. *Nat. Commun.* **2018**, *9*. [CrossRef] [PubMed]
88. Erb, T.J.; Jones, P.R.; Bar-Even, A. Synthetic metabolism: metabolic engineering meets enzyme design. *Curr. Opin. Chem. Biol.* **2017**, *37*, 56–62. [CrossRef] [PubMed]
89. Mantzouridou, F.; Tsimidou, M.Z. Observations on squalene accumulation in *Saccharomyces cerevisiae* due to the manipulation of HMG2 and ERG6. *FEMS Yeast Res.* **2010**, *10*, 699–707. [CrossRef] [PubMed]
90. Fischer, M.J.C.; Meyer, S.; Claudel, P.; Bergdoll, M.; Karst, F. Metabolic engineering of monoterpene synthesis in yeast. *Biotechnol. Bioeng.* **2011**, *108*, 1883–1892. [CrossRef] [PubMed]
91. Drew, D.P.; Andersen, T.B.; Sweetman, C.; Møller, B.L.; Ford, C.; Simonsen, H.T. Two key polymorphisms in a newly discovered allele of the Vitis vinifera TPS24 gene are responsible for the production of the rotundone precursor α-guaiene. *J. Exp. Bot.* **2016**, *67*, 799–808. [CrossRef] [PubMed]
92. Schatz, M.C.; Witkowski, J.; McCombie, W.R. Current challenges in *de novo* plant genome sequencing and assembly. *Genome Biol.* **2012**, *13*. [CrossRef]
93. Claros, M.G.; Bautista, R.; Guerrero-Fernández, D.; Benzerki, H.; Seoane, P.; Fernández-Pozo, N. Why Assembling plant genome sequences is so challenging. *Biology* **2012**, *1*, 439–459. [CrossRef] [PubMed]
94. Van Wyk, N.; Trollope, K.M.; Steenkamp, E.T.; Wingfield, B.D.; Volschenk, H. Identification of the gene for β-fructofuranosidase from Ceratocystis moniliformis CMW 10134 and characterization of the enzyme expressed in *Saccharomyces cerevisiae*. *BMC Biotechnol.* **2013**, *13*. [CrossRef] [PubMed]
95. Wagner, J.M.; Alper, H.S. Synthetic biology and molecular genetics in non-conventional yeasts: Current tools and future advances. *Fungal Genet. Biol.* **2016**, *89*, 126–136. [CrossRef] [PubMed]

96. Jagtap, U.B.; Jadhav, J.P.; Bapat, V.A.; Pretorius, I.S. Synthetic biology stretching the realms of possibility in wine yeast research. *Int. J. Food Microbiol.* **2017**, *252*, 24–34. [CrossRef] [PubMed]
97. Pretorius, I.S. Solving yeast jigsaw puzzles over a glass of wine. *EMBO Rep.* **2017**, *18*, 1–10. [CrossRef] [PubMed]
98. Pretorius, I. Conducting wine symphonics with the aid of yeast genomics. *Beverages* **2016**, *2*. [CrossRef]
99. Moore, S.J.; Tosi, T.; Hleba, Y.B.; Bell, D.; Polizzi, K.; Freemont, P. A cell-free synthetic biochemistry platform for raspberry ketone production. *bioRxiv* **2017**. [CrossRef]

Permissions

All chapters in this book were first published in MDPI; hereby published with permission under the Creative Commons Attribution License or equivalent. Every chapter published in this book has been scrutinized by our experts. Their significance has been extensively debated. The topics covered herein carry significant findings which will fuel the growth of the discipline. They may even be implemented as practical applications or may be referred to as a beginning point for another development.

The contributors of this book come from diverse backgrounds, making this book a truly international effort. This book will bring forth new frontiers with its revolutionizing research information and detailed analysis of the nascent developments around the world.

We would like to thank all the contributing authors for lending their expertise to make the book truly unique. They have played a crucial role in the development of this book. Without their invaluable contributions this book wouldn't have been possible. They have made vital efforts to compile up to date information on the varied aspects of this subject to make this book a valuable addition to the collection of many professionals and students.

This book was conceptualized with the vision of imparting up-to-date information and advanced data in this field. To ensure the same, a matchless editorial board was set up. Every individual on the board went through rigorous rounds of assessment to prove their worth. After which they invested a large part of their time researching and compiling the most relevant data for our readers.

The editorial board has been involved in producing this book since its inception. They have spent rigorous hours researching and exploring the diverse topics which have resulted in the successful publishing of this book. They have passed on their knowledge of decades through this book. To expedite this challenging task, the publisher supported the team at every step. A small team of assistant editors was also appointed to further simplify the editing procedure and attain best results for the readers.

Apart from the editorial board, the designing team has also invested a significant amount of their time in understanding the subject and creating the most relevant covers. They scrutinized every image to scout for the most suitable representation of the subject and create an appropriate cover for the book.

The publishing team has been an ardent support to the editorial, designing and production team. Their endless efforts to recruit the best for this project, has resulted in the accomplishment of this book. They are a veteran in the field of academics and their pool of knowledge is as vast as their experience in printing. Their expertise and guidance has proved useful at every step. Their uncompromising quality standards have made this book an exceptional effort. Their encouragement from time to time has been an inspiration for everyone.

The publisher and the editorial board hope that this book will prove to be a valuable piece of knowledge for researchers, students, practitioners and scholars across the globe.

List of Contributors

Svetlana V. Kamzolova, Vladimir A. Samoilenko, Julia N. Lunina and Igor G. Morgunov
Federal Research Center, Pushchino Scientific Center for Biological Research of the Russian Academy of Sciences, G.K. Skryabin Institute of Biochemistry and Physiology of Microorganisms of the Russian Academy of Sciences, Prospect Nauki 5, Pushchino, 142290 Moscow, Russia

Yuki Yamamoto, Yuki Fujimaru, Yuki Kusaba and Hiroshi Kitagaki
Graduate School of Advanced Health Sciences, Saga University, 1, Honjo, Saga city, Saga 840-8502, Japan

Jennifer E. G. Gallagher
Biology Department, West Virginia University, 53 Campus Drive, Morgantown, WV 26506-6057, USA

Matthew J. Winans
Graduate School of Advanced Health Sciences, Saga University, 1, Honjo, Saga city, Saga 840-8502, Japan
Biology Department, West Virginia University, 53 Campus Drive, Morgantown, WV 26506-6057, USA

Dung Minh Ha-Tran
Molecular and Biological Agricultural Sciences Program, Taiwan International Graduate Program, Academia Sinica and National Chung Hsing University, Taipei 11529, Taiwan
Department of Life Sciences, National Chung Hsing University, Taichung 40227, Taiwan
Graduate Institute of Biotechnology, National Chung Hsing University, Taichung 40227, Taiwan

Trinh Thi My Nguyen
Department of Life Sciences, National Chung Hsing University, Taichung 40227, Taiwan

Chieh-Chen Huang
Department of Life Sciences, National Chung Hsing University, Taichung 40227, Taiwan
Innovation and Development Center of Sustainable Agriculture, National Chung Hsing University, Taichung 40227, Taiwan

Mesfin Haile
Department of Horticulture, Kangwon National University, Chuncheon 24341, Korea

Won Hee Kang
Department of Horticulture, Kangwon National University, Chuncheon 24341, Korea
Convergence Program of Coffee Science, Kangwon National University, Chuncheon 24341, Korea

Maria Papagianni
Department of Hygiene and Technology of Food of Animal Origin, School of Veterinary Medicine, Aristotle University of Thessaloniki, Thessaloniki 54006, Greece

Emmanuel M. Papamichael
Department of Chemistry, University of Ionnina, Ioannina 45500, Greece

José Juan Mateo and Sergi Maicas
Departament de Microbiologia i Ecologia, Universitat de València, E-46100-Burjassot, Spain

Esteban García-Romero and Adela Mena-Morales
Instituto Regional de Investigación y Desarrollo Agroalimentario y Forestal de Castilla-La Mancha (IRIAF), IVICAM, Ctra. Toledo-Albacete s/n., 13700 Tomelloso, Ciudad Real, Spain

Pedro Miguel Izquierdo-Cañas
Instituto Regional de Investigación y Desarrollo Agroalimentario y Forestal de Castilla-La Mancha (IRIAF), IVICAM, Ctra. Toledo-Albacete s/n., 13700 Tomelloso, Ciudad Real, Spain
Parque Científico y Tecnológico de Castilla-La Mancha, Paseo de la Innovación 1, 02006 Albacete, Spain

María Ríos-Carrasco
Department of Biomedicine, Biotechnology and Public Health, University of Cádiz, Puerto Real, 11003 Cádiz, Spain

José María Heras-Manso
Lallemand Península Ibérica, Tomás Edison 4, 28521 Madrid

Ronnie G. Willaert
Alliance Research Group VUB-UGent NanoMicrobiology (NAMI), IJRG VUB-EPFL NanoBiotechnology & NanoMedicine (NANO), Research Group Structural Biology Brussels, Vrije Universiteit Brussel, 1050 Brussels, Belgium
Department Bioscience Engineering, University Antwerp, 2020 Antwerp, Belgium

Mar Gil-Díaz, Margarita García, Juan Mariano Cabellos and Teresa Arroyo
Departamento de Agroalimentación, Instituto Madrileño de Investigación y Desarrollo Rural Agrario y Alimentario, Autovía A2 km 38.2, Alcalá de Henares, Madrid 28800, Spain

Gustavo Cordero-Bueso
Departamento de Agroalimentación, Instituto Madrileño de Investigación y Desarrollo Rural Agrario y Alimentario, Autovía A2 km 38.2, Alcalá de Henares, Madrid 28800, Spain
Departamento de Biomedicina, Biotecnología y Salud Pública, Universidad de Cádiz, Avenida de la República Saharaui s/n, Puerto Real, Cádiz 11510, Spain

Braulio Esteve-Zarzoso
Biotecnologia Enològica, Departament de Bioquimica i Biotecnologia, Facultat d'Enologia, Universitat Rovira i Virgili, Marcel li Domingo s/n, Tarragona 43007, Spain

Susan Krull, Malin Lünsmann, Ulf Prüße and Anja Kuenz
Thünen-Institute of Agricultural Technology, Bundesallee 47, 38116 Braunschweig, Germany

Fei Yang
Cool Climate Oenology and Viticulture Institute, Brock University, St. Catharines, ON L2S 3A1, Canada

Caitlin Heit
Cool Climate Oenology and Viticulture Institute, Brock University, St. Catharines, ON L2S 3A1, Canada
Centre for Biotechnology, Brock University, St. Catharines, ON L2S 3A1, Canada

Debra L. Inglis
Cool Climate Oenology and Viticulture Institute, Brock University, St. Catharines, ON L2S 3A1, Canada
Centre for Biotechnology, Brock University, St. Catharines, ON L2S 3A1, Canada
Department of Biological Sciences, Brock University, St. Catharines, ON L2S 3A1, Canada

Tingting Liu, Shuangcheng Huang and Anli Geng
School of Life Sciences and Chemical Technology, Ngee Ann Polytechnic, Singapore 599489, Singapore

Konstantin Bellut, Jonas J. Atzler, Andrea Hoehnel and Kieran M. Lynch
School of Food and Nutritional Sciences, University College Cork, College Road, Cork, Ireland

Maximilian Michel, Martin Zarnkow, Mathias Hutzler and Fritz Jacob
Research Center Weihenstephan for Brewing and Food Quality, Technische Universität München, Alte Akademie 3, 85354 Freising-Weihenstephan, Germany

Elke K. Arendt
School of Food and Nutritional Sciences, University College Cork, College Road, Cork, Ireland
APC Microbiome Ireland, University College Cork, College Road, Cork, Ireland

Carmela Garofalo, Mattia Pia Arena, Luciano Beneduce, Carmen Berbegal, Giuseppe Spano and Vittorio Capozzi
Dipartimento di Scienze Agrarie, degli Alimenti e dell'Ambiente, Università di Foggia, via Napoli 25, Foggia 71100, Italy

Barbara Laddomada, Maria Stella Cappello, Gianluca Bleve and Francesco Grieco
Istituto di Scienze delle Produzioni Alimentari, Consiglio Nazionale delle Ricerche, Unità Operativa di Supporto di Lecce, Lecce 73100, Italy

Heinrich Kroukamp
Department of Molecular Sciences, Faculty of Science and Engineering, Sydney, NSW 2109, Australia

Niël van Wyk
Department of Molecular Sciences, Faculty of Science and Engineering, Sydney, NSW 2109, Australia
Institut für Mikrobiologie und Biochemie Zentrum Analytische Chemie und Mikrobiologie, Hochschule Geisenheim University, 65366 Geisenheim, Germany

Isak S. Pretorius
Chancellery, Macquarie University, Sydney, NSW 2109, Australia

Index

www.ingramcontent.com/pod-product-compliance
Lightning Source LLC
LaVergne TN
LVHW081022210926
848485LV00004B/929